Green Energy

Green Energy

Edited by **Marrianne Fox**

SYRAWOOD
PUBLISHING HOUSE
New York

Published by Syrawood Publishing House,
750 Third Avenue, 9th Floor,
New York, NY 10017, USA
www.syrawoodpublishinghouse.com

Green Energy
Edited by Marrianne Fox

International Standard Book Number: 978-1-68286-372-5 (Hardback)

Printed in China.

Contents

Preface

Green energy has gained global significance with increasing demand and need for sustainable energy resources. There has been a significant focus on procuring energy from non-conventional, natural resources like sunlight, tides, plants, algae, etc. This book contains some path-breaking studies that explore various aspects of renewable energy. The topics included in this book on energy efficiency and energy security are of utmost significance and bound to provide incredible insights to readers. The book presents researches and case-studies performed by experts across the globe which emphasize on global socio-economic issues and different alternative energy resources. Those in search of information to further their knowledge will be greatly assisted by this book.

All of the data presented henceforth, was collaborated in the wake of recent advancements in the field. The aim of this book is to present the diversified developments from across the globe in a comprehensible manner. The opinions expressed in each chapter belong solely to the contributing authors. Their interpretations of the topics are the integral part of this book, which I have carefully compiled for a better understanding of the readers.

At the end, I would like to thank all those who dedicated their time and efforts for the successful completion of this book. I also wish to convey my gratitude towards my friends and family who supported me at every step.

Editor

Evaluation of portable water in five provinces of Zambia using a water pollution index

Kabunga NACHIYUNDE*, Hideo IKEDA, Kazuhiko TANAKA, Daisuke KOZAKI

Graduate School for International Development and Cooperation, Hiroshima University.

Five provinces with different environmental characteristics were evaluated for water pollution with the aim of identifying the major sources and area(s) most affected. The results indicate that the water in Lusaka Province is significantly high in nitrate and sodium concentration compared to other provinces considered in this study. The results further reveal that ground water is most affected by nitrate and sodium. On the contrary, dissolved organic carbon (DOC) was found to be significantly higher in surface water but no particular province registered a higher amount compared to others. The results further reveal that the pollution is primarily a function of human activities, social amenities and industrial activities in the study areas.

Key words: Nitrate, sulphate, groundwater, Lusaka, ion exchange/cation-exchange chromatography.

INTRODUCTION

Surface and ground water resources are threatened by pollution from release of dissolved substances including inorganic ions and organic compounds such as oils, into rivers, wetlands and ground water from industrial activity in Zambia. Cement waste, molasses and bagasse, textile sediment sludge, petroleum, paint, pesticides, fungicides, human waste, fertilizers from farms/gardens, and lime sludge from industries in Lusaka, Kafue and the Copperbelt, all continue to find their way into water systems through direct discharge, seepage or overflow to ground and surface water courses (Chundama, 2008). To mitigate some of these problems, it is strongly recommended that there be adequate supervision on what type of sanitation is used in different areas depending on the geology and hydrology of the area; solid waste should be seen as a serious threat to water resources and the solid waste hierarchy of reduce-reuse-recycle should be followed up in solid waste management activities (Nakamboa, 2005).

Many studies cite groundwater quality data as evidence of sewer – related contamination (Baird and Cann, 2008; Marquita, 2010). Common indicators cited include bacteria, nitrate, ammonium and various organic compounds (Misstear and Bishop, 1997). Furthermore, potassium and boron are also considered to be amongst the substances indicative of contamination by sewage (Whitehead et al., 1999).

In areas without mains sewage system, runoff water from washing and flushing is conducted to pit latrines and sometimes absorber wells, and again flows into groundwater which is hence a source of pollution (Khazai and Riggi, 1999). Foster et al. (1993) argues that, in such a situation, about 90% of the water supply will end up as recharge to groundwater. Septic tank systems are frequently reported sources of localised groundwater pollution and in some cases regional groundwater problems have also been recognised in areas of high septic tank density resulting in degradation of groundwater. One common reason for degradation of the above is that the capacity of the soil to absorb effluent from the tank has exceeded the limit, and the waste added to the system moves upwards. Many other pathogens, such as typhoid, cholera, streptococci, salmonella, poliomyelitis, and protozoans are transmitted by septic tank systems (Chandra et al., 1997). This view is further supported by Foster et al. (1999) who contends that urban [or rural] residential districts without, or with incomplete, coverage of mains sewage system, seepage from unsewered sanitation sys-

*Corresponding author. E-mail: bungaman@live.jp.

tems such as septic tanks, cesspits and latrines probably represents diffuse pollution sources. Ironically, Barret and Howard (2002) states that despite the importance of groundwater to urban and urbanising regions in Sub-Saharan Africa, there are few published studies of urban groundwater in this region, and a detailed scenario of the urban impacts on groundwater quality does not exist.

In Lusaka, there are three major types of sanitation services namely: water borne sewer system provided by Lusaka Water and Sewerage Company (LWSC); septic tanks; and pit latrines. The coverage of water borne sewer system is about 30% of the total area of which the LWSC supplies water [Environmental Council of Zambia and Zambia Environment Outlook Report 3 (ECZ and ZEOR3, 2008)]. With this meagre 30% coverage, and bearing in mind that an adult human being excretes around 4 kg of nitrogen per year (Barret and Howard, 2002), it is our contention that there is massive input of nitrogen into the soil, and eventually groundwater in Lusaka and other towns lacking adequate sanitation. Furthermore, urine stream is known to contain a substantial amount of nitrogen, about 8.5 and 2 mg/day of phosphorus (Green, 2003; Pickford, 1995) which is either potentially recoverable or a problem to treat. The fate of this nitrogen and phosphorus is usually in pit latrines and open ground or in stormy water runoff and subsequently combines with surface and groundwater. The consequences are obvious when it comes to health status of the general citizenry. In addition, the shallow geology and hydrogeological regime of Lusaka is of crucial importance when trying to predict the effect of urbanisation on groundwater. Furthermore, it is also important to acknowledge that nitrate contamination is the subject of extensive research because of its potential hazard resulting in mathaemoglobinemia (Fan and Steinberg, 1996), hypertrophy of the thyroid (van Maanen et al., 1994), and diabetes (Parslow et al., 1997). Zeb et al. (2011) argues that a regular monitoring of water bodies with required number of parameters in relation to water quality prevents the outbreak of diseases and occurrence of hazards.

Lusaka is built over a karstic dolomite aquifer and ground water accounts for 61% of the total water supply within Lusaka; ironically, there has been a registered increase in the amount of waste generated such that by 2006, it was estimated at 242,803 metric tons (ECZ and ZEOR3, 2008). The fate of this waste is usually in landfills, which consequently are sources of contamination for groundwater, and may result in disease-related problems, with associated risk to the environment (Magmedov and Yakovleva, 1997; Baird and Cann, 2008). This view is reinforced by Lerner et al. 1999) who contend that urban sources of nitrogen in groundwater include leaking sewers, leaking water mains, landfills and industrial chemical spillages. Even with a developed sewer system in the UK, Lerner et al. (1999) reported that sewers contribute about 13% of nitrogen loading, leaking water mains contribute about 36% and the remaining 50% of the nitrogen loading includes parks, gardens, landfills and industries. With an underdeveloped sewer system and the widespread use of illegal landfills in Lusaka and other Zambian towns, the nitrogen loadings and other pollutants from these sources can be assumed to be high and hence the need to do a comprehensive study not only in Lusaka but also in the various provinces. Zambia has an abundance of groundwater well distributed in many parts of the country of which the best known aquifer is nearly 25,000 km^2 in size extending from Lusaka to the Copperbelt region (ECZ and ZEOR3, 2008). Our premise is that unplanned urban development and especially development in recharge areas have resulted in increased pollution in this aquifer, which also happens to be the most populated. It is the source of groundwater for major cities and towns like Lusaka, Kabwe and Ndola.

In Zambia, most of the waste is disposed of in undesignated places, burnt, buried in the yard or recycled and there is no separation of the various types of waste. It can be argued that leachate from the waste has resulted into ammonium and other pollutants (Mocanu et al., 1997). And considering the fact that there is aged waste from old dumps and the on-going dumping or recent waste, it can be said with certainty that ground water is currently being affected by contamination from both modes, that is, from recent and aged waste leachate contamination.

The Kafue, one of the country's main rivers, is threatened by industrial activity and residue from agriculture run off. Pollutant accumulation within the Kafue ecosystem has been associated with various toxicological manifestations. The disappearance of Hippopotamus amphibious from the Kafue River in Chingola, the proliferation of water hyacinth and the bioaccumulation of heavy metals within wildlife liver tissue have been associated with pollutants in the Kafue River ecosystem (Choongo et al., 2005; Nakayama et al., 2010; von der Heyden and New, 2003).

No comprehensive research has been done in Zambia, covering a vast area, to accurately determine and compare the loadings of nitrogen and other pollutants in different areas and from different sources. The study's objective therefore was to carry out an evaluation of the portable water due to non-functioning social facilities, industries and general negative attitudes of the citizens to prudent management of their surroundings. The study further aimed at isolating areas of poor water quality and to sensitise stakeholders and interest groups and individuals on the state of the water quality in Zambia.

MATERIALS AND METHOD

The water samples were collected in September and October 2011, covering five of the ten provinces. The samples were collected in

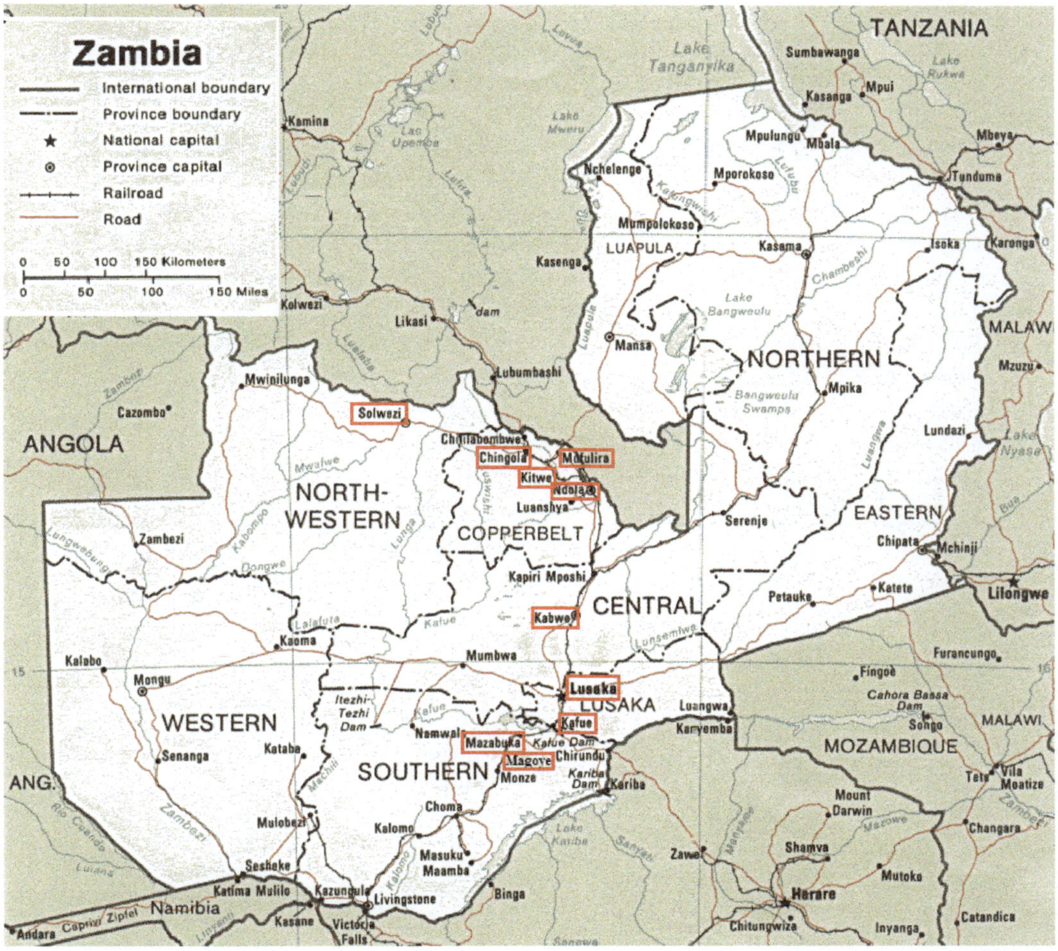

Figure 1. Map showing the ten locations where the 73 water samples were collected from. Source: Modified from Perry-Castañeda Library Map Collection (2001).

the driest period (dry season) to take care of diluting effect of rain water. Eight samples were from public water taps; 34 samples from rivers and/or streams; three samples from dams/water reservoirs; and 28 samples were from ground water sources such as wells and boreholes. Figure 1 shows the towns where the water samples were obtained.

Filtration of the samples was done through 0.45 μm filters to remove bacteria/suspended particles and to slow down sample degradation. A non-suppressed ion-exclusion/cation-exchange chromatography with conductimetric detection for the simultaneous determination of anions (SO_4^{2-}, Cl^- and NO_3^-) and cations (Na^+, K^+, NH_4^+, Mg^{2+} and Ca^{2+}) was performed on a polymethacrylate-based weakly acidic cation-exchange resin column and a mixed eluent consisting of succinic acid, tartaric acid and 18-crown-6 at pH 2.9; a Tosho IC – 2001 analyser was used. The separation mechanism was based on the ion-exclusion effect for the anions and the cation exchange effect for the cations. Under the optimised efluent conditions (26 mM succinic acid, 4 mM tartaric acid and 1 mM 18-crown-6 at 0.6mL/min), the simultaneous separation of the cations and cations was achieved in *ca.* 20 min. The other parameters: PO_4^{3-}, ClO^-, O_3, H_2O_2, CN^-, phenol, total hardness (TH), CH_2O, NO_2^-, B, SiO_2; pH; temperature; and dissolved organic carbon (DOC) were determined by diverse methods as shown in Table 1.

Creation of the water pollution index maps was achieved using ArcGIS version 9.3(ESRI Inc., USA). Further analysis of the data was done by factor analysis, T-test, spearman correlation and Tukey's honestly significant different (HSD) test following one-way analysis of variance (ANOVA). The statistical analyses were performed using SPSS statistical package version 17.0 (SPSS Inc., Chicago, Ill).

Part of the analysis involved use of the Nemerow–Sumitomo Water Pollution Index (*WPI*). In this study, 73 water samples from different locales around Zambia and 20 different parameters were analysed. However, the following parameters: CH_2O, TH, H_2O_2, O_3, ClO^-, pH and DOC were not included in the water pollution index (*WPI*) because there are no clear permissible values (*PVs)* in Zambia (or worldwide consensus on safe levels in drinking water) indicating the maximum or minimum permissible values.

The *WPI* was used to evaluate the pattern of portable water in the study areas. The function of this method was to standardize the concentrations ranges for the parameters such that the different concentrations ranges for each water parameter were rescaled by the equation to produce a relative value that lies within a comparable range (Nemerow and Sumitomo, 1970). The *WPI* is a function of relative values (C_i/L_i), where, C_i represents the concentration of parameter i and L_i represents the *PV* of parameter i defined by a regulation.

$$WPI = \text{a function of } (C_i/L_i)\text{'s} = f(C_1/L_1, C_2/L_2, C_3/L_3 \ldots C_n/L_n) \quad (i = 1, 2, 3 \ldots n) \tag{1}$$

Table 1. Analysed parameters and the methods/instruments used.

Analysed parameter	Method/Instrumentation)
Phosphate (PO_4^{3-})	Molybdenum blue
Hypochlorite ($C\ell O^-$)	N,N-diethyl-p-phenylenediamine sulphate
Ozone (O_3)	4-Aminoantipyrine with enzyme
Hydrogen Peroxide (H_2O_2)	4-Aminoantipyrine with enzyme
Cyanide (CN^-)	4-Pyridinecarboxylic acid
Phenol	4-Aminoantipyrine with enzyme
Total Hardness (TH)	Phthalein Complex
Formaldehyde (CH_2O)	N-methylbenzothiazolone hydrazone (MBTH)
Nitrate (NO_2^-)	Naphthylethylenediamine
Boron (B)	Azomethine H
Silica (SiO_2)	Molybdenum blue
pH and Temperature	As One 392R pH meter
Dissolved Organic Carbon (DOC)	Total Organic Carbon Analyser TOC-V CSN Shimadzu

Parameters 1 to 12 were determined by digital water analyser, DPM-MT.

Then, the *WPI* for a specific water use *j* (WPI_j) is further expressed by the following equation:

$$WPI = \sum_{i=1}^{n} \sqrt{\frac{(C_i/L_{ij})^2_{max} + (C_i/L_{ij})^2_{ave}}{2}} \tag{2}$$

Where, C_i is the measured concentration of parameter *i*, L_{ij} is the *PV* for the parameter *i* determined for water use *j* (e.g. drinking or irrigation), and $(C_i/L_{ij})_{max}$ and $(C_i/L_{ij})_{ave}$ are maximum and average values of C_i/L_{ij} for water use *j*, respectively.

For the water parameters for which the higher value represents a higher level of pollutions, such as nitrate and heavy metals, the values of C_i/L_{ij} obtained from the field measurements can be directly calculated using the above equation, with a prerequisite. The prerequisite is that if the value of C_i/L_{ij} obtained from measurement is greater than 1.0, then the C_i/L_{ij} value must be standardized by applying the following equation:

$$(C_i/L_{ij})_{new} = 1.0 + x \log (C_i/L_{ij})_{ave} \tag{3}$$

Where, *x* is a constant value (as a standard value for a relative comparison, 5.0 is arbitrary employed for *x* value in the application of the index for the existing pollution).

For the parameters where the lower value represents a higher level of pollution, such as dissolved oxygen (DO), the C_i/L_{ij} values obtained from the field measurements must be standardized by using the following equation:

$$(C_i/L_{ij})_{new} = \frac{C_{im} - C_i}{C_{im} - L_{ij}} \tag{4}$$

Where, C_{im} is the saturation value for any parameter at room temperature.

For parameters for which the *PV* (L_{ij}) is defined by a range of numbers, such as for pH, where the *PV* ranges from 6 to 9, a standardized value C_i/L_{ij} is required, which is calculated by the following equation:

If $C_i \leq$ average L_{ij},

$$(C_i/L_{ij})_{new} = \frac{C_i - (L_{ij})_{ave}}{(L_{ij})_{min} - (L_{ij})_{ave}} \tag{5}$$

If $C_i >$ average L_{ij},

$$(C_i/L_{ij})_{new} = \frac{C_i - (L_{ij})_{ave}}{(L_{ij})_{max} - (L_{ij})_{ave}} \tag{6}$$

Where, $(L_{ij})_{min}$ and $(L_{ij})_{max}$ are, respectively, the maximum and minimum values of L_{ij} (e.g., pH: min = 6, max = 9). The $(L_{ij})_{ave}$ is the average value of L_{ij} (e.g., pH: (6 + 9)/2 = 7.5).

Based on chemical loadings relative to their *PVs*, the results from the water samples were classified into four categories. The classification used in this study reflects the suitability of the water for human consumption because only *PVs* for drinking water were used. It did not cater for the suitability of the water for use on animal husbandry purposes or crops. In addition, the *WPI* did take into account of water contamination due to biological activities. So if the water meets the *PV* criteria, it may still need some form of treatment (e.g. chlorination and filtration) but at a far lower cost.

The *PVs* for this study are based on Zambia Bureau of Standards (ZBS)/Environmental Council of Zambia (ECZ) except for NH_4^+, Na^+ (WHO, 2006), and K^+, PO_4^{3-} (USEPA, 2009). Utilizing the *PVs* obtained from ZBS/ECZ, WHO and US-EPA, the *WPI* was classified into four categories expressing the portable water's suitability for human consumption; the categories are as shown below.

$0.0 \leq WPI \leq 1.0$ = clean water (meets the *PV* criteria)
$1.0 < WPI \leq 5.0$ = slightly polluted water
$5.0 < WPI \leq 10$ = moderately polluted water
$WPI > 10$ = highly polluted water.

RESULTS

The Copperbelt Province and North-western Provinces (CB/NW) were analysed as one entity because the latter did not have sufficient data to analyse separately. At face value, comparison of Tables 2 and 3 reveal that the Copperbelt region/North-western Province had the most

Table 2. The averages, maximum, minimum values and standard deviations of the twenty-two parameters in the three provinces and their corresponding *PVs*.

Parameter analysed	Copperbelt Province/North Western Province						Lusaka Province					
	PV (drinking)	PV (waste water)	Max.	Min.	SD	Average	PV (drinking)	PV (waste water)	Max.	Min.	SD	Average
pH		6 - 9	8.36	6.73	0.39	7.77		6-9	8.44	5.75	0.46	7.63
Temp			27.00	22.00	0.93	24.23			26.30	20.70	1.40	23.47
SO_4^{2-}	400.00	1500.00	1659.72	0.47	396.93	197.12	400.00	1500.00	105.33	0.47	24.74	29.48
Cl^-	250.00	800.00	598.24	0.09	138.22	61.23	250.00	800.00	72.92	0.63	17.10	19.15
NO_3^-	10.00	50.00	11.54	0.00	2.76	1.26	10.00	50.00	127.56	0.00	24.22	19.10
Na^+	200.00	nd	37.13	2.40	10.96	12.41	200.00	nd	108.77	0.53	25.68	35.18
NH_4^+	1.50	10.00	nd	nd	nd	nd	1.50	10.00	6.49	0.17	2.29	3.09
K^+	250.00	nd	51.32	0.01	9.97	6.84	250.00	nd	43.48	0.01	7.88	4.54
Mg^{2+}	150.00	500.00	341.96	0.40	71.48	43.38	150.00	500.00	71.36	0.21	15.92	32.06
Ca^{2+}	200.00	nd	93.21	4.45	25.57	54.21	200.00	nd	108.98	0.83	22.75	54.63
DOC	nd	nd	2.44	0.19	0.59	1.26	nd	nd	5.77	0.05	1.62	1.51
PO_4^{2-}	0.02	0.03	1.00	0.16	0.26	0.39	0.02	0.03	3.20	0.10	0.74	0.46
$C\ell O^-$	nd	nd	0.32	0.32	0.00	0.32	nd	nd	0.53	0.14	0.10	0.26
O_3	nd	nd	0.70	0.32	0.17	0.46	nd	nd	0.39	0.27	0.05	0.35
H_2O_2	nd	nd	0.25	0.25	0.00	0.25	nd	nd	0.22	0.11	0.04	0.18
CN^-	0.10	0.20	nd	nd	nd	nd	0.10	0.20	0.06	0.03	0.01	0.03
Phenol	0.00	0.20	2.18	0.77	0.39	1.36	0.00	0.20	1.03	0.21	0.30	0.52
TH	200.00	nd	2080.00	26.00	530.89	495.91	200.00	nd	410.00	68.00	79.50	188.39
CH_2O	nd	nd	nd	nd	nd	nd	md	nd	0.23	0.19	0.01	0.21
NO_2^-	1.00	2.00	0.28	0.02	0.08	0.06	1.00	2.00	1.92	0.02	0.44	0.24
B	nd	0.50	nd	nd	nd	nd	nd	0.50	1.57	0.72	0.37	1.23
SIO_2	nd	nd	nd	nd	nd	nd	nd	nd	41.20	3.30	7.37	15.17

nd, Not determined.

Table 3. The averages, maximum, minimum values and standard deviations of the twenty-two parameters in the two provinces and their corresponding *PVs*.

Parameter analysed	Central Province						Southern Province					
	PV (Drinking)	PV (Waste Water)	Max	Min	SD	Average	PV (Drinking)	PV (Waste Water)	Max	Min	SD	Average
pH		6 - 9	7.74	7.63	0.06	7.69		6 - 9	8.13	7.63	0.21	7.93
Temp			21.30	20.80	0.25	21.05			20.90	20.50	0.16	20.70
SO_4^{2-}	400.00	1500.00	11.01	4.14	3.43	7.58	400.00	1500.00	62.08	21.50	17.80	46.39
Cl^-	250.00	800.00	2.78	0.70	1.04	1.74	250.00	800.00	12.44	2.57	4.12	6.91
NO_3^-	10.00	50.00	10.16	4.46	2.85	7.31	10.00	50.00	48.41	0.00	19.86	25.58
Na^+	200.00	nd	5.92	3.55	1.18	4.73	200.00	nd	81.02	10.64	28.79	47.10
NH_4^+	1.50	10.00	nd	nd	nd	nd	1.50	10.00	nd	nd	nd	nd
K^+	250.00		0.01	0.01	0.00	0.01	250.00		9.66	2.90	2.77	6.14
Mg^{2+}	150.00	500.00	50.23	45.65	2.29	47.94	150.00	500.00	49.08	16.93	13.34	34.68
Ca^{2+}	200.00	nd	66.65	43.62	11.51	55.14	200.00	nd	58.19	31.00	12.31	48.35
DOC	nd	nd	0.39	0.27	0.06	0.33	nd	nd	5.39	0.36	2.06	3.00
PO_4^{2-}	0.02	0.03	nd	nd	nd	nd	0.02	0.03	0.25	0.10	0.08	0.18
$C\ell O^-$	nd	nd	nd	nd	nd	nd	nd	nd	0.15	0.15	0.00	0.15
O_3	nd	nd	nd	nd	nd	nd	nd	nd	nd	nd	nd	nd
H_2O_2	nd	nd	nd	nd	nd	nd	nd	nd	0.36	0.36	0.00	0.36
CN^-	0.10	0.20	nd	nd	nd	nd	0.10	0.20	nd	nd		nd
Phenol	0.00	0.20	nd	nd	nd	nd	0.00	0.20	0.44	0.21	0.12	0.33
TH	200.00	nd	94.00	66.00	14.00	80.00	200.00	nd	200.00	74.00	51.55	134.67
CH_2O	nd	nd	nd	nd	nd	nd	nd	nd	0.41	0.41	0.00	0.41
NO_2^-	1.00	2.00	nd	nd	nd	nd	1.00	2.00	6.00	0.02	2.99	3.01
B	nd	0.50	nd	nd	nd	nd	nd	0.50	nd	nd	nd	nd
SIO_2	nd	nd	nd	nd	nd	nd	nd	nd	nd	nd	nd	nd

nd, not determined.

parameters whose average concentrations exceeded the averages obtained from the other provinces. An in-depth discussion was done to determine if the observed mean differences for CB/NW were significantly different across the regions and to determine which parameter(s) was significantly higher compared to others. Furthermore, the result of the analysis of ground versus surface water pollution is reported in the subsequent sections.

Statistical treatment of the results

Results across all the provinces

A one-way between-groups analysis of variance was conducted to explore the impact of location on pollution as measured by different methods. There was a statisticcally significant difference at the $p<0.05$ level in nitrate levels: $F(3, 68) = 5.2$, $p = 0.02$. Despite reaching statistical significance, the actual difference in mean scores between the locations was not very big. The effect size, calculated using eta squared, was 0.2. Post-hoc comparisons using the Tukey HSD test indicated that the mean score for Lusaka (M = 19.5, SD=24.7) was significantly different from CB/NW (M = 1.21, SD = 2.78), Central Province (M = 7.31, SD = 4.03), Southern Province (M = 25.6, SD = 24.3).

Furthermore, there was a statistically significant difference at the $p<0.05$ level in sodium levels: $F(3, 68) = 8.6$, $p = 0.00$. The effect size, calculated using eta squa-red, was 0.3. Post-hoc comparisons using the Tukey HSD test indicated that the mean score for Lusaka (M= 36.7, SD=25.7) was significantly different from CB/NW (M = 12.0, SD = 11.2), Central Province (M = 4.7, SD = 1.7), Southern Province (M = 47.1, SD = 35.3). Furthermore it was found that Southern Province differed significantly from CB/NW.

In addition, there was a statistically significant difference at the $p < 0.05$ level in total hardness: $F(3, 68) = 8.6$, $p = 0.00$. The effect size, calculated using eta squared, was 0.2. Post-hoc comparisons using the Tukey HSD test indicated that the mean score for Lusaka (M = 187.9, SD = 80.9) was significantly different from CB/NW (M = 496.0, SD = 543.4). An independent-samples T-test was conducted to compare the pollution levels of ground and surface water. With reference to nitrate, there was a significant difference in mean scores for ground water (M=25.9, SD = 26.9) and surface water (M = 4.1, SD = 9.0); $t(30.1) = 4.149$, $p = 0.000$ (two tailed). The magnitude of the differences in the means was 21.8, and with reference to sodium, there was a significant difference between ground water (M = 39.3, SD = 28.6) and surface water (M = 18.6, SD = 17.8); $t(42.6) = 3.5$, $p = 0.01$.

Furthermore, there was a significant difference in DOC levels between the water obtained from the ground (M = 0.90, SD = 1.2) and surface water (M = 1.8, SD = 1.5); $t(67.5) = -3.0$, $p = 0.004$.

Results for parameters obtained from Lusaka and Copperbelt Provinces *(phenol, NO_2^-, H_2O_2, O_3, ClO^-, PO_4^{3-})*

An independent-samples T-test was conducted to compare the pollution levels of the water obtained from the Lusaka and Copperbelt/North-western Provinces. With reference to phenol, there was a significant difference in scores for Lusaka (M=0.54, SD = 0.30) and Copperbelt / North-western Province (M =1.36, SD = 0.41); $t(17) = -5.574$, $p = 0.000$ (two tailed). The magnitude of the differences in the means was 0.81. The remaining parameters were not significantly different across the two provinces.

An independent-samples T-test was conducted to compare the pollution levels of the water obtained from the ground and surface water. With reference to the parameters listed above, there was no significant difference between ground and surface water.

Principal component analysis (PCA)

Principal component analysis (PCA) was performed only on the parameters that were analysed in all the five provinces (some parameters were not analysed in some provinces as shown in Tables 2 and 3). The results show that there was no ambiguity in the component matrix. After analysing the rotated matrix, it was even clearer that four components emerged. As shown in Table 5, component 1 (F1) was comprised of Cl^-, Mg, TH and K. Component 2 (F2) was constituted of NO_3^- and Na. The third component (F3) was made up of SO_4^{2-} and Ca^{2+}. DOC constituted the last component (F4). Table 4 shows that the four parameters explained 82.9% of the variance.

Correlations, as shown in Table 6, were observed between several parameters and the significant ones were between Na^+ and NO_3^- ($r = 0.588$); between K^+ and Cl^- ($r = 0.573$); between PO_4^{3-} and Ca^{2+} ($r = 0.651$) and between Mg^{2+} and Cl^- ($r = 0.925$) among others. In addition, location (province) was correlated positively with H_2O_2 and NO_2^- at 0.810 and 0.488, respectively. Water source (either obtained from the ground or rivers/streams) correlated significantly with DOC ($r = 0.319$). Using Tukey HSD test, PCA and the correlation coefficients, it was found that NO_3^- and Na^+ are indeed closely associated suggesting that they may have the same origin; the increase or decrease of either parameter will affect the other in a predictable manner. In addition, significant correlations of Cl^-, TH, K^+ and Mg^{2+} reinforces component 1 (F1) extracted by PCA. However component 3 (F3) was not supported by correlation coefficient; there was no evidence supporting a close relation that an increase or decrease of SO_4^{2-} leads to corresponding changes in levels of Ca^{2+} even though they fall under one component, F3. Furthermore, DOC analysis correlates significantly and positively with location, this finding confers addition insight to t-test results which showed that DOC levels

Table 4. Total variance explained.

Component	Initial Eigen values			Extraction sums of squared loadings			Rotation sums of squared loadings
	Total	% Variance	Cumulative %	Total	% Variance	Cumulative %	Total
1	3.41	37.84	37.84	3.41	37.84	37.84	3.34
2	1.67	18.57	56.41	1.67	18.57	56.41	1.65
3	1.23	13.63	70.04	1.23	13.63	70.04	1.53
4	1.16	12.89	82.93	1.16	12.89	82.93	1.19
5	.82	9.09	92.02				
6	.34	3.81	95.83				
7	.31	3.49	99.32				
8	.05	.53	99.85				
9	.01	.15	100.00				

Table 5. Pattern and structure matrix.

Component	Pattern matrix				Component	Structure matrix			
	1	2	3	4		1	2	3	4
Cl^-	0.976				Mg	0.967			
Mg^{2+}	0.961				Cl	0.937			
TH	0.873				TH	0.911			
K^+	0.756				K	0.776			
NO_3^-		0.903			NO_3^-		0.889		
Na^+		0.868			Na		0.882		
SO_4^{2-}			0.827		SO_4^{2-}			0.831	
Ca^{2+}			0.737		Ca			0.739	
DOC				0.959	DOC				0.957

were significantly high in surface water compared to ground water. However, there are several other parameters which correlated significantly but no rational explanation was given for their correlations because they were not supported by other analyses methods and they also proved not to be statistically significant in other regions compared to others; they did not either prove higher when compared to ground and surface water.

The water pollution index (WPI)

The findings of the Zambian water quality using the *WPI* are summarised in Tables 7, 8 and 9. The results are arranged in ascending order with water samples that met the benchmark (0.0 ≤*WPI* ≤1.0) to the water samples that were highly polluted (*WPI*> 10).

It is apparent from the *WPI* tables, that the Zambian water is compromised in one way or another, apart from the nitrate issue. The *WPI* reveals that out of all the water collected from Lusaka Province, 39.1% were not compromised (the water was safe for household consumption with reference to the analysed parameters); out of the water samples collected from the Copperbelt/North-western Provinces, 20.9% were safe for household consumption and 28.6% of the samples collected from Central and Southern Provinces were not polluted. Further analysis of the water shows that water treatment is reasonably effective in Mufulira as the water derived from the tap CB22T is suitable for human consumption despite its source being the highly polluted Kafue River (CB40R). However, the other water sample obtained directly from the mine ground aquifer and supplied to the former mine residential area proved to be unsuitable for human consumption, despite whatever treatment procedure is performed on the water, as shown by sample CB20T and according to the *WPI* of 1.63, is moderately polluted. The Lusaka city (Figure 2) and Copperbelt region (Figure 3) data is further presented on maps, using ArcGIS, showing spatial distribution of water quality across the two regions. There was a general tendency of the groundwater to degrade on the outskirts (commonly referred to as the peri-urban) of the Lusaka city boundary; most of the unplanned settlements in Lusaka are also built on the outskirtsof the city. As for surface water, it was expected that more pollutants would be observed towards the north east as it is generally the flow of all surface water in Lusaka, but there is no genera

Table 6. Pearson correlation of the pollution data.

	SO_4^{2-}	Cl^-	NO_3^-	Na^+	K^{2+}	Mg^{2+}	Ca^{2+}	DOC	PO_4^{3-}	ClO^-	O_3	H_2O_2	PNL	TH	NO_2^-	SiO_2	Location
Cl^-	-0.108	1															
NO_3^-	-0.106	-0.015	1														
Na^+	-0.063	0.190	**0.588****	1													
K^{2+}	0.174	**0.573****	-0.050	0.248*	1												
Mg^{2+}	0.156	**0.925****	0.048	0.179	**0.658****	1											
Ca^{2+}	0.243	0.092	0.036	0.026	0.040	0.204	1										
DOC	-0.058	-0.022	-0.230	0.074	0.212	0.006	0.174	1									
PO_4^{3-}	-0.113	0.033	0.027	0.356	**0.509****	-0.049	**0.651****	0.461*	1								
ClO^-	0.132	0.241	-0.089	0.062	-0.065	-0.268	0.088	-0.263	-0.003	1							
O_3	-0.242	-0.124	-0.460	-0.426	-0.001	-0.650	-0.750	-0.111	-0.651	0.490	1						
H_2O_2	0.287	-0.191	0.436	0.482	**0.852***	0.024	0.118	**0.825***	-0.941	-0.114	-10.000**	1					
PNL	0.045	0.113	-0.563**	-0.549*	-0.197	-0.593*	-0.038	-0.366	0.409	**0.887****	**10.000****	0.121	1				
TH	0.435**	**0.813****	-0.124	0.046	**0.609****	**0.892****	0.320*	-0.014	0.063	0.155	-0.040	-0.148	0.162	1			
NO_2^-	-0.018	-0.056	0.197	0.163	-0.031	-0.022	0.045	**0.679****	-0.745*	-0.281	-10.000**	0.920	-0.507	-0.054	1		
SiO_2	-0.292	-0.234	-0.129	-0.012	-0.083	-0.204	-0.318	0.052	0.022	0.029	-0.038	-0.849	-0.293	-0.163	-0.256	1	
Location	0.148	0.072	-0.189	-0.237*	0.049	0.063	-0.065	0.038	-0.136	-0.152	0.437	**0.810***	0.049	0.146	**0.488****	-	1
Water source	0.086	0.112	-0.511**	-0.412**	0.097	-0.077	0.154	**0.319****	0.299	0.137	0.437	-0.153	0.374	0.194	-0.145	-0.030	0.298*

**Correlation is significant at the 0.01 level (2-tailed). *Correlation is significant at the 0.05 level (2-tailed).

Observable trend. However, there is a general observable trend that the surface water which meets *WPI* clean water criteria is at or near the centre (e.g. LSK16R) of the city and but also intermingled with some of the worst polluted groundwater (e.g. LSK68G).

DISCUSSION

The water in Lusaka city revealed mixed results, that is, both acceptable chemical loadings and unacceptably high chemical loadings (to pollution levels) with nitrate being more widespread and predominant in high-density residential areas, like sites; LSK2R, LSK8R, LSK13T, LSK14G, LSK48G, LSK49T, LSK54G, LSK55G, LSK57G, LSK59G and LSK63G among others. The locations that revealed high levels of nitrates in tap water derive their water from boreholes and wells. A similar trend is observed in Magoye (site MG71G and MG74G), a typical rural area with a very low population density and no industrial activities; the sampled water was extracted from a borehole. The source of nitrate is mostly from decaying organic matter, sewage, fertilizers, manure, and nitrate in the soil or natural deposits, animal waste and septic tanks. The areas which tested for high nitrate have also been a source of perpetual cholera outbreaks; relentless cholera outbreaks have been reported over the past years with the first reported cases being in 1977 (Sasaki et al., 2009; WHO, 2011).

Analysis of the data shows that nitrate pollution is a major problem in Lusaka city in contrast to the CB/NW and Central Provinces, where none of the samples tested for nitrates beyond what is naturally expected. The low levels of nitrates prevailing in the CB/NW and Central Provinces can be attributed to the well organised and planned settlements which are serviced by sewer pipelines. This fact is quite opposite to Lusaka which is comprised of a large number of unplanned settlements (shanty townships, slums etc.), which are not serviced by a sewer system and garbage collection by the city council is way beyond their capacity.

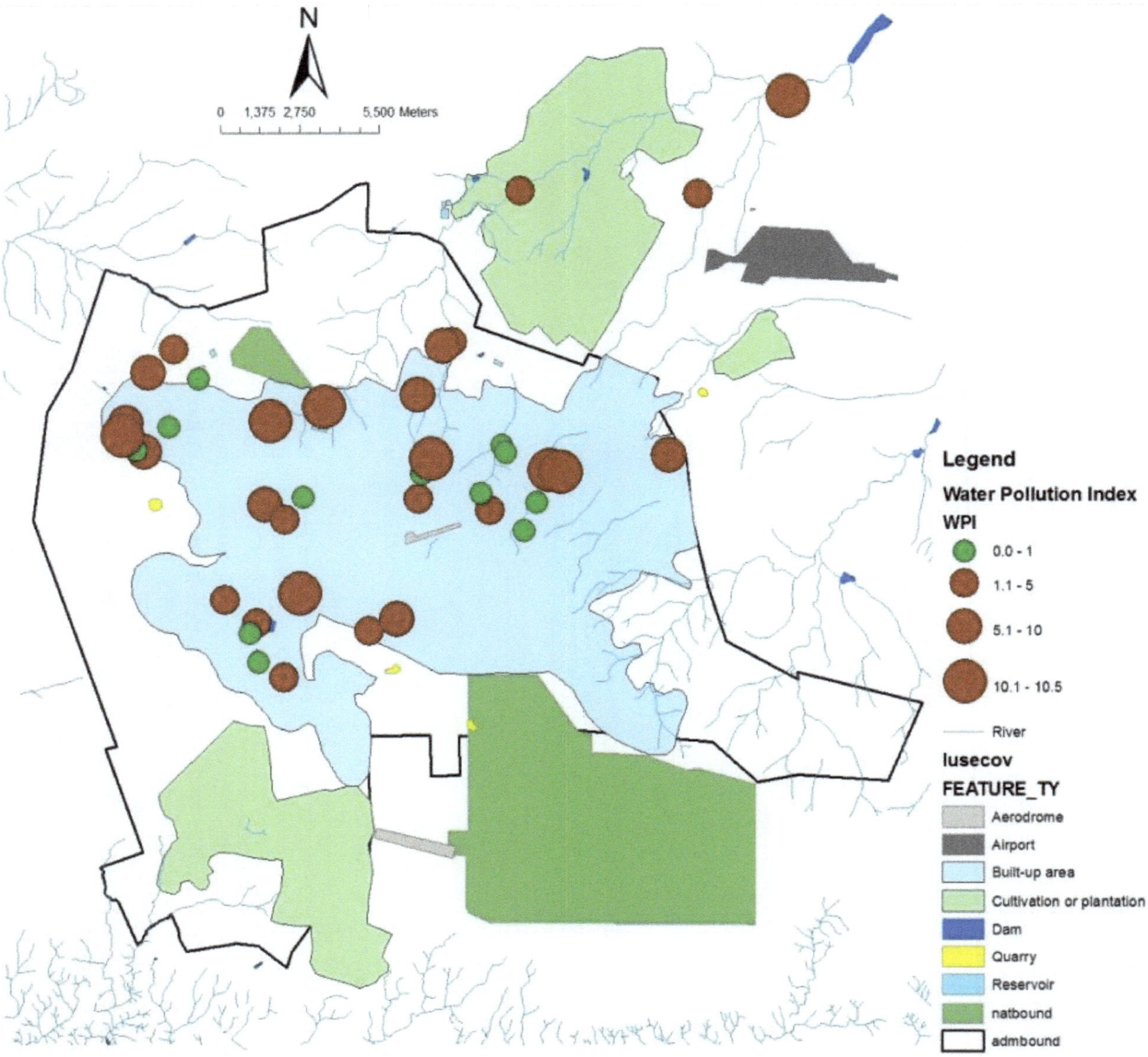

Figure 2. Lusaka city Data using the water pollution index.

This has resulted in many illegal/backyard dumpsites. The situation in Lusaka is also found in most rural areas like Magoye, where there has never existed any sewer system (except for *in-situ* sanitation), or any garbage collection system.

The high sulphate content (Table 2) is restricted to the Copperbelt region especially, near the effluent water discharged from the copper mines in Chingola and Mufulira towns. The mining activities are a probable explanation of the high levels of sulphates in the drinking water (e.g. site CB20T). The water in the Butondo River (CB21R, receiving drainage water from the mines) equally revealed high sulphate, an indication of how negatively the mine has affected the environment. "Zambia: MCM Acid Spillage" (2008) reports that Mopani Copper Mines (MCM) in Mufulira has disclosed that part of its ground water table has been contaminated following an acid spillage into the main domestic water supply system that pumps water to households in former mine townships.

Part of the tap water used in Lusaka is obtained from the Kafue River at Kafue town. Nevertheless, groundwater abstraction in the city has increased because of the construction of public and private boreholes. LWSC has approximately 50 boreholes around the city which draw water from the Lusaka and Cheta aquifers. For surface water derived from the Kafue River, there is a conventional treatment facility at the Ioland Water Treatment Plant. As for LWSC groundwater obtained from public boreholes, there are simple on-site chlorination facilities at borehole sites (LWSC, Personal Communication,

Table 7. Water quality in Lusaka Province using the water pollution index.

Code	Site name	WPI
LSK47G	Chainama Lusaka General Hospital	0.19
LSK16R	Roma/Garden Township Stream	0.24
LSK1D	Mutendere Dam	0.29
ND43R	Kafubu Upstream	0.32
LSK15T	Matero Petroda Station Tap Water (LWSC)	0.44
LSK56G	Mutendere Near Chainama Golf Club Well Water	0.44
LSK4R	Mumana Pleasure Resort (stream)	0.46
LSK51G	UNZA Tap Water - Groundwater	0.51
LSK65G	Zeko Camp Borehole	0.87
LSK17D	Blue Water Dam	1.05
LSK7R	Mutendere/Maplot midstream	1.36
LSK52G	Femag Garden John Laing Tap Water from the Ground	1.56
LSK58G	Entrance of Chunga - Ground Water	1.60
LSK50G	Vera Chiluba Basic School Tap from tap ground	2.17
LSK14G	John Laing Tap (ground water)	2.19
LSK9R	Ngombe Basic school Stream	2.31
LSK8R	Malimba Ngwerere Stream	2.43
LSK57G	Chunga, Africa Methodist Episcopal Church Ground Water Borehole	2.60
LSK49T	Libala Basic School Tap from the Ground	2.75
LSK63G	Ndeke Village	3.09
LSK13T	Chawama Compound Tap Water (LWSC)	3.15
LSK54G	Gospel Outreach Tap Water-ground	3.19
LSK53G	Kanyama Ground Water-Borehole-Filling Station	3.56
LSK68G	Thornpark Construction Centre Borehole	3.68
LSK60G	Avondale River Side Street Tap Water from the Ground	4.19
LSK10G	New N'gombe Basic Tap (underground)	4.50
LSK55G	Zingalume Police Tap Water from the Ground	4.72
LSK67G	Villa Elizabetha	4.73
LSK62G	George Compound Borehole	5.02
LSK64G	Chisengalumbwe Basic School	5.25
LSK61G	Ngome tap Water from Ground Aquifer	5.56
LSK59G	Twikatane Area Ground Water	5.65
ND46R	Kafubu Downstream	7.24
LSK3R	Mutendere River upstream	7.41
LSK48G	Kamwala Remand Prison Borehole Water	8.34
LSK72G	Zingalume borehole water	8.60
LSK2R	Chipata/Marapodi at Garden Park (stream)	9.05
LSK66G	NRDC Borehole Tap	9.22
LSK6R	Ngwerere Stream (near Kasisi)	9.28
LSK12R	Lumumba/Great North Road Junction (stream)	10.19
LSK70D	Goma lakes	10.47
ND31R	Kafubu River	10.93
ND45T	Laka Petroleum Filling Station Tap Water - Ndola	11.26

LSK, Lusaka city; ND, Ndola city; G, groundwater; T, tap water; R, river water; D, dam/water reservoir.

Table 8. Water quality in Copperbelt and North-western Province using the Water pollution Index.

Code	Site name	WPI
CB30R	Malembeka Stream into Kafue (1st site)	0.02
SL73G	Solwezi well water	0.08
CB23R	Boating Club Dam (Mufulira)	0.16
CB28R	Chingola/Chililabobwe Road Kafue River	0.20
CB22T	Tap Water - Mufulira Town Centre (Mulonga Water)	0.22
CB19R	Kafue River Mufulira/Kitwe Road	0.25
CB24R	Kafue River, Nkana East Waterworks	0.47
CB25R	River Near Sabina	0.62
CB26R	Before Tailings Damp (Butondo Stream)	0.67
CB20T	Mine Water Mulonga Tap (Entebbe)	1.63
CB27R	Tailings Damp Effluent)	2.10
CB21R	West Shaft Effluent Water	2.54
CB29R	Chingola Mine Effluent	4.77
CB34R	Ndola Kitwe Dual Carriage Kafue River	10.00
CB39R	Kawama stream - Mufulira	10.17
CB36R	Chingola River Upstream	10.55
CB33R	Butondo Stream at Road Near Butondo High	10.58
CB32R	West Mine Shaft - Butondo Stream (at Bridge)	10.70
CB40R	Kafue Water Treatment Plant - Mulonga WSC	10.77
CB44R	Kasuswa Stream (at Mufulira/Kitwe Road)	10.92
CB37R	Tailings Damp (TD 11)	11.09
CB38G	Eagle High School Tap Water (Groundwater)	11.17
CB35R	Malembeka Stream into Kafue (2nd site)	11.57

CB, Copperbelt; SL, Solwezi town; G, groundwater; T, tap water; R, river water.

Table 9. Water quality in central and southern province using the Water pollution index.

Code	Site name	WPI
KB41T	Kabwe City Council Tap Water (Lukanga WSC)	0.33
KB42T	Jack and Jill School - Lukanga WSC	0.75
MG74G	Magoye borehole water- school	2.34
KF11R	Kafue River at Motor Bridge	5.07
MZ69T	Mazabuka Tap water	8.04
KF18R	Train Bridge (Kafue River)	8.85
MG71G	Magoye borehole water - main compound	9.21

KB, Kabwe town; KF, Kafue town; MG, Magoye town; MZ, Mazabuka town; G, groundwater; T, tap water; R, river water.

January 11, 2013). And for private boreholes, the onus is on the individual to treat or not to. Inevitably these uncontrolled, unmonitored and untreated shallow and deeper boreholes can pose serious health hazard

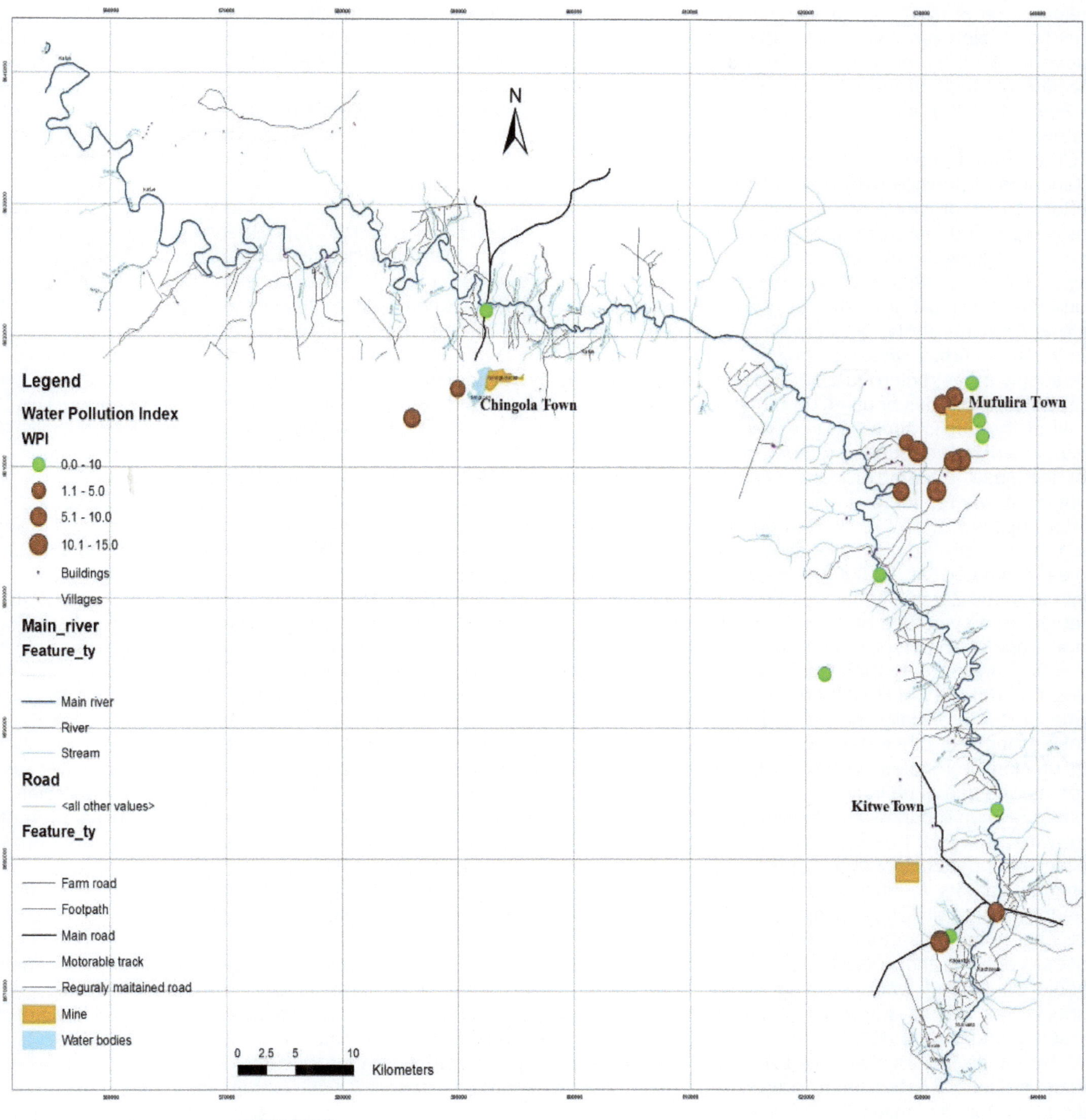

Figure 3. Copperbelt region Water Pollution Index.

Nitrate concentrations are similar in urban and rural areas and frequently exceed the drinking water limits set by the Zambia bureau of standards and environment council of Zambia. There is unequivocal evidence of pollution by sewer system as this problem was more predominant in the sites with a rudimentary sewer system, both urban and rural. Rural areas may have an addition loading from fertilizer application from the nearby farms but there is no strong evidence linking the two. For example site SL73G, found in a typical farming area did not test for nitrates beyond what is expected in nature.

It can be deduced from the locations sampled that the

main source of nitrate is from human excrement (pit latrines) and biomass decay (garbage). This fact is further supported by the presence of ammonia, and nitrite (intermediate products in the nitrogen cycle) in both the rural (MZ69T, MG71G and MG74G) water and Lusaka city water.

Groundwater protection is best achieved by defining "standards of practice" which, for example would exclude certain types of development and land use activity in areas specified as hygrogeologically sensitive. Recent hydrogeological and hydrochemical studies at the University of Toronto confirm that urbanization represents a serious threat to local groundwater quality (Howard, 1997). Due to its vulnerability, groundwater use may be restricted to certain areas where geology is favourable or where the supply infrastructure has not been installed, or it may only be certain types of users. In Birmingham, UK, virtually all of the public supply is surface water, with groundwater restricted to industries (Howard, 1997), but this is not the case in Zambian (and many other developing countries) where, for example, 58% (130,000 m^3/d) of water supplied by LWSC is from ground sources, excluding the sources from satellite systems and private boreholes estimated at 80,000 m^3/d (ECZ.ZEOR3, 2008). Data from the Zambian Ministry of Health (MoH) indicates that diarrhoea was one of the cause of hospital visitation, an average of 75 persons per 1,000 population from 2006 to 2008, ranking only third from malaria and respiratory infections (Ministry of Health and European Development Fund, 2010). Ironically, one of the interventions by the MoH, acting with other stakeholders, is to advocate sinking of more boreholes on top of the existing ones and promote the use of pit latrines. However, it is important to note that when groundwater is used as a source of drinking water, it is vital that sanitation methods and garbage collection and its disposal are of the highest efficiency to eliminate threats to human safety (Khalid et al., 2011; Zaadnoordijk et al., 2004), which is not the case in Zambia and numerous other developing countries. A risk assessment into the potential for sewer-related pollution should be carried when locating a new borehole site, especially if this is within or close to an urban area (Misstear and Bishop, 1997).

Some of the factors associated with environmental problems in Zambia is that there are no clear promulgated standards for the minimum or maximum groundwater depths, making evaluation of drainage situation difficult and although for some chemicals, quality standards are present, but the picture is not complete (for example DOC, TOC, pH, $C\ell O^-$, TH, B and PO_4^{3-}). Currently, there is no systematic priority list available on which pollutants are most serious in Zambia and what should be emphasized at household, industry and/or government level. Furthermore, there is a lack of an integral approach, by this we mean, ground water, stormwater, surface water, sewers and drinking water supply, all represents different dimensions of the same water system within a given locality. Nonetheless, these activities have been studied and planned separately. Hardly is the issue of groundwater given the gravity that it deserves and how it relates to human settlements and town planning. It was the purpose of this study to highlight some of the important water related aspects that would be worth giving a serious consideration.

One parameter worth noting is the amount of water TH in Lusaka city and Copperbelt region; it is important to note that guidelines for hardness are based on aesthetic, rather than health concerns. "Safe Drinking Water Foundation" (2006) has stated that levels greater than 200 ppm are considered poor but can be tolerated and levels in excess of 500 ppm are normally considered unacceptable. Of the Lusaka water samples, 33% were beyond 200 ppm, but all were below 500 ppm. This is in contrast to the Copperbelt region water (especially near the mining towns), where 13.6% of the samples' TH concentrations were greater than 200ppm; 31.8% with concentrations well beyond 500 ppm. The other sites assessed from miscellaneous regions around Zambia revealed normal TH with occasional undetectable levels of TH below 20 ppm.

The other parameters (ClO^-, H_2O_2, CH_2O, pH, DOC, O_3 and SiO_2), even though lack agreed upon standards, tested positive in some of the water samples and hence it would be appropriated to embark on research to establish the effects on human health from such chemicals as is common in the Zambian water. For instance, SiO_2 was detectable in all the samples tested from Lusaka city and 38% of all the samples tested positive for ClO^-; further discussion was not done for lack of Zambian *PV* standards to warrant an in-depth discussion. Another parameter highly considered in public discourse but with widely varying standards is pH; the ZBS/ECZ has not set any guidelines for drinking water. By and large, the pH values were within range of 6.5 to 8.5 (USEPA standards) except for one sample (LSK13T) from Chawama compound in the capital city of Zambia, Lusaka, which was slightly acidic at 5.75. However, it is important to use the unregulated contaminant monitoring program where data is collected for contaminants that were present in water samples, but do not have health-based standards set under the safe drinking water regulations. If safety of water is to be sustained, surface and groundwater protection process must be an integral part of the ministry in charge of environmental issues (Ministry of Tourism, Environment and Natural Resources in case of Zambia).

Conclusion

After analysing water from the five Zambian Provinces with different characteristics, it is indisputable that the pollution is anthropogenic in nature and is a consequence of mining on the Copperbelt, as revealed by TH levels, and the rampart unplanned residential areas in Lusaka Province as revealed by PCA, Tukey HSD test and correlation coefficients which show NO_3^- and Na^+ to be significantly higher in Lusaka compared to other regions. These parameters, among others, are a typical signature of human induced (human excrement) pollution. Furthermore, the levels of NO_3^- and Na^+ in ground water were significantly higher than in surface water. On the other hand, DOC levels were significantly higher in surface water compared to ground water and were not specific to any province. The users of the water in Zambia should pay particular attention when using ground water because of NO_3^- and Na^+ (that is, especially for Lusaka and Southern Provinces) and DOC pollution when using surface water irrespective of province. The residents of the Copperbelt Province should however be cautious of sulphate, phenolic compounds and excessive water TH especially when using the water obtained from the mine ground aquifers.

It is clear from this study that living in Lusaka (or any location heavily dependent on in-situ sanitation and backyard dumpsites like Magoye town) is particularly hazardous, given the high prevalence of nitrate pollution in tap water derived from the ground aquifers. Due to many people obtaining their drinking water from ground sources, it is important to set regulatory standards for more parameters and to monitor supplies regularly to ensure that potential health risk to humans are avoided. In Zambia, it is a popular advice (through private and public media and MoH) to tell the general public to either chlorinate and/or boil their drinking water, but merely boiling water will increase rather than decrease the contaminant concentrations. It is correct to argue that by boiling their drinking water, most Zambians are exposed to elevated pollutant concentrations than what was elucidated in this study. This study, therefore, suggests that water especially in Lusaka should not be boiled as a way of disinfecting it, but rather chlorinate because boil-ing will not remove the nitrates and other pollutants but instead increase their concentration.

ACKNOWLEDGEMENT

The authors would like to earnestly express their gratitude to the Global Environmental Leaders (GEL) program of Hiroshima University for the generous financial support which enabled the researchers to collect water samples from various locations in Zambia.

REFERENCES

Baird C, Cann M (2008).Environmental Chemistry. W.H. Freeman, San Francisco, USA.

Barret MH, Howard AG (2002). Urban groundwater and sanitation – Developed and Developing Countries. In: Howard KWF, Israfilov RG, Eds. Current problems of hydrogeology in urban areas, urban agglome-rates, and industrial centres. Kluwer Academic publishers, Netherlands. pp. 36-56

Chandra DS, Kirk S, Watkins J (1997). Groundwater pollution threat to public water supplies from urbanisation. In: Chilton J et al., Eds. Groundwater in the urban environment: problems, processes and management. Rotterdam, Balkema Publishers. pp. 31-40

Choongo KC, Syakalima MS, Mwase M (2005). Coefficient of condition in relation to copper levels in muscle of Serranochromis fish and sediment from the Kafue River, Zambia. Bull. Environ. Contam. Toxicol. 75(4) 645–651.

Chundama M (2008).An overview of the state of the environmental movement in Zambia: A status report. Retrieved from http://www.utviklingsfondet.no/filestore/EMiS_zambia_country_assessment.pdf

Environmental Council of Zambia and Zambia Environment Outlook Report 3 (2008). A report prepared by Environmental Council of Zambia. ECZ, Lusaka, Zambia.

Fan AM, Steinberg VE(1996). Health implications of nitrate and nitrite in drinking water: an update on Methemoglobinemia occurrence and reproductive and developmentaltoxicity. Regul. Toxicol. Pharmacol. 23(1): 35–43.

Foster F, Morris B, Lawrence A, Chilton J(1999). Groundwater impacts and issues in developing cities - An introductory review. In: Chilton J et al.,Eds. Groundwater in the urban environment: selected city profiles, pp. 3-15.

Foster SSD, Morris BL, Lawrence AR (1993). Effects of urbanization on groundwater recharge. In: Wilkinson WB, Ed. Groundwater problems in urban areas, proceeding of institution of civil engineers, London, pp. 43-63.

Green CH (2003).The handbook of water economics: principles and practice. John Wiley and Sons.

Howard KWF (1997). Incorporating policies for groundwater protection into the urban planning process. In: Chilton J et al, Eds. Groundwater in the urban environment: problems, processes and management. Balkema Publishers, Rotterdam. pp. 31-40

Khalid A, Malik AH, Waseem A, Zahra S, Murtaza G (2011). Qualitative and quantitative analysis of drinking water samples of different localities in Abbottabad district, Pakistan. Intern. J. Phys. Sci. 6(33): 7480 – 7489.

Khazai E, Riggi MG(1999). Impact of urbanization on the Khash aquifer, an arid region of south east Iran. In:Ellis JB, Ed. Impacts of urban growth on surface water and groundwater quality: proceedings of an international symposium held during IUGG 99, the XXII General Assembly of the International Union of Geodesy and Geophysics, at Birmingham, UK 18-30 July.IAHS, pp. 211-217.

Lerner ND, Yang Y, Barret MH,Tellam JH(1999).Loadings of non-agricultural nitrogen in urban groundwater. In:Impacts of urban growth on surface water and groundwater quality. Proceeding of IUUGG 99 Symposium HS5, Birmingham, July 1999. IAHS Pbli. no. 259, pp. 117-123

LWSC, Personal Communication, January 11, 2013

Magmedov VG,Yakovleva LI (1997). The problem of groundwater contamination at waste disposal sites in Ukraine. In: Chilton, J. et al.,Eds. Groundwater in the urban environment: problems, processes and management. Balkema Publisher, Rotterdam. pp. 481-486

Marquita H (2010). Understanding environmental pollution. Cambridge University Press, Cambridge New York.

Ministry of Health and European Development Fund (2010). Annual Health Statistical Bulletin 2008. Intercopy Enterprise, Lusaka.

Misstear BD, Bishop PK(1997). Groundwater contamination from sewers: experience from Britain. In: Chilton, J. et al.,Eds. Groundwater in the urban environment: problems, processes and management. Balkema Publishers, Rotterdam. pp. 491-496.

Mocanu VD, Mirca VD, Albu M (1997). Risk assessment of groundwater contamination from the Southeastern Bucharest landfill. In Chilton, J. et al.,Eds. Groundwater in the urban environment: problems, processes and management. Balkema Publishers, Rotterdam. pp. 491-496

Nakamboa S (2005). Review of the integration of water and sanitation in the PRSP. Partnership for Africa's water development (PAWD) project. Retrieved from http://www.gwpsa.org/pawd/docs/Review%20of%20integration%20of%20water%20in%20Zambias%20%20PRSP.pdf.

Nakayama S, Ikenaka Y, Muzandu K, Choongo K, Oroszlany B, Teraoka H, Mizuno N,Ishizuka M (2010).Heavy metal accumulation in lake sediments, fish (Oreochromis niloticus and Serranochromis thumbergi), and crayfish (Cherax quadricarinatus) in Lake Itezhi-tezhi and Lake Kariba, Zambia. Arch. Environ. Contam. Toxicol.59(2): 291–300.

Nemerow NL, Sumitomo H (1970). Benefits of water quality enhancement. Report No 16110 DAJ, prepared for the US Protection Agency. Syracuse University, Syracuse, NY.

Parslow RC, McKinney PA, Law GR, Staines A, Williams R,Bodansky HJ(1997). Incidence of childhood diabetes mellitus in Yorkshire, Northern England, is associated with nitrate in drinking water: an ecological analysis. Diabetologia 40(5): 550–556.

Perry-Castañeda Library Map Collection (2001). Zambia (Political). Retrieved from http://www.lib.utexas.edu/maps/zambia.html

Pickford J (1995). Lowcost sanitation. A survey of practical experience. Intermediate Technology Publications, London.

Safe Drinking Water Foundation. Total Hardness Analysis (Elementary School)(2006). Retrieved from http://www.safewater.org/PDFS/owd/OWDElementaryTotalHardnessAnalysiswithpgnumbers.pdf

Sasaki S, Suzuki H, Fujino Y, Kimura Y, Cheelo M (2009).Impact of drainage networks on cholera outbreaks in Lusaka, Zambia. Am. J. Public Health. 99(11): 1982-1987.

USEPA (2009).Drinking Water Contaminants. Retrieved from http://water.epa.gov/drink/contaminants/index.cfm

van Maanen JMS, van Dijk A, Mulder K, de Baets MH, Menheere PCA, van der Heide D, Mertens PLJM, Kleinjans CSJ(1994). Consumption of drinking water with high nitrate levels causes hypertrophy of the thyroid. Toxicol. Lett. 72(1–3): 365–374.

von der Heyden CJ,New MG(2003). Natural wetlands for mine effluent remediation? The case of the Copperbelt. In: Bernard T, Mosepele K, Ramberg L, Eds, Environmental monitoring of tropical and subtropical wetlands, Harry Oppenheimer Okavango Research Center of the University of Botswana and H.T. Odum Center for Wetlands of the University of Florida, Maun, Botswana. pp. 444–457

Whitehead E, Hiscock K, Dennis P (1999). Evidence for sewerage contamination for the Sherwood Sandstone aquifer beneath Liverpool, UK. In:Impacts of urban growth on surface water and groundwater quality (Proceedings of IUGG 99 Symposium HS5, Birmingham, July 1999). IAHS Publ. no. 259. pp. 179-185

WHO (2006).Guidelines for drinking-water quality [electronic resource] : incorporating first addendum. Vol. 1, Recommendations. – 3rd ed. Retrieved from http://www.who.int/water_sanitation_health/dwq/gdwq0506.pdf.

WHO (2011). Global taskforce on cholera control. Cholera country profile: Zambia. Retrieved from http://www.who.int/cholera/countries/ZambiaCountryProfile2011.pdf

Zaadnoordijk WJ, van den Brink C, van den Akker C,Chambers J(2004). Values and functions of groundwater under cities. In: Lerner DN, Ed. urban groundwater pollution, A. A. Balkema, Lisse. pp. 1-28

Zambia: MCM acid spillage pollutes Mufulira's domestic water supply (2008, January 2). Lusakatimes. Retrieved from http://www.lusakatimes.com/?p=1814

Zeb SB, Malik HA, Waseem A, Mahmood Q (2011). Water quality assessment of Siran river, Pakistan. Int. J. Phys. Sci. 6(34): 7789 – 7798.

2

Transesterification of oil extracted from different species of algae for biodiesel production

Farooq Ahmad, Amin U. Khan and Abdullah Yasar

GC University, Sustainable Development Study Centre, Lahore, Pakistan.

In the current study, biodiesel production efficiency of *Chlorella vulgaris, Rhizoclonium hieroglyphicum* and mixed algae culture was measured by transesterification process. Growth rate of algal species was measured on the basis of increase in their dry matter in various media. Protein, carbohydrates and lipids in all selected algae were measured on dry matter basis. Extracted oil was analyzed for water contents, iodine value, saponification value, acid number and fatty acid composition. Transesterification of algal oil was performed by using sodium methoxide as a catalyst. Yield of biodiesel from extracted oil was calculated for *C. vulgaris* (95%), *R. hieroglyphicum* (91%) and mixed algae culture (92%). Produced biodiesel was analyzed for kinematic viscosity (4.9, 5.0 and 4.7 mm^2/s), flash point (160, 156 and 155°C), specific gravity (0.91, 0.914 and 0.912 g/ml), cetain number (51, 49 and 47 min), iodine value (47, 53 and 49 mg/g), acid number (0.49, 0.5 and 0.46 mg.KOH/g), carbon residue (0.01, 0.02 and 0.01 mass%), sulfated ash (0.007, 0.003 and 0.004 mass%), sulphur (0.013, 0.012 and 0.01 wt%) and water contents (15, 23 and 17 mg/kg) for *C. vulgaris, R. hieroglyphicum* and mixed algae culture, respectively. Properties of biodiesel were compared with ASTM standards and it was found with high quality biodiesel.

Key words: Biodiesel, oil extraction, algae, transesterification.

INTRODUCTION

Fuel production from biomass is getting more importance these days due to scarcity, pollutants emission and increase in cost of conventional fossil fuels (Sensoz et al., 2000). Earlier studies revealed the fuel properties of algae and the effect temperature on the yield of algal biofuel (Demirbas, 2006). All algae contain proteins, carbohydrates, lipids and nucleic acids in varying proportions. While the percentages vary with the type of algae and the cultivation conditions, some algae types could accumulate 40% or more of their overall mass by fatty acids (Becker, 1994). Algae have the potential to produce more oil per acre than any other feedstock being used to make biodiesel (Demirbas, 2009).

Biodiesel are monoalkyl esters of long chain fatty acids which are transesterified from vegetable oil or animal fat (Ma and Hanna, 1999). Biodiesel has been commonly used in United States and Europe with annual production of 57 to 67 million L and 500 million to 1 billion L, respectively (Gerpan, 2005). The main obstacle in commercialization of biodiesel is high cost of oil feedstock. Research efforts are made to minimize oil feedstock cost (Zhang et al., 2003; Noureddini et al., 2005; Dhaermadi et al., 2006).

The most significant distinguishing characteristic of algal oil is its yield and hence its biodiesel yield. According to some estimates, the yield (per acre) of oil from algae is over 200 times the yield from the best-performing plant/vegetable oils. Many species exhibit rapid growth and high productivity and many microalgal species can be induced to accumulate substantial quantities of lipids, often greater than 60% of their dry biomass (Sheehan et al., 1998).

There are several methods used to extract oil from algae that includes mechanical systems, chemical, thermal, plasma and microwave techniques. Most of the traditional methods do not offer the long-term solutions. Algal oil can be extracted by using suitable solvents (Xu et al., 2006). Transesterification is the most common method and leads to monoalkyl esters of algal oil, vegetable oils and fats. The methyl ester produced by transesterification of vegetable oil has low viscosity and improved heating value as compared to those of pure vegetable oil which results in shorter ignition delay (Demirbas, 2010).

The current study was conducted to assess the biodiesel production efficiency of various species of algae. A comparison was made to find out the algal species with high oil contents and biodiesel production efficiency. Growth rate of selected algae was compared by growing in various culture media. Harvested biomass from these algal species was dried, ground and oil was extracted by Soxhlet extractor using hexane as a solvent. The extracted oil was transesterified to biodiesel using sodium methoxide as a catalyst. Biodiesel produced from these oils was analyzed and compared with ASTM standards.

MATERIALS AND METHODS

Algal sampling and identification

Samples of *Chlorella vulgaris, Rhizoclonium hieroglyphicum* and mixed algae culture were collected from fish forms and laboratory of Punjab Fisheries Department, Lahore Pakistan. These algal species were identified using standard methods described by Zarina et al. (2005a, b).

Culture media preparation

Aquatic cultures of algae were prepared in artificial ponds having dimensions of 0.3 x 0.3 x 0.15 m with final capacity 13.5 L. The rate of growth of these algal species was measured by the estimation of increase in its biomass for 6 days. Experiments were conducted in the month of November with light intensity of 25.5 K lux and 30°C of average temperature. The following three types of culture media were used for algal culture.

Wastewater

Raw sewage water collected from municipal drain was used for the growth of all selected algae in separate ponds.

Agar solution

This medium was prepared by dissolving 10 g of analytical grade agar in 1 L of a filtered tap water to measure the growth rate of selected algal species in ponds of same dimensions as mentioned earlier.

Synthetic medium

For the preparation of synthetic medium, six stock solutions were prepared in a distilled water as follows: (a) 30 g sodium nitrate per liter, (b) 2 g potassium phosphate per liter, (c) 3 g iron sulfate, 3 g citric acid, 1.5 g boric acid and 1 g manganese chloride per liter, (d) 0.22 g zinc sulphate, 0.79 g copper sulphate, 1.5 g ammonium molybdate, 0.23 g ammonium vanadate, 2.5 g ethylene diamine tatra acetic acid and 0.12 g cobalt chloride per liter, (e) 0.007 g vitamin B_{12} per liter and (f) 3 g ethylene diamine tatra acetic acid, disodium salt per liter. Synthetic medium for algal growth was prepared by mixing 10, 5, 1, 0.1, 20 and 20 ml per liter of each stock solution from serial (a) to (f) respectively (Hur, 2008).

Biodiesel production processes

Oil extraction from algae

In the current study, solvent extraction method was used because solvent used was recycled, reducing processing cost. Three hundred milliliters of n-hexane was used for 40 to 60 g of dried algae for the oil extraction. The extraction was carried out in a Soxhlet extractor (UNE-EN 734-1, 2006) for 4 h in order to determine the algae oil contents. All the experiments were conducted using a 0.5 L round-bottomed glass flask. The resultant solution was separated from solvent by distillation. The solvent was reused in the next batch of extraction.

Transesterification of algal oil

The extracted oil was transesterified to methyl ester (biodiesel) by sodium methoxide. The mixture was heated at 62°C and methanol (10 wt% dry basis) having sodium metal (0.1 wt% dry basis) previously dissolved was added to the reactor (Charoenchaitrakool and Juthagate, 2011). Reaction was conducted at the same temperature for 1 h with constant stirring at 110 rpm. The reaction mixture was cooled to room temperature after this process. Then, the solid phase was separated using a separating funnel under vacuum condition. Finally, the bottom layer of glycerin was separated from the mixture biodiesel and hexane layer (top layer), which was then washed with water to remove the methanol excess and the traces of catalyst (Karaosmanoglu et al., 1996; Lang et al., 2001; Chen et al., 2012). In order to obtain the crude biodiesel, it was necessary to remove the solvent by distillation.

Analysis of algal biomass and extracted oil

Algal biomass was analyzed for protein, carbohydrates and lipids. Protein content was determined by the block digestion method and lipid contents by solvent extraction (Boccard et al., 1981). The fatty acid profile of extracted oil was analyzed by GC-MS in a gas chromatograph (Shimadzu GC-14-A). Acid value, free fatty acids, iodine value and saponification value were determined by methods of Raie (2008).

Biodiesel analysis

The final product of methyl ester identified with the help of thin layer chromatography (TLC) and analyzed for kinematic viscosity, flash point, specific gravity, cetain number, carbon residue, sulfated ash, sulphur and water contents by ASTM standard methods and iodine value, acid number were by methods suggested by Raie (2008). In thin layer chromatography, thin layer chromatogram (0.25 mm of thick) having length of 20 cm and width of 20 cm was prepared by using water and silica gel. The plate was air dried and activated through heating at 105°C in an oven for one hour. Diethyl ether and n-hexane (20:80) were used as solvent system, biodiesel was dissolved in a solvent and 2,7-dichlorofluorescein was used as non-destructive locating agent to see color bands (purple-yellow) under ultra violet light of 366 nm wavelength.

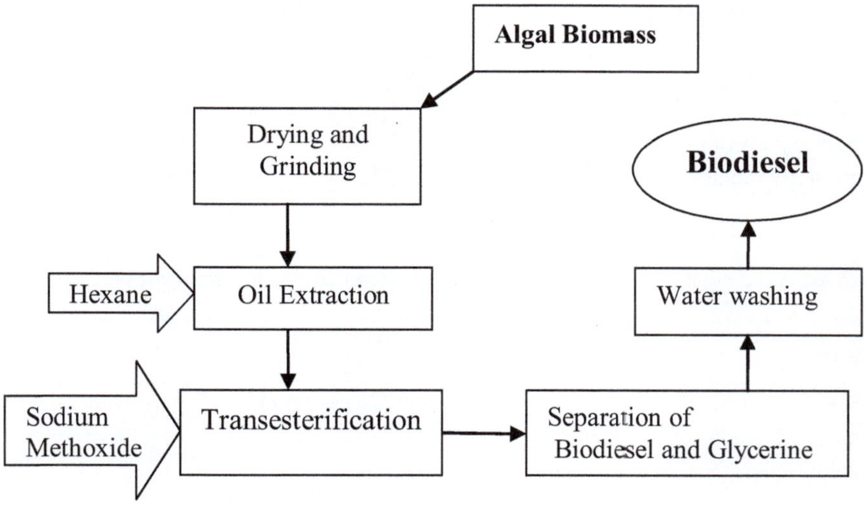

Figure 1. Steps involved in the biodiesel production from algal biomass.

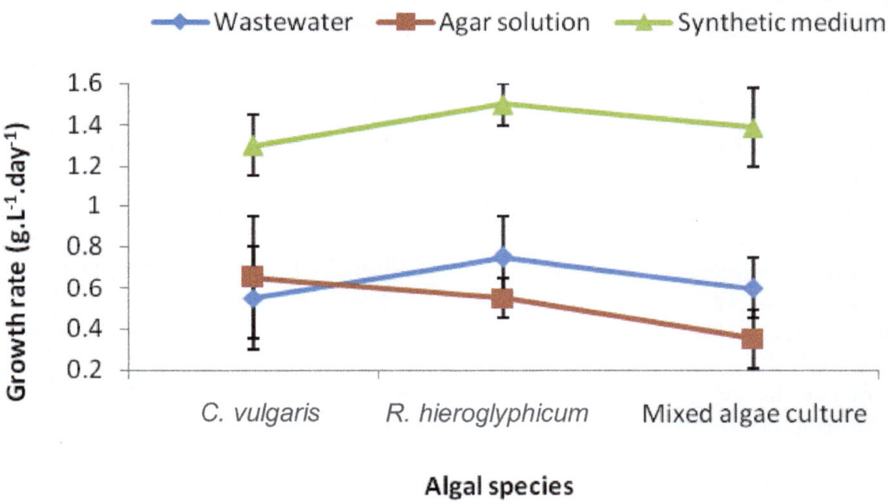

Figure 2. Growth rate of algal species in three different types of nutrient media.

RESULTS AND DISCUSSION

Processing algae for biodiesel

Algae were processed to biodiesel by extracting oil from dried and ground biomass of algae using hexane as solvent. It was transesterified to biodiesel by sodium methoxide catalyst. Biodiesel produced was separated from glycerin by separating funnel and washed with water to get pure biodiesel (Figure 1).

Growth rate measurement

Growth rate of all three types of algae *(C. vulgaris, R. hieroglyphicum* and mixed algae culture*)* was measured by growing these species in three types of nutrient media (wastewater, agar solution and synthetic medium) and it was observed that all species showed maximum growth in synthetic medium as it contained all the necessary nutrients its required amount. *R. hieroglyphicum* showed growth rate of 1.5 $g.L^{-1}.day^{-1}$ in synthetic medium which was higher than *C. vulgaris* and mixed algae culture. The growth rate of these species in wastewater and agar solution was almost similar with *C. vulgaris* which showed more growth in agar solution while *C. vulgaris* and mixed algae culture showed more growth in wastewater (Figure 2) but it was decreased with time (over 6 days) due to deficiency of nutrients if no fresh medium was added.

Ruiz-Marin et al. (2010) performed many experiment under batch culture condition in which microalgae showed high growth rates in initial days but growth and chlorophyll contents were decreased after four cycles of culture

Table 1. Percentage chemical composition of selected algal species on dry matter basis.

Sample specie	Protein	Carbohydrate	Lipid	other
C. vulgaris	35±3	16±1.5	45±1.9	4±0.4
R. hieroglyphicum	38±2.1	19±1.8	34±0.9	9±0.7
Mixed algae culture	28±1.6	29±3.4	35±2.8	8±1.3

Table 2. Analysis of extracted algal oil for various parameters before converting it to biodiesel.

Analytical parameter	Oil extracted from *C. vulgaris*	Oil extracted from *R. hieroglyphicum*	Oil extracted from mixed algae culture
Water content (%)	0.07±0.01	0.08±0.009	0.05±0.015
Iodine value (mg/g)	63±2	75±1	69±2
Saponification value (mg/g)	173.12±3.4	168.3±4.1	182.6±1.5
Free fatty acid (%)	2.6±0.3	3.6±0.7	3.3±0.1
Acid number (mg KOH/g)	31.46±2.1	34.5±1.8	37.3±2.8

indicating collapse of the culture due to nutrient deficiency.

Algal biomass analysis

Algal biomass was harvested by filtration and gravity sedimentation methods. It was dried and analyzed for protein, carbohydrates and lipid contents. *C. vulgaris* showed higher lipid contents (45%) than *R. hieroglyphicum* (34%) and mixed algae culture (35%) on the basis of their dry mass whereas protein and carbohydrates were observed to be more in *R. hieroglyphicum* (Table 1). An integrated approach of biodiesel production from heterotrophic *C. protothecoides* showed that the lipid content reached 46.1, 48.7 and 44.3% of cell dry weight in samples from 5, 750 and 11,000 L bioreactors, respectively (Li et al., 2007).

Analysis of extracted algal oil

Oil extracted from algal species was analyzed for water contents, iodine value, saponification value, free fatty acids and acid number. It was deduced from the results that there was no much difference in the above mentioned parameters in oil extracted from all three species of algae (Table 2). Properties of oil were comparable with that of the study conducted by Li et al. (2007) in which water content in the extracted oil was found to be less than 1% which means that oil feasible for methyl ester production as transesterification level decreased if the water content increased to 10% of the oil quantity. Saponification value of 189.3 mg KOH g^{-1} and acid value of 8.97 mg KOH g^{-1} were also measured in lipid extracted from heterotrophic chlorella cells (Li et al., 2007).

Fatty acids profile (14:0, 16:0, 16:1, 16:2, 16:3, 16:4, 18:0, 18:1, 18:2, 18:3, 18:4, 20:0, 20:1, 20:2, 20:3 and 20:4) of algal oils was also analyzed in algal oils of selected types of algae and it was observed that unsaturated fatty acids percentage was more in *C. vulgaris* (78.09%) than *R. hieroglyphicum* (77.5%) and mixed algae culture (77.99%), while saturated fatty acids were higher in *R. hieroglyphicum* (Table 3). Feasibility of biodiesel production from algal consortium grown in treated wastewater was checked by Chinnasamy et al. (2010), about 126.7 g (144 mL) of crude algal oil was extracted from 2.3 kg of dry algal biomass and extracted oil was examined for the fatty acid profile using GC. The profile showed 14:0, 16:0, 16:1, 18:0, 18:1, 18:2, 18:3, 20:0, 20:1, 20:2, 20:3, 20:4, 20:5, 22:5 and 22:6 fatty acids. These results are also in conformity with the observations made by Gouveia and Oliveira (2009) that microalgal lipids derived from *C. vulgaris, Scenedesmus maxima, Nannochloropsis oleabundans, Scenedesmus obliquus* and *Dunaliella tertiolecta* were mainly composed of unsaturated fatty acids (50 to 65%).

Yield of biodiesel

Hundred grams algal biomass of each species was used to measure the production of oil and biodiesel. Maximum quantity of oil was obtained from *C. vulgaris* biomass (45 ml/100 g), it was higher than *R. hieroglyphicum* (35 ml/100 g) and mixed algae culture (33 ml/100 g). Similarly yield of biodiesel was 94, 92 and 91% by *C. vulgaris, R. hieroglyphicum* and mixed algae culture, respectively (Table 4). Li et al. (2007) showed that 98.15% of the oil was converted to monoalkyl esters of fatty acids in 12 h by transesterification of the microalgal oil, catalyzed by immobilized lipase from *Candidia* sp.

Identification of biodiesel by thin layer chromatography

The oil extracted from algal species, biodiesel produced from *C. vulgaris* (Sample 1), *R. hieroglyphicum* (Sample 2) and mixed algae culture (Sample 3) were analyzed and compared with standards of fatty acids and methyl

Table 3. Percentage composition of fatty acids of oil extracted from *C. vulgaris, R. hieroglyphicum* and mixed algae culture algae.

Fatty acid profile	Oil extracted from *C. vulgaris*	Oil extracted from *R. hieroglyphicum*	Oil extracted from mixed algae culture
C14:0	2.26	0.35	1.35
C16:0	18.9	20.89	19.33
C16:1	9.53	11.39	15.01
C16:2	0.04	0.02	n.d.
C16:3	6.29	5.34	4.36
C16:4	7.62	10.03	8.94
C18:0	0.65	1.26	1.15
C18:1	19.58	15.65	14.75
C18:2	11.15	10.16	15.6
C18:3	22.17	23.73	18.07
C18:4	n.d.	0.07	0.02
C20:0	0.1	n.d.	0.18
C20:1	0.91	0.65	0.57
C20:2	0.79	0.43	0.67
C20:3	n.d.	0.03	n.d.
C20:4	0.01	n.d.	n.d.
Saturated	21.91	22.5	22.01
Unsaturated	78.09	77.5	77.99

*n.d. = not detected.

Table 4. Quantification of yield of biodiesel and by product synthesized from algal biomass.

Products and by products	*C. vulgaris*	*R. hieroglyphicum*	Mixed algae culture
Total biomass used (g)	100	100	100
Residual biomass (g)	57±2	69±3.5	71±2.6
Quantity of oil extracted (ml)	45±0.9	35±0.6	33±1.1
Yield of biodiesel (%)	94	92	91
Glycerin and other by products (%)	6	8	9

ester by thin layer chromatography. Figure 3 show that biodiesel produced from all three species was pure methyl ester as there were no traces of glycerides found in all three samples.

Biodiesel analysis

Biodiesel produced by transesterification of algal oil was analyzed for various parameters and results showed that quality of biodiesel obtained from all species was comparable to ASTM D-6751~02 standards. It was also observed that there was no significant difference in the properties of biodiesel produced from *C. vulgaris, R. hieroglyphicum* and mixed algae culture (Table 5). Li et al. (2007) obtained similar results in which properties of biodiesel produced from *Chlorella* sp. were comparable to conventional diesel fuel and comply with the US Standard (ASTM 6751) for Biodiesel.

Conclusions

In the current study, three algal species were used to extract oil and its conversion to biodiesel. The study revealed that algae are fast growing and effective organism for biodiesel production as these can be grown in wastewater as well as in artificial media. Raw municipal wastewater can be used as a medium of growth for algae as growth rate of *C. vulgaris, R. hieroglyphicum* and mixed algae culture were comparable with artificial nutrient media. Oil extracted from harvested biomass of these algae was transesterified to biodiesel using sodium methoxide as a catalyst. Resultant biodiesel was analyzed and compared with ASTM standards. It was found that properties of biodiesel were in accordance with standards limits so it can be blended with fossil fuels or can be used individually.

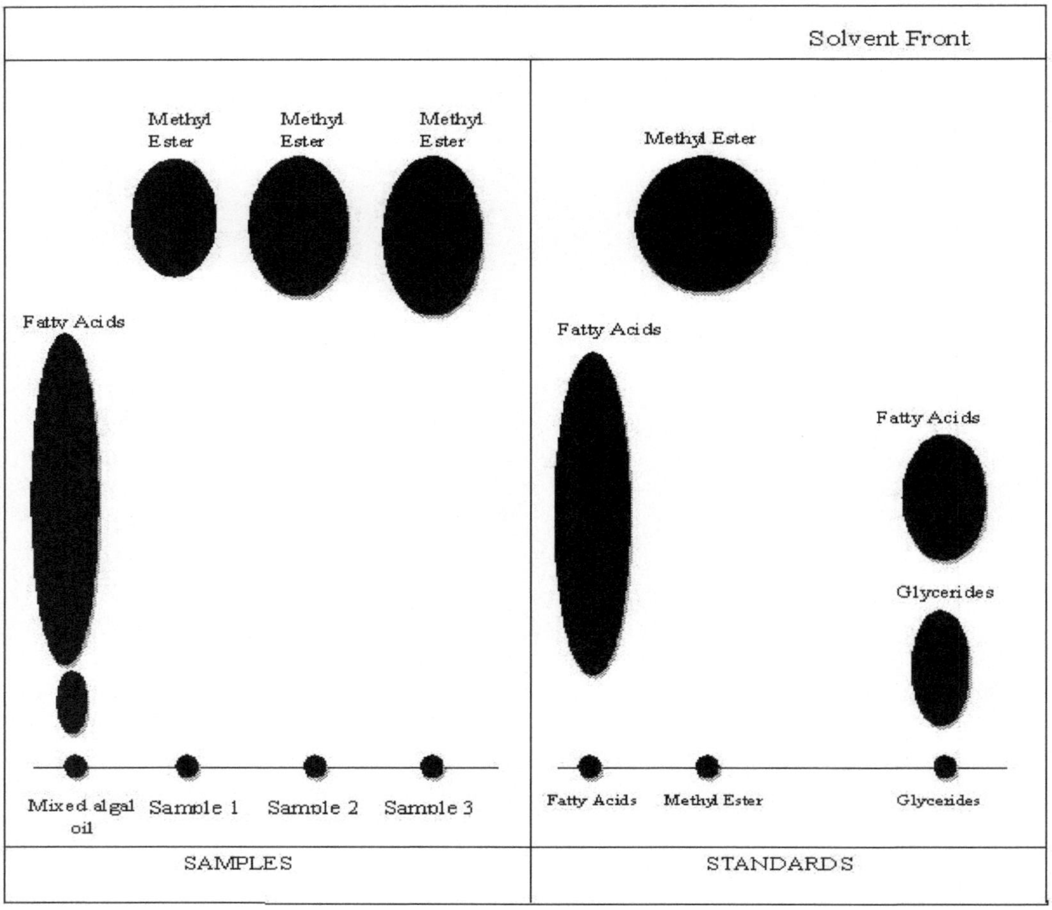

Figure 3. Identification and comparison of biodiesel (methyl ester) with standards using solvent system of n-hexane to diethylether ratio of 80:20.

Table 5. Analysis of biodiesel produced from selected algae and its comparison with international standards.

Properties	Unit	*C. vulgaris*	*R. hieroglyphicum*	Mixed algae culture	ASTM D-6751~02 Standard
Kinematic viscosity at 40°C	mm^2/s	5.2	5.0	4.8	1.9–6.0
Flash point	°C	145	146	140	>130
Specific gravity at 28°C	g/ml	0.916	0.914	0.912	0.88
Cetain number	min	53	51	49	>47
Acid number	mg.KOH/g	0.49	0.5	0.46	0.8 max.
Water contents	%vol	0.04	0.05	0.07	0.05 max.
Calorific value (gross)	MJ/Kg	41.2	35.3	37.2	-----

ACKNOWLEDGEMENTS

The authors acknowledge GC University Lahore for providing facilities for the current study. The authors are also indebted to Dr. Muhammad Zeeshan and Ikram Hussain Arain at PCSIR and SEPCOL Lahore, respectively for assistance in oil and biodiesel testing. The authors also acknowledge Mr. Tariq Rashid and Director, Punjab Fisheries Department for providing algal samples and guidance during the course of the work.

REFERENCES

Becker EW (1994). Microalgae: biotechnology and microbiology. Cambridge University Press, New York, USA. pp. 177-195.

Boccard R, Buchter L, Casteels E, Cosentino E, Dransfield E, Hood DE, Joseph RL, Macdougall DB, Rhodes DN, Schön L, Tinbergen BJ, Touraill C (1981). Procedures for measuring meat quality characteristics in beef production experiments. Livest. Prod. Sci. 8: 385–397.

Charoenchaitrakool M, Juthagate T (2011). Statistical optimization for biodiesel production from waste frying oil through two-step catalyzed process. Fuel Process. Technol. 92: 112–118.

Chen L, Liu T, Zhang W, Chen X, Wang J (2012). Biodiesel production from algae oil high in free fatty acids by two-step catalytic conversion. Biores. Technol. 111: 208-14.

Chinnasamy S, Bhatnagar A, Hunt RW, Das KC (2010). Microalgae cultivation in a wastewater dominated by carpet mill effluents for biofuel applications Biores. Technol. 101: 3097–3105

Demirbas A (2006). Oily products from mosses and algae via pyrolysis. Energy Sources 28: 933–940.

Demirbas AH (2009). Inexpensive oil and fats feedstocks for production of biodiesel. Energy Edu. Sci. Technol. 23: 1–13.

Demirbas MF (2010). Microalgae as a feedstock for biodiesel. Energy Edu. Sci. Technol. 25: 31–43.

Dhaermadi Y, Murarka A, Gonzalez R (2006). Anaerobic fermentation of glycerol by *Escherichia coli*: A new platform for metabolic engineering. Biotechnol. Bioenergy 94: 811–829.

Gerpan JV (2005). Biodiesel processing and production. Fuel Process. Technol. 86: 1097–1107.

Gouveia L, Oliveira AC (2009). Microalgae as a raw material for biofuels production. J. Ind. Microbiol. Biotechnol. 36: 269–274.

Hur SB (2008). Korea Marine Microalgae Culture Center - List of Strains. Algae 23(3): 1-68.

Karaosmanoglu F, Cigizoglu KB, Tuter M, Ertekin S (1996). Investigation of the refining step of biodiesel production. Energy Fuels 10: 890-895.

Lang X, Dalai AK, Bakhshi NN, Reaney MJ, Hertz PB (2001). Preparation and characterisation of biodiesels from various bio oils. Biores. Technol. 80: 53-62.

Li X, Xu H, Wu Q (2007). Large-scale biodiesel production from microalga *Chlorella protothecoides* through heterotrophic cultivation inbioreactors.Biotechnol.Bioeng. 98(4):764-71.DOI10.1002/bit.21489.

Ma FR, Hanna MA (1999). Biodiesel production: A review. Biores. Technol. 70: 1–15.

Noureddini H, Gao X, Philkana RS (2005). Immobilized Pseudomonas cepacia lipase for biodiesel fuel production from soybean oil. Biores. Technol. 96: 769–777.

Raie MY (2008). Oil, Fats and Waxes. 1st Edition, National Book Foundation, Islamabad, (Pakistan).

Ruiz-Marin A, Mendoza-Espinosa LG, Stephenson T (2010). Growth and nutrient removal in free and immobilized green algae in batch and semi continuous cultures treating real wastewater. Bioresour. Technol. 101: 58–64.

Sensoz S, Ang n D, Yorgun S (2000). Influence of particle size on the pyrolysis of rapeseed (Brassica napus L.): fuel properties of bio-oil. Biomass Bioenergy 19: 271–279.

Sheehan J, Dunahay T, Benemann J, Roessler P (1998). A Look Back at the U.S. Department of Energy's Aquatic Species Program Biodiesel from Algae. National Renewable Energy Laboratory (NREL) Report: NREL/TP-580-24190. Golden, CO.Xu H, Miao X, Wu Q (2006). High quality biodiesel production from a microalgae *Chlorella protothecoides* by heterotrophic growth in fermenters. J. Biotechnol. 126: 499–507.

Zarina A, Masud-ul-Hasan, Shameel M (2005a). Taxonomic study of Vaucheriophyta Shameel from certain areas of the Punjab and NWFP, Pakistan. Int. J. Phycol. Phycochem. 1: 159-166.

Zarina A, Masud-ul-Hasan, Shameel M (2005b). Taxonomic study of the order Ulotrichales (Chlorophyta) from north-eastern areas pf Pakistan. Pak. J. Bot. 37: 797-806.

Zhang Y, Dube MA, Mclean DD (2003). Biodiesel production from waste cooking oil: Process design and technological assessment. Bioresour. Technol. 89: 1–16.

3

Bio-ethanol production from poultry manure at Bonga Poultry Farm in Ethiopia

Asrat Gebremariam Woldesenbet[1] , Gizachew Shiferaw[1] and Bhagwan Singh Chandravanshi[2]

[1]School of Chemical and Bio-Engineering, Addis Ababa Institute of Technology, Addis Ababa University, P.O. Box 385, Addis Ababa, Ethiopia.
[2]Department of Chemistry, Faculty of Sciences, Addis Ababa University, P.O.Box 1176, Addis Ababa, Ethiopia.

This study aimed to determine bio-ethanol potential of poultry manure. The manure was hydrolyzed by dilute H_2SO_4 (0.2, 0.4, 0.6, 0.8 and 1 M) at temperature of 100 and 130°C for 30 min. Maximum glucose (115.6 g/l) was obtained from hydrolysis of 100 g poultry manure by 0.8 M H_2SO_4 at 100°C for 30 min. The study shows that treatment of poultry manure releases more total reducing sugar using 0.8 M H_2SO_4. Refractometry analysis of the fermented hydrolysate gave maximum ethanol concentration of 5% w/v with 0.8 M H_2SO_4 and 4, 4, 3 and 2.5% of ethanol with 1, 0.6, 0.4 and 0.2 M H_2SO_4 hydrolysis at 130°C for 30 min, respectively. Treatment with 0.8 M acid is an effective way that results to maximum yield and treatment with 0.2 M acid is the least effective. The optimum fermentation time is 24 h.

Key words: Bioethanol, bio-mass, fermentation, conversion, ethanol fuel.

INTRODUCTION

The high solubility of poultry manure in water provides a strong possibility for the occurrence of water pollution. The application of excessive amounts of poultry manure can result in the leaching of nutrients through the soil and into the local groundwater. Efficient manure management ensures that manure or its constituents cannot gain access to rivers, streams, lakes or water supplies (Akpan et al., 2008).

One of the greatest challenges for the growing society in this century is to meet the energy demand for transportation, heating, lighting and industrial processes, which have significant impact on the environment. These challenges demand an urgent need to carry out research to find out the viable alternative fuels (Akpan et al., 2008).

The Energy Independence and Security Act (EISA) mandate that at least 16 billion gallons (61 billion liters) per year of cellulosic fuel be in production by the year 2022. To meet this ambitious goal, many feedstocks, with appropriate conversion technologies, will be required for fuel production (Budsberg et al., 2012).

One of the promising alternative fuels to gasoline in the transport sector is bio-ethanol. Fuel ethanol is known as bio-ethanol, since it is produced from plant materials by biological processes. Bioethanol is mainly produced by fermentation of sugar containing crops like corn, maize, wheat, sugar cane, sugar beet, potatoes, sorghum and cassava (Akpan et al., 2008).

There are lots of other alternative fuels such as methanol, methane, natural gas, propane, hydrogen, etc. Nevertheless, the remarkable characteristics of ethanol distinguish it as the best alternative fuel for automobile. It has high latent heat of vaporization, high octane number and rating, and emission of toxic compounds on its combustion is low (Akpan et al., 2008).

Approximately 80% of the ethanol produced in the world is still obtained from fermentation (Gunasekaran, 2007); the remainder comes largely by synthesis from the petroleum product ethylene (Tanaka, 2006). All the ethanol used for fuel and green house gas and they are responsible for global warming and climate change. Nowadays, research on alcoholic drinks, and most industrial ethanol, is made by this process (Licht, 2001). In 2002, world ethanol production was projected at 34 million m^3 (Licht, 2007).

Fuel ethanol is the largest market by far, accounting for 60% of total ethanol production worldwide (Licht, 2001). This share is likely to increase over the coming years as many countries set up fuel ethanol programmes. Industrial ethanol accounts for 20% of the market and beverages for about 15%; both these markets are growing comparatively slowly.

According to Licht (2007), the total world production of ethanol was about 40,700 million litres(L), of which 73% was used as vehicles fuel, 17% for production of beverages and 10% for other industry needs (Anuj et al., 2007). Brazil and USA are by far the largest producers of fuel alcohol. Ethanol produced from biomass is considered as one of the modern forms of biomass energy that has the potential to be a sustainable transportation fuel for gasoline engines.

Ethanol is a clear, colorless, flammable, oxygenated hydrocarbon with the chemical formula C_2H_5OH. Ethanol can be used as a transport fuel in at least four forms: anhydrous ethanol (100% ethanol), hydrous ethanol (95% ethanol and 5% water), anhydrous ethanol-gasoline blends (10 to 20% ethanol in gasoline) and as raw material for ethyl *t*-butyl ether (ETBE).

Bio-ethanol has a number of advantages over conventional fuels. Some of these are: it comes from a renewable resource, that is, crops and not from a finite resource, and the crops it derives from can grow well (like cereals, sugar beet and maize). The road transport network accounts for 22% of all greenhouse gas emissions and through the use of bio-ethanol, some of these emissions will be reduced as the fuel crops absorb the CO_2 they emit through growing.

Blending bio-ethanol with petrol will reduce heavy reliance on oil producing nations. Bio-ethanol is biodegradable and far less toxic than fossil fuels. Ethanol has much higher latent heat of vaporization (855 MJ/kg) than petrol (293 kJ/kg). Ethanol has a higher octane number (99) than petrol (80 to 100) as a result pre-ignition does not occur when ethanol is used: ethanol burn more completely, so that hydrocarbon emission is drastically lower as compared to petrol. Ethanol is much less likely to catch fire and explodes in case of fuel leakage. Using bio-ethanol in older engines can help to reduce the amount of carbon monoxide produced by the vehicle thus improving air quality (Attygalle, 2008). The production and use of biofuels has the potential to reduce greenhouse gas emissions, particularly with the development of bioethanol from agricultural wastes or lignocellulosic crops (Scott et al., 2013).

The use of fossil fuel in cars releases carbon dioxide, carbon monoxide, sulfur dioxide, oxides of nitrogen and heavy metals. These gases are the main constituents of non-food crops and cellulosic materials is getting great attention worldwide because they are cheap, easily available, and profitable as compared to food crops and also reduces inflation of the cost of food crops used for bio-ethanol production.

Animal manures have been used effectively as fertilizers for centuries. Poultry manure has long been recognized as perhaps the most desirable of these natural fertilizers because of its high nitrogen content. In addition, manures supply other essential plant nutrients and serve as a soil amendment by adding organic matter. Organic matter persistence will vary with temperature, drainage, rainfall and other environmental factors. Organic matter in soil improves moisture and nutrient retention. The utilization of manure is an integral part of sustainable agriculture.

Livestock manures contain lignocelluloses, polysaccharides, proteins and other organic materials. The conversion of these waste materials to value-added products has been recognized as an attractive alternative waste management solution (Industry Canada, Bio-Products Canada, 2004). However, as a feedstock, this organic (lignocellulosic) waste has a complex physical and chemical composition and, to date, the limited efforts to convert waste biomass to higher-value chemicals and energy on a commercial scale have been unsuccessful (Attygalle, 2008).

Preliminary research using organic waste materials as lignocellulosic feed stocks for the production of bio-ethanol and other value-added products has shown great promise and is capturing the interest of the bio-products industry. However, further detailed research is critical to investigate its ultimate application beyond the laboratory scale (Demirbas, 2005).

The Food and Agriculture Organization of the United Nations estimated that in 2002 there were nearly sixteen billion chickens in the world; counting a total population of 15,853,900,000. The figures from the Global Livestock Production and Health Atlas for 2004 were as follows: China (8,860,000,000), Mexico (540,000,000), United States (1,970,000,000), India (425,000,000), Indonesia (1,200,000,000), Japan (286,000,000), Brazil (1,100,000,000), Iran (280,000,000), Russia (340,000,000) and Turkey (250,000,000),

Ethiopia is believed to have the largest livestock population in Africa. This livestock sector has been contributing considerable portion to the economy of the country, and still promising to rally round the economic development of the country. Livestock provides farmyard manure that is commonly applied to improve soil fertility and also used as a source of energy. Rural poultry production in Ethiopia represents a significant part of the national economy in general and the rural economy in particular, and contributes 98.5 and 99.2% of the national egg and poultry meat production, respectively (Canettieri et al., 2007), with an annual output of 72,300 metric tones of meat and 78,000 metric tones of eggs (Choge et al., 2007).

Total poultry population in Ethiopia is estimated to be about 38.13 million. Cocks and cockerels are also estimated separately, and are 4.01 and 1.78 million, respectively. The others are non-laying hens that make up about 3.99% (1.52 million) of the total poultry population in the country. With regard to breed, 96.4, 3.06 and 0.53% of the total poultry were reported to be indigenous, hybrid and exotic, respectively.

Small and large scale chicken farms are rapidly growing in Ethiopia. Some of the poultry farms are found in Addis Ababa, Debre Zeit, Awassa, Wolaita Sodo, Benshagul Gumuz Regional State and Bonga (the study area). Furthermore, private farms (Alema and Genesis Farms from Debre Zeit, four small-scale poultry farms in Addis Ababa), two Poultry Multiplication and Distribution Centers in Awassa and Wolaita Sodo, and backyard farms in Wolaita Sodo and Benshangul Gumuz Regional State.

This research was conducted to investigate bioethanol production potential of poultry manure and provide environment friendly alternatives to make use of poultry manure. Thus, the objectives of the present study were: (i) to determine the quantity of reducing sugars produced from the H_2SO_4 hydrolysis of poultry manure at different acid concentrations and at different temperatures by liquid chromatography and optical rotation and (ii) to determine the quantity of ethanol produced from the fermentation of reducing sugars at different fermentation times by refractometry.

MATERIALS AND METHODS

Materials and chemicals

Fermentation bottles (Erlenmeyer flask, 250 to 500 ml), yeast (*Saccharomyces cerevisiae*), sulfuric acid (98%, sd fine-chem. limited, Mumbai), poultry manure, high performance liquid chromatography (HPLC) (2053E00X, ECOM Ltd, Czech Republic), Polarimeter (AUTOPOL IV, Rudolph, USA), vacuum filter, sealed plastic bucket, centrifuge, oven (VO200, memmert, Germany) and grinder were used throughout the experiment.

Methods

Alcohol was produced by preparing sugar-containing raw material for fermentation by yeast. The experimental work was designed to determine the ethanol content of poultry manure. Ethanol was produced from the manure by the hydrolysis and sugar fermentation processes. In order to produce sugars from the manure, it was pre-treated with acid and the hemicellulose portions was broken down (hydrolyzed) by dilute acids into different sugars that was then fermented into ethanol.

Feedstock material collection and preparation

Poultry manure with no bedding material was collected from Bonga Poultry Farm (Ethiopia). The solid manure was transported in a sealedplasticbucket.Priortotreatment,manuresamples were thawed in a refrigerator at 4°C for 24 h and then mixed. The raw poultry manure feedstock was washed with distilled water and the sand was removed through sedimentation, the remaining solid fraction was employed in the experimental investigation and the liquid portion which may have negative effect on the environment is kept for treatment before disposal. It needs characterization of the liquid before discharging to the environment. The washed poultry manure was neutralized to a pH of 7.0 with 0.5 M H_2SO_4. The neutralized manure solution was then centrifuged for 10 min at 4,700 rpm. The solid fraction was refrigerated at 4°C in plastic bottles prior to acid hydrolysis.

Dilute acid hydrolysis

The poultry manure (100 g) was hydrolyzed (pretreated) with 250 ml of different concentrations of sulfuric acid (0.2, 0.4, 0.6, 0.8 M and 1 M) in 500 ml Erlenmeyer flask and separately heated at 100 and 130°C for 30 min using oil thermostat. After hydrolysis, the liquid fraction was cooled, filtered and determined for glucose concentration using HPLC and polarimeter. The acid hydrolysates were adjusted to pH 5-6 by adding concentrated sulfuric acid and 1 M sodium hydroxide, and the solutions were filtered and prepared for fermentation (Dawson and Boopaty, 2008).

Fermentation process

The hydrolysis process breaks down the cellulose part of the manure into sugar solutions and yeast was added to the solution, which was then heated. Bach fermentations of hydrolysates were carried out in flask incubated with 5 g/l yeast (*S. cerevisiae*) at 30°C, as described in the literature (Thuesombat et al., 2007). The hydrolysates of 0.2, 0.4, 0.6, 0.8 and 1 M sulfuric acid treated hydrolysates are fermented with 5 g/l yeast. The overall process is shown in Figure 1.

RESULTS AND DISCUSSION

Reducing sugar concentration versus acid concentration

The experimental results showed that the sample (poultry manure) has good reducing sugar potential. The results also showed that the amount of total reducing sugar in the sample increases with increase in acid concentration used for hydrolysis and temperature of hydrolysis. The results are given in Table 1.

Table 1 shows that all expected reducing sugars are detected by hydrolysis of 0.8 M H_2SO_4. It can also be seen that some of the reducing sugars (fructose, glucose and sucrose) are detected from breaking down of cellulosic material in the sample on hydrolysis using 1 and 0.6 M H_2SO_4. While only one of the expected reducing sugars (fructose) is detected on hydrolysis using 0.4 M H_2SO_4, which indicates the concentration of the acid used for hydrolysis of the sample is not able to break cellulosic materials in the sample completely. In the similar manner, only one of the components of expected reducing sugars (fructose) is detected with much reduced peak area on hydrolysis using 0.2 M H_2SO_4.

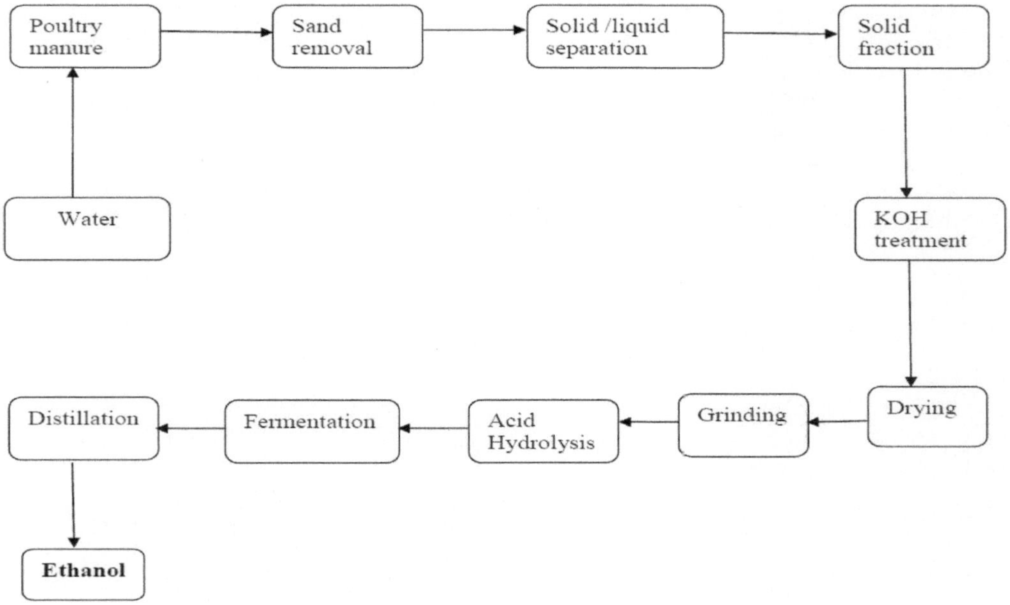

Figure 1. General process flow diagram.

Table 1. Chromatographic results of the samples hydrolyzed by 1.0, 0.8, 0.6, 0.4 and 0.2 M H_2SO_4.

Component	Solution hydrolyzed by 1 M H_2SO_4 (%)	Solution hydrolyzed by 0.8 M H_2SO_4 (%)	Solution hydrolyzed by 0.6 M H_2SO_4	Solution hydrolyzed by 0.4 M H_2SO_4	Solution hydrolyzed by 0.2 M H_2SO_4
Fructose	13.2	16.9	1.0	100.0	100.0
Glucose	75.7	29.8	85.6	-	-
Sucrose	11.1	17.6	13.5	-	-
Maltose	-	6.2	-	-	-
Lactose	-	29.6	-	-	-
Total	100.0	100.0	100.0	100.0	100.0

In general, the above results from HPLC indicate the presence of reducing sugars in the sample, and the amount of sugar depends on the concentration of acid used for hydrolyzing the sample. The total amount of sugar increases with increase in concentration of acid used for hydrolysis up to 0.8 M H_2SO_4 and decreases after while.

Similar results were also obtained by polarimetry (Table 2) indicating that maximum amount of total sugar (115.6 g/l) is obtained from the sample of 0.8 M acid hydrolysate at 130°C.

Ethanol concentration

The amount of ethanol in the sample was determined using refractometer which was calibrated with different percentages of standard ethanol before measurements of the samples. The data are listed in Tables 3 and 4.

In Table 3, the highest refractive index is obtained for 0.8 M hydrolysate after 24 h fermentation. As can be seen, the refractive index increases until 24 h fermentation and decreases after 24 h for all concentrations of hydrolysate.

In the same manner, maximum refractive index is obtained for 0.8 M hydrolysate after 24 h fermentation. But here the refractive index is relatively greater than the results in Table 3. This is because the temperature in Table 4 for hydrolysis, which is 130°C, is greater than 100°C.

Ethanol concentration as a function of acid concentration and fermentation time

The amount of ethanol in the sample was determined using refractometry. The amounts of ethanol by volume in the sample hydrolyzed with acids of different concentration for different fermentation time are given in Tables5 and 6. It is determined from the refractive index of the

Table 2. Optical rotation and concentration of glucose formed after 30 min hydrolysis at 100 and 130°C.

Sample hydrolyzed by H_2SO_4 (M)	α_{obs} /(at 100°C)	Glucose conc. /g/l (at 100°C)	α_{obs} /° (at 130°C)	Glucose conc. / g/l (at 130°C)
1.0	0.53	72.6	0.711	100.3
0.8	0.62	91.1	0.850	115.6
0.6	0.46	67.4	0.670	93.4
0.4	0.41	56.9	0.493	70.2
0.2	0.38	49.4	0.436	59.3

Table 3. Refractive index of sample solution hydrolyzed at 100°C for 30 min at different concentration of sulfuric acid and at different fermentation time.

No.	Conc. of acid (M)	Refractive index of the sample hydrolyzed and fermented at different time					
		12 h	24 h	36 h	48 h	60 h	72 h
1	1.0	28	32	31	28	25	21
2	0.8	30	36	33	31	28	23
3	0.6	27	31	30	26	23	19
4	0.4	22	25	23	22	20	18
5	0.2	18	21	19	18	18	17

Table 4. Refractive index of sample solution hydrolyzed at 130°C for 30 min at different concentration of sulfuric acid and at different fermentation time.

No.	Conc. of acid (M)	Refractive index of the sample hydrolyzed and fermented at different time					
		12 h	24 h	36 h	48 h	60 h	72 h
1	1.0	30	32	31	29	27	23
2	0.8	32	39	36	33	30	26
3	0.6	29	33	31	28	25	20
4	0.4	23	26	25	23	22	19
5	0.2	20	22	21	19	18	18

sample which is compared with the calibration curve (volume of ethanol versus refractive index of the standard solution).

Table 5 shows maximum volume of ethanol (13.5%) obtained from 0.8 M acid hydrolysate after 24 h fermentation. The table also shows the volume of ethanol in the sample increases with the concentration of acid used for hydrolysis. Therefore, it can be concluded that the percentage volume of ethanol increases until 24 h fermentation, and decrease after fermentation of 24 h. Decrease in concentration of ethanol might be decrease in microbial mass.

As can be seen in Tables 5 and 6, the percentage volume of ethanol in the sample increases with increase in concentration of the acid used for hydrolysis of the sample (poultry manure). The tables also show that the of hydrolysis, that is, maximum amount of ethanol is obtained at 130°C, because the temperature is optimum to break cellulosic material. The experiment shows the amount of ethanol increases until 24 h fermentation, but it decreases after 24 h fermentation because of decrease in microbial mass.

Generally, the experiment showed that the maximum percentage volume of ethanol (14.5%) is obtained at the concentration of 0.8 M acid hydrolyzed sample at 130°C and after the fermentation of 24 h.

ACKNOWLEDGEMENTS

The authors would like to appreciate the financial support from SNNPR Mines and Energy Agency. We also thank the Head of Bonga Poultry farm, for his kind cooperation during sample collection.

Table 5. Percentage volume of ethanol in the sample hydrolyzed at 100°C for 30 min at different concentration of acid and fermentation time.

Conc. of acid (M)	Refractive index of the sample hydrolyzed and fermented at different time					
	12 h	24 h	36 h	48 h	60 h	72 h
1.0	8	12	11	8 5	6	3
0.8	10	13.5	12.5	10.5	8.5	4
0.6	7	10.5	10	6	4	2
0.4	3.5	5	4	3.5	2.5	1.5
0.2	1.5	3	2	1.5	1.5	1

Table 6. Percentage volume of ethanol in the sample hydrolyzed at 130°C for 30 min at different concentration of acid and fermentation time.

Conc. of acid (M)	Refractive index of the sample hydrolyzed and fermented at different time					
	12 h	24 h	36 h	48 h	60 h	72 h
1.0	11	13	11	10	8.5	4
0.8	12	14.5	13.5	12.5	10	6
0.6	9	12.5	10.5	8.5	5	2.5
0.4	4	6	5	4	3.5	2
0.2	2.5	3.5	3	2	1.5	1.5

REFERENCES

Akpan UG, Alhakim AA, Ijah UJ (2008). Production of ethanol fuel from organic and food wastes. J. Pract. Technol. 13(3):1-11.

Anuj KC, Ravinder R, Lakshmi MN, Rao V, Ravindra P (2007). Economic and environmental impact of bioethanol production technology. Biol. Technol. Mol. Biol. 2(2):14-32.

Attygalle A (2008). Instrumental analysis. Institute of Technology, Diaz publication, London.

Budsberg E, Rastogi M, Puettmann ME, Caputo J, Balogh S, Volk TA, Gustafson R, Johnson L (2012). Life-cycle assessment for the production of bioethanol from willow biomass crops via biochemical conversion. Forest Prod. J. 62(4):305-313.

Canettieri EV, Rocha GJ, Carvalho JA, Silva JB (2007). Optimization of acid hydrolysis from the hemicellulosic fraction of eucalyptus grandis residue using response surface methodology. Biores. Technol. 98(2):422-428.

Choge SK, Pasiecznik NM, Harvey M, Wright J, Awan SZ, Harris PJ (2007). Prosopis pods as human food, with special reference to Kenya. Forest Res. Inst. 33(3):323-335.

Dawson L, Boopaty R (2008). Cellulosic ethanol production from sugar-cane bagase without enzymatic saccharification. Biores. 3(2):452-460.

Demirbas A (2005). Bioethanol from cellulosic materials: A renewable motor fuel from biomass. Energy Sources 27(5):327-337.

Gunasekaran P, Chandra KR (2007). Ethanol fermentation technology: Zymomonas mobilis. Madurai Kamary University, Madurai, India. pp. 1-22.

Licht FAO (2001). World ethanol market: Analysis and outlook. Agra Europe, London.

Licht FAO (2007). World ethanol production: To hit new record, World Ethanol and Biofuels Report.

Scott F, Quintero J, Morales M, Conejeros R, Cardona C, Aroca G (2013). Process design and sustainability in the production of bioethanol from lignocellulosic materials. Bio.Technol. 16(3):1-7.

Tanaka YL (2006). Ethanol fermentation from biomass resources: current state and prospects. Appl. Microbiol. Bio. Technol. 69(6):627-642.

A study of the physico-chemical water quality, hydrology and zooplankton fauna of Opa Reservoir catchment area, Ile-Ife, Nigeria

EO Akindele[1] and IF Adeniyi[2]

[1]Department of Biological Sciences, Bowen University, Iwo, Nigeria.
[2]Department of Zoology, Obafemi Awolowo University, Ile-Ife, Nigeria.

This study was aimed at relating the zooplankton fauna of Opa reservoir catchment area (Rivers Obudu, Opa and Esinmirin) to their physico-chemical water parameters and some aspects of their hydrology. Basin perimeter, basin width, total length, number of streams, river order and discharge showed significant positive correlation with species occurrence, while compactness ratio showed significant positive correlation with abundance. Conductivity, total dissolved solids, biological oxygen demand (BOD_5) and major ions were generally higher in the dry season than in the rainy season. On the other hand, flow velocity, discharge, total suspended solids, apparent colour, true colour, turbidity, dissolved oxygen (DO), DO saturation, phosphate and organic matter were higher in the rainy season than in the dry season. Cluster analysis indicated a strong relationship among some physico-chemical water parameters. pH, Na^+, Mg^{2+}, SO_4^{2-}, DO, DO saturation, BOD_5, Cu, Mn, Cr, Co, Pb, As and PO_4^{3-} all showed significant correlation ($p < 0.05$) with zooplankton occurrence and/or abundance.

Key words: Abundance, basin, hydrology, rivers, water parameters, species occurrence.

INTRODUCTION

Most limnological studies have concentrated on the biology and the physico-chemistry of lakes and rivers with little or no attention on their hydrology. A complete assessment of a water body is however based on appropriate monitoring of its hydrology, physico-chemistry and biology (Meybeck and Helmer, 1992). In river systems, nutrients, sediments and organisms move according to the water speed. This flow if permanent is a controlling factor in the distribution and abundance of organisms (Neiff, 1996). Variation in hydrologic regime are undoubtedly of great importance, and responsible for main changes in a great number of environmental factors related to each other and to the zooplankton, such as current velocity, turbidity and suspended solids (Casanova and Henry, 2004).

Zooplankton are an important component of the aquatic biota and play a pivotal role in the food web by forming a link between the lower trophic level organisms (phytoplankton) and the higher trophic level organisms such as macroinvertebrates and fish. They are susceptible to variation in a wide number of environmental factors including water temperature, light, chemistry (particularly pH, oxygen, salinity, toxic contaminants) and food availability (Paterson, 2001). Zooplankton, like all organisms have a range of environmental conditions to which they are adapted. The optimum environment for one species may be barely tolerable to another (Parker and Corbilt, 1993). Animals also live within a certain range of environmental condition called the tolerance range for any environmental factor, and at either limit of

Table 1. Site description of the sampling stations.

Station Ref. code	River system	Longitude (E)	Latitude (N)	Altitude (m)	Depth range (m)
OB1	Obudu	004°37.142^1	07°33.281^1	335	0.10-0.45
OB2	Obudu	004°37.394^1	07°34.256^1	333	0.12-0.40
OB3	Obudu	004°36.140^1	07°33.757^1	314	0.15-.0.65
OB4	Obudu	004°34.680^1	07°32.783^1	287	0.35-1.12
OB5	Obudu	004°32.760^1	07°31.432^1	269	0.50-1.50
OB6	Obudu	004°36.215^1	07°33.421^1	313	0.25-0.90
E2	Esinmirin	004°36.154^1	07°28.670^1	313	0.15-0.50
E3	Esinmirin	004°35.476^1	07°28.672^1	271	0.20-0.65
E6	Esinmirin	004°34.205^1	07°29.866^1	265	0.35-0.95
E8	Esinmirin	004°33.033^1	07°30.799^1	256	0.45-1.25
MK1	Esinmirin	004°36.675^1	07°30.353^1	ND	0.15-0.45
MK2	Esinmirin	004°36.190^1	07°29.150^1	ND	0.17-0.52
OP2	Opa	004°34.568^1	07°31.231^1	267	0.70-1.05
OP4	Opa	004°33.345^1	07°31.112^1	263	0.70-1.15
OP5	Opa	004°35.433^1	07°31.684^1	274	0.65-0.95
OP6	Opa	004°32.862^1	07°31.056^1	262	0.84-1.75

the tolerance range, one or more essential functions cease (Miller and Harley, 1999).Previous limnological surveys of Opa Reservoir (Akinbuwa ,1992; Akinbuwa and Adeniyi, 1991; 1996) accounted for the taxonomic composition of zooplankton fauna as well as their relationship with the physico-chemical water quality of the reservoir. Ogunfowokan et al. (2011) also gave an account of the trophic status and physico-chemical water parameters of Opa Reservoir and two other reservoirs (Asejire Reservoir and Ede Reservoir) in south west Nigeria. Zooplankton studies in other Nigerian fresh-waters include the works of Iloba (2002) on Ikpoba Reservoir, Ayodele and Adeniyi (2006) on River Oshun impoundments, Ibrahim (2009) on Challawa River, Bwala et al. (2010) on NIFFR reservoirs, as well as Achionye-Nzeh and Ismaikaiye (2010) on the University of Ilorin reservoir. This study was aimed at making a complete assessment of the three river systems that flow into Opa Reservoir based on their hydrology, water physico-chemistry and an aspect of their biology (zooplankton), and relating the findings to previous limnological accounts on the reservoir.

MATERIALS AND METHODS

Study area

The study was based on the catchment area of Opa reservoir, covering parts of Atakumosa West, Ife Central and Ife East Local Government Areas of Osun State, Nigeria. The area lies between Latitudes 07°28^1-07°35^{1}N and Longitudes 004°32^1-004°38^{1}E. Coordinates and site descriptions of the sampling stations are provided in Table 1. Sixteen (16) sampling stations were established in the study area covering the upper reach, middle reach and the lower reach of each river system (Figure 1). The dry season is shorter and usually lasts for about five months (November-March) while the rainy season prevails for the remaining seven months of the year (April-October). The annual regime of rainfall shows two peaks (one in June/July and the other in September) separated by a short dry spell in August.

Sample collection and analyses

Sampling was carried out every other month between September 2004 and July 2005 on the sixteen established stations. Ambient air and water temperature were determined by a mercury-in-glass thermometer. Flow velocity was determined by placing a float in the water and measuring the time taken to travel a predetermined distance (Jones and Reynolds, 1996). Discharge was estimated from the product of the flow velocity and the cross-sectional area of the river (that is basin width and water depth) (Chapman and Kimstach, 1996). pH was determined by using a standardized pH meter while conductivity was determined by using a standardized Jenway conductivity meter. Samples for physico-chemical and zooplankton analyses were collected from the surface layers in all the sampling stations. Dissolved oxygen (DO) and biological oxygen demand (BOD_5) samples were collected in oxygen bottles (250 ml reagent bottles). DO samples were fixed *in situ* with Winkler's reagents A (Manganous sulphate) and B (Alkali iodide). BOD_5 samples were incubated in a dark cupboard at room temperature for five days and afterwards fixed with Winkler's reagents.

Zooplankton samples were filtered through 45 µm plankton net and preserved in 5% formalin solution. Further analysis of zooplankton was done in the laboratory by viewing under a compound light microscope and identification was done based on standard identification guides (Egborge and Chigbu, 1988; Egborge, 1994; Alekseev, 2002; Fernando, 2002; Kutikova, 2002). Species diversity was determined using the Margalef index. Water samples for physico-chemical analyses were collected in new and uncontaminated 2 liters plastic bottles. Colour, turbidity, total solids (TS), total dissolved solids (TDS) and total suspended solids (TSS) were all determined using the appropriate laboratory methods described by APHA et al. (1995). DO and BOD_5 were determined by

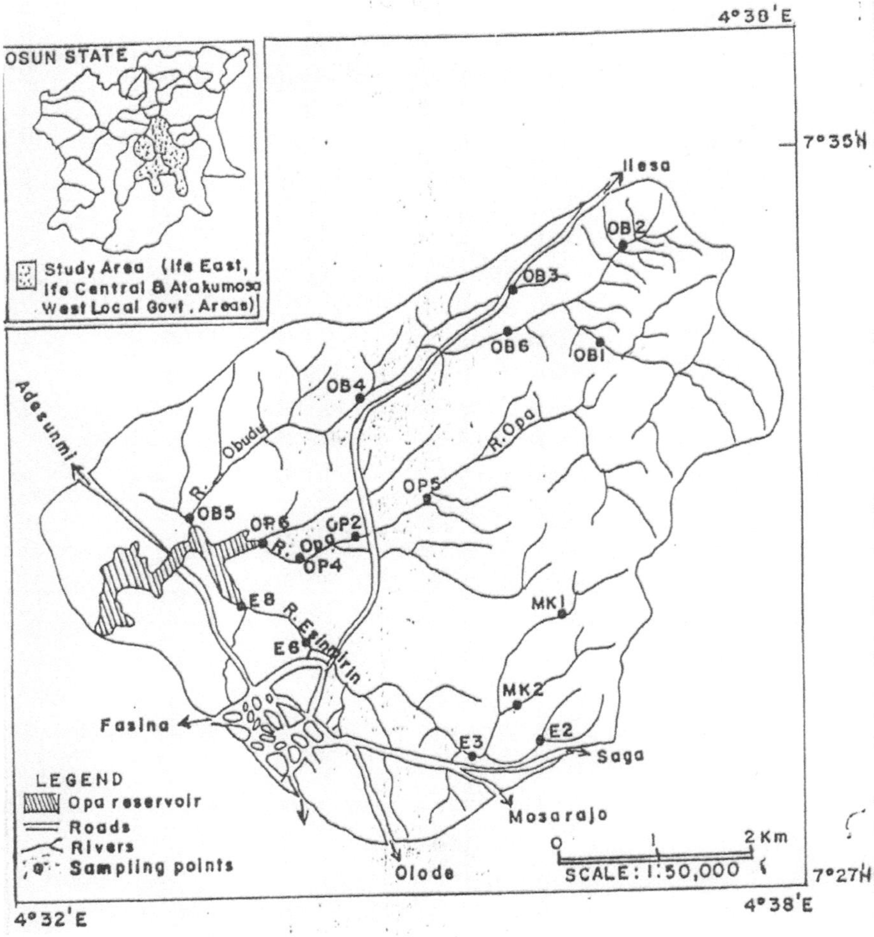

Figure 1. The Catchment basin of Opa Reservoir showing the water sampling stations (Inset: Map of Osun State showing the study area).

Winkler's titration method while alkalinity and acidity were determined by acid-base titration method (APHA et al., 1995). Mohr titration method was used to analyze chloride ion (Cl^-) while complexiometric titration method was used for calcium and magnesium (APHA et al., 1995). Sulphate (SO_4^{2-}) was determined by turbidimetric/spectrophotometric method while Organic matter was analysed by chromic acid digestion (APHA et al., 1995). Spectrophotometric method was used for both nitrate and phosphate (APHA et al., 1995) while atomic absorption spectrophotometry (AAS) was used for sodium (Na^+), potassium (K^+) and all the investigated trace metals (Pb, Zn, Fe, Cu, Mn, Ni, Co, Cr, Cd, and As) at their respective wavelengths (APHA et al., 1995). The data obtained were subjected to statistical analyses using descriptive statistics, cluster analysis and regression analysis.

The various definitions and determination methods used to determine the hydrological parameters are given in Table 2.

RESULTS

Physico-chemical parameters

The general descriptive statistics of physico-chemical water parameters in Opa Reservoir catchment basin are given in Table 3. For hydro-physical parameters; air temperature ranged from 25.0 to 36.0°C, while water temperature ranged from 22.0 to 33.2°C. Apparent colour and true colour ranged from 52.0 to 874.3 Pt.Co and 8.8 to 355.0 Pt.Co respectively. Turbidity ranged from 4.6 to 69.7 NTU. Total solids and total suspended solids ranged from 90 to 300 mgL^{-1} and 10 to 60 mgL^{-1} respectively. Flow velocity ranged from 0.05 to 0.44 ms^{-1}, while discharge ranged from 0.002 to 3.53 m^3s^{-1}.

For the general chemical characteristics of water quality; pH ranged from 6.70 to 8.11, while conductivity ranged from 64.3 to 421.4 μScm^{-1}. Alkalinity and acidity ranged from 7 to 210 $mgCaCO_3L^{-1}$ and 12 to 102 $mgCaCO_3L^{-1}$ respectively, while total dissolved solids ranged from 40 to 300 mgL^{-1}.

For major ions; Calcium and magnesium ranged from 2.4 to 34.8 mgL^{-1} and 0.5 to 27.5 mgL^{-1} respectively. Sodium ranged from 5.62 to 11.45 mgL^{-1}, potassium ranged from 1.03 to 9.23 mgL^{-1}. Cl^- ranged from 3.6 to 47.8 mgL^{-1}, while HCO_3^- and SO_4^{2-} ranged from 8.4 to 252 mgL^{-1} and 0.2 to 7.2 mgL^{-1}, respectively.

For nutrient compounds and oxygen parameters; nitrate and phosphate ranged from 0.61 to 2.48 mgL^{-1}

Table 2. Determination of hydrological aarameters.

Parameter	Definition/method of determination
Basin length (L_B)	Straight line distance from outlet to the point on the basin divide used to determine the main channel length lire
Basin width (W_B)	Average width of the basin determined by dividing the area, A, by the basin length
Basin perimeter(P_B)	The length of the line that defines the surface divide of the basin
Basin shape (SH_B)	A measure of the shape of the basin computed as the ratio of the length of the basin to its average width
Compactness ratio (CR_B)	The ratio of the perimeter of the bas n to the circumference of a circle of equal area. Computed from A and P_B
Main channel length ((L_C)	The length of the main channel from :he mouth to the basin divide
Sinuosity ratio	The ratio of the channel length to the basin length
Total length	The length of the basin from the point of outlet to discharge
Drainage density	The ratio of the total length of the basin to the basin area
Stream frequency	The ratio of basin area to number of stream segments
Drainage index	The ratio of the stream frequency to the basin drainage density
Drainage type	The pattern of flow of the rivers and streams in the basin

Wetzel and Likens (2000).

Table 3. General descriptive statistics of physico-chemical water parameters in Opa Reservoir catchment basin.

Parameter	Minimum	Maximum	Mean	Standard deviation	Median	Mode
Air Temp. (°C)	25.0	36.0	29.7	2.7	29.5	32.0
Water Temp. (°C)	22.0	33.2	26.3	1.7	26.0	27.0
Apparent Colour (Pt.Co)	52.0	874.3	278.2	202.9	210.7	153.0
True Colour (Pt.Co)	8.8	355.0	64.1	80.6	37.6	8.8
Turbidity (NTU)	4.6	69.7	22.0	16.2	16.7	13.7
TS (mgL^{-1})	90	300	142	54	120	100
TSS (mgL^{-1})	10	60	28	13	25	20
Flow Velocity (cms^{-1})	5.0	44.4	20.2	11.3	20.0	33.3
Discharge (m^3s^{-1})	0.002	3.53	0.43	0.61	0.16	0.02
pH	6.70	8.11	7.38	0.36	7.37	7.25
Conductivity (μScm^{-1})	64.3	421.4	161.9	76.5	136.9	120.2
Alkalinity ($mgCaCO_3L^-$)	7	210	58	40	44	22
Acidity ($mgCaCO_3L^-$)	12	102	29	18	22	16
TDS (mgL^-)	40	300	111	53	95	80
Ca^{2+} (mgL^{-1})	2.4	34.8	14.6	8.2	12.9	7.7
Mg^{2+} (mgL^{-1})	0.5	27.5	3.9	4.8	2.4	0.5
Na^+ (mgL^{-1})	5.62	11.45	8.94	1.90	9.1	8.82
K^+ (mgL^{-1})	1.03	9.23	4.36	2.52	3.19	3.12
Cl^- (mgL^{-1})	3.6	47.8	18.4	11.8	15.1	8.2
HCO_3^- (mgL^{-1})	8.4	252	72	47.9	55.2	26.4
SO_4^{2-} (mgL^{-1})	0.2	7.2	1.5	1.4	1.0	0.2
NO_3^- (mgL^{-1})	0.61	2.48	1.24	0.47	1.17	1.28
PO_4^{3-} (mgL^{-1})	0.08	3.17	0.49	0.51	0.31	0.23
SiO_2 (mgL^{-1})	1.2	47.0	16.1	10.8	12.1	25.0
Organic Matter (mgL^{-1})	0.46	9.2	4.20	2.52	3.93	3.72
DO (mgL^{-1})	1.2	7.6	5.1	1.3	5.2	6.0
DO Saturation (%)	15.0	94.4	64.0	15.7	64.7	85
BOD_5 (mgL^{-1})	0.4	4.4	1.8	0.9	1.6	1.2
Cu (mgL^{-1})	0.01	1.58	0.69	0.45	0.68	0.01
Fe (mgL^{-1})	0.05	41.2	1.7	6.04	0.9	0.3

Table 3. Contd.

As (mgL^{-1})	0.05	0.7	0.26	0.18	0.20	0.20
Mn (mgL^{-1})	0.01	2.36	0.12	0.33	0.07	0.01
Co (mgL^{-1})	0.01	0.03	0.02	0.01	0.01	0.01
Pb (mgL^{-1})	0.03	16.74	6.17	4.39	5.73	0.06
Ni (mgL^{-1})	0.01	0.05	0.2	0.1	0.1	0.1
Cr (mgL^{-1})	0.01	0.14	0.04	0.03	0.04	0.06
Zn (mgL^{-1})	0.01	0.05	0.02	0.01	0.02	0.01
Cd (mgL^{-1})	0.004	0.872	0.306	0.280	0.030	0.208

and 0.08 to 3.17 mgL^{-1}, respectively. Organic matter ranged from 0.46 to 9.2 mgL^{-1}. DO and DO saturation ranged from 1.2 to 7.6 mgL^{-1} and 15.0 to 94.4% respectively, while BOD_5 ranged from 0.4 to 4.4mgL^{-1}.

For heavy metals; Cu ranged from 0.01 to 1.58 mgL^{-1}, Fe ranged from 0.05 to 41.2 mgL^{-1}, As ranged from 0.05 to 0.7 mgL^{-1}, Mn and Co ranged from 0.01 to 2.36 mgL^{-1} and 0.01 to 0.03 mgL^{-1} respectively. Pb ranged from 0.03 to 16.74 mgL^{-1}, Ni and Cr ranged from 0.01 to 0.05 mgL^{-1} and 0.01 to 0.14 mgL^{-1}, respectively. Zn ranged from 0.01 to 0.05 mgL^{-1}, while Cd ranged from 0.004 to 0.872 mgL^{-1}.

Seasonal and intra-basin variations of hydro-physical parameters

Apparent colour, true colour, turbidity, total suspended solids (TSS), flow velocity and discharge were all generally higher in the rainy season (304.6±30.6 Pt.Co, 83.5±14.1 Pt.Co, 23.3±2.4 NTU, 29±2 mgL^{-1}, 0.23±0.02 ms^{-1}, and 0.57±0.10 m^3s^{-1} respectively) than in the dry season (238.2±32.6 Pt.Co, 35.8±4.1 Pt.Co, 20.1±3.0 NTU, 27±2 mgL^{-1}, 0.12±0.02 ms^{-1}, and 0.10±0.03 m^3s^{-1}). Apparent colour decreased from the upper reach to the lower reach in River Opa (383.9±201.8, 156.7±85.5 and 122.8±64.8 Pt.Co), while there were no definite patterns in the other two rivers. Like in apparent colour, turbidity decreased downstream of River Opa (30.3±15.3 NTU, 11.0±2.8 NTU, 9.6±3.7 NTU) but showed no patterns in the other two rivers. Flow velocity showed a definite pattern in River Obudu only, with an upward trend downstream of the river (0.15±0.09 ms^{-1}, 0.23±0.09 ms^{-1}, 0.27±0.11 ms^{-1}). Discharge on the other hand showed a distinct pattern in River Esinmirin only (0.05±0.03, 0.07±0.05 and 0.80±0.62 m^3s^{-1}), although highest value in River Obudu was also recorded in the lower reach. TSS was lowest in the upper reaches of River Obudu and River Esinmirin and this showed a significant difference ($p < 0.05$) from the middle and the lower reaches.

Seasonal and intra-basin variations of general chemical characteristics of water quality

Conductivity, alkalinity, acidity and total dissolved solids (TDS) were generally higher in the dry season (195.1 μScm^{-1}, 70 $mgCaCO_3L^{-1}$, 34 $mgCaCO_3L^{-1}$, 131 mgL^{-1}, respectively) than in the rainy season (140.0 μScm^{-1}, 49 $mgCaCO_3L^{-1}$, 25 $mgCaCO_3L^{-1}$, 99 mgL^{-1}). There was a slight increase in pH downstream of River Opa (7.22, 7.24, and 7.41). Conductivity increased downstream of Rivers Obudu (116.5, 124.8 and 126.6 μScm^{-1}) and Opa (127.9, 140.4 and 196.0 μScm^{-1}). Total dissolved solids and conductivity were highest at the lower reach of each river.

Seasonal and intra-basin variations of major ions

All the major ions (Ca^{2+}, Mg^{2+}, Na^+, K^+, Cl^-, HCO_3^- and SO_4^{2-}) except Na^+ were higher in the dry season than in the rainy season. Sulphate showed a significant difference ($p < 0.05$) between the two seasons. Na^+ increased downstream in River Esinmirin (5.81, 8.62 and 10.02 mgL^{-1}). K^+ increased downstream in the Obudu River (3.23, 4.28 and 4.38 mgL^{-1}) and likewise was Cl^- (13.7 mgL^{-1}, 15.0 mgL^{-1}, 17.2 mgL^{-1}). In River Opa, HCO_3^- increased downstream (40.7, 54.9 and 83.3 mgL^{-1}) and likewise was SO_4^{2-} (0.9 mgL^{-1}, 1.7 and 1.9 mgL^{-1}). K^+, Cl^-, HCO_3^- and SO_4^{2-} were all highest at the lower reaches of the three rivers, although only HCO_3^- showed a significant difference ($p < 0.05$).

Seasonal and intra-basin variations of nutrient compounds and oxygen parameters

The investigated nutrient compounds and oxygen parameters were nitrate (NO_3^-), phosphate (PO_4^{3}), organic matter, dissolved oxygen (DO), DO saturation and biological oxygen demand (BOD_5). Phosphate, organic matter, DO and DO saturation were generally higher in the rainy season (0.50, 4.10 and 5.4 mgL^{-1} and 67%) than in the dry season (0.48 mgL^{-1}, 3.71 mgL^{-1}, 4.7 mgL^{-1} and 59.7%, respectively), while nitrate and BOD_5 were generally higher in the dry season (1.32 mgL^{-1} and 2.0 mgL^{-1}) than in the rainy season (1.19 mgL^{-1}, 1.7 mgL^{-1}%, respectively).

DO showed a significant difference ($p < 0.05$) between the two seasons. There was a decrease in organic matter

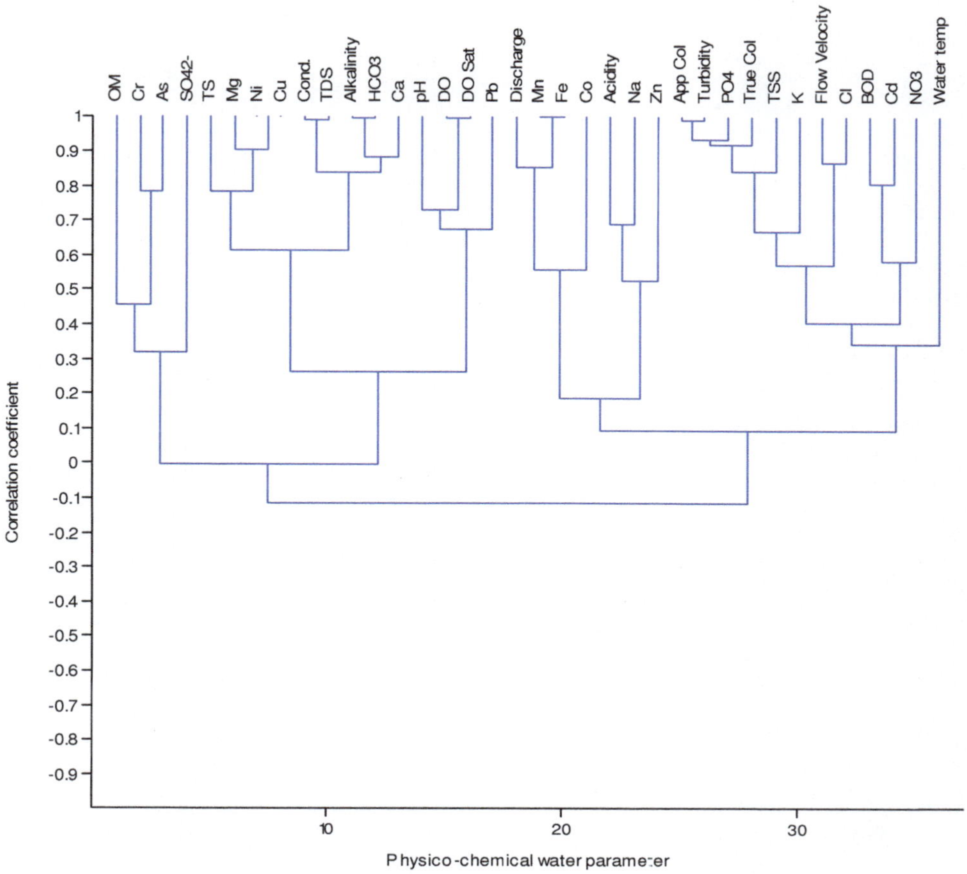

Figure 2. Cluster analysis showing the relationship among physico-chemical water parameters in River Obudu ($p < 0.05$; $r = 0.7067$; $n = 6$).

concentration from the upper reach to the lower reach of River Esinmirin (4.92, 3.83, and 3.20 mgL^{-1}). DO decreased down-stream of River Obudu (5.5, 5.4 and 5.1 mgL^{-1}) and River Esinmirin (5.8, 5.5 and 4.0 mgL^{-1}). DO saturation also decreased downstream of Rivers Obudu (68.4, 68.0 and 63.2%) and Esinmirin (70.6, 69.9 and 50.8%).

Seasonal and intra-basin variations of trace/heavy metals

Cu, As, Co and Ni were higher in the rainy season (0.76 mgL^{-1}, 0.30 mgL^{-1}, 0.02 mgL^{-1}, 0.20 mgL^{-1}) than in the dry season (0.54 mgL^{-1}, 0.24 mgL^{-1}, 0.01 mgL^{-1}, 0.13 mgL^{-1}), while Cd, Fe, Mn, Pb, and Cr were higher in the dry season (0.33, 2.30, 0.20, 6.28 and 0.05 mgL^{-1}) than in the rainy season (0.29, 0.80, 0.07, 6.07 and 0.04 mgL^{-1}). Only nickel however showed a significant difference ($p < 0.05$) between the two seasons. Cu increased downstream of the Obudu River (0.55 mgL^{-1}, 0.89 mgL^{-1}, 1.08 mgL^{-1}) and so also were Fe (0.70 mgL^{-1}, 0.90 mgL^{-1}, 11.0 mgL^{-1}) and Mn (0.07 mgL^{-1}, 0.08 mgL^{-1}, 0.47 mgL^{-1}). In River Esinmirin, Mn increased downstream (0.02, 0.08 and 0.09 mgL^{-1}) and likewise was Pb (0.05, 5.00 and 8.29 mgL^{-1}). Cr decreased downstream of River Opa (0.05, 0.04 and 0.03 mgL^{-1}).

Relationship among the physico-chemical parameters

In River Obudu, a number of clusters were observed among the physico-chemical water parameters at $p < 0.05$. A major cluster was formed among conductivity, TDS, alkalinity, HCO_3^- and Ca while another one was formed among apparent colour, turbidity, PO_4^{3-}, true colour and TSS (Figure 2). pH, conductivity, TDS, HCO_3^-, alkalinity, Ca^{2+}, TS, Cl^-, K^+, acidity and SO_4^{2-} formed a big cluster at $p < 0.05$ in River Opa. There was another major cluster among TSS, DO, DO saturation, true colour, Cd, apparent colour, turbidity, flow velocity and phosphate (Figure 3). The clustering pattern of physico-chemical water parameters in River Esinmirin was similar to that of River Opa (Figure 4), with TS, TDS, conductivity, Ca^{2+}, discharge, Cl^-, HCO_3^-, SO_4^{2-}, K^+, acidity and flow velocity forming a big cluster at $p < 0.05$. TSS, Zn, Mn, apparent colour, turbidity, NO_3^- and PO_4^{3-} formed another big cluster in R. Esinmirin.

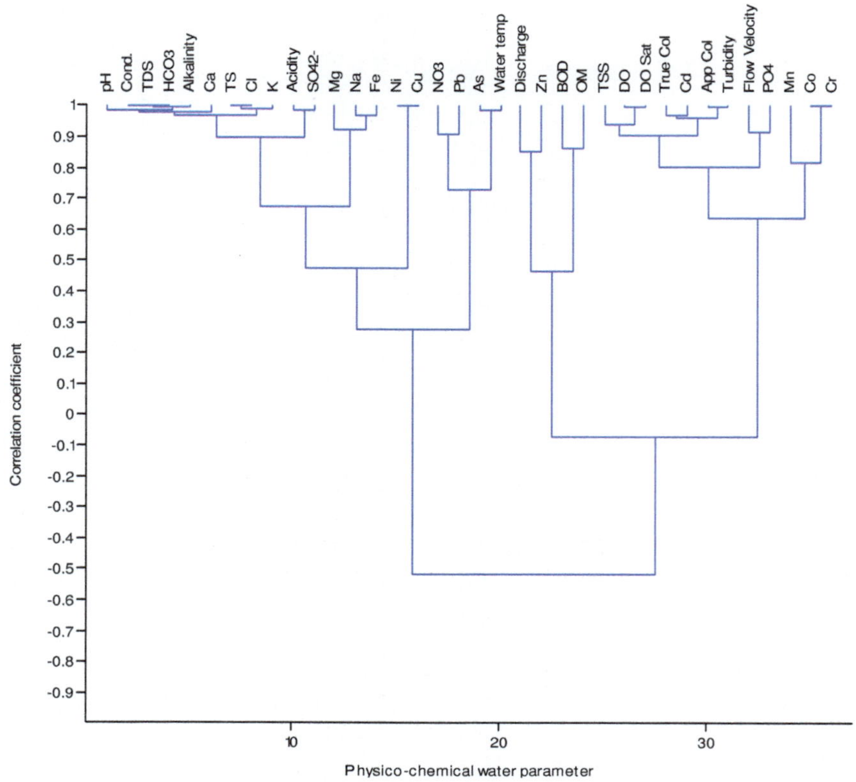

Figure 3. Cluster analysis showing the relationship among physico-chemical water parameters in River Opa ($p < 0.05$; $r = 0.8114$; $n = 4$).

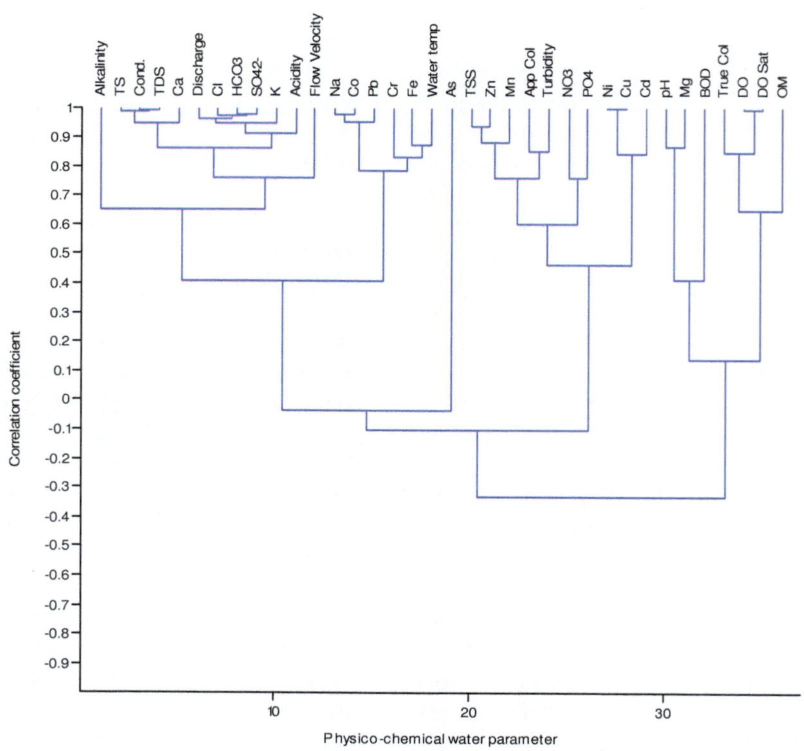

Figure 4. Cluster analysis showing the relationship among physico-chemical water parameters in River Esinmirin ($p < 0.05$; $r = 0.7067$; $n = 6$).

Hydrology characteristics

Discharge was highest at the lower reach in R. Obudu and increased towards the downstream in R. Esinmirin ($0.05 \pm 0.03\ m^3s^{-1}$, $0.07 \pm 0.05\ m^3s^{-1}$, $0.80 \pm 0.62\ m^3s^{-1}$). It was also generally higher in the rainy season ($0.57 \pm 0.10\ m^3s^{-1}$) than in the dry season ($0.10 \pm 0.03\ m^3s^{-1}$) and this showed a significant difference ($p < 0.05$). The drainage pattern in the three rivers was generally dendritic. Area, perimeter, basin width, main channel length and basin length were all highest at the lower reaches of the three rivers (OB5, OP6 and E8). Drainage density decreased downstream in the three rivers.

Zooplankton fauna

The zooplankton fauna of the rivers comprised two phyla, three classes, four orders, fourteen families and forty-six species. A checklist of the species is given in Table 4. Rotifers were the most dominant with nine families and thirty-five species, while the crustacean zooplankton were represented by only five families and eleven species. The relative occurrence and abundance of zooplankton in the sampling stations are also given in Table 5.

Relationship between zooplankton and physico-chemical parameters

The relationships between zooplankton and investigated physico-chemical parameters were established using the regression and correlation analysis. pH, Na^+, SO_4^{2-}, biological oxygen demand (BOD_5), Cu, Mn and Cr all showed significant correlation ($p < 0.05$) with zooplankton occurrence. pH and BOD_5 showed significant inverse correlation ($p < 0.05$) with occurrence, while Na^+, SO_4^{2-}, Cu, Mn and Cr all showed significant positive correlation ($p < 0.05$) with the occurrence of zooplankton.

Phosphate (PO_4^{3-}), DO, DO saturation and Co showed significant correlations ($p < 0.05$) with the abundance of zooplankton. PO_4^{3-} showed a significant positive correlation ($p < 0.05$) with abundance, DO and DO saturation showed highly significant inverse correlation ($p < 0.01$) with abundance, while Co showed a highly significant positive correlation ($p < 0.01$).

Mg^{2+}, Na^+, SO_4^{2-} and Cu showed highly significant positive correlation ($p < 0.01$) with species diversity of zooplankton. As, Co and Pb also showed significant positive correlation ($p < 0.05$) with species diversity.

Relationship between hydrological parameters and zooplankton

Discharge, river order, area, perimeter, basin width, number of streams, total length and discharge all showed significant positive correlation ($p<0.05$) with the occurrence of zooplankton. Total length showed a highly significant positive correlation with occurrence ($p<0.01$). All other hydrological parameters showed insignificant positive correlation ($p>0.05$) with occurrence.

There was a significant positive correlation ($p<0.05$) between compactness ratio and abundance. River order, sinuosity ratio, stream frequency and drainage density showed insignificant inverse correlation ($p>0.05$) with the abundance of zooplankton. A highly significant positive correlation ($p<0.01$) occurred between river order and species diversity, while a significant positive correlation ($p<0.05$) occurred between number of streams and species diversity.

DISCUSSION

Apparent colour, true colour, turbidity, total suspended solids, flow velocity and discharge were all higher in the rainy season than in the dry season. This seasonal variation pattern is common for most tropical inland ecosystems as a result of allochthonous run-off into river basins during the rainy season. Turbidity is a measure of suspended solids in water, and the amount of these solids is largely determined by the velocity of the water (Moore, 1989; Michaud, 1991). This is obvious in the clustering of turbidity with apparent colour, true colour and flow velocity in River Opa, and in the clustering of turbidity with apparent colour and total suspended solids in River Obudu. The increase in flow velocity and discharge in the rainy season can be attributed to rainfall which further increased their values. Discharge, volume of water per unit time (Allaby, 1999) was highest at the lower reaches of River Obudu and River Esinmirin since the volume of water passing down increases as water winds its way along a stream or river from its source (Chapman and Reiss, 1995).

Conductivity, total dissolved solids and virtually all the major ions were higher in the dry season than in the rainy season. Conductivity is a measure of dissolved ions or solids in water, and tends to increase with evaporation of water from the basin, and consequent concentration of the dissolved ions. The direct relationship between conductivity and dissolved solids or ions is evidenced in the clustering of conductivity with the major ions and total dissolved solids. Conductivity and total dissolved solids were also observed to be highest at the lower reach of each river. This may be associated with anthropogenic activities like farming and dumping of wastes in the downstream stations. The connection of the downstream stations with the reservoir was another probable factor for relatively high conductivity and total dissolved solids.

Phosphate and organic matter were also higher in the rainy season than in the dry season. This seasonal pattern can easily be linked to allochthonous inputs from the nearby terrestrial surfaces in the rainy season, thereby increasing their concentrations. The water bodies

Table 4. Taxonomic checklist of zooplankton in Opa Reservoir catchment area.

Taxon				
Phylum	**Class**	**Order**	**Family**	**Species**
Rotifera	Eurotatoria	Ploima	Brachionidae	*Brachionus calyciflorus*
				Brachionus quadridentatus
				Brachionus budapestinensis
				Brachionus falcatus
				Brachionus bennini
				Brachionus angularis
				Brachionus plicatilis
				Branchionus sp.
				Anuraeopsis racenensis
				Keratella tropica
				Keratella lenzi
				K. cochlearis cochlearis
				K. cochlearis macracantha
				Macrochaetus longipes
				Squatinella mutica
				Notholca sp.
				Platyias leloupi
			Colurellidae	*Lepadella patella*
				Lepadella uncinata
			Lecanidae	*Lecane quadridentata*
				Lecane luna
				L.luna presumpta
				L. leontina
				L.lunaris
				Lecane sp.
				Monostyla bulla styrax
				Monostyla lunaris
			Trichocercidae	*Trichocerca cylindrica*
				Trichocerca sp.
			Notommatidae	*Cephalodella mucromata*
				Enteroplea lacustris
			Gastropodidae	*Ascomorpha ovalis*
			Asplanchnidae	*Asplanchna priodonta*
			Euchlanidae	*Beauchampiella eudactylota*
		Flosculariacea	Filinidae	*Filinia terminalis*
Arthropoda	Brachiopoda	Cladocera	Chydoridae	*Camptocercus rectrirostris*
			Moinidae	*Moina micrura*
				Moinodaphnia macleayi
			Sididae	*Diaphanosoma excisum*
			Daphnidae	*Scapholeberis kingi*
				Simocephalus vetulus
	Copepoda	Cyclopoida	Cyclopidae	*Thermocyclops neglectus*
				Halicyclops troglodytes
				Microcyclops varicans
				Eucyclops macrurus
				Halicyclops korodiensis

Table 5. Species occurrence, mean abundance (org/m^3) and species diversity of zooplankton in the stations.

River system	Station	Species occurrence	Mean Abundance (org/m^3)	Species diversity
River Opa	OP2	8	240	0.16
	OP4	25	910	0.17
	OP5	21	430	0.25
	OP6	27	1240	0.27
River Obudu	OB1	16	200	0.27
	OB2	15	340	0.25
	OB3	18	2170	0.19
	OB4	18	320	0.25
	OB5	27	1200	0.27
	OB6	16	370	0.24
River Esinmirin	E2	4	200	0.10
	E3	5	100	0.14
	E6	10	890	0.18
	E8	13	370	0.20
	MK1	8	300	0.08
	MK2	1	80	0.10

might be considered meso-eutrophic as a result of the relatively high concentrations of phosphorous and nitrogenous nutrients. The overall mean concentration of NO_3^- (1.24 mgL^{-1}) fairly exceeded the typical concentration of freshwaters (<1.0 mgL^{-1}) though far less than the recommended maximum level of 5 mgL^{-1} (Chapman and Kimstach, 1996). PO_4^{3-} (0.49 mgL^{-1}) also exceeded the recommended maximum level (0.1 mgL^{-1}) of phosphates in lakes and streams (Moore, 1989; Michaud, 1991). Fertilizer run-off into the basins could have contributed to the elevated levels of these nutrients (Chapman and Kimstach, 1996) since most of the sampled stations were close to farmlands. Dissolved oxygen (DO) and DO saturation were higher in the rainy season owing to the fact that the solubility of oxygen in water increases with decrease in temperature (Lowe-Connel, 1987), which is characteristic of the rainy season. Biological oxygen demand was generally higher in the dry season than in the rainy season, most likely as a result of increase in ambient and water temperature and the consequent rise in the metabolic rate of the aquatic organisms in the dry season. Increased BOD_5 in the dry season could also be associated with increased biological activities (respiration and decomposition of organic matter) which in turn must have been catalyzed by the rise in temperature. Generally, the seasonal pattern in the physico-chemical water parameters of these three rivers indicated that the onset of rains signals a radical change in physico-chemical characteristics of tropical rivers (Lowe-McConnel, 1987; Chapman and Kramer, 1991).

The significant correlation shown between some selected hydrological parameters and the community structure of zooplankton further confirms the relevance of hydrology in limnological studies. River order, area, perimeter, basin width, total length and discharge had a direct relationship with occurrence. A direct relationship was observed between compactness ratio and abundance, and also between river order and species diversity. All these parameters are indices of carrying capacity in a lotic freshwater environment, which in turn determines the community structure of organisms.

Discharge, area, perimeter, basin width, main channel length and basin length were all highest at the lower reaches of the three rivers. High occurrence and abundance of zooplankton were also recorded at the downstream stations. It could therefore be inferred from this study that these hydrological parameters play a vital role in the occurrence and population density of zooplankton. This also indicates that the parameters were good indices of the carrying capacity of the rivers.

Water discharge is one of the parameters affecting zooplankton seasonal variation in rivers (Saunders and Lewis, 1988a, b; Brown et al., 1989, Pace et al., 1991; Van Dijk and Van Zanten, 1995; Vranovsky, 1995), as confirmed from this study with high discharge and high occurrence of zooplankton species in the rainy season. This observation is further supported by the beliefs of Canfield and Jones (1996) that high water discharge provides a larger water volume where animals can develop, and Pace et al. (1991) that small rivers with low discharge have correspondingly lower zooplankton abundance. High discharge was also observed to have coincided with high nutrient loading in the rainy season, which could have in turn stimulated plankton growth as evident in the positive correlation between PO_4^{3-} and zooplankton abundance. The relationship between phosphate and zooplankton abundance is a strong indication that the abundance and biovolume of both phytoplankton and zooplankton are largely regulated by the resource base and tend to increase with the trophic state of freshwaters (Canfield and Jones, 1996). The richness of the water bodies in nitrates and phosphates as well as the moderate presence of *Brachionus spp.*, *Keratella* spp., *Asplanchna priodonta* and *Filinia terminalis* suggests that the rivers were meso-eutrophic (Attayde and Bozelli, 1998).

DO and DO saturation showed highly significant inverse correlation with zooplankton abundance in contrast to previous surveys of the reservoir (Akinbuwa and Adeniyi, 1991; Akinbuwa, 1992). This is probably because dissolved oxygen is a more sensitive environmental factor in a reservoir (lentic environment) than in a lotic river, where the water is constantly being oxygenated by turbulent actions. It is also expected that oxygen consumption through respiration should increase with increase in abundance of organisms, thereby lowering the concentration of dissolved oxygen in a water body. Many of the investigated parameters did not however show significant correlation with the community structure of zooplankton. This is probably because many environmental factors affect zooplankton only at extreme levels (example toxic contaminants, oxygen) and will not be important in all freshwaters (Paterson, 2001).

Conclusion

This study reveals that the physico-chemical and biological nature of inland water bodies is closely linked to their hydrology, the surrounding terrestrial surfaces, as well as the anthropogenic activities in and around the basin. Oxygen and hydrological parameters as well as nutrient compounds played a major role in the occurrence and distribution of zooplankton fauna in the reservoir catchment basin. High concentrations of nu-trients brought into the basin through anthropogenic acti-vities and run-off, as well as the presence of zooplankton indicator species of eutrophication were an indication of a eutrophic freshwater system.

REFERENCES

Achionye-Nzeh CG, Isimaikaiye A (2010). Fauna and flora composition and water quality of a reservoir in Ilorin, Nigeria. Int. J. Lakes Rivers 3(1):7-15.

Akinbuwa O (1992). A preliminary study of diurnal vertical distribution of rotifers in Opa Reservoir, Nigeria. J. Aquat. Sci. 7:19-28.

Akinbuwa O, Adeniyi IF (1996). Seasonal variation, distribution and interrelationships of rotifers in Opa Reservoir, Nigeria. Afr. J. Ecol. 34:351-263.

Akinbuwa O, Adeniyi IF (1991). The rotifera fauna of Opa reservoir, Ile-Ife,Nigeria. J. Afr. Zool. 105(5):383-391.

Alekseev VR (2002). Copepoda. In: Fernando CH (ed), A Guide to Tropical Freshwater Zooplankton; Identification, Ecology and Impact on Fisheries. Backhuys Publishers, Leiden, pp.123-187.

Allaby M (1999). Oxford dictionary of ecology, Oxford University Press, p. 440.

APHA, AWWA and WEF (1995). Standard methods for the examination of water and wastewater, 19th ed., Washington.

Attayde JL, Bozelli RL (1998). Assessing the indicator properties of zooplankton assemblages to disturbance gradients by Canonical Correspondence Analysis. Can. J .Fish. Aquat.Sci. 55(8):1789-1797.

Ayodele HA, Adeniyi IF (2006). The zooplankton fauna of six impoundments on River Osun, Southwest, Nigeria. The Zoologist. 1(4):49-67.

Brown AV, Limbeck RL, Schram MD (1989). Trophic importance of zooplankton in rivers and streams with alluvial riffle and pool geomorphometry. Arch. Hydrobiol. pp. 349-367.

Bwala RL, Ovie SI, Ajayi O, Haruna A (2010). Prelimniary study on the limnology and plankton abundance in relation to fish production in some NIFFR Reservoirs. Report Opinion 2(6):7.

Canfield TJ, Jones JR (1996). Zooplankton abundance, biomass and size distribution in selected Midwestern water bodies and relation with trophic state. J. Freshwat. Ecol. 11:171-181.

Casanova SMC, Henry R (2004). Longitudinal distribution of copepoda populations in the transition zone of Paranapanema river and Jurumirin reservoir (S.Paulo, Brazil) and interchange with two lateral lakes. Braz. J. Biol. 64(1):11-26

Chapman D, Kimstach V (1996). Selection of water quality variables. In: Chapman D (ed.), Water quality assessments, Chapman and Hall, London. pp.65-122.

Chapman JL, Kramer DL (1991). Limnological observations of an intermittent tropical dry forest stream. Hydrobiol. 226:153-166.

Chapman JL, Reiss MJ (1995). Ecology; Principles and Applications. Cambridge University Press, Low price edition. pp. 222-223.

Egborge ABM (1994). Water Pollution in Nigeria, Vol. 1, Biodiversity and Chemistry of Warri River, Ben Miller Books Nig. Ltd. p.331.

Egborge ABM, Chigbu P (1988). The Rotifers of Ikpoba River, Bendel State. The Nigerian Field 53:117-132.

Fernando CH (2002). A Guide to Tropical Freshwater Zooplankton; Identification, Ecology and Impact on Fisheries. Backhuys Publishers, Leiden. p. 291.

Ibrahim S (2009). A survey of zooplankton diversity of Challawa River, Kano and evaluation of some of its physico-chemical conditions. Bayero J. Pure Appl. Sci. 2(1):19-26.

Iloba KI (2002). Vertical distribution of rotifera in the Ikpoba Reservoir in Southern Nigeria. Tropical Freshwater Biology 11:69-89.

Jones JC, Reynolds JD (1996). Environmental variables. In: Sutherland WJ (ed.), Ecological census techniques, First ed., Cambridge University Press. pp. 281-316.

Kutikova LA (2002). Rotifera. In: Fernando CH (ed.), A Guide to Tropical Freshwater Zooplankton; Identification, Ecology and Impact on Fisheries. Backhuys Publishers, Leiden pp. 23-68.

Lowe-McConnel RH (1987). Ecological studies in fish communities, Cambridge. pp. 3-33.

Meybeck M, Helmer R (1992). An introduction to water quality. In Chapman D(ed.), Water quality assessment. UNESO/WHO/UNEP. p. 21.

Michaud JP (1991). A citizen's guide to understanding and monitoring lakes and Streams, Publ. #94-149, Washington State Dept of Ecology, Publications office, Olympia, WA, USA, 360, pp.407-7472.

Miller SA, Harley JB (1999). Zoology, 4th ed, McGraw-Hill Companies, Inc. pp. 192-223.

Moore ML (1989). NALMS management guide for lakes and reservoirs, North American Lake Management Society (*http: www.nalms.org*).

Neiff JJ (1996). Large rivers of South America; toward a new approach. Verh. Internat. Verein. Limnol. 26:167-180.

Ogunfowokan AO, Akanni MS, Ajibola RO, Ayinde FO (2011). Trophic status and physico-chemical parameters of three reservoirs in Osun State, Nigeria. Ife J. Sci. 13 (1):27-44.

Pace ML, Findlay SEG, Lints D (1991). Zooplankton in advective environments. The Hudson River community and a comparative

analysis. Can. J. Fish Aquat. Sci. 49:1060-1069.
Parker SP, Corbilt RA (1993). Encyclopedia of Environmental Science and Technology, 3rd edition, McGraw-Hill, Inc., pp. 711-715.
Paterson M (2001).Protocols for measuring biodiversity; zooplankton in freshwaters.www.eman-resecaleman/ecotools/protocols/freshwater/zooplankton.
Saunders JF, Lewis WM (1988a). Zooplankton abundance in the Caura River, Venezuela. Biotropica 20:206-214.
Saunders JF, Lewis WM (1988b). Zooplankton abundance and transport in a tropical white-water river. Hydrobiologia 162:147-155.
Van Dijk GM, Van Zanten B (1995). Seasonal changes in zooplankton abundance in the lower Rhine during 1987-1991. Hydrobiologia 162:147-155.
Vranovsky M (1995). The effect of current velocity upon the biomass of zooplankton in the River Danube side arms. Biologia Bratislava 50:461-464.
Wetzel RG, Likens GE (2000). Limnological Analyses, Third edition, Springer, New York. p. 429.

5

Quantification of the potential for biogas and biogas manure from the selected fruit wastes in Addis Ababa city, Ethiopia

Getachew Dagnew Gebreeyessus[1] and Berhanu Assefa Demissie[2]

[1]Department of Environmental Health Sciences, Haramaya University, P.O. Box 235, Harar, Ethiopia.
[2]School of Chemical and Bio-Engineering, Addis Ababa Institute of Technology, Addis Ababa University, P.O. Box 385, Addis Ababa, Ethiopia.

In this paper, the overall potential of biogas and biogas manure from the selected fruit wastes in the city of Addis Ababa was estimated (quantified). The prediction is based on the characterization and biogas yield results in a related study, using structured questionnaire and checklists during field survey. This study has shown a survey of the major sources of the fruit wastes and compiled the number of registered service providers in this particular sector by category. Consequently, this study has shown that the wasted biogas potential of the city from the selected 164.51 tons of daily fruit wastes alone was 20896 m^3 per day. This wasted energy material is equivalent to 9000 L of diesel fuel that currently would cost 9389 US dollars. Further, soil nutrients that could be obtained from use of anaerobic digestion of the selected fruit wastes are 366 kg/day of total phosphorus, 798 kg/day of total nitrogen and 565 kg/day of potassium, which are also lost in vain. Therefore, the study shows the importance of proper organic waste management for the recovery of clean energy and nutrients, while at the same time avoiding expenses to buy fossil fuel as source of energy and reduce the associated stress on our environment and more.

Key words: Biogas potential, fruit waste, quantification, prediction, biogas manure.

INTRODUCTION

The so-called wastes that we discard and suffer with the consequences of improper management are of course partly huge energy and fertilizer sources that can support energy demands of cities greatly. Urban waste disposal is a serious challenge in all cities in the developing world, and its accumulation is an additional health hazard (Young, 2012).

Food waste problem is not only in this part of the world, India or Africa. It is a problem in the whole world. In big and medium cities, the problem intensifies due to socio-economic factors. For example, consumption of fruits and vegetables along with the resulting waste increase with development mainly due to awareness and capability of the public (Kumar, 2005; Belloso and Fortuny, 2011)

Addis Ababa city inhabiting closer to four million people generates solid waste at a rate of 0.4 kg/head/day and 60% of the municipal solid waste is organic in nature (Tesema, 2010). The existing solid waste collection system of the city is inadequate and covers only 60% of the waste generated. This is due to lack of transportation facilities and financial matters as well as geographic inconvenience for collection signifying the unreliability of the system for waste items like easily decaying fruit and vegetable wastes (FVW) (Regassa et al., 2011).

In relation to this, Ethiopia's horticulture has become one of big cash crop production. The production of fruit and vegetable (FVs), including root crops, was 2.16 million tones. The total FV production in 2008/2009 for the season comprises about 351 thousand tones of fruits (16%), 600 thousand tons of vegetables (28%), and 1.2 million tons of root crops (56%) (Ethiopian Horticultural Development Agency, 2011).

Tesema (2010) estimated that the city of Addis Ababa generates 23.1 tones of FVWs daily, in a study conducted earlier and it looks very much under estimated as it is done on household basis. Therefore, it could be simply additional to what 'point sources' discharge as the same waste. Moreover, these figures could increase with the growth of the horticulture industry year after year.

The sources for FVWs are diverse (Peavy et al., 1985; AbdAlqader and Hamad, 2012). FVWs in the city emanate from general store, open markets, FV markets, transporters, supermarkets, fruit juice houses, 'mobile' fruit sellers and consumer households. These wastes unnecessarily join municipal waste stream mostly in waste containers that are picked up using hauling and are transported to a local open dumpsite.

An important step towards a sustainable waste management system is to augment the waste reduction, reuse and recycle paradigm with processes that actually reduce solid waste accumulation (Pinderhughes, 2004). One of the compelling processes in this category is anaerobic digestion where little is done in this regard (Mshandete and Parawira, 2009). When fruit wastes (FWs) are collected at the outset, the operation would be easy to deal with.

FVWs are known to contain highly biodegradable compounds (70 - >80% volatile solid) and moisture strongly supporting their use for anaerobic biogas digesters' input. These wastes therefore holdup signifi-cant energy (Hawkins and Rains, 2008) and manure resource. However, these resources are wasted and left to decompose where a related problem arises in other areas (Gustavsson et al., 2011). Due to the decomposition, the wastes emit huge amount of CH_4 and CO_2 (green house gasses) to the atmosphere together with NH_3 and H_2S-noxious gases, causing environmental and health problems. Moreover, the wastes affect aquatic systems by depleting dissolved oxygen due to the biochemical strength of the wastes if washed with floodwater.

The reason for considering environment and methane gas in the management of such wastes can be explained (Esposito et al., 2012) considering mainly three factors. First, is the need to apply a process to dispose organic solid wastes in a more environmentally friendly manner along with useful soil conditions.

Secondly, the opportunity to obtain from this process a renewable fuel called biogas alternative to fossil fuel. Thirdly, the advantage of relatively low costs in starting up and managing this process as it mainly does not require oxygen supply, as well as, its feasibility of being even commercialized.

Though converting these wastes to a renewable energy/biogas and soil manure is proved to be magnificent, little is known on the quantity of the biogas potential of such wastes in the city. Therefore, this study is aimed at estimating the quantity of biogas and prediction of biogas manure that can be recovered from FW generated by Addis Ababa city. Thus, this finding could help many stakeholders interested in an Integrated Solid Waste Management (ISWM) activity that is geared towards city sustainability in terms of energy and environmenta management.

METHODS AND MATERIALS

Description of the study area

Addis Ababa, the capital of Ethiopia, which is by far the largest in the country, is located almost at the geographical centre of the national territory. The city covers an area size of 530.21 square kilometres and a population of 3,059,000 million persons (CSA, 2006). Addis Ababa has a history of a little more than 100 years. In this short span of time, the city grew from a small traditional Ethiopian town to one of Africa's metropolises.

Astronomically, Addis Ababa is located at 09° 02' N latitude and 38° 44' E longitude. It is LOCATEDon the steep escarpment of Mt. Entoto in the North (2900 m) to the South with an average altitude of 2400 m. The central FVM is located in Arada sub city near the municipality ard on the way from Piassa to Mercato, the largest commercial area (Figure 1).

Sampling

The study area covered the entire city where there are fruit retails, 'mobile' fruit sellers, fruit wholesales, juice houses, super markets including the major FV distribution centre, which is located at the heart of the city, where this study was focused. This study excluded FWs from households. The commercial facilities related to FWs are sampled randomly in the city. Up-to-date information was obtained on the number of registered FV related business firms from Trade and Investment Bureau of the City Administration.

Moisture resistant and baskets with cover were used to collect the selected FWs that were weighed daily to determine the average wastage in those sampling points for about two weeks as indicated in Figure 1.

Data collection and analysis

The quantification (estimation) was made based on supplementary data collected using structured questionnaire administered in a face-to-face interview and fieldwork made using observation checklists following letter communications with all respondents and daily weighted selected FWs.

Official letter communication was also made to the relevant government agency to obtain data on the number of the service providers. The TS and biogas yield calculators were obtained from a previous study by the same authors. The result was analyzed using Excel 2007 and discussed with related findings, journals and relevant literatures.

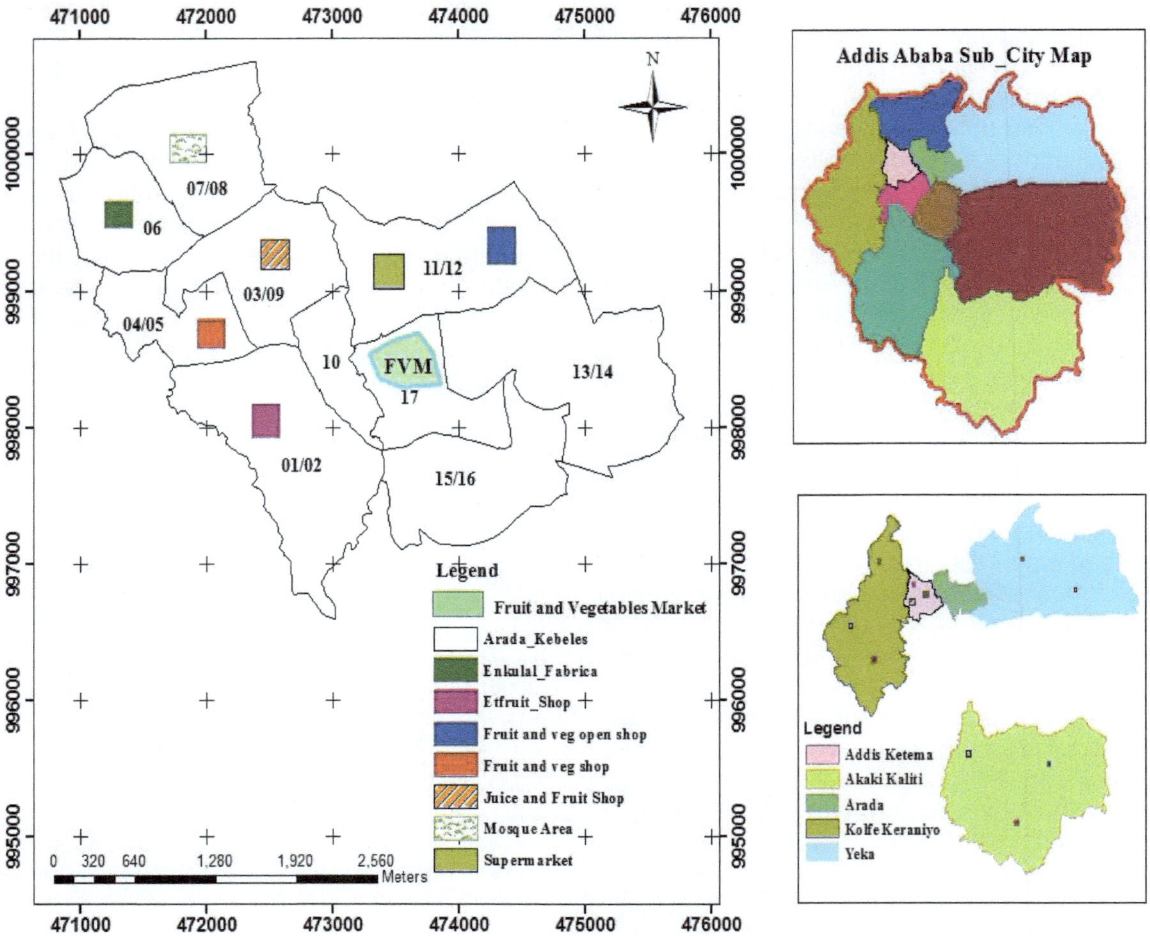

Figure 1. Map of the study area (GIS-2011).

Table 1. The summary of the number of fruit and vegetable juice, supermarkets and the subtotal in sub city in Addis Ababa city (data are up to the year 2010).

Business field	Sub city								
	Addis ketema	Akaki	Arada	Bole	Gulele	Kirkos	Kolfe	Lideta	Yeka and N/silk
FV juice	14	6	16	13	15	9	15	24	2
Supermarkets	157	19	54	114	23	121	94	37	87
Sub-total	171	25	60	117	28	130	109	41	89

Source: Extracted from Addis Ababa Trade and Industry Bureau (2010).

RESULTS AND DISCUSSION

Basics

A related laboratory scale study on the potential of the selected FWs in the city was made for biogas and relevant results were brought (Gebreeyessus and Demissie, 2013), which triggered this further study. In fact, the magnitude of the resource wasted has to be quantified in order to show the relevance of related interventions to recover such resources.

General

Sources of such FWs are diverse. In this regard, is not only the formal sector that is running the FV business but also the informal people are involved. The Ethiopian government's agency in control of registering this group is also operating behind the existing distribution of the sector.

The distribution of the fruit juice houses in the city is uneven (Table 1) ranging between 2 and 24 in number among sub cities. The huge quantity is generated by

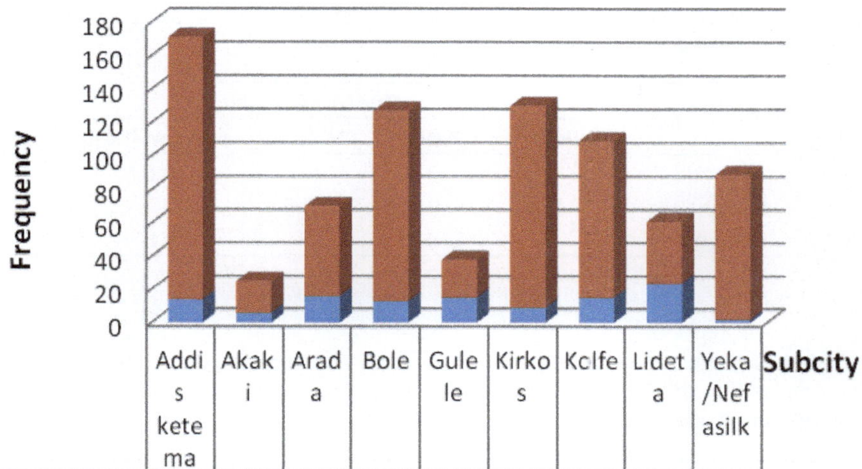

	Addis ketema	Akaki	Arada	Bole	Gulele	Kirkos	Kclfe	Lideta	Yeka /Nefasilk
Supermarkets	157	19	54	114	23	121	94	37	87
FV juice	14	6	16	13	15	9	15	24	2

Figure 2. Distribution of supermarkets and fruit and vegetable juice houses in Addis Ababa City, 2010.

"Atikilt tera" which is followed by fruit juice houses, based on personal observation.

Addis ketema sub city is the highest to have those super markets and FV juice houses followed by kirkos sub city. The least is Akaki sub city, as it is located at the out skirt of the city (Table 1 and Figure 2) (Trade and Industry Bureau of Addis Ababa, Ethiopia, letter communication, 2012).

In a related fact, the operation and waste generation is also variable temporally. The reasons for fluctuation of FVW generation are mainly the seasonality of supply of fruits in the market and the consumers factor which shows peak consumption in fasting days than others as well as location of the houses. In general, auditing of input versus wasted (locally called "marsh") is almost uncontrolled by the private business makers except the state owned one (Etfruit).

Quantification of overall FVWs from major contributors

Rough estimation of the total FWs was made based on the number of firms involved in this trade field (registered and informal). The responses of focal persons and the approximate weight of the wastes in a given volume as well as approximate composition of the wastes were studied. However, it should be considered that supply and discharge are variable within days of a week and seasons of the year as mentioned earlier.

The grand share of FVWs in general and FWs in particular is generated by the central FV market-"Atikilt tera" as marked earlier. The reasons for wastage in particular to the site are partly due to physical damage, rotten fruits and improper temperature application for heating fruits to ripen the row. There divisions of private and shared FV companies in this zone are made based on types of FV commodities; mainly banana, mango and avocado, vegetables and root crops.

The tomato-dominated corner disposes nearly 1500 kg of waste daily. Regarding banana waste, there are 12 compounds in the two areas; one "Atikilt tera" and the other to the north in front of "Atikilt tera" commonly called "Talian sefer". The hugest amount of FWs contributed is banana, including its holding stem and little rolling leaf 8 m^3 of FW is discharged every day from each store totalling 12 containers of the waste.

Within "Atikilt tera", the mango FW discharge is also comparable to banana, however mango is more seasonally available as it is supplied for about only 8 months in a year to market. It is estimated that 4 to 7 number of an 8 m^3 container mango wastes are discarded daily from "Atikilt tera".

The avocado FW is more contributed by juice houses and retailers than in "Atikilt tera". In fact, over four 8 m^3 volume waste containers are emptied to dumpsite daily, and more during the supply season.

According to the Ethiopian FV Market Share Company (Etfruit) there are over 60 fruit shops within Addis Ababa city that are relatively evenly distributed. These fruit shops receive 60 kg of fruits daily on average. These shops unlike the private shops have to return almost 20 kg of avocado, banana, and mango FWs daily on average back to storage centre due to defect for disposal (Etfruit Ethiopia, personal communication, 2012).

In all other contributors (except "Atikilt tera" and mobile fruit sellers) the primary FWs discharged are avocado and mango followed by banana all accounting to 2/3 of

Table 2. Summary of generation rate (in tons/day) for the selected FWs by the identified contributors.

Fruit waste	Contributor sector				
	Atikilt tera	Etfruit	FV Juice	supermarket	Mobile
Avocado	43	0.3	30.	0.8	-
Banana	110	0.4	8	0.1	1
Mango	62	0.2	18	0.6	-

Table 3. Mean value of TS (g/l), VS (g/l) removal, and biogas productivity and average methane for the various conditions evaluated in batch experiment of 35 days of anaerobic digestion time.

System (digester)	TS Initial	VS Initial	TS Final	VS final	VS (%) removal	Biogas (m^3/kg VS)	Average CH_4 (%)
A_0	7.7	6.2	4.1	3.157	49	0.48	55.56
B_0	7.6	6.06	3.8	2.998	50.5	0.57	58.16
M_0	7.8	5.96	3.6	2.495	58.1	0.53	51.64
A_1	7.9	6.36	3.1	2.3	63.8	0.76	54.68
B_1	7.8	6.22	3.2	2.32	62.7	0.82	56.38
M_1	7.9	6.05	3.2	2.22	63.3	0.82	52.14

Source: Gebreeyessus and Demissie, 2013. TS, Total solids; VS, volatile solids.

Table 4. Summary of important information and prediction values for biogas (methane) per day.

Total FWs (kg)	TS average (%)	VS average (%)	VS total (kg)	Average biogas(m^3/kg VS)	Total biogas (m^3)	Average CH_4 (m^3/kg VS)	Total CH_4 (m^3/kg VS)
164510	20.2	78.6	26120	0.8	20896	0.44	11493

Source: Gebreeyessus and Demissie, 2013.

total. Roughly, over 144 fruit, vegetable and juice houses discard 0.3 m^3 FVWs daily dominated by mango and avocado (2/3) in the city.

Giving observatory check to those informal mobile sellers that range from 100-200 in number in the city, about 5 - 11 kg of banana waste is discarded daily by each seller. In addition to that, the low rate of discard is by the supermarkets, 706 in number, as most of them do have better facilities for storage, 3 kg of FWs per day each.

The amount of selected FWs reported in unit of volume can be expressed in unit of weight using the average respective densities of the selected FWs for subsequent calculations. The weight per day and the contributors for the selected FWs is summarized in Table 2.

The total quantity of these selected FWs when added amounts to 164.51 tons/day. Even this amount is exclusive of the same discharge from households. Thus, this huge resource should not be lost. Therefore, the objective of this paper was to predict possible recoverable biogas out of these resources thereby supporting decision makers and further studies from applying such recovery technologies in efficient manner in the following sections.

Estimation of possible biogas from these selected FWs

According to a recent bench scale experimental study, the biogas as well as methane yield obtainable from these FWs was found to be promising (Table 3).
Based on these data, rough estimate of the possible clean energy obtainable from the selected FWs in the city is calculated below. The necessary information to predict was the amount of such waste discarded daily, those FWs characterization results (moisture, TS and VS), and total biogas as well as methane yield from the preceding experimental work (Table 4).

The TS and VS values are average values for the selected FWs combined and are calculated the same way as shown in Table 3. The biogas volume estimated

Table 5. Summary of the total determinant nutrients quantification on daily basis.

Nutrient	Average gm/kg TS	Total amount (kg)
Total phosphorus	11	366
Total nitrogen	24	798
Potassium	17	565

recoverable by proper management of these wastes in the city is about 20896 m^3/day. This, with average methane content of 54.4% is estimated to give a total of 11496 m^3/day CH_4 (methane).In a related report, the fuel value of 0.24 m^3 biogas with 40% CH_4 content was reported to be equivalent to 0.1 L of diesel fuel (Department of Environment and Resource Management, Australia, 2011). This implies the city of Addis Ababa is wasting an equivalent fuel value of 9000 L of diesel fuel, which is about 9389 US dollar in terms of money. In other terms of comparison, the same amount of daily biogas yield could have be converted to 11.2 mega watt of electricity had it been managed using the anaerobic digesters.

Currently, interests in the use of biogas are increasing by the sub-Saharan African cities due to its economic and environmental benefits (Austin et al., 2011). The recovery of such huge resource needs attention by the various concerned parties. Basic biogas technology therefore can make a big difference (Environmental Health Perspect, 2006).

The issue does not end with only the lost of such resource but also the inevitable accompanying environmental stress the wastes bring in; emission to atmosphere and mainly water sources pollution. On the other hand, our dependence on fossil fuel as energy source and the synthetic fertilizer production adds much green house gas emissions in addition to the costs to buy the items (Young, 2012).

Obviously, the collection and transportation to dump site is not free as well. Based on the information from "Atikilt tera" merchants, 4.5 US dollar was spent to discard 1 m^3 of the wastes. However, collection and transportation of the wastes to gas centres will also not be free.

The biogas manure prediction

The advantage of applying biogas technology for management of such wastes is better than obtaining the clean and renewable energy. It can also provide opportunity for soil nutrient to recycle far better than applying the row waste to farm and more.

Based on the daily TS wasted which are 33, 231 kg and the amount of the soil fertility indicators determined in in a related study, the maximum obtainable nutrient is projected (Table 5).

Given the variability in the nutrient content among the different FWs, the average biogas manure could contain 366 kg/day of total phosphorus, 798 kg/day of total nitrogen and 565 kg/day of potassium. Had such nutrients been synthesized artificially, the cost and environmental implication would have been huge. In Cuba for instance, use of chemical fertilizers is prohibited in cities rather, the use of organic composting using a peri-urban horticulture imitative is encouraged (Young, 2012).

Undoubtedly, this amount is obtainable excluding other components of FVWs. The content of nutrient by the other vegetable wastes would probably be higher in this regard. Therefore, this prediction also adds to the need to recover such nutrients from the waste resources in addition to the clean energy obtainable.

In relation to this, if such waste is simply dumped on soils, it will give little humus due to consumption of the organic matter by microorganisms for their growth (Memon et al., 2012). Contrary to this, the biogas manure interrupts succession of plant parasites from year to year in addition to the best nutritive value it gives which can be considered also as an indirect benefit. Therefore, it is imperative to pay attention to issue for action not only to meet environmental demands but also for economic advantage as well.

Conclusion

Even though energy is a need in cities of developing countries, it is being disposed and unrecovered. Fruit and vegetable wastes are dumped in open field, discarded in adjoining streams and in cities including Addis Ababa in some locations. These known source wastes however hold huge energy potential. The quantity of fruit wastes discarded in the city of Addis Ababa is so huge (~164.51 tons/day) that its indiscriminate disposal or open dumping poses serious public and environmental health problem.

Nevertheless, using some technology these wastes can be converted to useful materials; energy and soil nutrients while at the same time avoiding the undesirable consequences of open dumping. Conversely, use of fossil fuels as energy sources would result in economic and environmental implications.

The renewable energy that would have been recovered is equivalent to 11.2 mega watt of electricity daily. The benefit from proper management is even greater. The anaerobic digestion of such wastes can also result to sustainability of soil nutrients, replacing significantly the synthetic fertilizers.

Thus, the various stakeholders in an integrated solid waste management should focus on to unhide such huge energy and nutrient potential in order to sustain cities; however, further studies have to be done in this regard.

REFERENCES

AbdAlqader A,Hamad J(2012). Municipal solid waste composition determination supporting the integrated solid waste management in

Gaza strip. Int. J. Environ. Sci. Dev. 3(2):172-177.
Austin G, Bob O, Jecinta M (2011). The potential of biogas digesters to improve livelihoods and long term sustainability of ecosystem services in Sub-Saharan Africa. Available at www.abdu.ac.uk.
Belloso OM, Fortuny RS (2011). Food Preservation Technology: advances in fresh-cut fruits and vegetables processing. pp.1-5. Tylor and Francis Group, USA.
Central Statistical Agency(CSA) (2006). Population and Housing Census Report-Country - 2006, Ethiopia.
Environmental Health Perspect (2006). Biogas: A bright idea for Africa. 114(5):300-303.
Esposito G, Luigi F, Flavia L, Antonio P, Francesco P (2012). Bio-Methane Potential Tests to Measure the Biogas Production from the Digestion and Co-Digestion of Complex Organic Substrates. Open Environ. Eng. J. 5:1-8.
Ethiopian Horticultural Development Agency (2011). Assessment of development potentials and investment options in the export-oriented fruit and vegetable sector, online document, (http://www.africa-do-business)
Gebreeyessus GD, Demissie BA (2013). Effect of Adding Urea on Biogas Potential of Selected Fruit Wastes, Addis Ababa, Ethiopia.
Gustavsson J, Cederberg C, Sonesson U, Otterdijk R, Meybeck A (2011). Global food loss and food waste: extent, causes and prevention, Food and Agriculture Organization of the United Nations, Rome, Italy.
Hawkins GL, Rains G (2008). Assessing the physical and chemical properties of fruit and vegetable waste for use in anaerobic digestion. ASABE, annual international meeting. USA.
Kumar A (2005). Verms and Vermitechnology. 129-131. New Delhi, India.
Letter communication with Trade and Industry bureau of Addis Ababa City, Ethiopia, 2012.
Memon M, Memon KS, Mirani S, Jamro GM (2012). 'Comparative evaluation of organic wastes for improving maize growth and NPK Content'. Afr. J. Biotechnol. 11:39:9343-9349.
Mshandete AM, Parawira W (2009). Biogas technology research in selected sub-Saharan, African countries- A review. Afr. J. Biotechnol. 8:116-125.
Office of Environment and Resource Management, Queen's land Government (2011). Use of banana waste as fuel source-online document, Australia.
Peavy HS, Rowe DR, Tchobanoglous G (1985). Environmental Engineering, McGraw-Hill.
Personal communication with Ethiopian Fruit and Vegetable Market Share Company (etfruit) (2012). Addis Ababa, Ethiopia.
Pinderhughes R (2004). Alternative Urban Future: Planning sustainable development in cities through the world. 87-90. Rowman and Littlefield, USA.
Regassa N, Sundaraa RD, Seboka BB (2011). Challenges and opportunities in municipal solid waste management: The case of Addis Ababa city, central Ethiopia. J. Human Ecol. 33(3):179-190.
Tesema H (2010). Overview of Addis Ababa city solid waste management system, Division for Sustainable Development, Department of Economic and Social Affairs, United Nations, Addis Ababa, Ethiopia.
Young EM (2012). Food and Development. 259-269. Routledge, USA.

6

Appraising the combustion of biogas for sustainable rural energy needs

A. M. Stanley[1] , D. M. Stanley[2], Dadu, D. W.[1] and A. M. Abah[1]

[1]Department of Building, Ahmadu Bello University, Zaria, Kaduna State, Nigeria
[2]Department of Crop Protection, University of Maiduguri, Borno State, Nigeria.

This paper shows the combustion of biogas in rural households' appliances. Biogas has been known since 1800s as an odourless and colourless gas with high combustion rate. Its use is beginning to gain ground in most developing countries like Nigeria due to its availability, ease of generation and environmental friendliness. Developing countries are characterized by poor infrastructural development, inadequate energy and water supply, poor health delivery system, etc. which hinders economic and social development. Most sources of rural households' energy are firewood, animal dung, crop residue and kerosene which are associated with negative environmental impacts. The study was carried-out by articulation of past literatures on biogas combustion and consumption in household's appliances and internal combustion engines. The study ascertains from the past studies high efficiency of biogas compared with natural gas and liquid petroleum gas (LPG) on stove-top burner, oven and two panel flue heater. It was observed that, biogas consumption is higher in all the appliances under investigation as compared to natural gas and LPG. The study recommended public enlightenment on biogas technology and its associate benefits to rural areas. The government and NGOs should encourage the application of this technology through financing of pilot projects in community leaders' households which will extend to the populace. The technology should also be embraced because it is associated with environmental hygiene.

Key words: Efficiency, hygiene, sustainability, developing countries, biogas.

INTRODUCTION

The rate of energy consumption and waste generation in developing countries necessitates the adoption of technologies that promote renewable energy and the conversion of waste into viable commodity. Biogas technology is one of such systems, it is cost effective and environmentally friendly (Brown, 2003). According to Matthew (1998), after the advent of the oil boom of the 1970s in Nigeria, there has been a growing national concern on shortage of energy supply. Similar problem has been envisaged around the globe (Oslaj and Mursec, 2010; Preston and Murgueitio, 1992; Mattocks, 1984). In the United States of America for instance, several million gallons of fuel is consumed on daily basis for transportation and other activities to meet their daily needs (Habmigern, 2003). These fuels are obtained from non-renewable sources which will not guarantee production in the nearest future. In addition, their production is associated with negative impacts on the environment. Sustainability is a current global trend which addresses issues concerning environmental impacts. It was defined as development that meets the needs of the present generation without compromising the ability of future generations to meet their needs (World Commission on Environment and Development, 1987). To achieve sustainability, the use of sustain-

Table 1. Composition of biogas.

Substance	Symbol	Percentage
Methane	CH_4	50 - 70
Carbondioxide	CO_2	30 - 40
Hydrogen	H_2	5 - 10
Nitrogen	N_2	1 - 2
Water vapour	H_2O	0.3
Hydrogen sulphite	H_2S	Trace

Source: Yadava and Hesse (1981).

able energy sources (such as biogas, solar photovoltaic and wind) (Pick et al., 2012; Oslaj and Mursec, 2010) must be adopted in developing countries.

Biogas has been known since early 1800s in different parts of the world to complement most household energy needs (Stanley, 2006). Rapaport (1995), defined biogas as 'a mixture of methane and carbon dioxide produced by anaerobic bacteria as they digest organic materials'. It is composed of methane, carbondioxide and other small amounts of gases as shown in Table 1.

Biogas is lighter than air and has an ignition temperature of approximately 700°C as compared to diesel oil of 350°C; petrol and propane of about 500°C (Brown, 2003). The temperature of the flame is 870°C (Brown, 2003). In rural areas, biogas are used in modified form for stove, refrigerator, water heater, diesel engine, gas lamps and indeed any piece of equipment that uses liquid petroleum gas (LPG) (Engineers Without Borders, 2004).

Rural areas are characterized by inadequate provision of essential amenities like power supply, water, road, hospital and schools. Biogas production reduces the risk of exposure to smoke from firewood which causes respiratory illness and promotes deforestation. It reduces shallow ground water contamination from nearby pit latrines and soakaway. It also saves money, replacing electricity or liquid gas as a source of energy for cooking (Aquilar, 2001). Interest in biogas as a viable energy resource has spread throughout the globe in the past two decades (Stanley, 2006). Biogas digester operates throughout Asia, with more than 100,000 reported in India, 30,000 in Korea and several million in China (Mattocks, 1984). Biogas produced from a town septic tank was used for street lighting in Exeter England in the late 1800s and today used as a source of energy for operations in sewage treatment plants in most parts of the world (Rapaport, 1995). Also in Pnatna, India, 24 seat pour-flush latrines serve several thousand people and generate sufficient energy to light a 4 km road (Li, 1984). Also, Vijay (2006) outlined the use of biogas in cars, buses, tractors, auto rickshaws, irrigation pumps and in rural industries. The calorific value of the gas is about 6 kWh/m^3 which corresponds to about a litre of diesel oil (Brown, 2003). These associated advantages necessitate the adoption of biogas technology in rural areas of the developing countries.

The major disadvantage in using the gas is the presence of hydrogen sulphide. The hydrogen sulphide combines with water vapour to form a corrosive acid. Water heating appliance, utensils and refrigerators are particularly at risk. However, according to Brown (2003), biogas can be rid of the sulphide using iron oxide filters, as given in the following equations:

$$FeO + H_2S \rightarrow FeS + H_2O$$

$$2FeS + O2 \rightarrow 2FeO + 2S$$

ANAEROBIC DIGESTION

Biogas is produced from a process known as anaerobic digestion. Anaerobic digestion is the process of bacterial decomposition in the absence of oxygen. It is a natural phenomenon that occurs in everyday life, that is, plants and animals die and are recycled by nature. Swamp areas, landfill wastes, septic and soakway tanks, farm wastes, animal dung, poultry wastes, abattoir wastes, etc. are potential sources of anaerobic digestion (Pick et al., 2012). The major by-products of the anaerobic digestion are: Biogas and a semi-solid effluent or sludge (Brown, 2003; Mattocks, 1984).

Biogas

Biogas is a mixture of gas produced by methanogenic bacteria while acting upon biodegradable materials such as raw cow manure, human excreta, vegetable waste and water in an anaerobic condition (SDdimension, 1997).

Biogas is about 20% lighter than air and has an ignition temperature in the range of 650 to 750°C as compared to diesel oil, 350°C; petrol and propane of about 500°C; it is an odourless and colourless gas that burns with clear flame at a temperature of 870°C (Brown, 2003; Sathianathan, 1975). Biogas is known by many names: swamp gas, marsh gas, "will o' the wisp" and gober gas (Mattocks, 1984).

Biogas can be used in the same way as many other combustible gases. It can be used in a modified gas stove, refrigerator, water heater, diesel engine, gas lamp and in any piece of equipment that uses liquid petroleum gas (LPG) (Engineers Without Borders, 2004).

Effluent

Effluent (called slurry or sludge) is the residue of inputs that comes out of the compensation chamber after the substrate is acted upon by the methanogenic bacteria in an anaerobic condition inside the digester. The effluent is the second by-product that comes out of the digester after

Figure 1. Oil drum biodigester.

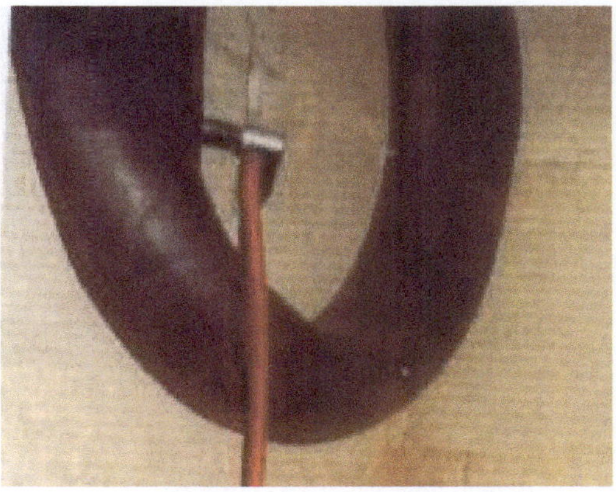

Figure 2. Flotation tube.

Figure 3. Cooker burner using biogas.

Figure 4. Water boils by biogas.

after extracting biogas. It is a stabilized manure almost pathogen free and has proved to contain high quality organic materials for plant nutrition and fish production (Karki and Gautam, 2000). It has no odour and neither dissemination of diseases/weeds and comes out from the compensation chamber in one of the following forms (SDdimension, 1997):

1. A high rather solid fraction, mainly fibrous material, which float on the top forming the scum.
2. A very liquid and watery fraction remaining in the middle layer of the digester.
3. A viscous fraction below which is the real sludge.
4. Heavy solids, mainly sand and soils deposited at the bottom.

The above separation can be less in the slurry if the organic matters fed into the digester are homogenous, using appropriate ratio for urine, water and excrement; and intensive mixing before feeding the digester leads to homogenous slurry (SDdimension, 1997).

BIOGAS APPLICATIONS

Figure 1 shows oil drum biodigesters used for the production of the biogas in Ahmadu Bello University, Zaria, Nigeria, while Figure 2 shows flotation tube used for collection and storage of the biogas produced from the biolatrine system (Stanley, 2006).

Biogas can be stored in tractor tubes or in a flotation container as observed in Stanley (2006) and Forst (2002). This can be adopted in the rural areas for the storage of the biogas produced for immediate use. In an experiment conducted by Stanley (2006), biogas was used in cooker burner to boil water as shown in Figures 3 and 4.

This was also articulated by Brown (2004), ITDG (2004) and Shannon (1997), that biogas combust effecttively on cooker burner. Figures 5, 6 and 7 shows biogas burning from a gas valve with visible flame enhanced by wood and paper (Stanley, 2006).

The use of wood and piece of paper were to enhance the visibility of the combustion which is odourless and colourless that burns with clear flame at a temperature of 870°C (Brown, 2003; Sathiarathan, 1995) (Figure 8).

Some studies (Stanley, 2006; Brown, 2003; Shannon, 1997) have compared the combustion of biogas different on equipment and households appliances. Shannon (1997) compared the combustion of biogas, natural gas and LPG on stove-top burner, oven, two panel heater and flued heater. It revealed that biogas consumption is high-

Figure 5. Wood enhanced the visibility of biogas combustion.

Figure 6. Paper enhanced the visibility of biogas combustion.

Figure 7. Blue flame burning on biogas.

Figure 8. Flame of fire from biogas.

er in all the appliances under investigation as com-pared to natural gas and LPG as shown in Table 2.

Natural gas and LPG are obtained from non-renewable sources which are associated with environmental and health hazards. Biogas on the other hand, is environmental sustainable in the sense that it is renewable and obtained from renewable sources. This makes biogas environmentally sound.

Biogas has high efficiency as observed in ITDG (2004), 1 m^3 of biogas can make three (3) meals for a family of five to six. In BSP (2001), the efficiency of biogas stove was studied under perfectly controlled, semi-controlled and uncontrolled external conditions. The results show that the efficiency of biogas on the stove was 49.44, 43.8 and 32.26%, respectively, which are high. In another study by CES (2001), the overall efficiency of biogas was 57.4% higher than that of LPG, kerosene and wood stove with frequencies of 53.6, 49.6 and 22.8%, respectively. These shows that biogas has high efficiency of combustion in stove. Table 3 also shows the consumption ption rates of biogas and time taken in cooking using modified gas stove.

Table 4 and Figure 9 show various households equipment and the quantity of biogas consumed in litre per hour for lighting and cooking. Household gas burner consume 200 to 450 L/h of biogas to cook 350 to 1000 g of pulse as shown in Table 3, while a gas lamp of 60 W bulb equivalent consume 120 to 150 L/h of biogas, etc. Brown (2003) and Forst (2002) have measured the quantity of consumption and performance of biogas on different household equipment which proved excellent. Figure 10 shows the storage bag for biogas as observed in Forst (2002).

Internal combustion engine are improvised to use biogas. Engine carburetors and injectors are the major parts which need partial or no modification to burn

Table 2. Typical consumption figures - domestic appliance.

Appliance	Approximate consumption (m^3/h)		
	Biogas	Natural gas	LPG
Stove-top Burner (9 Mj)	0.5	0.25	0.1
Oven (8.5 - 10 Mj)	0.40 - 0.60	0.20 - 0.30	0.08 - 0.12
Small, two-panel heater (11 Mj)	0.55	0.30	0.11
Large, flued heater (44 Mj)	2.20	1.10	0.44

Source: Shannon, (1997).

Table 3. The consumption rates of biogas in cooking using modified gas stove.

Amount cooked	Time (min)	Gas (L)
1 L water	10	40
5 L water	35	165
500 g rice	30	140
1000 g rice	37	175
350 g pulse	60	270
700 g pulse	70	315

Source: Brown (2003).

Table 4. Consumption of biogas by various equipment.

Equipment	Amount of gas (L/h)
Household burners	200 - 450
Industrial burners	1000 - 3000
Refrigerator, 100 L depending on outside temperature	30 - 75
Gas lamp, equivalent to 60 W bulb	120 - 150
Biogas/diesel engine per bhp	420
Generation of 1 kWh of electricity with biogas/diesel mixture	700

Source: Brown (2003).

Figure 9. Modified gas stove and pressure lamp attached to biogas storage container. Source: Forst (2002).

Figure 10. Flotation storage tank. Source: Forst (2002).

Figure 11. Biogas packed in cylinder for car consumption. Source: Vijay (2006).

Figure 12. Packed biogas for household application. Source: Vijay (2006).

smoothly on biogas as dual or whole. In ITDG (2004), it was observed that 1 m^3 of biogas generate 1.25 kWh of electricity, or power a one horse power internal combustion engine for 2 h which is equivalent to 2.5 L of petrol (Vijay, 2006). In Quenum (2007), it was also observed that 1 m^3 of biogas burned on electric generator to produce 2 kWh of electricity. This figure is projected to 4,612,320 kWh electricity per year for a pig farm with flock of 3,206. In another study by Siripornakarachai and Sucharitakul (2007), a Hino K-13CTI 13,000cc 24 valves diesel bus engine was modified to operate on biogas. The engine was coupled to a 3-phase 4-pole induction motor to produce electricity at 50 Hz. The study established optimal engine efficiency at 28.6%, while the generator power output was 134.20 kW with emission of CO and NOx at 1,154 and 896 ppm, respectively. It was observed that emission of CO and NOx were significantly low as compared to same emission from a 2.3 kW petrol generator of 916 and 1.8 ppm, respectively (Stanley, 2011).

Vijay (2006), articulated packaging of biogas for cars and home use in cylinders (Figures 11 and 12).

It was observed that one (1) cylinder filled with 120 m^3 biogas bottle equals to eight (8) of 6 kg CNG cylinders and the 6 kg CNG cylinder equals 6 L of petrol (Vijay, 2006). The gas filled in these cylinders equal to 6 multiple by 8 which is 48 L of petrol per day.

King and Mintner (1998) ascertained the removal of impurities present in biogas which is fed into an internal combustion engine. Biogas supplied to the engine flows through a filter which removes hydrogen sulphide and merceptans and also through a coalescer to remove water vapour.

In Mehta (2002), the production of biogas per cow and the quantity of electricity generated were compared. The electricity generated per cow by European generator was 0.15 kW on a continuous basis, while in America it was 0.2 kW. The difference in the generation was owed to animal size and feed. Mehta (2002) estimated the potential of electric power generation that can be produced by number of cows in farm (Table 5).

In PREGA (2006), the production of biogas and the electric power output in summer and winter were studied. It was observed that 40 m^3 of biogas was generated per day in summer, while 80 m^3 in winter with electricity generation of 80 to 160 kWh/day. While GEDA (2003), included the estimation of plant capacity of biogas production per day, which range from 15 to 85 m^3 as shown in Table 6 and Figure 13 shows the China dome digester used for the biogas production. The corresponding cattle population, dung required, quantity of electric power generated size and the running hours were also estimated.

GEDA (2003) established that 15 m^3 of biogas per day is produced by 25 to 30 cows which can run a 3.5 KVA/3.0 kW diesel generator for 4 to 5 h per day. Subsequently, 220 to 250 cows produce 85 m^3 and run

Table 5. Maximum demand and electric power potential.

Number of cows	Electric power potential (kW)
30	6
60	12
200	200 20 40
400	80

Source: Mehta (2002).

Table 6. Plant capacity of biogas and electricity generation.

Plant capacity (m^3/day)	Cattle population required	Dung required on daily basis (kg)	Possible size of DG set to be coupled with biogas plant	Approximate DG running per day and total unit generation/day
15	25 -30	225	3.5 kva/3.0 kW (single phase genset)	4 - 5 h/day (12 to 15 units/day)
25	62 -70	625	3.5 kva/3.0 kW (single phase genset)	7 - 8 h/day (20 - 24 units/day)
35	85 - 95	875	7.5 kva/6.0 kW (Three phase genset)	5 - 6 h/day (30 - 36 units/day)
45	120 - 130	1130	7.5 kva/6.0 kW (Three phase genset)	7 - 8 h/day (45- 50 units/day)
60	150 - 160	1500	10 kva/8.0 kW (Three phase genset)	9 - 10 h/day (80 - 90 units/day)
85	220 - 250	2125	10 kva/8.0 kW (Three phase genset)	13 -14 h/day(120-130 units/day)

Source: GEDA (2003)

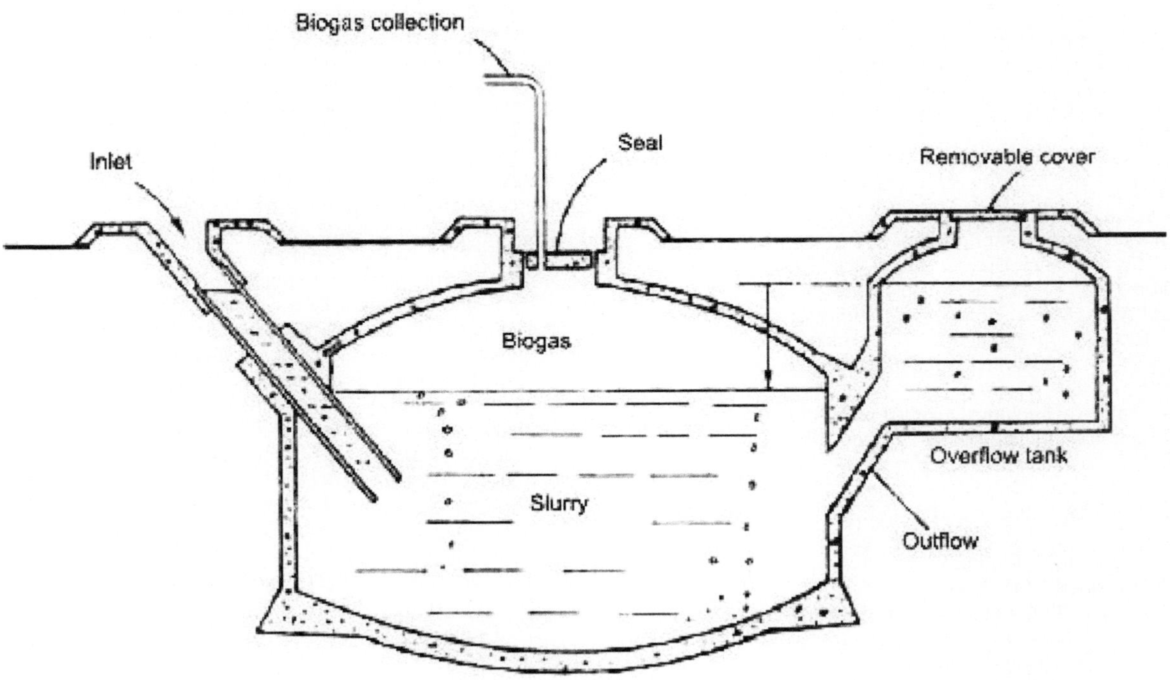

Figure 13. China dome digester. Source: Kangmin and Ho (2006).

10 KVA/8.0 kW diesel generator for 13 to 14 h per day.

Conclusion and recommendations

From the findings, biogas has been a viable source of energy which has been practiced for a very long time in China, India, Korea, etc. because of its cheapness, accessibility and sustainability. It has been observed that biogas production is a process of waste disposal (both solid and liquid) and environmental sanitation. Biogas has been tested and proved to perform well in household appliances (cooker burner, gas lamp, electric power generators and on cars). No environmental contaminant is associated with the production of biogas. The technology requires no or less expertise and this will suits the nature of the Nigerian rural populace. It will enhance the income generation and reduce spending of the poor. It was recommended that public enlightenment on biogas technology and its associated benefits should be disse-minated. The government and NGOs should encou-rage the application of this technology through financing of pilot projects in households of rural head which will be extended to the entire populace.

REFERENCES

Biogas Support Programme (BSP) (2001). A Studies Report on Efficiency Measurement of Biogas, Kerosene and LPG Stoves. Temp No. 71.

Brown J (2003). Wastewater treatment: The Experiences of the Scientific Research Council in Jamaica (SRC). Hope Gerdons, Kingston 6. 7(10):16-20.

Center for Energy Studies (CES) (2001). A Study Report on Efficiency Measurement of Biogas, Kerosene and LPG Stoves. Biogas Support Program Lalitpur, Nepal.

Engineers Without Borders (2004). The Biogas Digester - A Sustainable Energy Production Technology for Rural Developments of Sub - Sahara Countries. [Online] Available from http//:www.umanihba.ewb.ca/sdrc/2004/002BON.pdf (accessed 17th October, 2005).

Forst C (2002). Technologies Demonstrated at Echo: Horizontal Biogas Digester. AN ECHO CONCEPT PAPER [Online] Available from http://www.echonet.org/.

Gujarat Energy Development Agency (GEDA) (2003). Power Generation from Animal Dung / Alternative Feed Stock. [Online] Available from http://www.cybersurfindia.com/.

Habmigern (2003). Biogas Production: The Methane Digester for Biogas. Biogas Production. [Online] Available from http//:www.hamigern2003.info/biogas/methane-digester.html (accessed 15th August, 2005).

ITDG (Intermediate Technology Development Group) (2004). Biogas and liquid biofuels. In: Engineers Without Borders: The Biogas Digester - A Sustainable Energy Production Technology for Rural Developments of Sub - Sahara Countries. [Online] Available from http//:www.umanihba.ewb.ca/sdrc/2004/002BON.pdf (accessed 17th October, 2005).

Kangmin L, Ho M (2006). Biogas China. Available from http://www.i-sis.org.uk/full/BiogasChinaFull.php.

Karki AB, Gautam U (2000). Latrine - cum - Biogas Plant at Ward No. 1 of Pathari VDC of Morang Distict of Nepal. Internet Conference on Material Flow Analysis of Integrated Bio-Systems (March - October). Consolidate Management Services, New Baneshwor, Kathmandes, Nepal.

King JL, Mintner J (1998): Biogas driven generator set. [Online] Available from http://www.freepatentsonline.com/5724948.html.

Li (1984). In: WHO (1992) A Guide to the development of On-site Sanitation. Available from www.who.int/docstore/water_sanitation_health/onsitesan/begin.htm. (accessed 17th August 2005).

Matthew A (1998). Levels of Biogas Production with Different Subtrates. An Unpublished Undergraduate Project, Department of Microbiology, Ahmadu Bello University, Zaria.

Mattocks R (1984): Understanding Biogas Generation. Technical Paper. Published by VITA, Virginia, USA. [online] email: pr-info@vita.org (accessed 13th August, 2005).

Mehta A (2002). The Economics and Feasibility of Electricity Generation using Manure Digesters on Small and Mid-size Dairy Farms. Department of Agricultural and Applied Economics Energy Analysis and Policy Program. University of Wisconsin - Madison.

Oslaj M, Mursec B (2010). Biogas as a Renewable Energy Source. Technical Gazette Vol. 17, 109-114.

Pick D, Dieterich M, Heintschel S (2012). Biogas Production Potential from Economically Usable Green Waste. [Online] Available from www.mdpi.com/journal/sustainability. J. Sustain. 4:682-702.

Preston TR, Murgueitio E (1992): Strategy for Sustainable Livestock Production in the Tropics. and Amani, Tanzania. University of Aarhus, Denmark. [online] Available from www.husdyr.kul.dk/htm/php/tune16/16knie/sen.htm (accessed 20th September, 2005).

Promotion of renewable energy, Energy efficiency and greenhouse gas abatement (PREGA) (2006). Biogas Unit for Farm. A Feasibility Study Report

Quenum BM (2007). An Animal Breeding Operation Coupled with A Biodigester to Producing Biogas and Electricity. [Online] Available from http://businessafrica.net/africabiz//botproofemail1.php.

Rapaport D (1995). Sewage Pollution in Pacific Island Countries and How to Prevent It. A Report Produced for Greenpeace Pacific. Center for Clean Development, Greenpeace Pacific, Eugene, Oregon, USA. [online] Available from www.dev_pacis.htm (accessed 6th November, 2005).

Sathianathan MA (1975). Biogas Achievements and Challenges. In: SDdimensions (1997): A System Approach to Biogas Technology. Energy and Environmental Technology. [online] Available from www.fao.org/sd/EG-direct/EGe0022.htm (accessed 9th August, 2005).

SDdimensions (1997). A System Approach to Biogas Technology. Energy and Environmental Technology. [online] Available from www.fao.org/sd/EG-direct/EGe0022.htm (accessed 9th August, 2005).

Shannon R (1997). Biogas. Conference Proceeding. IPC - VI Designing for a Sustainable Future. [Online] Available from http://www.rosneath.com.au/ipc6/.

Siripornakarachai S, Sucharitakul T (2007). Modification and tuning of diesel bus engine for biogas electricity production. Maejo Int. J. Sci. Technol. 1(2):194-207.

Stanley AM (2006). An Experimental Biolatrine for a Sustainable Toilet System for Rural Communities in Nigeria. M.Sc. Thesis. Department of Building, Ahmadu Bello University, Zaria - Nigeria.

Stanley AM (2011). Air Pollutants Concentration and Noise Levels from Electric Power Generators. Seminar Presented to Faculty of Environmental Design, Ahmadu Bello University, Zaria - Nigeria.

Vijay VK (2006). Biogas Enrichment and Bottling Technology for Vehicular Use Center for Rural Development and Technology, Indian Institute of Technology Delhi, India.

Yadava LS, Hesse PR (1981): The Development and Use of Biogas Technology in Rural Areas of Asia. In: SDdimensions (1997): A System Approach to Biogas Technology. Energy and Environmental Technology. [online] Available from www.fao.org/sd/EG-direct/EGe0022.htm (accessed 9th August, 2005).

Earthworm-assisted bioremediation of petroleum hydrocarbon contaminated soil from mechanic workshop

Ameh A. O.[1], Mohammed-Dabo I. A.[1], Ibrahim S.[2] and Ameh J. B.[3]

[1]Department of Chemical Engineering, Ahmadu Bello University, Zaria 810261, Kaduna State, Nigeria.
[2]Department of Biochemistry, Ahmadu Bello University, Zaria, Kaduna State, Nigeria.
[3]Department of Microbiology, Ahmadu Bello University, Zaria, Kaduna State, Nigeria.

The use of earthworms (*Eudrilus eugenia*) for vermi-assisted bioremediation of petroleum hydrocarbon contaminated mechanic workshop soils was investigated. Parameters of interest where earthworm survival, microbial count and identification, carbon and nitrogen contents of the soil and total petroleum hydrocarbon content of the soil. Earthworm survival was found to be dependent on the concentration of petroleum hydrocarbon and nature of the mechanic workshop. The survival of the worms in the as-obtained mechanic workshop soil was much lower than the survival in mechanic workshop soil diluted with petroleum hydrocarbon free soil. Earthworm inoculation also resulted in lower soil C and N contents after 35 days of vermi-assisted bioremediation. After 35 days of treatment, earthworm inoculation effected a higher drop in total petroleum hydrocarbon contents as compared to the samples without worms, indicating that earthworms may be used as biocatalysts in the bioremediation process.

Key words: *Eudrilus eugeniae*, total petroleum hydrocarbon, mechanic workshop soil.

INTRODUCTION

The quality of life on earth is linked, inextricably, to the overall quality of the environment. Releases of persistent, bioaccumulative and toxic chemicals have a detrimental impact on human health and the environment. Contaminated lands generally result from past industrial activities when awareness of the health and environmental effects connected with the production, use, and disposal of hazardous substances were less recognized than today (Vidali, 2001). The problem is worldwide, but more severe in the developing countries where there were no effective regulatory policies on the environment. Petroleum contaminants are typical examples of these hazardous constituents. Soil contaminated with petroleum products can be defined as any earthen material or artificial fill that has human or natural alteration of its physical, chemical, biological or radiological integrity resulting from the introduction of crude oil, any fraction or derivative thereof (such as gasoline, diesel or motor oil), or oil-based product (Dhanasekaran et al., 2009; Chukwuma et al., 2010; Refaat, 2010).

The problems associated with used engine oil (UEO) are mainly as a result of poor management. In most developing countries, oil spills at auto-mechanic workshops have been left uncared for over the years and its continuous accumulation is of serious environmental concern because of the hazard associated with it. For instance, spent motor oil disposed improperly contains potentially toxic substances such as benzene (carcinogens), lead, arsenic, zinc and cadmium, which can seep into the water table and contaminate ground water (Igwe et al., 2008; Shah et al., 2009). It consequently results in serious health hazard such as anemia and tremors, which can cause death. If used motor oil and the contaminants it contains are disposed of inappropriately and released into the environment, they can harm humans, plants, animals, fish and shellfish. In water, oil is a visible pollutant, floating as a scum on the surface. This oil scum can stop sunlight and oxygen from getting into the water, affecting fish and water plants. It can kill fish, frogs and other animals that breathe from the water's surface. According to USEPA (1996), a liter of used engine oil is enough to contaminate one million gallons of freshwater.

Physicochemical technologies including thermal treatment, dig and dump method, chemical method, separation techniques and stabilization/solidification technology have been used for ages for the treatment or disposal of soils contaminated with petroleum products. Some of the limitations of these methods include: they are expensive to implement at full scale, they are not environmentally friendly, their technologies are complex and they lead to destruction of soil texture and characteristics (Zhang et al., 2009) and do not always result in complete neutralization of pollutants (Yerushalmi et al., 2003). Due to limitations of the physicochemical technologies, great deals of literature have reported that bioremediation technologies are alternatives and or supplements to these technologies. This is because of their cost effectiveness, environmental friendliness, simplicity in technology and conservation of soil texture and characteristics (Vidali, 2001; Yerushalmi et al., 2003; Adams and Guzman-Osorio, 2008; Fouépé et al., 2009). Bioremediation of petroleum hydrocarbons suffers mass transfer limitations as a result of the high hydrophobicity and low solubility of the contaminants in aqueous media (Li et al., 1995). This makes bioremediation a very slow process.

Earthworms are terrestrial annelids with bilateral symmetry. They are important factors in the development of soil structure, organic matter breakdown and nutrient cycling. Generally, the activities of earthworm are considered to improve soil health and theoretically can translate to improved bioremediation. In spite of the numerous ways in which the activities of earthworm promotes soil health, relatively few research have been undertaken to study the effects of earthworm on petroleum hydrocarbons contamination removal from soil (Hickman and Reid, 2008). Schaefer and Juliane (2007) concluded that earthworms may trigger the degradation process and might therefore, be applied in the remediation of oil contaminated soil with moderate TPH concentrations. Schaefer et al. (2005) reported that earthworms can contribute positively to bioremediation of oil-contaminated soil, but that the effect may be species-dependent. Ameh et al. (2011a) reported the survival of earthworms (*Eudrilus eugeniae*) in used engine oil (from automobiles) contaminated soil. They reported that used engine oil concentration level tolerable to earthworms was dependent on the source of the used engine oil. Ameh et al. (2011b) investigated the effect of earthworm on the bioremediation of simulated used-engine oil-contaminated soil and reported that the presence of earthworms can influence the rate of bioremediation of used engine contaminated soil, given right conditions.

The objective of this study was to examine the ability of earthworm (*E. eugeniae*) to enhance the bioremediation of spent engine oil contaminated soil, making a comparison between biostimulation and vermi-assisted bioremediation.

MATERIALS AND METHODS

Procurement and preparation of materials

Soil samples from two types of mechanic workshops were considered for this investigation: motorcar and motorcycle mechanic workshops (MM and MC, respectively). Large particles (metal, gravels and sticks) were removed from the soil samples. Two TPH contents of the soils were considered (92.73 and 31.61 g oil/kg soil for MM and 138.18 and 69.27 g oil/kg soil for MC). The larger portion of the TPH contaminated soil was soil as obtained from mechanic workshop whereas the smaller of the two TPH values for both mechanic workshop soils considered was obtained by diluting the mechanic workshop soil with petroleum hydrocarbon free soil obtained from Samaru, Zaria. The water holding capacity of the undiluted and diluted soil samples was determined thus: six (6) inches of soil was placed in percolation tube and compacted by gentle bouncing. Water was added until the water level stood at about 2 inches above soil level. The tube was covered and left to stand for 2 days. After this period, the top half-inch of soil was discarded and wet soil was weighed into a pre-weighed evaporating dish. The sample was placed in oven at 110°C for 24 h. It was then removed and weighed to obtain bone dry weight of the sample. Percent water holding capacity was calculated as (Srivastava and Thakur, 2006):

$$\%\,water\ holding\ capacity\ (WHC) = \frac{wet\ weight - bone\ dry\ weight}{bone\ dry\ weight} * 100 \qquad 1$$

Cassava tubers where obtained from Samaru market and identified in the Biological Science Department of Ahmadu Bello University, Zaria. The cassava used was identified as *Manihot esculenta* (Crantz). The tubers were peeled and the peels soaked in water for a period of one week to allow for detoxification of any cyanohydrin which might be present. After decantation, the fermented cassava peels were then dried in an oven at 30°C for four days. The dried peels were then crushed and sieved using 5 mm sieve mesh. Cassava peels was prepared for use as organic amendment.

Earthworms were obtained from Samaru village Zaria and identified as *E. eugeniae*. The worms were collected along with the soil in which they were found and kept for 24 h before use in clean plastic containers.

Preparation of samples

500 g of the mechanic workshop soil was placed in a plastic container and deionized water was added to 60% water holding capacity and then thoroughly mixed manually. 50 g of fermented cassava peels was then added and again mixed. A second sample

was similarly prepared to which 10 earthworm were in addition introduced. Both samples were covered with muslin cloth to prevent the worms escaping from the sample with worms. All samples were studied for 35 days. This procedure was repeated for the mechanic workshop soil diluted with petroleum hydrocarbon free soil. This was done for the motorcar mechanic workshop soil as well as the motorcycle mechanic workshop soil in turns. Control samples for each case (diluted and undiluted) were also prepared. The control sample had neither organic amendment nor earthworms. All samples were prepared in duplicates and maintained at 60% water holding capacity throughout the period of the study. Starting from day zero and then every week, the TPH of all samples was followed using a gravimetric method (Adesodun and Mbagwu, 2008): ten grams of soil was collected and dried at room temperature for 72 h. Five grams of the soil was placed in 200 milliliters beaker into which was added 150 milliliters of toluene. The mixture was stirred continuously for 30 min, left to stand in a fume cupboard for 2 h and then filtered using Whatman No 42 filter paper. The residue, soil, was allowed to dry in an oven at 50°C. TPH was computed as:

$$TPH\,(g/kg) = \frac{(initial\ weight\ of\ soil - weight\ of\ soil\ after\ solvent\ extraction)\ in\ g}{Initial\ weigh\ of\ soil\ in\ kg} \quad 2$$

The composition of the petroleum hydrocarbon contaminant was determined by mixing two grams of dry contaminated soil (at time, t = 0) with 5 ml of dichloromethane and analyzing the supernatant after settling using GCMS (GCMS-QP2010 Plus Shimadzu, Japan). The initial column oven temperature was kept at 60°C for five minutes and then raised by 20°C per minute to 140°C and then at 5°C per minute to 200°C. It was held at 200°C for 5 min and then raised by 4°C per minute to 280°C. The total run time was 49 min.

Earthworm metal uptake and elimination

Elemental analysis using EDXRF (MiniPal PW4025, Philips Analytical) of soil samples at the start and finish was conducted also on the earthworms samples at the start and after 35 days of inoculation (for the samples for which the worms were still alive). Procedure was coined from OECD (2010). On the set days, 0 and 35, the worms were removed from the soil and placed on a filter paper wetted with distilled water in a flat plastic bowl. The worms were thus kept, on the wet filter paper, for 48 h to void their gut content after which they were quickly killed by placing on ice and their carcasses dried in an oven at 35°C for 48 h. The dried carcasses were then analyzed for metal content using EDXRF. The elemental composition of the soil samples were also determined by EDXRF analysis of the dried soil (at 30°C for 72 h) samples.

Effect on soil carbon and nitrogen content

On day 35, 10 g of soil was collected from each sample and the carbon and nitrogen contents were determined using the Walkley-Black (AOAC, 1990) and the Macro-Kjedahl (AOAC, 1990) methods, respectively.

Microbial count and identification

Microbial count and identification was conducted on the soil samples obtained from mechanic workshops (before treatment with organic amendment or earthworms) as well as on all diluted samples (motorcycle and motorcar, worms treated and worms free). Twenty five grams (25 g) of soil sample was mixed with 225 ml of normal saline (NaCl + H2O) (1:10). Ten fold dilution were then prepared, that is 1 ml of the diluted sample (stock) was transferred into 9 ml of sterile normal saline (10^1) by using a sterile pipette, this was repeated to obtain 10^2, 10^3, 10^4 and 10^5 dilutions, respectively. Using a sterile pipette, 0.1 milliliters of 10^3, 10^4 and 10^5 dilution was aseptically transferred into sterile plate count agar (PCA) plates in duplicates. With the aid of a sterile bend glass rod, the inoculum was spread on the surface of the culture medium (PCA). The inoculated plates were incubated at 37°C for 24 h. The isolates were Gram stained and observed under the microscope to determine the gram reaction, morphology and the presence of spores. The organisms were identified and characterized. Gram-reaction, motility, shape and color of colony, catalase and oxidase activities were checked.

RESULTS

Table 1 presents the result of percentage TPH decrease and earthworm survival after 35 days of bioremediation for biostimulated and vermi-assisted systems.

Figures 1 and 2 presents the result of the fit of the bioremediation data obtained from the mechanic workshop soil samples (motorcar mechanic workshop soil and motorcycle mechanic workshop soil, respectively) to the 1st order bioremediation kinetics (Yeung et al., 1990; Adesodun and Mbagwu, 2008).

Figures 3 and 4 presents the GC chromatogram of the diluted motorcycle mechanic workshop soil at the beginning and end of the study, whereas Table 2 presents the result of the MS comparison. Table 3 presents the microbial count and identification of the samples on day 35. Table 4 presents the result of the analysis of carbon and nitrogen contents of the treated soil samples. Table 5 presents the result of metal uptake and elimination by the worms in the 35 day study.

DISUSSION

For soil from both types of mechanic workshop considered, the use of earthworm inoculation proved to have an effect on the rate of degradation of petroleum hydrocarbon as shown in Table 1. All the samples treated with worms had higher drops in TPH than the corresponding control samples (without worms). The least effect was observed for the MCU samples (undiluted soil from motorcycle mechanic workshop) in which earthworm survival drooped to 0% (all the worms died) in less than 24 h. There was only a marginal difference recorded between the MCU and the control (24.26461 and 24.56509, respectively).

It may also be observed that the highest drops in TPH for both mechanic workshop soil samples was recorded in the earthworm treated dilute samples (44.18 and 39.10% for MC and MM, respectively). Incidentally, these samples recorded an appreciable survival of earthworm over the duration of the study: on day 35, MCDW recorded 70% earthworm survival whereas MMDW recorded 80% survival of the earthworms.

For the undiluted soil samples (UD), MCUW recorded

Table 1. Effect of earthworm inoculation on the bioremediation of petroleum hydrocarbon contaminated soil from mechanic workshops.

	Initial TPH (g/kg)	TPH Decrease after 35 days (%)	Earthworm survival after 35 days (%)
Soil from motorcar mechanic workshop			
MMUNW	92.73	34.96694	
MMUW	92.73	36.3521	30
MMDNW	31.61	29.49738	
MMDW	31.61	39.09603	80
Soil from motorcycle mechanic workshop			
MCUNW	138.18	24.56509	
MCUW	138.18	24.26461	0
MCDNW	69.27	28.8435	
MCDW	69.27	44.17804	70

MMUW: Undiluted motorcar mechanic workshop soil treated with worms, MMUNW: undiluted motorcar mechanic workshop soil without worms, MMDW: diluted motorcar mechanic workshop soil treated with worms, MMDNW: diluted motorcar mechanic workshop soil treated without worms, MCUW: undiluted motorcycle mechanic workshop soil treated with worms, MCUNW: undiluted motorcycle mechanic workshop soil without worms, MCDW: diluted motorcycle mechanic workshop soil treated with worms, MCDNW: diluted motorcycle mechanic workshop soil treated without worms

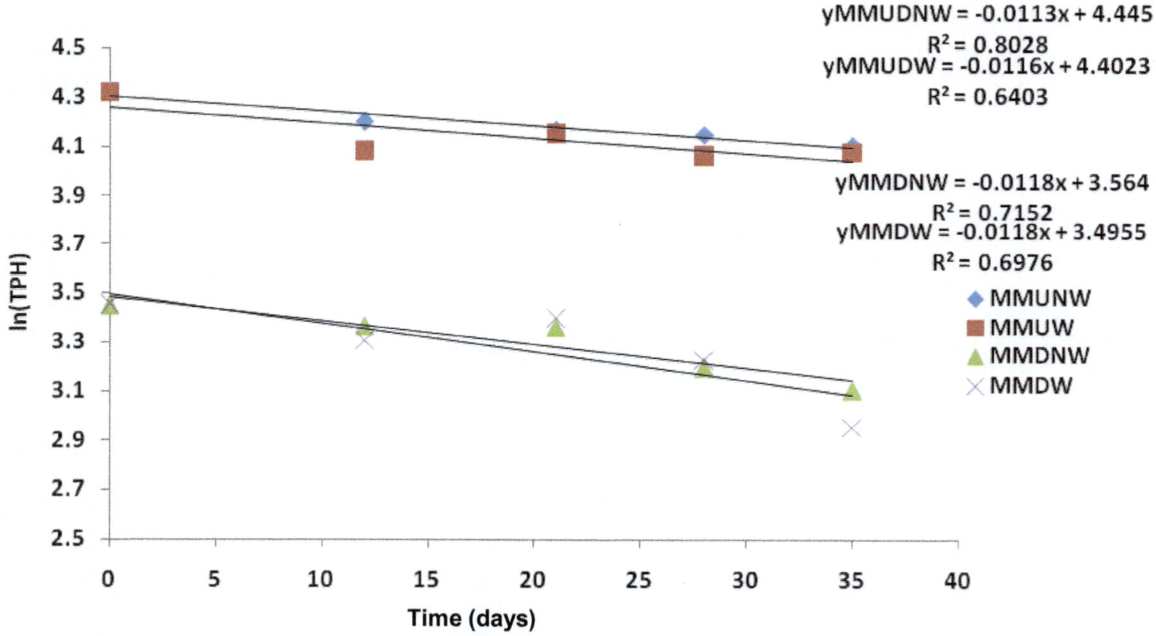

Figure 1. Effect of earthworm innoculation on the bioremediation kinetics of petroleum hydrocarbon contaminated soil from motorcar mechanic workshop.

100% death of the worms in less than 24 h. It was expected to have very similar bioremediation rate with the control (MCUNW) since all the worms died within 24 h after introduction. Table 1 confirmed this as percentage drop in TPH for MCUW and MCUNW which was found to be 24.26461 and 24.56509, respectively. MMUW recorded an earthworm survival rate of 30% at the end of the study (35 days) and also recorded a slightly higher TPH drop than the control (36.35 and 34.97, respectively). The death of the worms may be as a result of any or a combination of the following reasons: intolerable levels of petroleum hydrocarbon contamination, the pH, intolerable levels of heavy metal and carbon black (soot).

The results in Figures 1 and 2 indicate a poor fit to the first order bioremediation kinetics (R^2 generally less than 0.9). However, considering the duration of the study (only

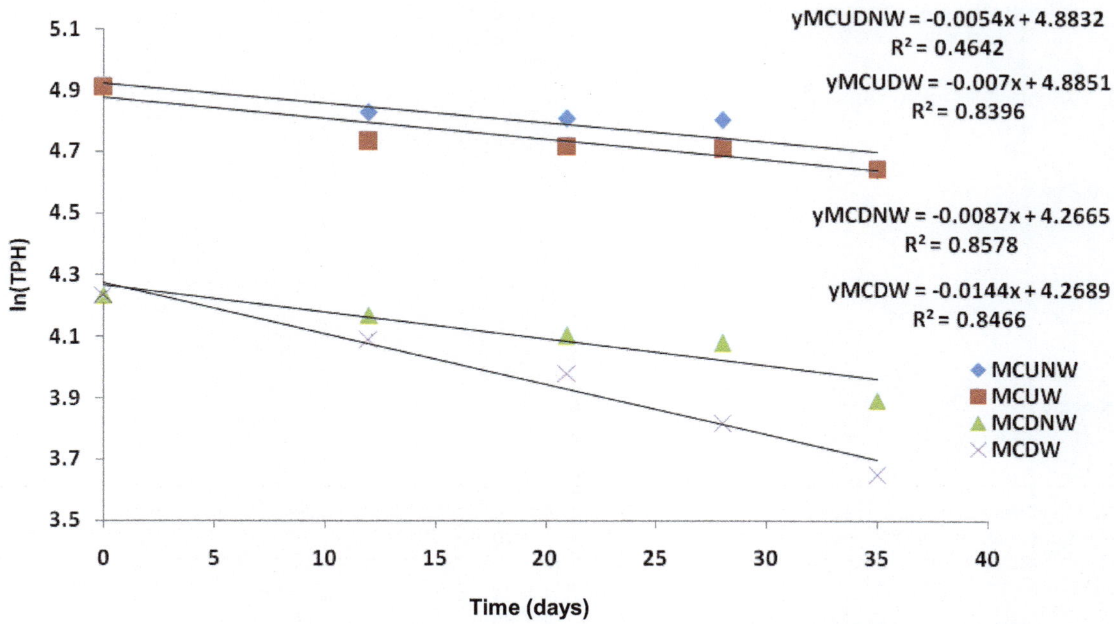

Figure 2. Effect of earthworm innoculation on the bioremediation kinetics of petroleum hydrocarbon contaminated soil from motorcycle mechanic workshop.

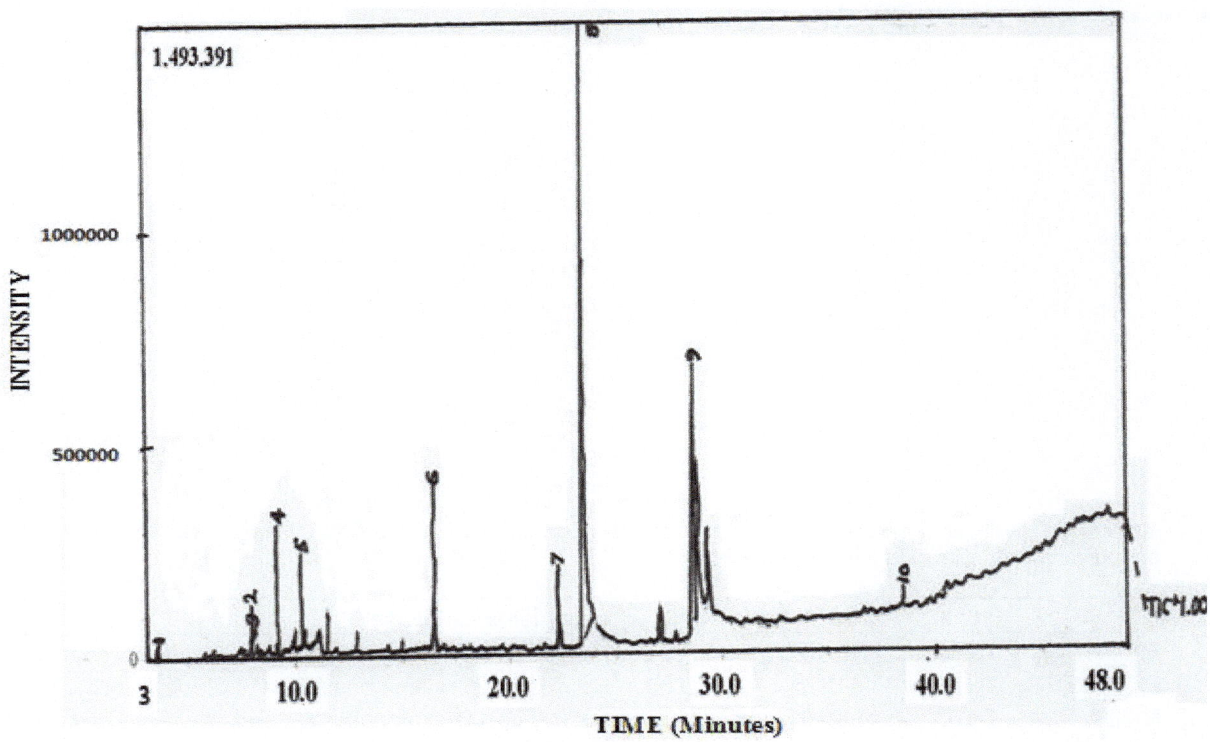

Figure 3. The GC chromatogram of the petroleum hydrocarbon contaminated soil from motorcycle workshop prior to bioremediation (MCO).

35 days) it should be possible to obtain R^2 value for longer study periods, or some other kinetics may be a better representation of the process.

Based on the result of this investigation, the use of earthworm for the bioremediation of petroleum hydrocarbon contaminated soil from mechanic workshop would

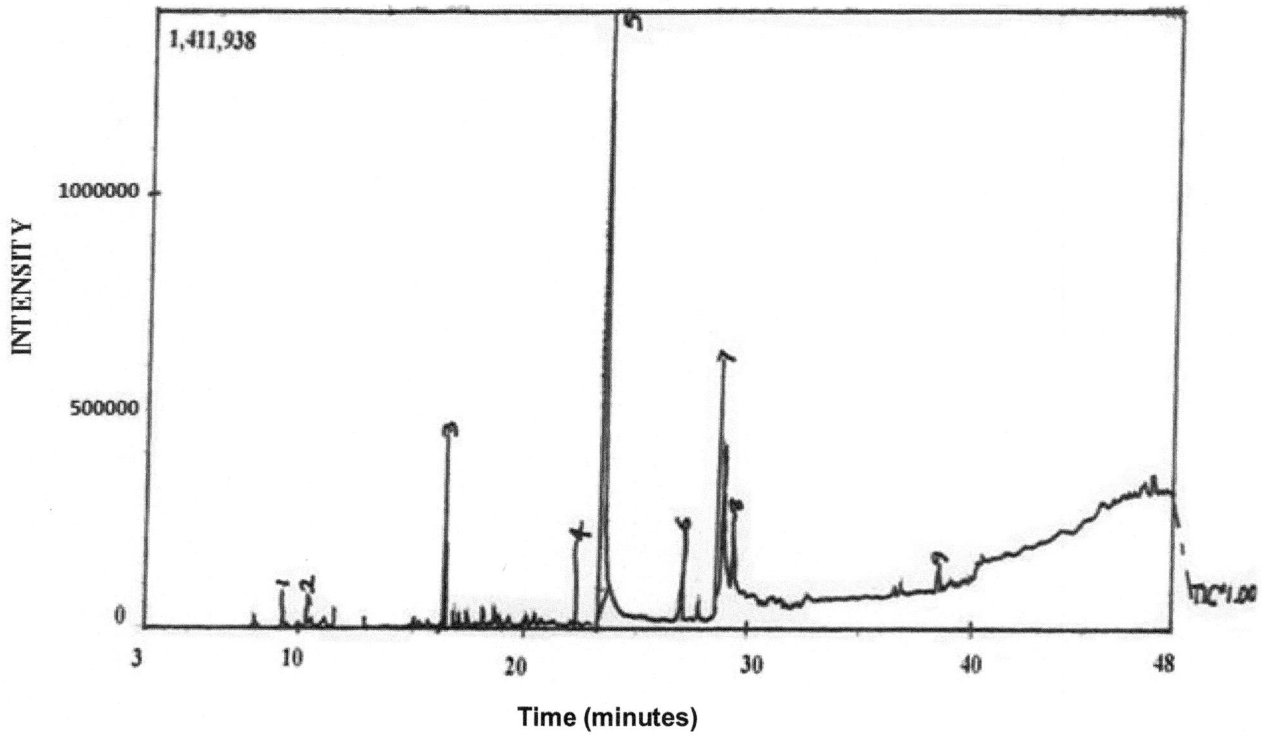

Figure 4. The GC chromatogram of the petroleum hydrocarbon contaminated soil from motorcycle workshop after 35 days of vermin-assisted bioremediation (MC35).

Table 2. Effect of vermin-assisted bioremediation on the composition of petroleum hydrocarbon content of motorcycle mechanic workshop soil.

Hydrocarbon	MCO	MC35
	Composition (%)	
C1 - C10	3.70	6.19
C11 - C20	95.47	92.31
C20 - C30	0.83	1.50

Table 3. Effect of vermi-assisted bioremediation on the microbial count and identification of mechanic workshop soil contaminated with petroleum hydrocarbon.

	10^3	10^4	10^5	Organisms isolated
MCDW	TNTC	3.18*10^5	8.15*10^5	*Bacillus* sp., *Pseudomonas* sp. and *Staphylococcus* sp.
MCDNW	TNTC	3.34*10^5	8.7*10^6	*Bacillus* sp., *Pseudomonas* sp. and *Staphylococcus* sp.
MMDW	TNTC	4.08*10^5	1.21*10^6	*Bacillus* sp., *Pseudomonas* sp. *and Staphylococcus* sp.
MMDNW	TNTC	3.39*10^5	9.4*10^6	Bacillus sp., Pseudomonas sp. and Staphylococcus sp.
MC (Raw)	5.0*10^3	1.0*10^3	NIL	*Pseudomonas* sp.
MM (Raw)	1.19*10^4	2.0*10^4	NIL	*Bacillus* sp and *pseudomonas* sp.

TNTC = Too numerous to count, MM (Raw): soil obtained from motorcar mechanic workshop (without organic amendment/earthworms), MC (Raw): soil obtained from motorcycle mechanic workshop (without organic amendment/earthworms) MMDW: diluted motorcar mechanic workshop soil treated with worms, MMDNW: diluted motorcar mechanic workshop soil treated without worms, MCUW: undiluted motorcycle mechanic workshop soil treated with worms, MCDW: diluted motorcycle mechanic workshop soil treated with worms, MCDNW: diluted motorcycle mechanic workshop soil treated without worms.

Table 4. Effect of vermi-assisted bioremediation on the carbon and nitrogen contents of mechanic workshop soil contaminated with petroleum hydrocarbon.

Sample	Nitrogen (%)	Carbon (%)
MMUW	0.49	11.88
MMUNW	0.53	9.77
MMDW	0.39	4.88
MMDNW	0.53	6.19
MCUW	0.53	11.72
MCUNW	0.49	11.72
MCDW	0.42	8.14
MCDNW	0.49	9.12

MMUW: undiluted motorcar mechanic workshop soil treated with worms, MMUNW: undiluted motorcar mechanic workshop soil without worms, MMDW: diluted motorcar mechanic workshop soil treated with worms, MMDNW: diluted motorcar mechanic workshop soil treated without worms, MCUW: undiluted motorcycle mechanic workshop soil treated with worms, MCUNW: undiluted motorcycle mechanic workshop soil without worms, MCDW: diluted motorcycle mechanic workshop soil treated with worms, MCDNW: diluted motorcycle mechanic workshop soil treated without worms.

Table 5. Elemental analysis of earthworms and soil.

Element	Earthworm sample				Soil sample				
	REW	MMUW	MMDW	MCDW	UCWS	MMUW	MMDW	MCDW	MCUW
Al	6.878	0.000	6.878	5.820	6.243	2.540	6.878	5.450	4.127
Si	10.930	4.036	12.425	16.676	25.410	22.468	33.197	31.669	29.708
S	5.240	8.680	4.320	4.720	0.376	0.560	0.000	0.744	2.580
K	8.058	11.453	7.627	6.108	3.718	4.067	2.465	3.494	2.896
Ca	8.864	17.441	8.435	6.004	4.796	8.077	1.501	1.573	3.031
Ti	0.689	0.276	0.893	0.743	1.031	0.303	0.566	0.893	0.605
V	0.030	0.0336	0.032	0.027	0.050	0.051	0.0308	0.0437	0.026
Cr	0.044	0.151	0.052	0.032	0.054	0.045	0.0226	0.047	0.099
Mn	0.085	0.070	0.147	0.132	0.093	0.077	0.058	0.108	0.148
Fe	11.106	7.518	10.576	9.712	15.666	10.873	7.276	9.52	9.569
Ni	0.035	0.110	0.039	0.030	0.039	0.028	0.008	0.033	0.035
Cu	0.066	0.104	0.058	0.036	0.061	0.064	0.013	0.053	0.058
Zn	0.321	0.2892	0.249	0.193	0.225	0.201	0.023	0.145	0.276
Ba	0.251	0.269	0.215	0.134	0.305	0.430	0.000	0.260	0.134

REW: Earthworm at t = 0, before introduction into mechanic workshop soil, MMUW: undiluted motorcar mechanic workshop soil, MMDW: diluted motorcar mechanic workshop soil, MCDW: diluted motorcycle mechanic workshop soil, MCUW: undiluted motorcycle mechanic workshop soil, UCS: petroleum hydrocarbon free soil used in diluting mechanic workshop soil.

be recommended.

From Figures 3 and 4 it may be seen that the numbers of peaks (corresponding to organic components of the contaminating oil) dropped from 10 to 9. The MS analysis in Table 2 indicated that the C11 - C20 fraction of the contaminating petroleum hydrocarbon was most affected by the bioremediation process, dropping from 95.47 to 92.31%. The C1 - C10 and the C20 - C30 fraction appeared to have increased. For the C1 - C10, this may be as a result of the breakdown of C11 - C20 resulting in the formation of more C1 - C10, whereas for the C20 - C30 increase may be a result of the reduction in TPH (by bioremediation which primarily affected the C11 - C20), giving the C20 - C30 content an apparent boost.

From the Table 3, it can be seen that the microbial count of the raw motorcar and motorcycle mechanic workshop soil (MM and MC, respectively) was very low. For instance, at a dilution of 10^5 no microbe was enumerated in both samples. This shows the negative effect petroleum hydrocarbon contamination has on soil microbial community. After amendment with fermented cassava peels with earthworm inoculation (in some samples) microbial activity picked up as shown at the end of the 35 day study.

The microbial count increased by a factor of at least 17, which is significant. In fact, for MC at dilution of 10^4, the count increased by more than 300 times. This increase in microbial count however may not directly be the reason for the reduction in TPH: the samples without worms both indicate slightly higher microbial counts than the corresponding samples with worms though the TPH analysis indicated that the samples with worms recorded lower values of TPH as compared to the corresponding samples without worms. Thus, increased microbial count may have contributed but other factors may have also influenced the TPH of the samples. The microorganisms isolated in all samples were similar: *Bacillus* sp, *Pseudomonas* sp and *Staphylococcus* sp. All these are known to produce biosurfactants of interest for application in petroleum hydrocarbon contaminated sites.

Comparing the results of the diluted samples (MMDW/MMDNW and MCDW/MCDNW) which recorded high earthworm survival in the 35 day study, the trend in the carbon/nitrogen contents followed the same pattern. The carbon and nitrogen contents were lower in the earthworm inoculated samples as compared to the earthworm free samples. Nitrogen is lost mainly through ammonia volatilization or denitrification, The process is performed primarily by heterotrophic bacteria (such as *Paracoccus denitrificans* and various pseudomonads). Carbon is lost in the form of CO_2 during respiration of earthworms/microbes. The results indicate heightened loss of nitrogen and carbon in the earthworm inoculated samples. The observed trend differs from the results of the findings of Li et al. (2002) which shows that earthworms had no significant influences on nitrogen content. It is possible that the organic amendment (fermented cassava peels) as well as the nature of the soil (contaminated with petroleum hydrocarbon over a long period of time which consequently affected the microbial community) used in the investigation offered a different condition that consequently affected the rate of mineralization of nitrogen.

Metal uptake and elimination

Elemental analysis of the soil from which the worms were obtained (UCWS) indicated that its iron (Fe) content was higher than that of all mechanic workshop soil considered as shown in Table 5. The composition of all other metals was close to or lower than the values recorded in the mechanic workshop soils. The earthworms prior to introductions into the mechanic workshop soil (REW) showed a high Fe content. The metals of interest were Zn, Cu, Cr, Ni, Ti and Mn. Generally, the order of abundance of Fe, Mn, Cu and Cr as shown in Table 5 (that is Fe> Mn> Cu> Cr>Ni) is similar to that reported by Nwachukwu et al. (2010).

Just as with the simulated used engine oil contaminated soil, there was an appreciable uptake of Ti and Mn in both samples that recorded high survival rate of earthworm (MMDW and MCDW). MMUW which recorded a 70% death rate of the worms however recorded elimination of both Ti and Mn. This may be because the unfavorable environment affected the feeding behavior of the worms leading to the elimination of these metals. For the metals Zn, Cu, Ni and Cr which recorded elimination in the study using simulated used engine oil contaminated soil: Zn recorded a decrease in earthworm body concentration of Zn for all samples, the concentration of Zn in the soil from which the worms were obtained was also found to be higher than all mechanic workshop soils considered. This suggests that a relation exist between the body concentration of the metal and the environment where the worms live. The concentration of Cu was found to be slightly higher in MMUW when compared with the UCWS but lower than MMDW and MCDW. MMDW and MCDW after the 35 day study recorded a drop in earthworm body concentration of Cu as was recorded in the simulation study. The anomalous behavior with MMUW may again be attributed to unfavorable environment which may have resulted in the preferential excretion of other materials by the worms, resulting in an increased Cu concentration. A definite relationship between earthworm body concentration and the concentration in the soil cannot be seen from the concentration in Ni and Cr. Heavy metals uptake/-elimination becomes important when earthworms are considered for use in selective removal of metals from contaminated soil or when the possibility of transferring harmful metals down the food chain exist. With this in mind, it is of note that the concentration of the following metals in worms introduced into the following samples increased sharply: Cr, Cu and Ni (MMUW), Mn (MMDW and MCDW).

Conclusions

The presence of earthworms (*E. eugeniae*) accelerated the bioremediation of petroleum hydrocarbon contaminated soil from mechanic workshops. Earthworm survival in the contaminated soil affects the degree to which the worms can influence bioremediation.

ACKNOWLEDGEMENTS

The authors thank the Science and Technology Education Post-Basic (STEP-B) Innovators of Tomorrow (IOT) and the University Board of Research (Ahmadu Bello University, Zaria) for their financial support.

REFERENCES

Adams RH, Guzmán-Osorio FJ (2008). Evaluation of land farming and chemico-biological stabilization for treatment of heavily contaminated sediments in a tropical environment. Int. J. Environ. Sci. Tech. 5(2):169-178.

Adesodun JK, Mbagwu JSC (2008). Biodegradation of waste lubricating petroleum oil in a tropical alfisol as mediated by animal droppings. Bioresour. Technol. 99:5659-5665.

Ameh AO, Mohammed-Dabo IA Ibrahim S, Ameh JB, Tanimu Y, Bello TK (2011b). Effect of earthworm inoculation on the bioremediation of used engine oil contaminated soil, Int. J. Biol. Chem. Sci. 6(1):493-503.

Ameh AO, Mohammed-Dabo IA, Ibrahim S, Ameh JB, Azienge CD, Yahuza T (2011a). Earthworm survival in used engine oil contaminated soil spiked with manure. Int. J. Biol. Chem. Sci. 5(3):923-929.

Association of Official Analytical Chemists (AOAC) (1990). Official Methods of Analysis, 14th Edition, Washington DC. Arlington, Virginia, USA

Chukwuma MC, Eshett ET, Onweremadu EU, Okon MA (2010). Zinc availability in relation to selected soil properties in a crude oil polluted eutric tropofluvent. Int. J. Environ. Sci. Tech. 7 (2):261-270.

Dhanasekaran D, Thajuddin N, Rashmi M, Deepika TL, Gunasekaran M (2009). Screening of biofouling activity in marine bacterial isolate from ship hull. Int. J. Environ. Sci. Tech. 6 (2):197-202.

Fouépé TA, Kengni TL, Gurunadha Rao VVS, Ndam NJR (2009). Transfer of moisture through the unsaturated zone in the tropical forest using the neutron probe. Int. J. Environ. Sci. Tech. 6 (3):379-388.

Hickman ZA, Reid BJ (2008). Earthworm assisted bioremediation of organic contaminants. Environ. Int. 34:1072-1081.

Igwe JC, Abia AA, Ibeh CA (2008). Adsorption kinetics and intraparticulate diffusivities of Hg, As and Pb ions on unmodified and thiolated coconut fiber. Int. J. Environ. Sci. Tech. 5 (1):83-92.

Nwachukwu MA, Feng H, Alinnor J (2010). Assessment of heavy metal pollution in soil and their implications within and around mechanic villages. Int. J. Environ. Sci. Tech. 7(2):347-358.

Organization for Economic Cooperation and Development (OECD) (2010). Guideline for testing of chemicals, Bioaccumulation in Terrestrial Oligochaetes, No. 317.

Refaat AA (2010). Different techniques for the production of biodiesel from waste vegetable oil. Int. J. Environ. Sci. Tech. 7 (1):183-213.

Schaefer M, Juliane F (2007). The influence of earthworms and organic additives on the biodegradation of oil contaminated soil. Appl. Soil Ecol. 36:53-62.

Schaefer M, Peterson SO, Filser J (2005) Effects of *Lumbricus chlorotica* and *Eisenia fetida* on microbial community dynamics in oil-contaminated soil. Soil Biol. Biochem. 37:2065-76.

Shah BA, Shah AV, Singh RR (2009). Sorption isotherms and kinetics of chromium uptake from wastewater using natural sorbent material. Int. J. Environ Sci. Tech. 6 (1):77-90.

Srivastava S, Thakur IS (2006). Evaluation of bioremediation and detoxification potentiality of *Aspergillus niger* for removal of hexavalent chromium in soil microcosm, Soil Biol. Biochem. 38:1904-1911.

USEPA (1996). Recycling Used Oil: What Can You Do? Cooperation Extension Services ENRI-317: 1-2.

Vidali M (2001). Bioremediation: An overview. J. Appl. Chem. 73(7):1163-1172.

Yerushalmi L, Rocheleau S, Cimpoia R, Sarrazin M, Sunahara G, Peisajovich A, Leclair G, Guiot RS (2003). Enhanced bioremediation of petroleum hydrocarbons in contaminated soil. Bioremed. J. 7 (1):37-51.

Yeung PY, Johnson RL, Xu JG (1990). Biodegradation of petroleum hydrocarbons in soil as affected by heating and forced aeration. J. Environ. Qual. 26:1511-1576.

Zhang H, Dang Z, Zheng LC, Yi XY (2009). Remediation of soil co-contaminated with pyrene and cadmium by growing maize (Zea mays L.). Int. J. Environ. Sci. Tech. 6 (1):69-76.

Hydrochemistry of shallow groundwater and surface water in the Ndop plain, North West Cameroon

Mengnjo J. Wirmvem[1], Takeshi Ohba[1], Wilson Y. Fantong[2], Samuel N. Ayonghe[3], Justice Y. Suila[4], Asobo Nkengmatia E. Asaah[5], Gregory Tanyileke[2], and Joseph V. Hell[2]

[1]Department of Chemistry, School of Science, Tokai University, Hiratsuka, 259-1211, Japan.
[2]Institute of Mining and Geological Research, P.O. Box 4110, Yaounde, Cameroon.
[3]Department of Environmental Science, University of Buea, Box 63, Buea, Cameroon.
[4]G.B.H.S. Atiela, Nkwen, P.O. Box 5147, Bamenda, Cameroon.
[5]Department of Earth and Planetary Sciences, Tokyo Institute of Technology, Tokyo, 152-8551, Japan.

This study was conducted on the hydrochemistry of shallow groundwater and surface water in the Ndop plain, North West Cameroon. The objectives were to determine the physico-chemical characteristics of water, controls on water chemistry and suitability for drinking and irrigation. Forty-six shallow groundwater and 26 surface water samples were investigated. Field measurements of physical parameters were preceded by chemical analyses of the samples for major ions and F^-. About 69% of the water sources had pH values below 6.5, classifying the water as barely acidic. Electrical conductivity (<282 µS/cm) and total dissolved solids (<183 mg/l) were low suggesting low-mineralised and freshwater. The relative abundance of major ions (meq/l) was $Na^{2+}>Ca^{2+}>Mg^{2+}>K^+$ for cations and HCO_3^- $>>NO_3 \geq Cl^- > SO_4^{2-}$ for anions. Major ion concentrations were low, and within the WHO guidelines for drinking water, but F^- concentrations were much lower (<0.39 mg/l) and below guideline value. Main water types and proportions were Na-HCO_3 (53 %), Ca-HCO_3 (35 %) and a mixed Na-Ca-HCO_3 (11%). Predominant processes influencing water chemistry were incongruent dissolution of silicate minerals and cation-exchange of Na^+ in rocks for Ca^{2+} in water. The low major ion concentrations indicated low water-rock interactions and short residence time. The analysed water was suitable for irrigation.

Key words: Water chemistry, hydrochemical controls, drinking-irrigation quality, Ndop plain, Cameroon.

INTRODUCTION

Groundwater and surface water geochemical studies can provide a better understanding of potential water quality variations due to geology and land use practices (Edmunds and Smedley, 1996; Deutsch, 1997; Appelo and Postma, 2005; Ramesh and Elango, 2012). In Africa, groundwater is a vital resource with nearly 80% of the continent's population relying on it as a main source of drinking water. However, in many parts of the continent, reaching basic health requirements is still a concern (Adelana and MacDonald, 2008; Adelana et al., 2011; Edet et al., 2011; Tay, 2012).

Cameroon is endowed with numerous water resources (Katte et al., 2003; Molua and Lambi, 2006), but an assessment of these resources, including their reliability, quality and sustainability has not been systematically carried out (Katte et al., 2003). Improved drinking water sanitation in semi-urban and rural communities is nearly the same from 37% in 1990 to 36% in 2010 (WHO/UNICEF, 2012). With increasing population, the high demand for water, especially in most poor rural communities, is met by shallow groundwater of generally unknown quality. This is usually the best option all year round as most surface water sources and springs are unreliable and run dry or reduce in size during the dry season, or become polluted by heavy floods in the rainy season.

Only few hydrochemical studies have been carried out to assess the geochemical characteristics of groundwater and its suitability for use in Cameroon (example, Njitchoua and Ngounou-Ngatcha, 1997; Katte et al., 2003; Fantong et al., 2009; Eneke et al., 2011; Ako et al., 2012). The Ndop plain is a semi-urban community where groundwater, as well as surface water, is a critical resource for human livelihood. The plain has an estimated population of over 200,000 people with ca. 70% involved in subsistence farming. Many ethnic groups have been attracted to the area as a result of fertile soils (Ndzeidze, 2008; Fonge et al., 2012). This has been followed by high demand for water and other natural resources (Fonge et al., 2012). Over 70% of the population depends solely on shallow groundwater, as well as surface water of poor microbial quality for drinking and other domestic purposes (Wirmvem et al., 2013). A hydrochemical study of water quality based on major ion composition has not been carried out. Since the development of the Bamendjin dam in the area in the 1970s, rice cultivation is the main cash crop using the dam and river waters. Unfortunately, there is no information on the suitability of the water sources for irrigation despite the on-going programme by the Cameroon government to revamp rice cultivation in the area. The chemical composition of water and its mineralization process are imperative in classifying and assessing drinking water quality (Deutsch, 1997; WHO, 2004; Kozisek, 2005) while irrigation water quality criteria can be used as a guideline by farmers for selecting appropriate management practice to overcome potential salinity hazard (Gupta et al., 2009; Ramesh and Elango, 2012).

The present study, therefore, presents the first hydrochemical baseline dataset of groundwater and surface water in the Ndop plain. The objectives were to (1) determine the physico-chemical characteristics of the water, (2) define the factors that control major ion chemistry, and (3) evaluate their suitability for drinking and irrigation. The data will provide a hydrochemical baseline data for water quality assessment, management, and monitoring of changes in the hydrochemical system from the prevailing human activities in the area. The study will also contribute to the limited knowledge on natural water composition in North West Cameroon. With the on-going programme to revamp rice cultivation in the area, an assessment of the suitability of the water sources for irrigation is invaluable.

The study area

The study area (Figure 1) is an inter-mountain basin (ca. 1,100 km^2) in the Western Highlands of Cameroon (WHC), along the Cameroon Volcanic Line (CVL). From the surrounding CVL rages, the relief drops steeply from an average altitude of ca. 1,899 m to ca. 1,200 m at the margins of the plain, and then gently to ca. 1,180 m above sea level in the south central portion. The flanks of the mountain chain are discharge points for numerous rivers and streams flowing dendritically into the plain and recharging the Bamendjin dam (Figure 1). Details on the drainage are described by Ndzeidze (2008). According to Neba (1999), the area has an Equatorial monsoon climate type characterized by high rainfall and relative humidity. This is attributed to the effect of the mountainous relief that characterises much of the WHC (Neba, 1999; Molua and Lambi, 2006). Two distinct seasons are differentiated, a long rainy season (mid-March to mid-November) and a short dry season (mid-November to mid-March). Mean annual rainfall and temperature, in 2012, were 1,540 mm and 26°C, respectively (Wirmvem et al., unpublished data), but with cooler climate in adjacent high elevations (Ndzeidze, 2008; Ndenecho, 2009). The heaviest rainfall occurs from July to September and often results in floods due to the poor drainage (Ndzeidze, 2008).

Geologically, the area is a shallow Cenozoic 'sedimentary' basin, underlain mainly by a consistent Precambrian granitic basement (Figure 1). This basement forms part of the tectonically inactive African shield consisting mostly of Precambrian Basement Complex rocks (Neba, 1999; Marzoli et al., 1999). Chemical weathering of the basement and surrounding volcanic rocks has produced thick unconsolidated sediments, mainly of clay to sand sizes. The basement is largely covered with these sediments, but outcrops in certain areas. The surrounding volcanoes are represented by voluminous Q-trachytes and minor rhyolitic ignimbrites with slight to moderate alkaline basalts and minor basanites (Marzoli et al., 1999). The voluminous Q-trachytes are enriched in alkali feldspars (Na_2O+K_2O). *In-situ* weathered basement and alkali-rich fluvial sediments constitute the aquifer material in the plain. From the margins of the plain towards the south-central portion, there is a progressive change from sandy to clayey sediments. The unconsolidated sandy-sediments form the groundwater aquifer system. This explains the concentric model of human settlement around the dam water (Ndzeidze, 2008) tapping groundwater through wells from this aquifer. The poor microbial quality of water with consequent water-borne infections (Wirmvem et al., 2013) has prompted the chlorination of water in wells by some members of the community.

MATERIALS AND METHODS

Water sampling and physico-chemical measurements

Water was collected from 72 water sources in January 2012 (dry season) within 13 communities (Figure 1). The sources comprised 32 open wells and 14 'boreholes' (hereafter termed groundwater); and 14 streams, 4 rivers, 5 dam waters, and 3 lakes (Lakes Bambili,

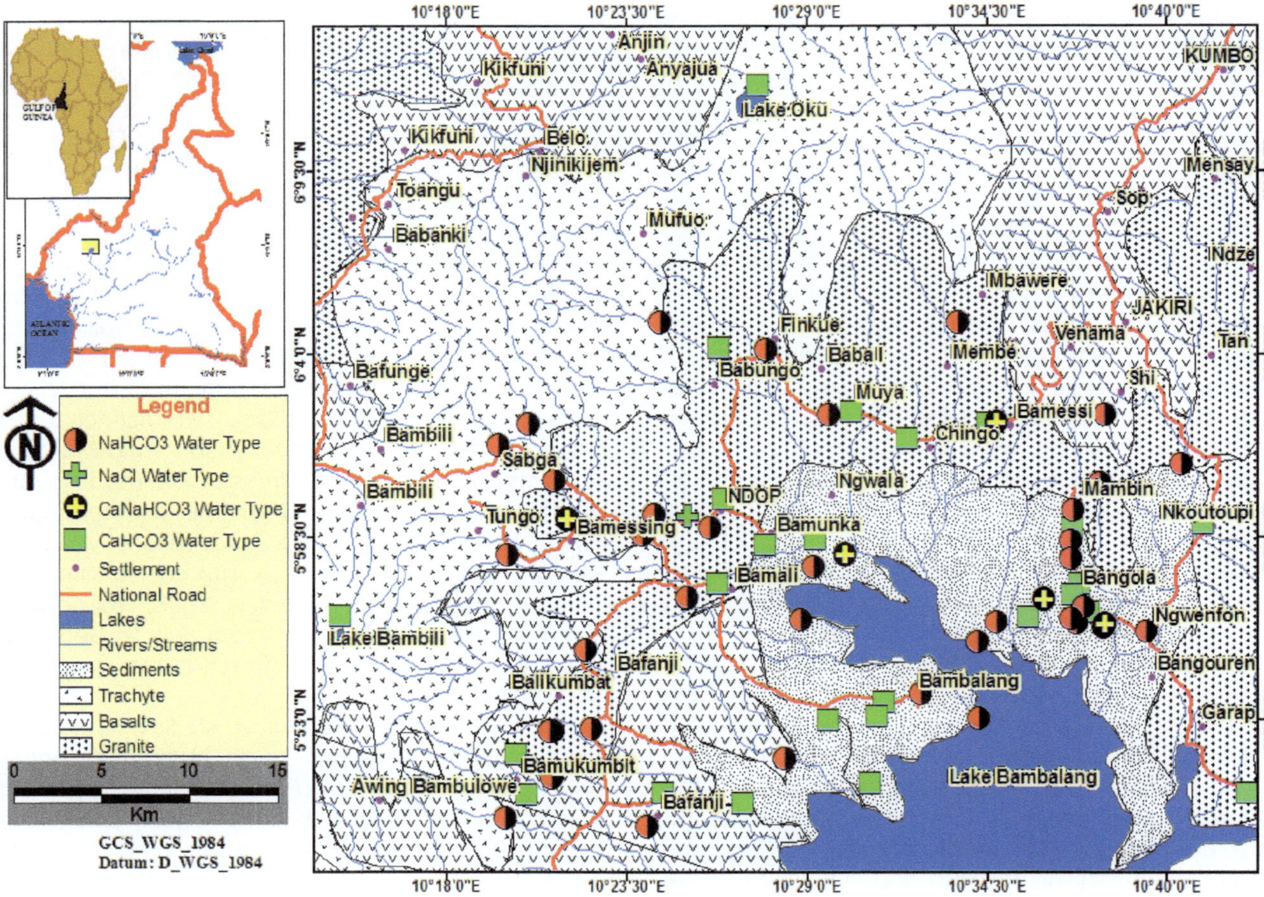

Figure 1. The geology and drainage map of Ndop Plain (modified after Ndzeidze, 2008) showing sample points/spatial distribution of water types.

Oku, and Ber) (hereafter called surface water). The sample location points (Figure 1) and altitudes were noted in the field by using a Garmin Vista CX GPS. Four sporadic rainfall events in the plain and its environs were also sampled. Field measurements and water sampling followed methods described by Karklins (1996). New 100 ml narrow mouth and low-density polyethylene (LDPE) bottles were properly rinsed with distilled water and dried in the Laboratory (in order to avoid contamination) and used for sampling the water. At each site, water was collected into a plastic bucket after thorough rinsing with water to be sampled. Groundwater from active wells was collected using a rope tied to a bucket and/or permanently installed hand-pumps (after several minutes of pumping). Surface water was sampled as far as possible from the edges of the water bodies and as deep as possible along the flow path. The water collected was then filtered directly into 2 LDPE bottles using a plastic syringe, fitted with a 0.20 µm filter after thorough rinsing with the same water that was sampled. Collected samples were properly capped and preserved in a cooler container containing ice blocks. Water for major ion analyses was put in one bottle and water for alkalinity determination in a separate bottle. The following physical parameters: activity of hydrogen ion (pH), electrical conductivity (EC), total dissolved solids (TDS), and water temperature were measured immediately from the remaining collected water using a multipurpose HI991300 hand held pH/EC/TDS/C Hanna Meter. The meter was calibrated before and during the field campaign using buffer solutions recommended by the manufacturer. The depths to standing water level and total depth of water column were determined with a water level indicator. Alkalinity/bicarbonate (HCO_3^-) was determined within 8 hours of sampling by acid titration from a volume of 0.02 M HCl, added to the sample and stirred with an automatic stirrer until the end-point, marked by a pH of 4.5.

Chemical analyses

Chemical analyses were carried out at Tokai University, Japan within two weeks of sample collection. Major cations: Na^+, K^+, Mg^{2+} and Ca^{2+} were determined by a Flame Technique in a High Resolution Continuum Source AAS (ContrAA 700) as described by Welz et al. (2006). Samples with EC values >100 µS/cm were diluted 5 or 10 times to acquire an absorbance within the range of the used standards for Na^+, K^+ and Ca^{2+}. The anions: F^-, Cl^-, NO_3^-, SO_4^{2-}, PO_4^{3-}, NO_2^- and Br^- were determined by using an Ion Chromatography (Dionex ICS-900) while dissolved Silica ($Si(OH)_4$), hereafter considered as SiO_2, was determined by using an ICP-AES. The reliability of the chemical analyses was verified by using an ionic balance error (IBE) equation (Appelo and Postma, 2005). The values were within ±5 %; hence, suitable for geochemical interpretations.

RESULTS AND DISCUSSION

Physico-chemical composition of groundwater and surface water

Table 1 presents results of the 72 water samples with a statistical summary shown in Table 2. A statistical summary of rain data is presented in Table 3. Mean depth to the water table (<27 m) and depth of the water column (<8.4 m) were low, suggesting a shallow aquifer system. The mean temperature of the water was closer to the mean atmospheric temperature of 26°C, suggesting present day climatic influence and recharge. Ground-water had a lower and narrow range in pH with a mean of 5.9, moderately acidic, while surface water had a relatively higher and wider range in pH with a mean of 6.8, slightly acidic. About 88% of the 72 samples had pH<7, suggesting an acidic aquifer system. The EC and TDS values of all water sources were low with mean values of 60 µS/cm and 39 mg/l, respectively suggesting low-mineralised and freshwater. Major ion concentrations, like TDS, were low (Table 2) and in the order (meq/l) Na^{2+}> Ca^{2+}>Mg^{2+}>K^+ for cations and HCO_3^->>NO_3^-≥Cl^->SO_4^{2-} for anions. The small range in their concentrations and higher dispersions around their respective mean values as reflected by the small STD values, except for NO_3^- and Cl^- (Table 2), suggests a unique source of most ions and a single aquifer system. Mean concentration of SiO_2 was higher than that of individual cations but less than the mean of HCO_3^- (Table 2). This suggests that SiO_2 was released more than the cations into solution. The high SiO_2 suggests the abundance of silicate minerals (Srinivasamoorthy et al., 2008) in agreement with the dominant granitic basement (Figure 1) and as observed in similar terrains by Njitchoua and Ngounou-Ngatcha (1997) and Fantong et al. (2009).

As expected, the mean TDS increased progressively from a minimum in rainwater (20 mg/l) to surface water (30 mg/l) and a maximum in groundwater (44 mg/l). Similarly, mean relative concentrations of dissolved ions showed a progressive increase from the lowest in rain-water, to surface water, and the highest in groundwater (Figure 2a). These results suggest chemical evolution from rain to surface water and groundwater. The relatively high TDS in groundwater indicated the influence of the unconsolidated sediments and granitic bedrock material in chemically enriching the water as it percolates through the under-saturated zone (or flows through the aquifer). The small difference in mean TDS of ground-water and rainwater by only 53 % is likely because these are early rains, usually enriched in dissolved ions (Takounjou et al., 2011). The low major ion concentrations in groundwater depict low water-rock interactions in the granitic basement, short residence time, the shallow nature of the aquifer and its acidic nature (Edmunds and Smedley, 1996). Similar low pH and major ions have been reported from granitic basement complex aquifers in Ghana (Adomako et al., 2011) and Nigeria (Edet et al., 2011).

Hydrochemical facies

The meq/l concentrations of major ions plotted on a pie chart (Figure 2b), showed that Na^+, HCO_3^- and to some extent Ca^{2+} were the major ionic species. Low K^+ may be due to its low geochemical mobility (Hem, 1989; Srinivasamoorthy et al., 2008) while low Mg^{2+} is possibly due its to low content in rocks of the study area. On a Piper diagram (computed using GW_Chart Version 1.23.7.0), most samples plotted in the field of alkali metals (Na+K) (Figure 3). This indicates the dominance of alkali metals in the groundwater aquifer system. The water types and their proportions are shown (Figure 3). The observed spatial distribution of the water types (Figure 1) showed no discernible pattern from Ca-HCO_3 to Na-HCO_3. Given the gentle relief in the plain, this suggests a mixed aquifer system as revealed by the mixed Na-Ca-HCO_3 water. Similarly, no clear pattern, in water type, was observed from surface water to ground-water (Figure 3). Nonetheless, there is chemical evolution from Ca+Mg+SO_4+Cl (rain), to Ca-HCO_3 (largely ground-water and lakes), and to Na-HCO_3 (basically groundwater, streams, rivers, and dam waters) as indicated by the dotted arrow (Figure 3). It is probable that the dissolution of Na^+ into surface waters from alkali-rich trachytic rocks in the surrounding highlands tends to mask the chemical evolution. The presence of Ca-HCO_3 water in ground-water within the plain, which is sporadic, in surface waters (streams and rivers) discharging from the high-lands suggests that the main recharge of groundwater may not necessarily occur at higher altitudes but rather through direct precipitation to the surface and subsequent percolation through the under-saturated zone into the aquifer.

Origin of solutes and hydrochemical controls

The major on chemistry of groundwater and compositional relations among ionic species can reveal the origin of solutes and processes that generated an observed water composition (Hounslow, 1995; Demlie et al., 2007; Cendon et al., 2011; Tay, 2012). Table 4 shows correlation coefficients between the ionic species in water. The good positive correlation of NO_3^- with Cl^- suggests their anthropogenic origin (Demlie et al., 2007; Fantong et al., 2009). Main potential sources of NO_3^- are numerous shallow pit toilets and oxidation of organic matter given the proximity (>15 m) to shallow ground-water and agricultural dominance, respectively. Fonge et al. (2012) identified relatively high concentrations of NO_3^- and PO_4^{3-} in soil water (<5 cm deep) from rice farms in the study area. The low content of NO_3^- and the absence of PO_4^{3-} in groundwater suggest their sorption on organic colloids (in the under-saturated source of zone) or the location of rice fields rather in low-lying swampy areas. The likely Cl^- is atmospheric inputs given the positive correlation with Na^+ and disinfection of wells through chlorination as noted in the field. Sulfur occurs in a reduced form in igneous rocks and is oxidized to SO_4^{2-} when weathered in contact with aerated water (Hem, 1989). The significant positive correlations of SO_4^{2-} with K^+, Ca^{2+} and HCO_3^- possibly suggest its igneous origin in

Table 1. Physico-chemical results of groundwater and surface water (n=72) in the Ndop plain (January-February, 2012).

ID	T	pH	EC	TDS	Na^+	K^+	Ca^{2+}	Mg^{2+}	HCO_3^-	NO_3^-	Cl^-	SO_4^{2-}	F^-	SiO_2	IB
OW06	25.4	5.7	15	9.8	1.0	0.9	0.2	0.4	6.3	0.1	0.1	0	0.03	9.9	2
OW07	24.3	5.9	28	18.2	2.0	1.0	0.6	0.8	10.6	1.4	0.2	0.02	0.04	12.1	1
OW09	24.4	6.2	25	16.3	0.4	0.9	1.7	0.3	7.9	0.2	0.1	0.54	0.02	7.7	2
OW10	24.3	5.5	19	12.4	1.3	0.6	0.4	0.7	8.2	0.5	0.1	0	0.02	10.8	3
OW18	23.8	5.6	9	5.9	0.2	0.5	0.3	0.1	2.5	0.1	0.1	0	0.01	3.8	7
OW19	24.0	5.8	52	33.8	6.1	1.0	0.4	0.6	21.0	0.6	0.1	0.04	0.07	3.7	1
OW20	23.6	5.8	59	38.4	1.4	2.0	5.0	0.8	20.4	0.1	1.1	0.77	0.1	9.8	5
OW21	19.5	5.9	56	36.4	4.0	1.4	2.9	0.7	20.5	0.0	2.1	0.06	0.01	3.1	2
OW22	20.4	5.3	73	47.5	2.8	8.6	2.0	1.1	21.9	7.6	0.9	0.18	0.01	2.7	2
OW28	25.0	6.0	45	29.3	1.1	1.5	2.4	0.6	13.5	0.3	0.2	0.7	0.06	12.1	1
OW32	25.0	5.6	44	28.6	2.9	1.9	0.9	0.9	11.0	5.7	0.4	0.02	0.03	9.8	2
OW37	24.2	5.9	65	42.3	6.1	1.8	0.3	0.7	19.7	3.1	0.1	0.06	0.13	3.6	2
OW39	23.5	5.9	88	57.2	5.1	5.0	0.9	2.1	18.6	0.1	6.2	3.17	0.08	2.7	1
OW49	19.5	6.4	230	149.5	1.8	24.7	9.4	6.7	91.1	0.1	1.2	8.42	0.03	3.2	4
OW50	22.6	5.2	14	9.1	0.9	1.1	0.2	0.1	4.6	0.4	0.2	0.08	0.01	5.9	4
OW56	24.3	6.6	71	46.2	2.7	3.3	4.1	0.3	23.1	0.6	0.2	0.35	0.03	8.1	0
OW59	23.0	6.5	45	29.3	5.4	2.3	0.2	0.2	13.7	1.7	0.2	2.06	0.22	2.3	4
OW62	23.1	5.7	26	16.9	2.5	1.2	0.3	0.4	8.5	0.1	0.5	0.64	0.06	13.4	2
OW64	24.0	5.5	34	22.1	2.5	1.6	0.5	0.6	10.5	1.3	0.6	0.27	0.05	8.0	5
OW66	22.4	5.6	16	10.4	1.7	1.3	0.1	0.2	5.1	0.9	0.2	0.38	0.05	14.9	3
OW67	23.4	6.1	38	24.7	4.7	2.9	0.6	0.7	18.4	1.0	0.5	0.29	0.05	16.4	1
OW72	23.0	6.1	16	10.4	1.1	0.9	0.3	0.3	5.7	0.4	0.2	0.02	0.02	6.5	1
OW74	24.0	6.7	145	94.3	3.3	9.4	12.2	0.9	61.4	0.0	0.2	1.31	0.06	10.2	3
OW82	23.8	5.6	24	15.6	0.7	1.0	0.6	1.1	9.0	0.0	0.4	0.26	0.05	7.3	4
OW85	23.7	5.4	61	39.7	5.3	3.9	1.1	0.9	15.3	0.0	4.7	2.07	0.17	13.7	1
OW86	24.2	5.5	135	87.8	21.5	3.3	1.3	0.4	4.4	15.5	26.9	0.17	0.1	3.1	3
OW88	24.3	5.4	26	16.9	1.4	2.0	0.6	0.3	9.4	0.0	0.3	0.09	0.03	9.5	4
OW90	23.9	6.6	158	102.7	5.8	19.8	3.4	4.7	62.5	0.0	3.1	4.72	0.11	10.5	2
OW125	23.1	6.0	48	31.2	3.1	1.7	1.6	1.3	9.5	7.2	2.6	0.13	0.02	3.1	9
OW126	23.2	5.8	42	27.4	4.4	3.1	0.5	1.0	17.5	0.8	0.3	0.159	0.08	19.0	2
OW127	24.3	5.8	61	39.7	3.0	1.6	2.7	0.9	20.0	1.0	0.3	0.43	0.06	12.1	1
OW129	23.4	5.8	126	81.9	7.5	1.0	3.7	4.7	14.5	27.9	6.6	1.56	0.08	15.3	2
BH02	23.2	5.8	40	26.0	3.0	1.1	1.6	0.5	14.5	0.7	0.1	0.09	0.19	13.1	1
BH11	24.5	5.8	43	28.0	2.7	0.9	1.0	1.0	14.6	1.0	0.2	0.08	0.05	18.8	1
BH29	22.4	6.4	131	85.2	5.6	3.5	3.5	3.1	38.9	3.3	1.0	0.77	0.39	2.2	1
BH41	24.5	6.2	272	176.8	3.8	2.2	1.6	2.3	11.1	12.4	3.0	0.58	0.11	2.2	0
BH52	25.1	6.3	45	29.3	1.4	1.3	2.7	0.2	13.4	0.7	0.2	0.27	0.02	6.6	5
BH53	24.0	5.4	34	22.1	1.5	3.3	1.4	0.3	10.6	1.4	0.7	0.16	0.03	7.2	3
BH55	25.5	5.7	43	28.0	0.9	1.0	3.2	0.3	12.8	0.4	0.4	0.18	0.02	6.5	1
BH60	23.0	6.0	56	36.4	3.3	3.0	2.8	0.2	15.9	0.0	0.1	4.37	0.32	17.5	-16
BH70	25.7	6.1	52	33.8	1.6	1.3	1.3	0.7	9.6	8.6	0.2	0.56	0.07	8.0	5
BH73	24.0	6.7	282	183.3	4.2	24.5	20.3	1.6	83.7	3.8	2.7	12.76	0.09	13.2	0
BH76	25.4	6.3	86	55.9	1.2	2.1	8.0	0.3	30.5	0.0	0.3	0.81	0.04	10.2	1
BH77	23.4	5.7	33	21.5	2.0	2.6	1.1	0.4	11.1	0.0	0.6	1.32	0.13	13.5	1
BH83	24.5	5.7	47	30.6	0.7	1.9	2.4	0.7	14.8	0.0	0.1	0.33	0.08	14.1	1
BH124	22.0	5.9	31	20.2	1.9	1.0	1.2	0.6	12.2	0.2	0.4	0.09	0.02	3.7	1
ST05	22.1	7.5	69	44.9	4.3	2.8	1.4	2.6	31.7	0.4	0.1	0	0.18	19.9	4
ST13	22.9	6.2	12	7.8	1.2	0.7	0.2	0.4	6.2	0.0	0.1	0	0.01	8.7	1
ST15	21.6	6.8	21	13.7	1.3	0.7	0.5	0.6	8.3	0.4	0.0	0	0.04	7.7	0
ST17	18.9	6.7	30	19.5	2.0	1.6	0.6	0.8	12.9	0.0	0.1	0	0.05	10.0	3
ST24	21.6	6.0	21	13.7	0.4	0.7	0.3	0.2	3.8	0.0	0.1	0.04	0	5.1	7

Table 1. Contd.

ST27	20.1	5.8	19	12.4	1.2	0.7	0.3	0.3	5.0	0.0	0.3	0.06	0.07	5.8	1
ST33	21.7	6.5	25	16.3	1.5	0.7	0.3	0.4	6.9	0.1	0.3	0.07	0.04	9.9	1
ST40	20.9	7.2	110	71.5	5.5	2.0	0.8	4.7	41.6	0.1	0.3	0.03	0.18	2.7	1
ST65	19.5	6.9	59	38.4	6.9	2.9	0.4	0.6	23.3	0.1	1.0	0.68	0.13	3.0	0
ST68	19.4	6.9	44	28.6	4.7	2.0	0.3	0.7	19.1	0.1	0.2	0.16	0.08	19.7	1
ST80	19.5	7.7	175	113.8	17.3	2.8	1.2	5.3	78.7	0.0	0.1	0.08	0.3	2.8	1
ST87	21.1	6.4	58	37.7	7.2	3.1	0.2	0.3	24.4	0.0	0.3	0.06	0.06	2.8	2
ST104	18.5	8.2	46	29.9	5.2	2.4	0.2	0.5	18.7	0.1	0.1	0.62	0.08	3.1	0
ST108	19.2	7.9	29	18.9	4.3	2.6	0.1	0.3	14.4	0.2	0.2	0.47	0.1	3.0	3
RW26	21.5	6.7	13	8.5	0.7	1.5	0.3	0.3	6.0	0.2	0.2	0.08	0.02	5.8	1
RW42	20.1	6.7	80	52.0	4.8	2.1	0.8	2.7	29.5	0.6	0.4	0.26	0.09	2.3	1
RW93	19.5	7.4	34	22.1	2.1	1.5	0.5	1.1	12.7	0.4	0.3	0.34	0.04	14.8	2
RW94	21.0	7.1	39	25.4	4.0	1.9	0.2	0.3	14.2	0.1	0.2	0.35	0.08	20.5	1
LK31	26.9	6.3	32	20.8	2.8	0.7	0.1	0.2	7.8	0.0	0.4	0.04	0.06	7.4	6
LK43	21.7	8.4	33	21.5	1.2	1.0	1.8	0.2	8.8	0.0	0.6	0.56	0.03	11.6	-1
LK123	20.2	7.8	44	28.6	2.1	1.4	0.8	1.4	14.8	0.0	1.1	0.096	0.037	9.4	0
DW16	25.3	6.5	47	30.6	3.7	2.6	0.4	0.9	16.5	0.0	0.9	0.01	0.07	15.9	3
DW51	20.2	5.9	46	29.9	3.7	5.2	0.7	0.0	15.7	0.0	1.9	0.02	0.07	13.4	3
DW54	22.7	5.5	24	15.6	2.9	2.0	0.4	0.8	14.4	0.0	0.3	0	0.06	12.2	4
DW75	20.1	6.8	45	29.3	2.9	2.8	0.7	0.9	17.4	0.0	0.5	0	0.11	13.2	1
DW128	27.3	5.8	58	37.7	1.7	0.9	1.0	1.1	12.9	16.1	0.8	0.413	0.03	9.1	-35

OW, Open well; BH, borehole; ST, stream water; RW, river water; LK, lake water; DW, dam water; EC, electrical conductivity (µS/cm); TDS, total dissolved solids (mg/l). All ions (mg/l); Br- was detected in only OW39 (0.02 mg/l); PO43-, in OW37 (0.43 mg/l), OW86 (0.1 mg/l) and LK43 (0.07 mg/l); NO2- (>0.3 mg/l) was detected in OW22, OW32, BH29, BH41 and LK16; IB, ionic balance.

Table 2. Statistical summary of physico-chemical data of groundwater and surface water in the study area.

Parameter	Surface and ground water (n=72)				Groundwater (n=46)				Surface water (n=26)			
	Min	Max	Mean	Std	Min	Max	Mean	STD	Min	Max	Mean	Std
Alt. (m)	1135	2272	**1235**	193	1163	1260	**1188**	19.24	1135	2272	**1317**	302
Dtw (m)	-	-	-	-	1.60	27.00	**8.81**	5.06	-	-	-	-
Dow (m)	-	-	-	-	0.30	8.40	**1.57**	1.36	-	-	-	-
Tw (°C)	18.5	27.30	**22.79**	2.04	19.50	25.70	**23.65**	1.32	18.5	27.3	**21.29**	2.22
pH	5.20	8.41	**6.23**	0.71	5.20	6.74	**5.90**	0.37	5.48	8.41	**6.83**	0.76
EC (µS/cm)	9.0	282	**59.75**	54.62	9.00	282	**67.16**	62.32	12.0	175	**46.65**	33.47
TDS (mg/l)	5.85	183.30	**38.84**	35.5	5.85	183.3	**43.65**	40.51	7.8	113.75	**30.33**	21.76
Na^+ (mg/l)	0.24	21.54	**3.37**	3.28	0.24	21.54	**3.21**	3.26	0.41	17.34	**3.67**	3.29
K^+ (mg/l)	0.47	24.69	**2.94**	4.48	0.47	24.69	**3.53**	5.45	0.66	5.24	**1.89**	1.05
Ca^{2+} (mg/l)	0.11	20.25	**1.80**	3.02	0.11	20.25	**2.50**	3.58	0.14	1.77	**0.56**	0.41
Mg^{2+} (mg/l)	0.0	6.74	**1.04**	1.31	0.12	6.74	**1.03**	1.31	0.0	5.34	**1.06**	1.31
HCO_3^- (mg/l)	2.45	91.12	**18.69**	17.53	2.45	91.12	**19.13**	18.80	3.83	78.65	**17.92**	14.97
NO_3^- (mg/l)	0.0	27.92	**1.80**	4.51	0.00	27.92	**2.41**	5.05	0.00	16.07	**0.73**	3.07
Cl^- (mg/l)	0.04	26.87	**1.13**	3.31	0.05	26.87	**1.53**	4.07	0.04	1.88	**0.42**	0.42
SO_4^{2-} (mg/l)	0.0	12.76	**0.77**	1.91	0.00	12.76	**1.12**	2.32	0.0	0.68	**0.17**	0.21
F^- (mg/l)	0.0	0.39	**0.08**	0.07	0.01	0.39	**0.08**	0.08	0.0	0.3	**0.08**	0.06
SiO_2 (mg/l)	2.18	20.54	**9.04**	5.10	2.18	18.97	**8.93**	4.80	2.26	20.54	**9.23**	5.58

Max, maximum, Min, minimum; STD, standard deviation; Alt, Altitude; Dtw, depth to water (m); Dow, depth of water (m); Tw, water temperature; EC, electrical conductivity; TDS, total dissolved solids.

agreement with the geology. Fluoride showed a distinct positive correlation with Na^+ relative to Ca^{2+} (Table 4). The likely source of F^- in the granitic basement is fluorite, CaF_2, the most common F^--bearing mineral (Edmunds

Table 3. Physico-chemical data of January 2012 sporadic rainfall events in the Ndop Plain and its Environs.

Location	pH	EC	TDS	Na^+	K^+	Ca^{2+}	Mg^{2+}	HCO_3^-	NO_3^-	Cl^-	SO_4^{2-}	F^-	Na/Ca
Bamenda	6.26	36	23.4	0.36	0.90	1.39	0.37	3.5	2.97	2.3	1.33	0.02	0.43
Ndop	6.57	30	19.5	0.26	0.88	1.62	0.43	3.77	2.58	1.56	1.25	0.03	0.27
Ndawara	6.45	28	18.2	0.29	0.87	1.30	0.36	3.97	2.47	1.43	1.12	0.01	0.37
Kumbo	6.53	32	20.8	0.23	1.09	1.37	0.27	4.12	3.73	1.22	2.03	0.03	0.28
Max.	**6.57**	**36**	**23.4**	**0.36**	**1.09**	**1.62**	**0.43**	**4.12**	**3.73**	**2.3**	**2.03**	**0.03**	**0.43**
Min.	**6.26**	**28**	**18.2**	**0.23**	**0.87**	**1.3**	**0.27**	**3.5**	**2.47**	**1.22**	**1.12**	**0.01**	**0.27**
Mean	**6.45**	**31.5**	**20.5**	**0.29**	**0.94**	**1.42**	**0.36**	**3.84**	**2.94**	**1.63**	**1.43**	**0.02**	**0.34**

EC = uS/cm, TDS and all ions = mg/l, Na/Ca

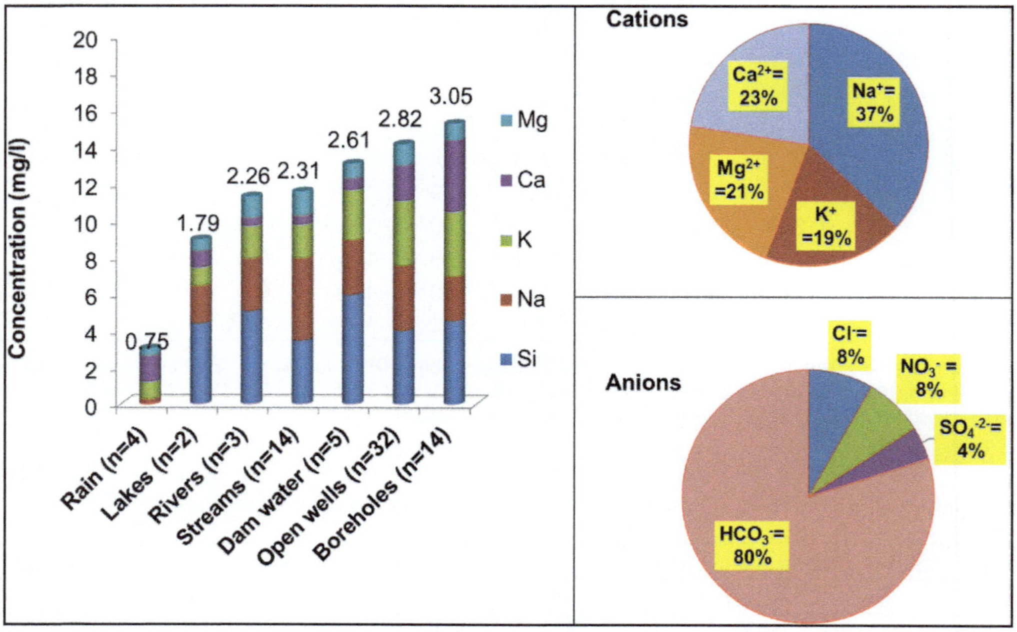

Figure 2. A, Relative mean concentrations of silica and cations (mg/l) in various water sources of the study area including rain. Si was not measured in rain; Si>Na>K>Ca>Mg. B, Pie chart of mean concentrations of ions (meq/l), showing high Na+ and dominant HCO3- in water sources of the study area.

and Smedley, 1996). However, the released Ca^{2+} is possibly exchanged for Na^+ on clay surfaces, resulting in a relative increase in F^- and a decrease in Ca^{2+}, hence, the weak correlation between Ca^{2+} and F^-.

High F^- concentrations in groundwater have been identified in North Cameroon, along the CVL, in a granitic-alkaline environment from fluorite, amphiboles and micas (Fantong et al., 2009). The low concentrations of F^- in groundwater of the study area may be due to its acidic nature which renders F^- immobile (Hem, 1989; Edmunds and Smedley, 1996). For example, the highest value of F^- (0.39 mg/l) was identified in BH29 ('borehole') with a pH of 6.4. The strong positive correlations of HCO_3^- with Ca^{2+}, Mg^{2+}, K^+ and SO_4^{2-} further indicate similar origin of ions. Given the Na-HCO_3 dominance (Figure 3), the weak positive correlation of Na^+ with HCO_3^- suggests that varied processes accounts for its concentration in solution.

Table 5 shows a summary of computed ratios between some ionic species. The observed Na^+/Cl^- molar ratios were high, relative to that of sea water, 0.86 (Millero et al., 2008). A Na^+/Cl^- molar ratio approximately equal to 1 is usually attributed to halite dissolution, whereas >1 is typically interpreted as reflecting Na^+ released from silicate weathering reactions (Fisher and Mullican, 1997; Cendon et al., 2010). Therefore, the excess Na^+ as further shown (Figure 4) is likely from silicate weathering, possibly feldspar mineral dissolution. On a plot of total cations, $Na^++K^++Ca^{2+}+Mg^{2+}$ (TC) versus alkalinity, most of the 72 samples (94 %) plotted on or near the 1:1 line (Figure 5a). The high mean $HCO_3^-/\sum$anions ratio in the water (Table 5) and the 1:1 relationship (Figure 5a) suggest silicate or carbonate weathering (Hounslow, 1995; Fisher and Mullican, 1997; Kim, 2002), but acid waters (as prevailed in the study area) usually results from dissolution of non-carbonate rocks such as granites (Edmunds and Smedley, 1996), the dominant rock in the

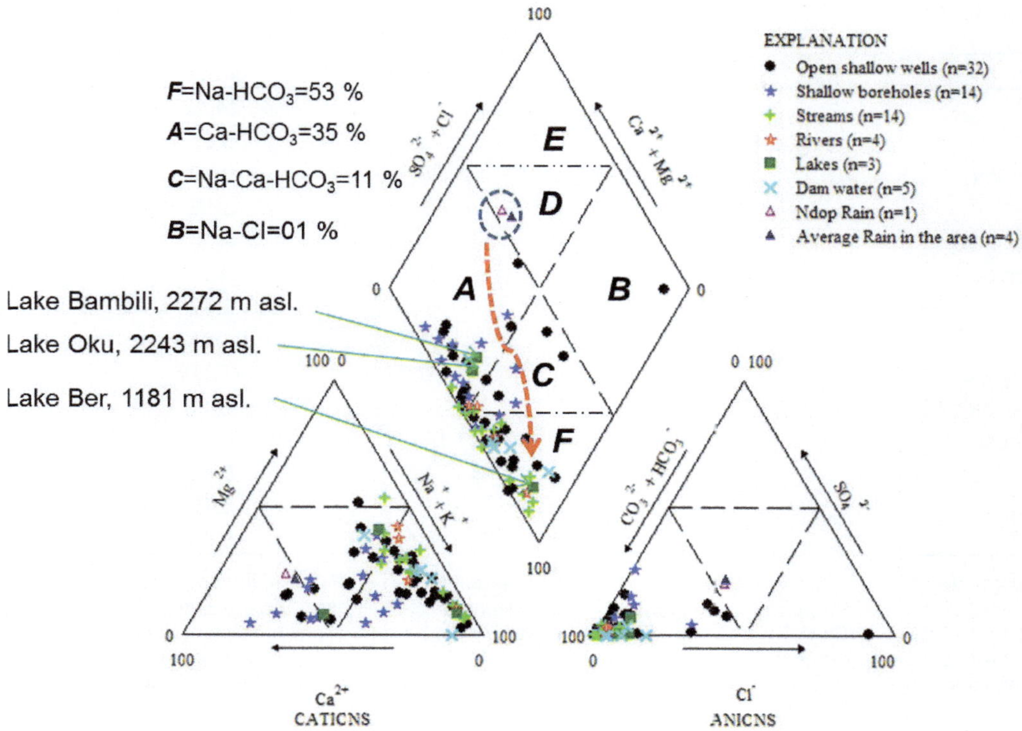

Figure 3. Piper's diagram showing the water types and proportions in the Ndop Plain. Alkali metals (Na+K) exceed alkali earth metals (Ca+Mg). Weak acids (CO_3+HCO_3) greatly exceed strong acids (SO_4+Cl). The dash line in the diamond section indicates the path of chemical evolution from Ca-Mg-SO_4-Cl (rain), to Ca-HCO_3 (groundwater, lakes and dam water), and to Na-HCO_3 (groundwater, streams and rivers). Ca-Na-HCO_3 represents a transition (mixing zone) from Ca-HCO3 to Na-HCO3 water.

Table 4. Pearson correlation coefficients of all analysed water sources (n=72).

Parameter	Na^+	K^+	Mg^{2+}	Ca^{2+}	HCO_3^-	Cl^-	NO_3^-	SO_4^{2-}	F^-	pH
Na^+	1.00									
K^+	0.12	1.00								
Mg^{2+}	0.35	0.52	1.00							
Ca^{2+}	-0.01	0.73	0.27	1.00						
HCO_3^-	0.33	0.81	0.72	0.70	1.00					
Cl^-	**0.69**	0.12	0.10	0.06	-0.03	1.00				
NO_3^-	0.34	-0.01	0.25	0.09	-0.08	**0.52**	1.00			
SO_4^{2-}	0.05	**0.87**	**0.40**	**0.77**	**0.68**	0.11	0.04	1.00		
F^-	**0.48**	0.07	0.36	0.04	0.35	0.07	0.01	0.17	1.00	
pH	0.17	0.10	0.23	0.08	0.33	-0.16	-0.21	0.08	0.21	1.00

area.

To further determine the likely influence of silicate dissolution, the molar ratio of HCO_3^-/SiO_2, saturation indices for calcite and aragonite, and $Ca^{2+}+Mg^{2+}$ versus $SO_4^{2-}+HCO_3^-$ relationship were examined. If HCO_3^-/SiO_2 ratio is greater than 10, (that is, $HCO_3^- >> SiO_2$), carbonate weathering predominates. However, if the ratio is low and especially <5, silicate weathering predominates (Hounslow, 1995). The computed HCO_3^-/SiO_2 ratios (Table 5) were 75% (<5), 19% (5-10) and 6 % (>10). Calculated saturation indices (SI) of the water samples (using AqQA software) showed under-saturation (SI<0) with respect to calcite and aragonite (Table 5) suggesting their absence in the formation and/or not enough time to interact. A charge balance will exist between the $Ca^{2+}+Mg^{2+}$ and $SO_4^{2-}+HCO_3^-$ if they are derived from basic dissolution of calcite, dolomite, and gypsum (Fisher and Mullican, 1997). On the contrary, there was a deficiency in $Ca^{2+}+Mg^{2+}$ re-

Table 5. A statistical summary of ratios between some ionic species in the analysed water.

Parameter	Minimum	Maximum	Mean	STD
Na^+/Cl^-	1.25	5345	**24.42**	63.92
Na^+/Ca^{2+}	0.13	27	**5.35**	6.66
$HCO_3^-/\sum$anions	0.06	68.34	**9.60**	10.93
HCO_3^-/SiO_2	0.34	28.11	**3.52**	5.31
SI of Calcite	-7.21	-1.41	**-4.31**	1.28
SI of Aragonite	-7.38	-0.17	**-4.39**	1.43
R_E Index	1.31	5.84	**3.82**	1.32
P_{CO2} (atm)	0.0000	0.1082	**0.0053**	0.01
Chloro-alkali index 1 (CAI 1)	-585	-0.59	**-32.59**	71
Chloro-alkali index 2 (CAI 2)	-2.37	-0.18	**-0.65**	0.30
Total Hardness	0.97	57.27	**8.77**	10.48
SAR	0.08	4.34	**0.64**	0.65
Kelly's Ratio	0.08	10.03	**1.70**	2.12
RSC (meq/l)	0.07	2.52	**0.48**	0.48

STD, Standard deviation; HCO3-/ Σ anions (meq/l); SI, Saturation Index; RE (weathering index) = 2(3K++3Na+2Ca2+-SiO2)/K++Na++Ca2+), RE was calculated using molar concentrations of ions; PCO2 (Partial pressure of CO2) in water; Na+/Ca2+ (meq/l); CAI 1 = Cl - (Na + K)/Cl; CAI 2 = Cl-(Na + K)/(SO4 + HCO3 + CO3 + NO3).Total hardness (as CaCO3 in mg/l) = 2.5(Ca2+)+ 4.1 (Mg2+); SAR (Sodium Adsorption Ratio) = Na+/{Ca2+ + Mg2+)/2}0.5; KR (Kelly's Ratio) = Na+/Ca2+ + Mg2+. RSC (Residual Sodium Carbonate) = (CO32-+HCO3-) + (Ca2++Mg2); CAI, SAR, KR, RSC were calculated from ionic concentrations in meq/l.

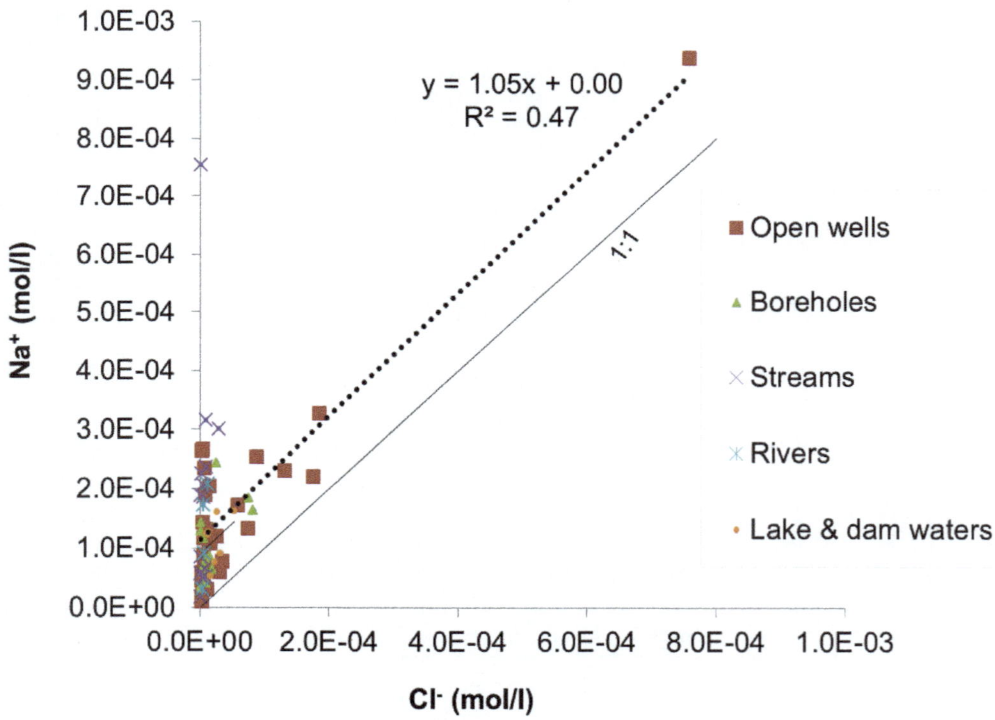

Figure 4. Plot of Na+ as a function of Cl- showing an excess of Na+ over Cl- in water sources.

lative to SO_4^{2-}+HCO_3^- (Figure 5b). Thus, the 75% dominance of samples with HCO_3^-/SiO_2 ratios <5, calcite and aragonite under-saturation, and excess of SO_4^{2-}+HCO_3^- relative Ca^{2+}+Mg^{2+} strongly suggest incongruent weathering of silicate minerals (water-rock interaction) as the main control on water chemistry in agreement with the geology. Similar observations of silicate dissolution in granitic formations (Njitchoua and Ngounou-Ngatcha, 1997; Fantong et al., 2009) and volcanic terrains (Tanyileke et al., 1996; Ako et al., 2012) along the CVL have been reported with the following generalized equation (Tanyileke et al., 1996):

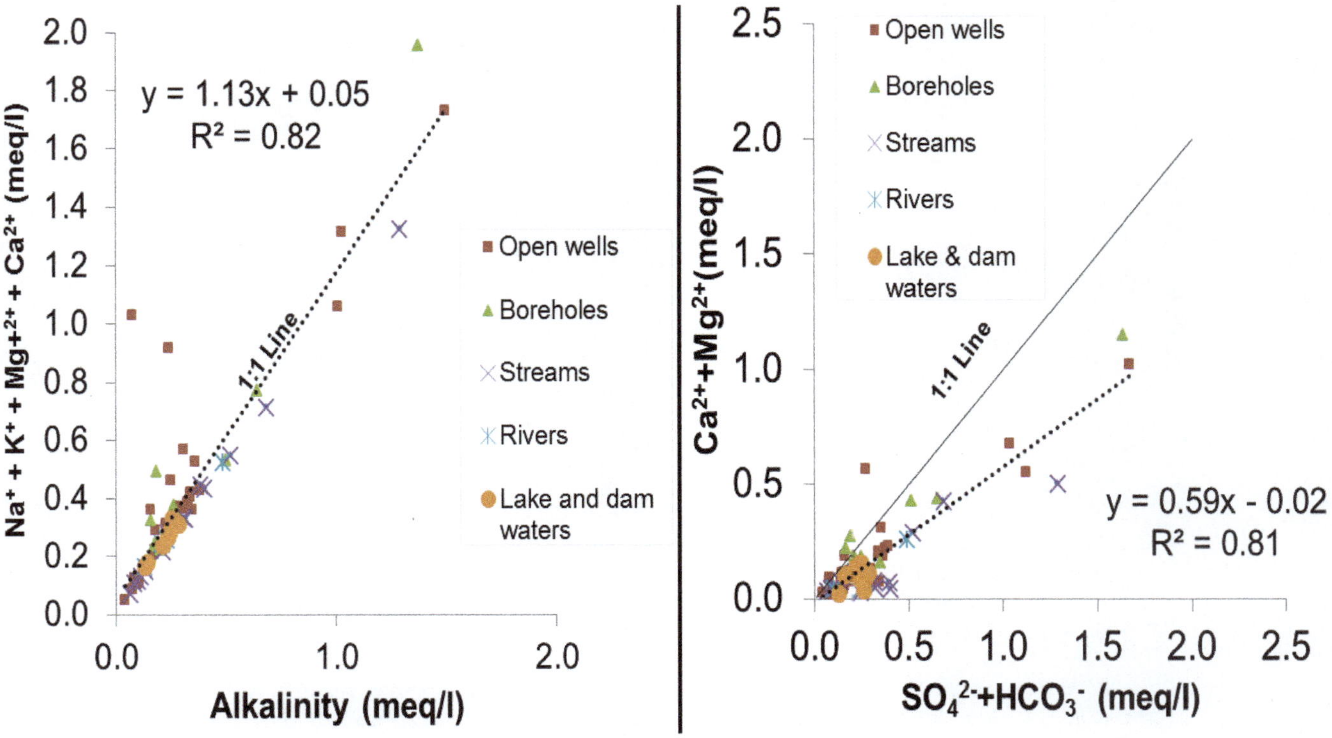

Figure 5. A, Plot of total cations as a function of alkalinity showing a 1:1 relationship in water sources of the study area. B, Plot showing deficiency in Ca^{2+}+Mg^{2+} in relation to SO_4^{2-}+HCO_3^- in water sources

Rocks + H_2O + CO_2 = Cations +H_4SiO_4 + HCO_3^- + solids (mostly clay minerals)... (1)

The HCO_3^- dominance and high SiO_2 are consistent with most natural waters along the CVL (Tanyileke et al., 1996).

Based on the geology and observed correlations between ionic species, the main mineral assemblage in the granitic basement under dissolution is possibly the plagioclase mineral series, with two end members, Albite ($NaAlSi_3O_8$) and Anorthite ($CaAl_2Si_2O_8$). In the dissolution process, which involves the release of SiO_2 and HCO_3^- into solution and the formation of clay minerals (Equation 1), Ca^{2+} is preferentially released into solution more than Na^+ (Clayton, 1988; Deutsch, 1997). As expected, Ca^{2+} would be greater in groundwater, but as earlier observed, Na-HCO_3 dominated Ca-HCO_3 waters. Calculated Weathering Index (R_E) using the formula of Tardy (1971) are shown in Table 5. About 88% of samples had R_E values <2, while 12% had values <2. Accordingly, the weathering process in the study area resulted mainly in montmorillonite clay (R_E>2) and minor kaolinite (R_E>2) (Tardy, 1971). Montmorillonite has a high cation-exchange capacity (Appelo and Postma, 2005) and will probably result in ion exchange of Ca^{2+} in solution for Na^+ in rocks given the freshwater aquifer system (Hem, 1989) as described in this equation (Njitchoua and Ngounou-Ngatcha, 1997):

In order to determine the significance of base-exchange in enriching the water chemistry, a plot of Ca^{2+}+Mg^{2+}-HCO_3^--SO_4^{2-} versus Na^+-Cl^- was examined. If cation exchange is a significant groundwater composition-controlling process, the relation between these two parameters will be linear with a slope of -1.0 (Fisher and Mullican, 1997; Adomako et al., 2011; Tay, 2012). About 89% of the 72 samples plotted along the line with a slope of -0.99 (Figure 6a). The 9 samples that diverted most from the line had relatively anomalous concentrations of cations or anions. A plot excluding these samples gave atrolling process, the relation between these two parameters will be linear with a slope of -1.0 (Fisher and Mullican, 1997; Adomako et al., 2011; Tay, 2012). About 89% of the 72 samples plotted along the line with a slope of -0.99 (Figure 6a). The 9 samples that diverted most from the line had relatively anomalous concentrations of cations or anions. A plot excluding these samples gave a slope of -1.03 with an improved linearity (Figure 6b), indicating that ion exchange reactions occurred in a greater portion of the aquifer. The nature of ion exchange was determined by calculating Chloro-Alkaline Indices (CAI) (Schoeller, 1965). All samples gave negative values from CAI 1 and CAI 2 (Table 5). This indicated that Mg^{2+}

$$2Na_{0.33}Al_{2.33}Si_{3.67}O_{10}(OH)_2 + 2Ca^{2+}_{(aq)} = 2Ca_{0.33}Al_{2.33}Si_{3.67}O_{10}(OH)_2 + 2Na^+_{(aq)} \quad (2)$$

Na-montmorillonite Ca-montmorillonite

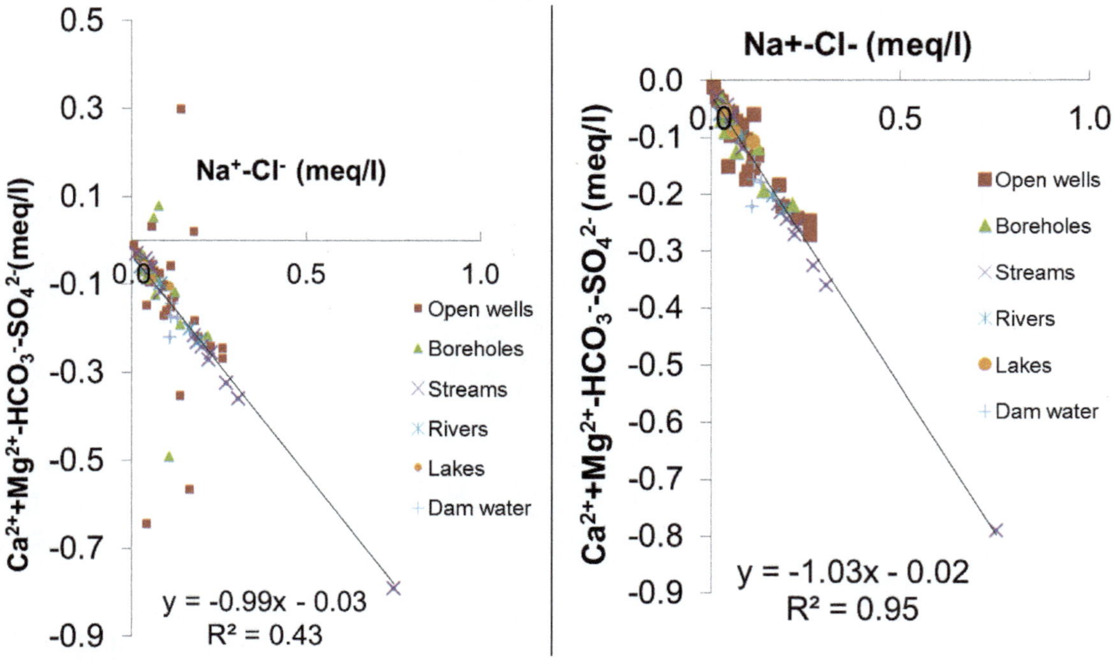

Figure 6. A, Relation of Ca^{2+}+Mg^{2+}-HCO_3+-SO_4^{2-} versus Na+-Cl- showing an increase in Na+ relative to a decrease in Ca^{2+}+Mg^{2+} or an increase in HCO_3^- and SO_4^{2-} with a slope close to -1.00 and a positive correlation (n=72). B, Relation of Ca^{2+}+Mg^{2+}-HCO_3+-SO_4^{2-} versus Na+-Cl- showing an increase in Na+ relative to a decrease in Ca^{2+}+Mg^{2+} or an increase in HCO_3^- and $SO4^{2-}$ with an improved linearity after excluding 9 anomalous samples (n=63).

and Ca^{2+} in groundwater were being exchanged with Na^+ and K^+ in rocks; hence, favouring cation-anion exchange reactions (chloro-alkaline disequilibrium) (Gupta et al., 2009). The Na^+ enrichment (addition) by ion exchange probably explains its observed weak positive correlation with other cations and SO_4^{2-}. Thus, besides the chemical dissolution of plagioclase feldspar minerals, cation-exchange reactions between water and montmorillonite in the aquifer also play a role in modifying the water composition.

The mean P_{CO2} in the groundwater (Table 5) was higher than the mean atmospheric value of 0.0003 atm (Appelo and Postma, 2005), suggesting that the major source of CO_2 governing the above reactions was probably from the oxidation of organic matter in the undersaturated zone (Karanth, 1987; Appelo and Postma, 2005). The high P_{CO2} in groundwater are consistent with findings along the CVL (Njitchoua and Ngounou-Ngatcha, 1997; Fantong et al., 2009; Ako et al., 2012), which has been attributed to the same process. Unlike in these related studies, groundwater in the study is comparatively acidic. The acidic nature probably results from reduction in acid buffering capacity of infiltrating rainwater and leaching of organic constituents from soils into the shallow groundwater.

Suitability for drinking and irrigation

The drinking water guidelines of the WHO (2004) were used to evaluate the quality of water in the study area for human consumption. Most water sources had a cream colour, likely due to suspended organic colloids. Regardless of the cream colour, the range of all major ions and TDS were below the maximum values of the guidelines for drinking water. However, the F^- content was low (0 to 0.39 mg/l) with a mean of 0.08 mg/l. At low concentrations (<0.2 mg/l) in drinking water, dental caries may result (Edmunds and Smedley, 1996). The water was soft, that is, low total hardness (Table 5) given the low Ca^{2+} and Mg^{2+}. Their concentrations were mostly below the minimum required: 10 mg/l (Mg^{2+}), 20 mg/l (Ca^{2+}) and 60 mg/l total hardness (Kozisek, 2005). Evidence is now available to confirm the existence of health risk from drinking water deficient in Ca^{2+} and Mg^{2+} (Rylander et al., 1991; Dissanayake and Chandrajith, 1999; Kozisek, 2005). Thus, the general water softness could be a health problem since soft water has been linked to cardiovascular ailments (Kozisek, 2005). About 69% of the water sources had pH values below 6.5. For effective disinfection of water with chlorine, the pH should preferably be less than 8, but lower pH water (<6.5) is likely to be corrosive (WHO, 2004). High trace-metal contents, a health problem, are also likely from such low pH waters (Edmunds and Smedley, 1996), hence, a need to assess potentially harmful trace element. The low TDS in groundwater are consistent with the observation that most exploited groundwater resources in the weathered Precambrian basement rocks of Africa area low in dissolved ions and generally suitable for consumption (Adelana et al., 2011). However, the microbial quality of groundwater in the tropical Africa is a serious health issue (Roche, 1993) as reported in the study area (Wirmvem et al., 2013) and elsewhere (Katte et al., 2003).

Suitability of water for irrigation has been assessed using various properties or ratios. These include EC, Sodium Adsorption Ratio (SAR), Kelley's Ration (KR) and Residual Sodium Carbonate (RSC) (Gupta et al., 2009; Ramesh and Elango, 2012). Electrical conductivity and SAR play a vital role in suitability of water for irrigation. The EC values in 99 % of the analysed samples were <250 µS/cm and SAR values were <5 (Table 5); thus, classifying the water into the low salinity hazard class (C1) and low sodium hazard class (S1), respectively (USSL, 1954; Gupta et al., 2009). Computed KR (Kelly, 1951) (Table 5) showed that 54% (KR<1), 25 % (KR, 1-2), and 21% (KR>2) were suitable, marginal and unsuitable, respectively, for irrigation. Calculated RSC (Eaton, 1950) classified 93% and 7% of the water as suitable (RSC≤1) and marginally suitable (RSC, 1.3-2.5), respectively, for irrigation.

Conclusion

A hydrochemical study, in the Ndop plain, to determine the physico-chemical characteristics, controls on water chemistry and suitability for drinking and irrigation has been carried out by analysing 72 water samples. The first physico-chemical baseline data of groundwater and surface water for water quality assessment, management and monitoring of changes in the hydrochemical system has been made available in the area. About 69% of the water sources had pH values below 6.5, classifying the water as barely acidic. Therefore, careful attention is necessary at all stages of pipe borne water treatment and supply in the area as this could lead to corrosion and scale formation in pipes. The low mean EC (60 µS/cm) and TDS (39 mg/l) classify the water as low-mineralised and freshwater. Major ion concentrations were low, and below maximum values of the WHO standards for drinking water, but the low Ca^{2+} and Mg^{2+} classify the water as soft (<<57 mg/l $CaCO_3$). This could be a health problem given the relationship between soft water and cardiovascular ailments. The low F^- content in the water sources (<0.39 mg/l) could lead to dental caries. In spite of these, the water sources are considered suitable for drinking as per WHO standards.

The relative abundance of major cations and anions in water (meq/l) is $Na^+ > Ca^{2+} > Mg^{2+} > K^+$ and $HCO3^- >> NO_3^- \geq Cl^- > SO_4^{2-}$, respectively. The main water types and proportions are Na-HCO_3 (53%), Ca-HCO_3 (35%) and mixed Na-Ca-HCO_3 (11%). Alkali metals (Na+K) exceed alkali earth metals (Ca+Mg) indicating their dominance in the aquifer system. The major controls on water chemistry are incongruent dissolution of silicate minerals (mainly plagioclase feldspars) and cation-exchange of Na^+ in rocks for Ca^{2+} in solution. The dissolution is governed by biogenic CO_2 and results to montmorillonite clay, which enhances the ion exchange. The low NO_3^- and general absence PO_4^{3-} in groundwater suggest their sorption by organic matter in the under-saturated zone. The low EC (TDS) and major ions in groundwater depict low solubility of the silicates, short residence time of groundwater, shallow nature of the aquifer and its acidic nature. The water sources are suitable for irrigation. Potentially harmful trace elements should be investigated given the acidic nature of the water. Hydrochemical investigations in the rainy season should also be carried out, to determine the seasonal controls on water chemistry.

ACKNOWLEDGEMENTS

This paper constitutes part of an on-going PhD research of the corresponding author, sponsored by the Japanese Government under the MONBUKAGAKUSHO Scholarship from the Ministry of Education, Culture, Sports, Science and Technology (MEXT). Material support is being provided by Japan Science and Technology Agency (JST) and Japan International Cooperation Agency (JICA) under the Science and Technology Research Partnership for Sustainable Development (SATREPS) project titled: Magmatic Fluid Supply into Lakes Nyos and Monoun, and Mitigation of Natural Disasters in Cameroon. The manuscript was improved by review comments from Dr Dion I. Cendon of Australian Nuclear Science and Technology Organization and an anonymous reviewer.

REFERENCES

Adelana S, Fantong W, Nedaw D, Duah A (2011). Groundwater and health: Meeting the unmet needs in Sub-Saharan Africa. In: Jones JAA (ed.) Sustainable groundwater resources. International Year of Planet Earth. Springer Science + Business Media B.V. pp. 21-33. DOI: 10.1007/978-90-481-3426-7_2.

Adelana SAM, MacDonald AM (2008). Groundwater research issues in Africa. In: Adelana SAM, MacDonald AM (eds.) Applied groundwater studies in Africa. Taylor and Francis, London. pp. 1-8.

Adomako D, Osae S, Akiti TT, Faye S, Maloszewski P (2011). Geochemical and isotopic studies of groundwater conditions in the Densu River Basin of Ghana. Environ. Earth Sci. 62:1071–1084.

Ako AA, Shimada J, Hosono T, Kagabu M, Akoachere RA, Nkeng GE, Eneke GT, Fouepe AT (2012). Spring water quality and usability in the Mount Cameroon area revealed by hydrogeochemistry. Environ. Geochem. Health 34 (5): 615-39.

Appelo CAJ, Postma D (2005). Geochemistry, groundwater, and pollution (2nd ed.). Balkema, Amsterdam p. 635.

Cendon D I, Larsen JR, Jones BG, Nanson GC, Rickleman D, Hankin SI, Pueyo JJ, Maroulis J (2010). Freshwater recharge into a shallow saline groundwater system, Cooper Creek floodplain, Queensland, Australia. J. Hydrol. 392 (2-4): 150-163.

Clayton JL (1988). Some observations on the stoichiometry of feldspar hydrolysis in granitic soil. J. Environ. Qual. 17:153-157.

Demlie M, Wohnlich S, Wisotzky F, Gizaw B (2007). Groundwater recharge, flow and hydrogeochemical evolution in a complex volcanic aquifer system, central Ethiopia. Hydrogeol. J. 15: 1169-1181.

Deutsch WJ (1997). Groundwater geochemistry: Fundamentals and applications to contamination. CRC Press, Florida p. 221.

Dissanayake CB, Chandrajith R (1999). Medical geochemistry of tropical environments. Earth-Sci. Rev. 47: 219-258.

Eaton FM (1950). Significance of carbonates in irrigation waters. Soil Sci. 39: 123-133.

Edet A, Nganje TN, Ukpong AL, Ekwere AS (2011). Groundwater chemistry and quality of Nigeria: A status review. Afr. J. Environ. Sci Technol. 5 (13): 1152-1169.
Edmunds WM, Smedley PL (1996). Groundwater geochemistry and health: An overview. In: Appleton JD, Fuge R, McCall GJH (eds.) Environmental geochemistry and health with special reference to developing countries. Geological Society Special Publication No.113, pp. 91-105.
Eneke GT, Ayonghe SN, Chandrasekharam D, Ntchancho R, Ako AA, Mouncherou OF, Thambidurai P (2011). Controls on groundwater chemistry in a highly urbanised coastal area. Int. J. Environ. Res. 5 (2): 475-490.
Fantong WY, Satake H, Ayonghe SN, Aka FT, Asai K (2009). Hydrogeochemical controls and usability of groundwater in semi-arid Mayo Tsanaga River Basin, Far-North Cameroon. Environ. Geol. 58 (6): 1281-1293.
Fisher RS, Mullican WF (1997). Hydrochemical evolution of sodium-sulfate and sodium-chloride groundwater beneath the northern Chihuahuan Desert, Trans-Pecos, Texas, USA. Hydrogeol. J. 5: 4-16.
Fonge BA, Egbe EA, Fongod AN, Tening AS, Achu RM, Yinda GS, Mvondo ZEA (2012). Effects of land use on macrophyte communities and water quality in the Ndop wetland plain, Cameroon. J. Agric. Soc. Sci. 12: 41-49.
Gupta S, Dandele PS, Verma MB, Maithani PB (2009). Geochemical assessment of groundwater around Macherla-Karempudi area, Guntur District, Andhra Pradesh. J. Geol. Soc. India 73: 202-212.
Hem HD (1989). Study and interpretation of the chemical characteristics of natural waters (3rd ed). U.S. Geological Survey Water-Supply Paper 2254.
Hounslow AW (1995). Water quality data: Analysis and interpretation. Lewis Publishers, Boca Raton.
Karanth KR (1987). Groundwater assessment, development and management. Tata McGrawl-Hill, New Delhi.
Karklins S (1996). Groundwater sampling desk reference. Wisconsin Department of Natural Resources Bureau of Drinking Water and Groundwater, PUBL-DG-037 96.
Katte VY, Fonteh MF, Guemuh GN (2003). Domestic water quality in urban centres in Cameroon: A case study of Dschang in the West Province. Afr. Water. J. Pilot ed.: 43-54.
Kelly WP (1951). Alkali soils. Their formation, properties and reclamation. Reinhold Publication, New York.
Kim K (2002). Plagioclase weathering in the groundwater system of a sandy, silicate aquifer. Hydrol. Process. 16: 1793-1806.
Kozisek F (2005). Health risks from drinking demineralised water. In: Nutrients in drinking water. World Health Organization, Geneva. pp. 148-163.
Marzoli A, Renne PR, Piccrilo EM, Francesca C, Bellieni G, Melfi A, Nyobe JB, N'ni J (1999). Silicic magmas from the continental Cameroon Volcanic Line (Oku, Bambouto and Ngaoundere): ^{40}Ar-^{39}Ar dates, petrology, Sr-Nd-O isotopes and their petrogenetic significance. Contrib. Mineral. Petr. 135: 133-150.
Millero FJ, Feistel R, Wright DG, McDougall TJ (2008). The standard composition of seawater: The Definition of a Reference Salinity Scale. Deep-Sea Res. I. 55 (1): 50-72.
Molua EL, Lambi CM (2006). Climate hydrology and water resources in Cameroon. CEEPA.
Ndenecho EN (2009). Cropping systems and post-cultivation vegetation succession: Agro- Ecosystem in Ndop Cameroon. J. Hum. Ecol. 27 (1): 27-33.
Ndzeidze SK (2008). Detecting changes in a wetland using multi-spectral and temporal Landsat in the Upper Noun Valley drainage basin-Cameroon. MSc. Thesis, Geography. Oregon State University.
Neba A (1999). Modern geography of the Republic of Cameroon (3rd ed.). Neba Publishers, Bamenda p. 235.
Njitchoua R, Ngounou-Ngatcha B (1997). Hydrogeochemistry and environmental isotope investigations of the north Diamare plain northern Cameroon. J. Afr. Earth Sci. 25 (2): 307-316.
Ramesh K, Elango L (2012). Groundwater quality and its suitability for domestic and agricultural use in Tondiar river basin, Tamil Nadu, India. Environ. Monit. Assess. 184:3887-3899.
Roche MA (1993). Water quality Issues in the humid tropics. In: Bonell et al. (eds.) Hydrology and water management in the humid tropics: Hydrological research issues and strategies for water management. UNESCO. Cambridge University Press, Cambridge. pp. 344-366.
Rylander R, Bonevik H, Rubenowitz E (1991). Magnesium and calcium in drinking water and cardiovascular mortality. Scan. J. Work. Environ. Health 17 (2): 91-94.
Schoeller H (1965). Qualitative evaluation of ground water resources. In: Methods and techniques of groundwater investigation and development. Water Resourc. Series No. 33. UNESCO. pp. 44-52.
Srinivasamoorthy K, Chidambaram S, Prasanna MV, Vasanthavihar M, Peter J, Anandhan P (2008). Identification of major sources controlling groundwater chemistry from a hard rock terrain- A case study from Mettur taluk, Salem district, Tamil Nadu. J. Earth Syst. Sci. 117: 49-58.
Takounjou AF, Ngoupayou JRN, Riotte J, Takem GE, Mafany G, Marechal JC, Ekodeck GE (2011). Estimation of groundwater recharge of shallow aquifer on humid environment in Yaounde, Cameroon using hybrid water-fluctuation and hydrochemistry methods. Environ. Earth Sci. 64 (1): 107-118.
Tanyileke GZ, Kusakabe M, Evans WC (1996). Chemical and isotopic characteristics of fluids along the Cameroon Volcanic Line, Cameroon. J. Afr. Earth Sci. 22 (4): 433-441.
Tardy Y (1971). Characterization of the principal weathering types by the geochemistry of waters from some European and African crystalline massifs. Chem. Geol. 7: 253-271.
Tay CK (2012). Hydrochemistry of groundwater in the Savelugu-Nanton District, Northern Ghana. Environ. Earth Sci. 67: 2077-2087.
US Salinity Laboratory Staff USSL (1954). Diagnosis and improvement of saline and alkali soils. Agricultural Handbook 60, USDA. US Government printing office, Washington DC.
Welz B, Becker-Ross H, Florek S, Heitmann U (2006). Front Matter, in High-Resolution Continuum Source AAS: The better way to do Atomic Absorption Spectrometry, Wiley-VCH Verlag GmbH & Co. KGaA, Weinheim, FRG. Doi: 10.1002/3527606513.fmatter.
WHO (2004). Guidelines for drinking-water quality (3rd ed.). Geneva, Switzerland.
WHO/UNICEF (2012). Joint monitoring programme for water supply and sanitation. Coverage estimates: Improved sanitation, Cameroon. www.wssinfo.org (Accessed: November 2012).
Wirmvem MJ, Fantong WY, Wotany ER, Ohba T, Ayonghe SN (2013). Sources of bacteriological contamination of shallow groundwater and health effects in the Ndop plain, Northwest Cameroon. J. Environ. Sci. Water Resourc. 2 (4): 127-132.

9

Hydrogeochemical and anthropogenic influence on the quality of water sources in the Rio del Rey Basin, South Western, Cameroon, Gulf of Guinea

Engome R. Wotany[1] , Samuel N. Ayonghe[2], Wilson Y. Fantong[3], Mengnjo J. Wirmvem[4] and Takeshi Ohba[4]

[1]Department of Geology, Faculty of Science, University of Buea, P.O. Box 63, Buea, Cameroon.
[2]Department of Environmental Science, Faculty of Science, University of Buea, P.O. Box 63, Buea, Cameroon.
[3]Institute of Mining and Geological Research, P.O. Box 4110, Yaounde, Cameroon.
[4]Department of Chemistry, School of Science, Tokai University, Hiratsuka, 259-1211, Japan.

This study which focuses on the hydrogeochemical and anthropogenic influence on water sources in the Rio del Rey Basin has provided an insight into the sources of major ions, water mineralisation processes, and the suitability of the water sources for drinking and agricultural purposes. Fifty-one (51) samples (38 groundwater, 9 surface water, 1 tap, 1 reservoir and 2 rainfall events) were analysed for physico-chemical and microbial quality. The pH of groundwater samples ranged from 5.5 to 8. Sixty percent (60%) of the groundwater samples were slightly acidic (5.5-6.4) and were different from the WHO (2004) standard for drinking water. Electrical conductivity values (3-274 µS/cm) and total dissolved solids (7.66 to 248.30 mg/l) for groundwater sources were low and signified low mineralisation and freshwater conditions. The relative abundance of major ions (meq/l) was $Na^+ > K^+ > Mg^{2+} > Ca^{2+}$ for cations and $HCO_3^- > NO_3^- > SO_4^{2-} > Cl^-$ for anions. The main water types were: Na-HCO_3 (45%), Ca-HCO_3 (35%), Na-Ca-HCO_3 (15%), Ca-Mg-Cl (2%) and NaCl (2%). The processes that influenced the water chemistry were incongruent silicate dissolution and cation exchange. The chemical constituents were low and within the WHO (2004) guidelines for drinking water except for high concentrations of NO_3^- (77.28 mg/l) and Br^- (0.08 mg/l) in one open well. Water quality index calculated using the adjusted form of CCME (2001) based on the major and chemical parameters (TDS, pH, F^- and Br^-) for groundwater and surface water data indicated that 90, 6, 2 and 2% provided excellent, good, marginal and poor water quality, respectively for drinking purposes. The evaluation of results showed that 90% of groundwater sources were suitable for agricultural purposes. Total coliform counts indicated that 97% of the groundwater sources were polluted.

Key words: Groundwater chemistry, microbial quality, anthropogenic influence, Rio del Rey Basin.

INTRODUCTION

In sub-Saharan Africa, almost all communities in coastal areas depend on groundwater for drinking and domestic purposes (Edet, 2008). This is because it is available and is relatively cleaner than surface water sources which are exposed to pollution from domestic wastes and runoff especially during the rainy season. Lastly, groundwater does not need extensive treatment before use (Yidana, 2010). The concept of water quality has been evaluated in recent years owing to greater understanding of water mineralization processes and greater concern about its origin (Chenini and Khemiri, 2009; Nwankwoala and Udom, 2011). Some of the factors that control groundwater quality and it's chemistry along the African coast are enhanced Na^+, Ca^{2+}, Mg^{2+} and HCO^-_3, due to silicate and carbonate weathering and enhanced NO^-_3 from indiscriminate waste disposal from human activities (Edet, 2008; Olufemi et al., 2010; Nduka and Orisakwe, 2011). Following the works of several authors (Chebotarev, 1955; Gibbs, 1970; Palmer and Cherry, 1984; Lawrence et al., 2000; Ellis and Rivett, 2007; Eneke et al., 2011), it is generally accepted that the chemical quality of groundwater is controlled by natural processes such as precipitation, mineral weathering and evaporation-crystallisation and also significantly influenced by anthropogenic processes (Lerner and Barrett, 1996). Variation in levels of water constituents (major ions) may be attributed to the differential derivation from source rocks, discharge of untreated effluents from industries, agricultural and domestic waste (Edet, 2009; Nduka and Orisakwe, 2011). The demand for potable water is rising due to population increase in developing countries. Such areas are expected to experience imbalances of potable water demand and supply due to climate change (Vairavamoorthy et al., 2008). The chemical constituents of groundwater is known to cause some health risks, so supply cannot be said to be safe if specific information on water quality which is needed for sustainable resource development and management is lacking (Nwankwoala and Udom, 2011).

The majority of health problems faced by African countries are water related: typhoid, diarrhoea, cholera, dysentery (Hamer et al., 1998; Kuitcha et al., 2008; WHO Regional Office for Africa, 2012). Managing the microbial aspects of water quality control is a major issue in the humid tropics as it has a direct effect on human health (Roche, 1993). For human consumption, the most important water quality parameter is total coliform (Younger, 1994). Studies carried out in different parts of Cameroon indicate that most domestic water sources have an alarming level of microbiological pollution (Katte et al., 2003; Mpakam, 2009; Wirmvem et al., 2013). Cases of water-borne diseases such as typhoid, cholera and amoebic dysentery are recurrent in most localities in Cameroon (Kuitcha et al., 2010) and these have been traced to the use of water from shallow, unprotected hand-dug wells (Nola, 1996).

Regrettably, the identification and evaluation of the quality of potential fresh water resources in rural and remote areas, as a precondition for sustainable management of water resources, has not been systematically carried out in Cameroon (Katte et al., 2003). Less than 40% of the population of Cameroon has access to pipe borne potable water (Tanawa et al., 2002). The vast majority is thus compelled to turn to alternative sources like springs, wells and streams whose chemical and microbial qualities are greatly compromised by their proximity to both point and diffuse sources of pollution.

Unlike the Douala Basin and other Basins along the coast of the Gulf of Guinea (e.g. the Niger Delta in the west of the Rio del Rey Basin), where hydrogeological and hydrogeochemical studies have been carried out (Edet, 2009; Williams and Benson, 2010; Eneke et al., 2011; Nduka and Orisakwe, 2011), little or no comprehensive data on the hydrogeochemistry and anthropogenic influence on the quality of water sources has been generated within the Rio del Rey Basin. Information on the quality of water sources used for household and agricultural purposes within the Bakassi Peninsular and its environs is non-existent; hence, the need to assess the suitability of these water source for users.

The present study accordingly aimed to evaluate the physico-chemical characteristics, sources of ions and mineralisation processes, the microbial quality and the suitability of the water sources for drinking and agricultural uses. The results from this study is useful towards the generation of a baseline hydrogeochemical and microbial data in the area in general and within the remote Bakassi Peninsular in particular for the first time; and will provide suitable information for harnessing potable water for the inhabitants.

The study area

The Rio del Rey Basin is located in the South Western Coast of Cameroon at the western end of the Gulf of Guinea (Figure 1). The basin is an extension of the western margin of the Niger Delta of Nigeria (Regnoult, 1986), and one of the Equatorial Atlantic marginal basins in the Gulf of Guinea. Together with the Douala and Kribi-Campo Basins, it was formed during the Aptian-Albian times (Agyingi et al., 2006). It is bordered in the west and northwest by the Niger Delta Basin, in the south it is limited by the Gulf of Guinea (Atlantic Ocean), in the north by the Rumpi Hills and to the east by the Cameroon Volcanic Line which separates the basin from the Douala Basin. The drainage pattern is dendritic (Figure 1) dominated by Rivers Ndian, Moko, Meme, Mungo and Akwayafe which have watersheds from high altitudes (800-1050 m) from the Rumpi Hills, the Manengouba mountains and the highlands of the Korup National Park.

It has a varied geological setting comprising of Cretaceous limestones, Tertiary and Quaternary sediments which are essentially clastics consisting of sands, sandstones, conglomerates, limestones, shales, clays and alluvium which are terminated landward by basaltic lava flows from the Rumpi Hills and by Precambrian basement rocks composed of gneisses, micaschists and quartzites (Dumort, 1968; Obenesaw et al., 1997; Njoh and Petters, 2008). The lithologic description made by Njoh and Petters (2008) describes

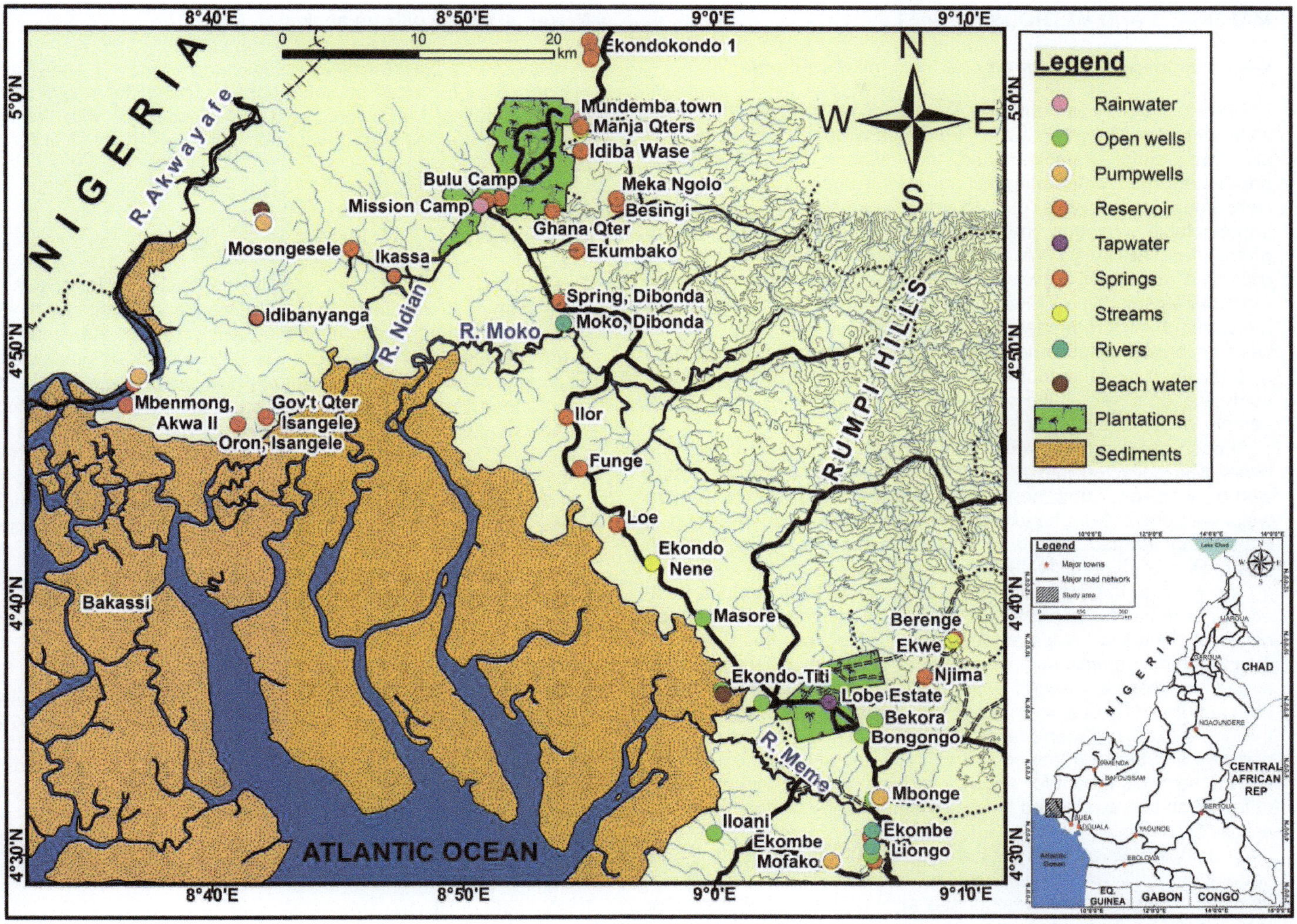

Figure 1. Geo-referenced map showing the location, drainage and sampling sites in the Rio del Rey Basin. Inset: Cameroon map showing study area (shaded rectangle).

the Cretaceous sedimentary rocks in this area as poorly sorted conglomeritic sandstones, dark micaceous siltstones, mudstones and shales. Directly overlying the Cretaceous sediments offshore is the lowest Tertiary unit known as the Isongo Formation, which is equivalent to the Akata Formation in the Niger Delta (Njoh and Petters, 2008).

This unit is composed mainly of dark grey marine shales ranging from Paleocene to Recent. The Agbada Formation is characterized by a lower shale section and an upper section, composed of an alternation of sandstones and shales. It ranges from Oligocene, Miocene to Recent. The Benin Formation forms the upper continental sequence in this basin and is composed of a few intercalations of shales towards the base, and rich in sands and sandstones.

The climate in this area is an Equatorial Cameroonian type (Etia, 1980) influenced by the meteorological equator. Two distinct seasons exist: a long rainy season of more than 8 months (March-October) and a dry season of four months (November-February; Gabche and Smith, 2002).

The study area is characterized by a high annual mean rainfall of over 5,000 mm (Zimmermann, 2000) subdivided into: 86 mm per month during the dry season and 326 mm per month during the rainy season (Gabche and Smith, 2002). The average annual variable temperature is 27°C (Ndian Meteorology, 2012). The coolest month typically occurs in August with mean monthly maximum temperatures of 25°C, while February is the hottest month with monthly mean maximum temperatures of 33°C.

Soils in the study area are ferralitic (Gavaud and Muller, 1980), yellowish in colour, and varying from clayey, silty, sandy to lateritic clay sub soils. Agricultural rich areas have been extensively cultivated for at least the last 200 years (Gartland, 1986). Large oil palm plantations of some 60 km^2 extent and belonging to the parastatal - PAMOL plantations Ltd - are found around Mundemba and Ekondo-Titi (Lawson, 1993).

MATERIALS AND METHODS

Water sampling and physico-chemical measurements

Fieldwork (January to February 2012) involved the collection of 51 water samples (38 groundwater, 9 surface water, 1 tap, 1 reservoir and two rainfall samples) and the measurement of physical parameters (pH, temperature, EC) *in situ*. The physical parameters were measured on-site due to their unstable nature to avoid unpredictable changes in characteristics as per the standard procedures (APHA/AWWA, 1998). The pH, temperature and total dissolved solids (TDS) were measured using the Hanna 98128 multi parameter type. Electrical conductivity (EC) values were measured using the Model 3301 conductivity meter. These meters were calibrated before and during fieldwork using buffer solutions as prescribed by the manufacturers. GPS Explorist 100 was used at each location to record the coordinates and altitudes to produce the sample location map (Figure 1).

At each sampling site, water from springs, streams, beaches and rivers was collected into a plastic collector after thorough rinsing with distilled water and the water to be sampled. Water from open wells was drawn from hand dug wells without any cement lining at depths that ranged from 3 to 10 meters using buckets tied with ropes. Hand dug pump wells were pumped before sampling for 5 to 10 min in order to purge any aquifer stagnant water to acquire fresh aquifer samples for analysis. Surface water was sampled from the edges of the water bodies as deep as possible along the flow path. Two sporadic rainfall samples were collected in different rainfall events in vessels cleaned with distilled water directly from the atmosphere. All samples meant for ion analyses were filtered through a 0.45 µm membrane filter which allowed the removal of particles and bacteria that could modify the contents (Atteia, 2005). For the samples meant for cation analysis, 1% nitric acid with a pH of less than two was added to keep metal ions in solution (Rodier et al., 2005).

All samples were put in plastic bottles (100 and 1500 ml in two sets) thoroughly rinsed with distilled water and the water to be sampled. These bottles were rinsed several times to ensure that they were free from contaminants. The samples were preserved airtight in order to minimize oxygen contamination and the escape of dissolve gases. The samples were stored in ice-chest containers at about 4°C prior to laboratory analysis to exclude microbial activity and unwanted chemical activity. The 1500 ml of samples were transported to the Soil and Water Laboratory of the Institute of Agronomic Research for Development (IRAD) Ekona for Alkalinity test and to the Microbiology Laboratory of the Faculty of Science of the University of Buea for coliform counts test within 24 h. The 100 ml containers were transported to the Chemistry Laboratory, Tokai University, Japan for major ions analysis. Total alkalinity as bicarbonate (HCO_3^-) was measured within 8 h of sampling through titration using a volume of 0.02 MHCl to the sample and stirred with an automatic stirrer until endpoint, marked by a pH of 4.5.

Chemical analyses

Cations: sodium (Na^+), potassium (K^+), calcium (Ca^{2+}) and magnesium (Mg^{2+}) were determined by the Flame Technique in a High Resolution Continuum Source AAS (ContrAA700). Details of the method used are described in Welz et al. (2006). Samples with EC values >100 µS/cm were diluted 5 or 10 times, to acquire an absorbance within the range of the used standards for Na^+, K^+ and Ca^{2+}. Measurements of the anions: chloride (Cl^-), phosphate (PO_4^{2-}) sulfate (SO_4^{2-}), nitrate (NO_3^{2-}), fluoride (F^-) and bromine (Br^-) were done using an Ion Chromatography (Dionex ICS-900). The charge balance for reliability of chemical measurements calculated after Domenico and Schwartz (1990) was within the acceptable limit of ±5%.

Microbial (total coliform analysis)

Microbial (total coliform count as bacteria indicator) analysis was performed by the most probable number (MPN) technique (APHA, 1992). Tubes of McConkey broth were inoculated with the water samples appropriately and incubated for 24-48 h at 37°C for total coliforms.

Data analysis

The SPSS 17.0 for windows statistical tool was used for Spearman correlation test, scattered plots and factor analysis to highlight indicators with high contribution that best explained the variability of state of water quality sampled. The WHO (2004) guidelines for water quality were used to evaluate the parameters analyzed for the purpose of drinking water. The water quality index (WQI) for the samples analysed was calculated using the methodology of the Canadian Department of Environment and Conservation (CCME, 2001) as described in Yidana et al. (2010). This method was used to characterize the usability of water resources for drinking purposes. The CCME (2001) was based on 3 factors: i) the scope which measures the percentage of the variables whose values deviate from the WHO (2004) value, ii) the frequency which measures the percentage of the samples that do not meet the standards and iii) the amplitude which measures the extent to which the non compliant parameters exceed the standards as described in Yidana (2010).

Agricultural suitability of groundwater samples was evaluated based on the following: Electrical conductivity (EC), the sodium percentage (Na %; equation 1), Wilcox diagram (Wilcox, 1955; Na % against EC), residual sodium carbonate (RSC; equation 2), sodium adsorption ratio (SAR; equation 3) and the United States Salinity Laboratory Classification (USSL) (Richard, 1954):

$$\text{Na}\ \% = \text{Na} / (\text{Na} + \text{Ca} + \text{Mg} + \text{K}) \times 100 \quad (1)$$

$$\text{RSC} = (CO_2 + HCO_3) - (\text{Mg} + \text{Ca}) \quad (2)$$

$$\text{SAR} = \text{Na} / \sqrt{(\text{Ca} + \text{Mg})} \quad (3)$$

RESULTS AND DISCUSSION

Physico-chemical composition of the groundwater sources

Results of the groundwater samples are presented in Table 1. The pH of springs, opens wells and pump wells sources ranged from 5.5 to 8. Sixty percent (60%) of these water samples were slightly acidic (5.5-6.4) and were different from the WHO (2004) standard for drinking water. EC values ranged from 3-274 µS/cm with a mean of 421.54 µS/cm. TDS of all groundwater sources ranged from 7.66 -248.30 mg/l with a mean of 54.30 mg/l. The chemical constituents analysed were low and were within the WHO (2004) guideline values except for high concentrations of NO_3^- (77.28 mg/l) and Br^-(0.08) in an open well source in the south east (Bekora) of the study area. The relative abundance of cations and anions (Figure 2a-d) were as follows:

$Na^+ > Mg^{2+} > K^+ > Ca^{2+}$ and $HCO_3^- > NO_3^- > Cl^- > SO_4^{2-}$ for

Table 1. Statistical summary of physico-chemical data of groundwater in the study area.

Parameter	Spring (n = 26)				Open well (n = 8)				Pump well (n = 4)				Summary of ground water samples (n = 38)		
	Mean	Min	Max	Std.	Mean	Min	Max	Std.	Mean	Min	Max	Std.	Min	Max	Mean
Water temperature (°C)	26.19	21.0	29.0	1.6	27.68	26.0	28.0	0.57	27.83	26.7	29.0	1.25	21.30	29.30	27.20
pH	6.72	5.64	7.61	0.55	6.12	5.5	7.2	0.54	6.78	5.87	8.0	0.99	5.5	8.00	6.60
EC	21.65	3.00	122.00	25.57	79.11	12.00	274.	90.23	95.63	7.00	171.5	82.04	3.00	274	41.54
Na^+ (mg/l)	1.93	0.33	5.28	1.47	6.21	1.19	20.8	6.65	5.09	2.19	9.06	3.06	0.33	20.80	3.17
K^+ (mg/l)	0.97	0.13	2.47	0.68	4.92	0.59	12.47	4.46	3.03	0.75	8.19	3.48	0.13	12.47	2.02
Ca^{2+} (mg/l)	0.31	0.03	1.66	0.34	1.53	0.02	6.88	2.22	2.85	0.16	8.6	3.87	0.02	8.60	0.83
Mg^{2+} (mg/l)	0.94	0.04	11.56	2.24	2.34	0.14	12.87	4.35	1.96	0.06	6.48	3.03	0.04	12.87	1.34
F^- (mg/l)	0.02	0.00	0.11	0.03	0.01	0.00	0.05	0.02	0.09	0.00	0.33	0.16	0	0.33	0.03
Cl^- (mg/l)	0.70	0.13	3.00	0.66	9.35	0.50	38.38	12.62	1.91	0.99	3.39	1.04	0.13	38.38	2.65
NO_2^-(mg/l)	0.00	0.00	0.00	0.00	0.01	0.00	0.08	0.03	0.01	0.00	0.04	0.02	0	0.08	0.007
Br^- (mg/l)	0.00	0.00	0.04	0.01	0.02	0.00	0.08	0.03	0.01	0.00	0.03	0.01	0	0.08	0.01
NO_3^- (mg/l)	1.10	0.00	8.10	1.79	17.67	0.00	77.28	25.35	3.57	0.00	8.5	3.80	0	77.28	4.85
PO_4^{2-} (mg/l)	0.04	0.00	0.39	0.11	0.01	0.00	0.05	0.02	0.00	0.00	0.00	0.00	0	0.39	0.02
SO_4^{2-} (mg/l)	0.82	0.20	2.39	0.69	2.33	0.19	11.20	3.66	2.44	1.29	5.58	2.09	0	11.00	1.31
HCO_3^- (mg/l)	19.22	4.88	118.34	21.75	91.94	75.56	106.20	9.80	52.92	10.98	101.26	47.44	4.88	118.34	38.08
TDS (mg/l)	26.07	7.55	144.52	26.32	136.34	98.62	248.30	49.29	73.90	19.69	124.01	56.43	7.55	248.30	54.32
SAR	0.4	0.09	1.40	0.30	0.76	0.22	1.99	0.69	0.53	0.36	0.83	0.20	0.09	1.99	0.56
Na/Cl	5.36	1.00	22.46	4.52	1.65	.84	4.78	1.29	5.87	1.49	14.06	5.82	0.84	22.46	4.29
Na/Ca	23.90	2.01	142.87	31.58	26.76	2.71	121.97	41.83	10.35	1.18	23.35	9.98	1.18	142.87	20.34
Na%	76.12	33.46	95.27	17.32	76.16	50.80	97.45	17.20	70.63	58.87	92.87	15.85	33.46	97.45	74.30
RSC	0.22	0.05	0.91	0.17	1.24	0.60	1.60	0.35	0.56	0.08	1.07	0.50	0.05	1.60	0.67
CAI1	-6.82	-26.8	-1.55	5.38	-2.13	-5.83	-0.05	1.71	-7.69	-15.8	-1.73	6.87	-26.8	-0.05	-5.55
CAI2	-0.27	-0.62	-0.05	0.15	-0.07	-0.15	-0.03	.05	-0.25	-0.26	-0.23	0.02	-0.62	-0.03	0.20
SI of Anhydrite	-6.26	-7.03	-4.86	0.52	-5.52	-6.58	-4.18	0.83	-4.95	-5.95	-3.71	0.93	-7.03	-3.71	-5.96
SI of Aragonite	-4.46	-7.27	-1.61	1.26	-4.91	-6.94	-2.39	1.38	-3.27	-5.39	-0.58	2.27	-6.94	-0.58	-4.37
SI of Calcite	-4.31	-7.12	-1.46	1.27	-4.77	-6.79	-2.25	1.38	-3.13	-5.24	-0.44	2.28	-6.79	-0.44	-4.23
SI of Dolomite	-7.33	-11.38	-1.71	2.46	-9.12	-12.25	-3.22	2.87	-6.19	-10.54	-1.47	4.58	-12.25	-1.47	-8.14
SI of Fluorite	-6.39	-7.92	-4.94	0.72	-6.43	-8.31	-4.55	2.66	-4.00	-4.41	-3.59	0.58	-8.31	-3.59	-5.22
SI of Gypsum	-6.04	-6.82	-4.65	0.51	-5.31	-6.36	-3.97	0.83	-4.75	-5.74	-3.51	0.92	-6.36	-3.51	-5.12
SI of Halite	-10.64	-11.37	-9.32	0.54	-9.32	-10.62	-7.66	1.04	-9.65	-9.93	-9.48	0.20	-10.62	-7.66	-9.43
PCO_2 (atm)	-2.77	-3.22	-2.29	0.26	-2.52	-2.69	-2.36	0.11	-2.50	-3.04	-2.22	0.37	-3.04	-2.22	-2.52

SAR = Sodium absorption ratio, TDS = total dissolve solids, Na % = sodium percent, RSC = residual sodium carbonate, CAI = chloroalkaline indices, SI = saturation index, PCO_2 = partial pressure of carbondioxide in water.

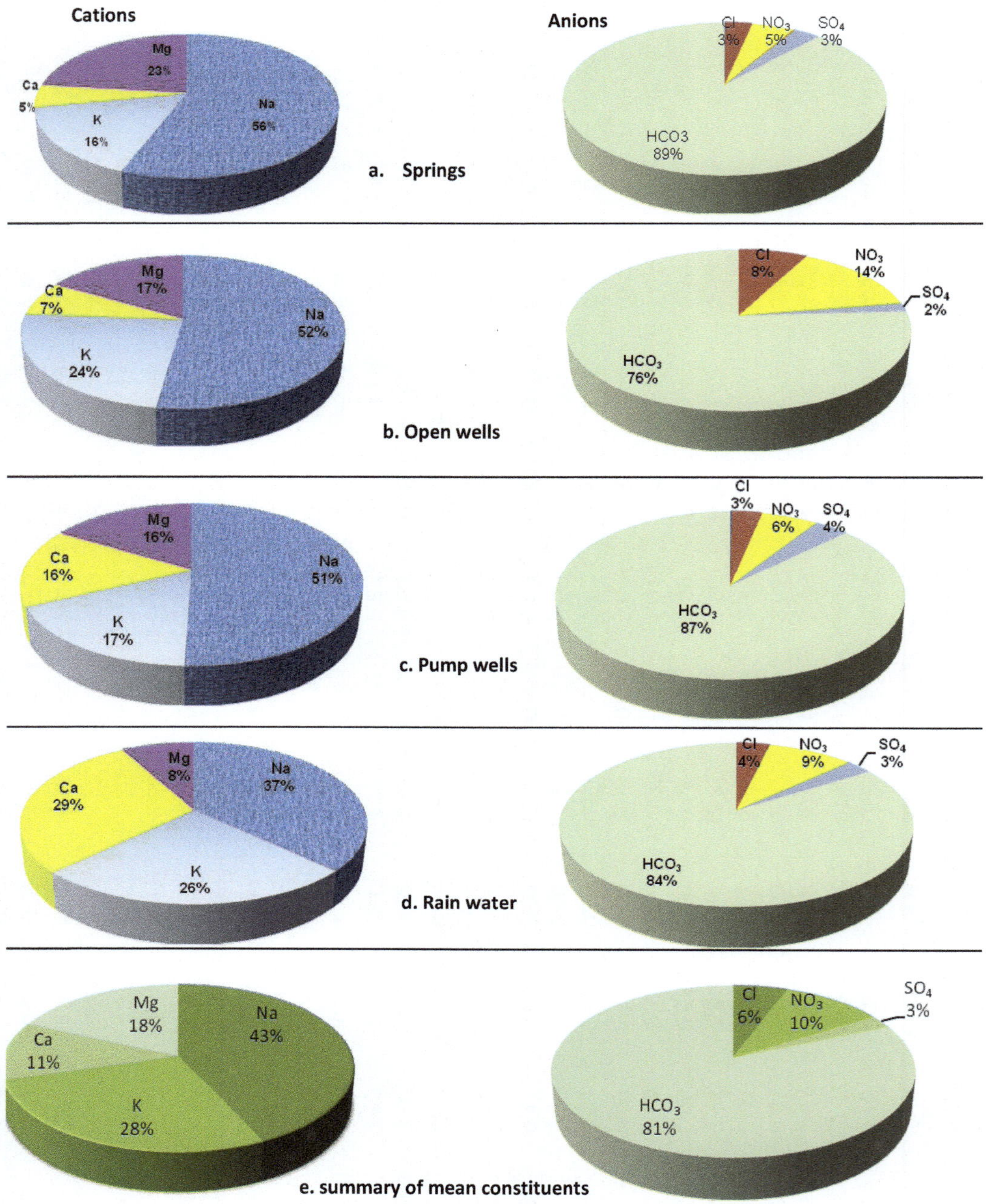

Figure 2. Pie charts of mean concentrations of ions (meq/l) for a) springs, b) open wells, c) pump wells and d) in rain water and e) summary of mean constituents showing dominant Na^+ and HCO_3^- in ground water sources.

springs; $Na^+ > K^+ > Mg^{2+} > Ca^{2+}$ and $HCO_3^- > NO_3^- > Cl^- > SO_4^{2-}$ for open wells; $Na^+ > Mg^{2+} > K^+ > Ca^{2+}$ and $HCO_3^- > NO_3^- > SO_4^{2-} > Cl^-$ for pump wells; $Na^+ > Ca^{2+} > K^+ > Mg^{2+}$ and $HCO_3^- > NO_3^- > Cl^- > SO_4^{2-}$ for rain water.

Sodium was the dominant cation and HCO_3^- the dominant anion (Figure 2e). This study has investigated the physico- (pH, temperature, EC) chemical (major: Na^+, K^+, Ca^{2+}, Mg^{2+}, Cl^-, HCO_3^-, SO_4^{2-}, NO_3^- and minor (F^-, Br^- ions) properties of some water sources (springs, open wells, pump wells) used by the local population as sources of drinking water and for agricultural purposes. According to Nduka and Orisakwe (2011), such surveillance of water quality to ensure safety is a vital public health function especially in developing countries.

The slightly acidic nature of the water sources are linked to the formation and dissolution of minerals and also influenced by biochemical processes in solution (Freeze and Cherry, 1979; Williams and Benson, 2010; Nduka and Orisakwe, 2011). The slight acidic nature of some of the potable water sources (Table 1) may possibly be attributed to the lateritic soils and organic litters from mangroves in the area (Williams and Benson, 2010). These processes involve groundwater reaction with lateritic soils and humic acids from the decomposition of organic matter in the infiltration zones leading to acidification from dissolved carbondioxide (CO_2). This is supported by the higher calculated partial pressures of carbondioxide (PCO_2), with which waters equilibrated in the infiltrating water ($10^{-3.04}$ - $10^{-2.22}$atm) (Table 1) compared to atmospheric air ($10^{-3.5}$ atm) which suggests that the waters have gained CO_2. The slightly acidic nature of the groundwater sources may also lead to the breakdown of HCO^-_3 releasing free hydrogen and carbondioxide.

Ako et al. (2011) and Fantong et al. (2009) obtained similar findings in groundwater sources along the Cameroon Volcanic Line and attributed the presence of dissolved CO_2 to root respiration, decay of organic matter or dissolution of carbonate minerals. According to Chapman (1996), marked spatial variations of groundwater temperature are largely the result of groundwater flow pattern and particularly the extent to which vertical movement occurs.

The lower mean groundwater temperature (27.20°C) (Table 1) relative to the air temperature (28.4°C) suggests that the depth of circulation (3-10 m deep) is short as in Local Systems and in close proximity to recharge areas with quick infiltration (Chapman, 1996). Studies carried out by Gnazou et al. (2011) in coastal aquifers in Togo (along the Gulf of Benin) indicated higher water temperatures (29-34°C) to relative lower air temperature (27°C) to confined aquifers with depths of 50 to ≥ 200 m. Following this, the lower mean water temperatures in the ground-water sources relative to the mean air temperatures as observed are associated with the unconfined nature of the aquifers. This fact is supported by evidence from field observations where aquifers are typically shallow, unconfined and characterised by depths of 3 to 10 m.

The EC values (3.00-274 us/cm) (Table 1) represent waters experiencing slight mineralization (Gnazou et al., 2011). The EC for open wells (Table 1) where slightly higher than those for springs and pump wells, probably due to increased value of nitrate (77.28 mg/l) from fertilizers in some of the open wells like those in Bekora. There is a relationship that exists between EC and nitrate (Das et al., 1999; De Neve et al., 2000). A positive correlation (r^2 = 0.69) (Table 2) which indicates a relationship exists between EC and NO_3^-. The EC values recorded are slightly lower than those obtained (27.80. to 899. µS/cm) in some localities in Douala (Eneke et al., 2011). In Togo where EC values in coastal aquifers were shown to exceed 1000 µS/cm in some areas (Gnazou et al., 2011), this was attributed to high sulphate content released from evaporites (gypsum) present in the surrounding geological formations. Low sulphate values (Table 1) obtained from groundwater sources in study area confirm low mineralization and the non-contribution from evaporites. According to Freeze and Cherry (1979), TDS values <1000 mg/l represent fresh water. Based on Table 1, TDS values for groundwater sources were below 248 mg/l, thus considered as fresh water.

Nitrate levels for groundwater sources varied from 3.80-77.28 mg/l (Table 1). There is a strong correlation between NO3- and Cl- (r2 = 0.97) (Table 3) which suggests an anthropogenic source of NO3- from the use of agrochemicals in the surrounding plantations and inputs from domestic waste that infiltrates into the groundwater sources. The ionic constituents of Ca2+ and K+ are slightly higher in the sporadic rain samples with lower values of Na+ and Mg2+ than the groundwater samples. The ionic constituents of Ca2+ and K+ are higher in rain samples because these were early rains which have shown to be enriched in major ions in tropical areas (Freeze and Cherry, 1979). The early rains enriched in dissolved ions would lead to the leaching of the base forming ions (Ca2+, Na2+, Mg2+ and K+) from the soils chemically enriching the water during infiltration in the subsurface (Apello and Postman, 1993; Feng et al., 2005; Eneke et al., 2011). The Na+ concentration in the aquifer system probably arises from the gneisses, basalts, sediments and clay units that exist in the study area.

Ternary plots (Piper, 1944) for major cations (Mg2+, Ca2+, Na+ and K+) and anions (SO42-, HCO3- and Cl-) in all the water samples revealed different water types with different proportions (Figure 3). According to Xu and Usher (2006), when fresh water is enriched by Ca2+ and depleted in Na+, an initial sodium chloride facies becomes calcium chloride in nature as in the study area (Ca-Mg-Cl) in a process of inverse cation exchange. Such changes in facies suggest freshwater contamination by sea water (Banton and Bangoy, 1997). The negative values of CAI and CAI2 (Table 1) also indicate that Na+ + K+ in rocks is exchanged for Ca2+ + Mg2+ in water. This facie (Ca-Mg-Cl) was highly mineralized with TDS content that reflects pollution due to anthropogenic activities in the area.

Main sources of ions and water mineralisation processes

The molar ratios of Na/Cl for groundwater sources ranged from 0.84 to 22.5 (Table 1). Meybeck (1987) used Na/Cl molar ratios to study silicate weathering reactions and showed that Na/Cl molar ratio >1 reflects Na^+ released from silicate (feldspar) weathering in the process of exchange of magnesium and calcium in water with sodium and potassium in rock (cation-anion exchange reaction) (Equation 4) (Utom et al., 2013):

$$2NaAlSiO_3O + Ca^{2+} \leftrightarrow CaAlSiO_3O + 2Na^+ \quad (4)$$

Values of $Na^+/Cl^- < 1$ (0.84 mg/l) recorded in an open well (Table 1) implies that another source is contributing

Table 2. Correlation coefficient matrix between chemical variables for groundwater in the Rio del Rey Basin.

	Temperature (°C)	pH	EC	TDS	Na^+	K^+	Ca^{2+}	Mg^{2+}	F^-	Cl^-	NO_2^-	Br^-	NO_3^-	PO_4^{2-}	SO_4^{2-}	HCO_3^-
T°C	1															
pH	-0.14	1														
EC	0.26	0.17	1													
TDS	0.37	-0.06	**0.84**	1												
Na^+	0.26	-0.05	**0.89**	**0.80**	1											
K^+	0.37	0.04	**0.89**	**0.81**	**0.84**	1										
Ca^{2+}	0.31	0.19	**0.78**	**0.64**	**0.65**	**0.74**	1									
Mg^{2+}	0.12	0.36	**0.59**	**0.55**	0.35	0.39	0.23	1								
F^-	-0.10	0.21	0.39	0.14	0.30	0.04	0.14	0.20	1							
Cl^-	0.24	-0.31	**0.73**	**0.73**	**0.89**	**0.80**	**0.57**	0.16	0.00	1						
NO_2^-	0.26	0.06	0.19	0.27	0.07	0.31	0.05	**0.55**	-0.13	0.06	1					
Br^-	0.25	-0.11	**0.60**	**0.60**	**0.65**	**0.78**	0.34	0.47	-0.10	**0.69**	**0.65**	1				
NO_3^-	0.21	-0.33	**0.69**	**0.69**	**0.87**	**0.74**	**0.60**	0.09	0.00	**0.97**	0.02	**0.62**	1			
PO_4^{2-}	-0.13	0.49	0.25	0.18	0.14	0.04	0.02	**0.51**	-0.04	-0.04	-0.01	0.00	-0.08	1		
SO_4^{2-}	0.33	0.30	0.46	0.43	0.22	**0.56**	0.40	**0.65**	0.00	0.10	**0.71**	**0.57**	0.03	0.06	1	
HCO_3^-	0.35	0.03	**0.65**	**0.91**	**0.53**	**0.58**	0.46	**0.59**	0.16	0.39	0.28	0.37	0.35	0.24	0.44	1

Bold values indicate highly correlated and shows evidence of observed water composition.

chloride to the groundwater (Edet et al., 2011?). Water containing significant amount of chloride (38.38 mg/l) (Table 1) in the open well at Bekora whose drainage has less salty rocks or evaporite is considered to be derived from the atmosphere (Al-Mikhlafi et al., 2003). Ninety six percent of Na/Cl molar ratio of the analysed groundwater and surface water samples were greater than that of seawater (0.86) and 4% was less than 0.86 possibly indicating sea water intrusion due to its proximity to the Gulf of Guinea (Yidana et al., 2010).

According to Rogers (1989), if silicate weathering is the possible source of Na^+, then these water sources will have HCO_3^- as the most abundant anion. This is because of the reaction of silicate minerals with carbonic acid in the presence of water, which releases HCO_3^- (Elango et al., 2003) as in Equation 5 (Tay, 2012):

$$2NaAlSi_3O_8 + 2CO_2 + 11H_2O = Al_2Si_2O_5(OH)_4 + 2Na^+ + 2HCO_3^- + 4H_4SiO_4 \quad (5)$$

(Albite) (Kaolinite)

This confirms the dominance of HCO_3^- in the groundwater sources as a result of weathering reactions on rocks in the study area by percolating groundwater. Where silicate mineral weathering is the major controlling process, concentrations of the major physico-chemical parameters are relatively low as with the case of the study area (Yidana et al., 2010). Yidana et al. (2010) indicated that the wells within the weathered gneisses in the Keta Basin, delivered relatively fresh groundwater whose quality is controlled by the weathering of silicate minerals and present the fresh water types.

Correlations between major ions in groundwater within the study area were carried out using Spearman's correlation analysis (Table 2). The results showed statistical, regression analysis between TDS-HCO_3, EC-Na, EC-K, Na-Cl with very strong positive relations with r^2 values of 0.91, 0.89, 0.89 and 0.89, respectively suggesting that the different aquifer chemistry is controlled by these parameters (Odoh et al., 2012). The correlation between Cl^--NO_3^- (r^2=0.97) and K-NO_3 (r^2 = 0.74) suggests the application of fertilizers and poor sanitation conditions. Other moderately inter-relationships exists as shown in Table 2.

Factor analysis was further used to establish the

Table 3. Factor analysis after varimax rotation on the physico-chemical data of groundwater sources.

Parameter	Factor			
	1	2	3	4
Water T°C	0.40	-0.01	-0.39	0.44
pH	0.02	0.68	0.46	0.02
EC	0.93	0.04	0.30	0.07
TDS (mg/l)	0.93	-0.01	0.07	0.02
Na^+ (mg/l)	0.90	-0.28	0.24	-0.06
K^+ (mg/l)	0.93	-0.08	-0.07	0.01
Ca^{2+} (mg/l)	0.74	-0.10	0.21	0.26
Mg^{2+} (mg/l)	0.58	0.69	0.06	-0.14
F^- (mg/l)	0.15	0.09	0.58	0.54
Cl^- (mg/l)	0.81	-0.51	-0.01	-0.21
NO_2^- (mg/l)	0.38	0.53	-0.64	-0.04
Br^- (mg/l)	0.78	0.04	-0.42	-0.26
NO_3^- (mg/l)	0.77	-0.57	0.03	-0.19
PO_4^{2-} (mg/l)	0.15	0.50	0.47	-0.57
SO_4^{2-} (mg/l)	0.56	0.61	-0.37	0.20
HCO_3^- (mg/l)	0.72	0.23	0.08	0.13
Eigen values	7.30	2.59	1.86	1.10
% total variance	45.63	16.17	11.64	6.86
Cumulative %	45.63	61.80	73.44	80.30

relationship between variables for the physicochemical data of groundwater sources (Table 3). Four factors were found to be responsible for the variation of groundwater quality. Factor 1 accounted for 45.63% of the total variance. It included high values of EC, TDS, Na^+, K^+, Ca^{2+}, Cl^-, Br^-, NO_3^- and HCO_3. Factor 2 accounted for 16.17% of the total variance. It included high values of pH, Mg^{2+} and SO_4^{2-}. Factor 3 accounted for 11.25% of the total variance and factor 4, 6.86% of the total variance and were associated with F^-. Factors 1 and 2 suggest mineralization of dissolved ions as a result of chemical weathering (Equation 5) and cation exchange processes at soil water interface due to the contribution of the major cations (Na^+, K^+, Ca^{2+} and Mg^{2+}) and Na^+-bearing minerals (Singh et al., 2005; Ganyaglo et al., 2012); where Na^+ is expected to exchange for Ca^{2+} (Utom et al., 2013) as in Equation 4.

The major (Cl^-, SO_4^{2-}) anions may have resulted from atmospheric inputs. Factors 3 and 4 are attributed to geogenic sources. Chloro-alkaline indices, CAI 1 = Cl-(Na+K) / Cl and CAI 2 = Cl - (Na+K) / SO_4+HCO_3 + CO_3 + NO_3 (Schoeller, 1965) were used to constrain the possible ion exchange reactions between groundwater and the host aquifers.

All groundwater samples had negative values for both indices (Table 1). CAI values suggest that Mg^{2+} and Ca^{2+} from water are exchanged with Na^+ and K^+ in rock favouring cation-anion exchange reactions (chloro-alkaline disequilibrium) (Gupta et al., 2009). The CAI ratio is negative when $Na^+$and K^+contents are high. This situation occurs when groundwater has been strongly in contact with minerals able to yield these interchangeable cations easily (Abderamane et al., 2013). This exchange is common with alumino-silicated clays formed by layers whose cohesion is ensured by the existence of interlayer cations and water. The surface of the layers is negatively charged thus promoting the possibility of cation exchange with those of groundwater.

In unconsolidated sediments of an aquifer system, there is always a considerable amount of clay minerals that adsorbs some of the cations in their pore spaces. This allows some species of cations in the groundwater to be replaced by other cations from the aquifer (Ma et al., 2013).

When there is an exchange between Na^+ and K+ in water sources with Mg^{2+} or Ca^{2+} in the aquifer material, both chloro-alkaline indices (CAI_1 and CAI_2) are positive, indicating reverse ion exchange (Jankowski et al., 1998). If the exchange takes place between Ca^{2+}or Mg^{2+} in groundwater with Na^+ or K^+ in the aquifer material, the indices will be negative, indicating ion exchange (Schoeller, 1967; Kumar et al., 2007) as observed in the study area.

Based on the geology of the area, the results of saturation indices (SI) of carbonates, sulfates, fluorite, halite minerals and partial pressures of CO_2 (P_{CO2}) calculated using PHREEQC, are presented in Table 1. Saturation indices express the extent of chemical

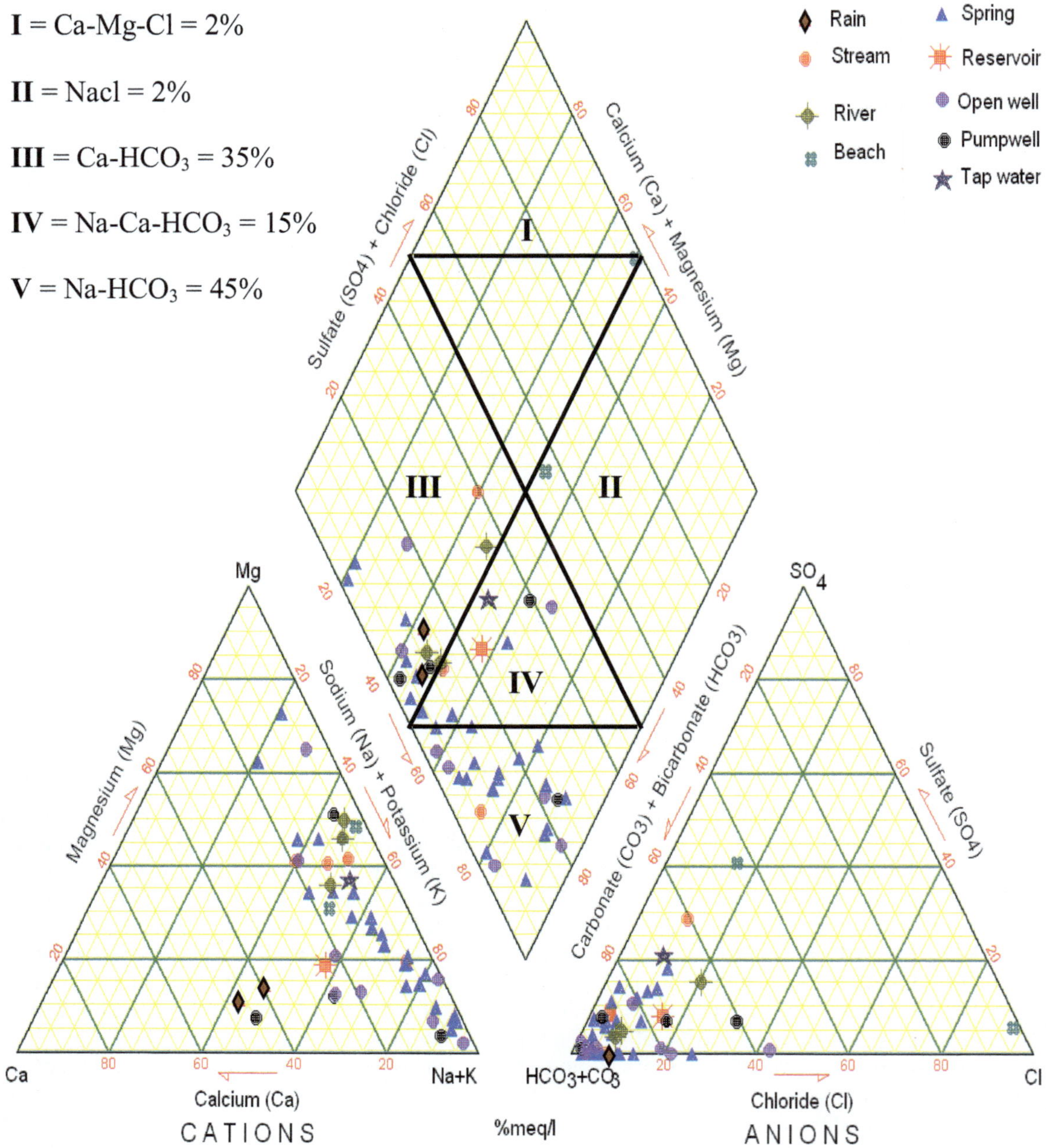

Figure 3. Piper diagram showing the water types with proportions in the Rio del Rey Basin. Alkali metals (Na^+ + K^+) exceed alkali earth metals (Ca^{2+} + Mg^{2+}). Weak acids (CO_3 + HCO_3) greatly exceed strong acids (SO_4 + Cl^-).

equilibrium between water and mineral phases in the matrix of the aquifers (Yidana et al., 2010) and could be used to evaluate the contribution of mineral dissolution and/or precipitation processes relating to the water-rock interaction (Drever, 1997; Ako et al., 2011). An index of SI<0, indicates that groundwater is under-saturated with respect to the particular mineral (Appelo and Postman, 1993). From the results obtained, all the groundwater sources were under-saturated with respect to calcite, dolomite, aragonite, gypsum, anhydrite, fluorite and halite.

The negative values registered for all the water samples could reflect the character of water from a formation with insufficient amount of the mineral in solution. The under-saturation of the sulphate minerals suggests that soluble sulphate mineral phases are absent in the host rocks (Gnazou et al., 2011). The influence of these sulphate minerals, on the chemistry of the groundwater was also less significant.

Gibbs (1970) was used to provide information on the mechanisms that actually controlled the groundwater chemistry. Figure 4a and b showed that the groundwater

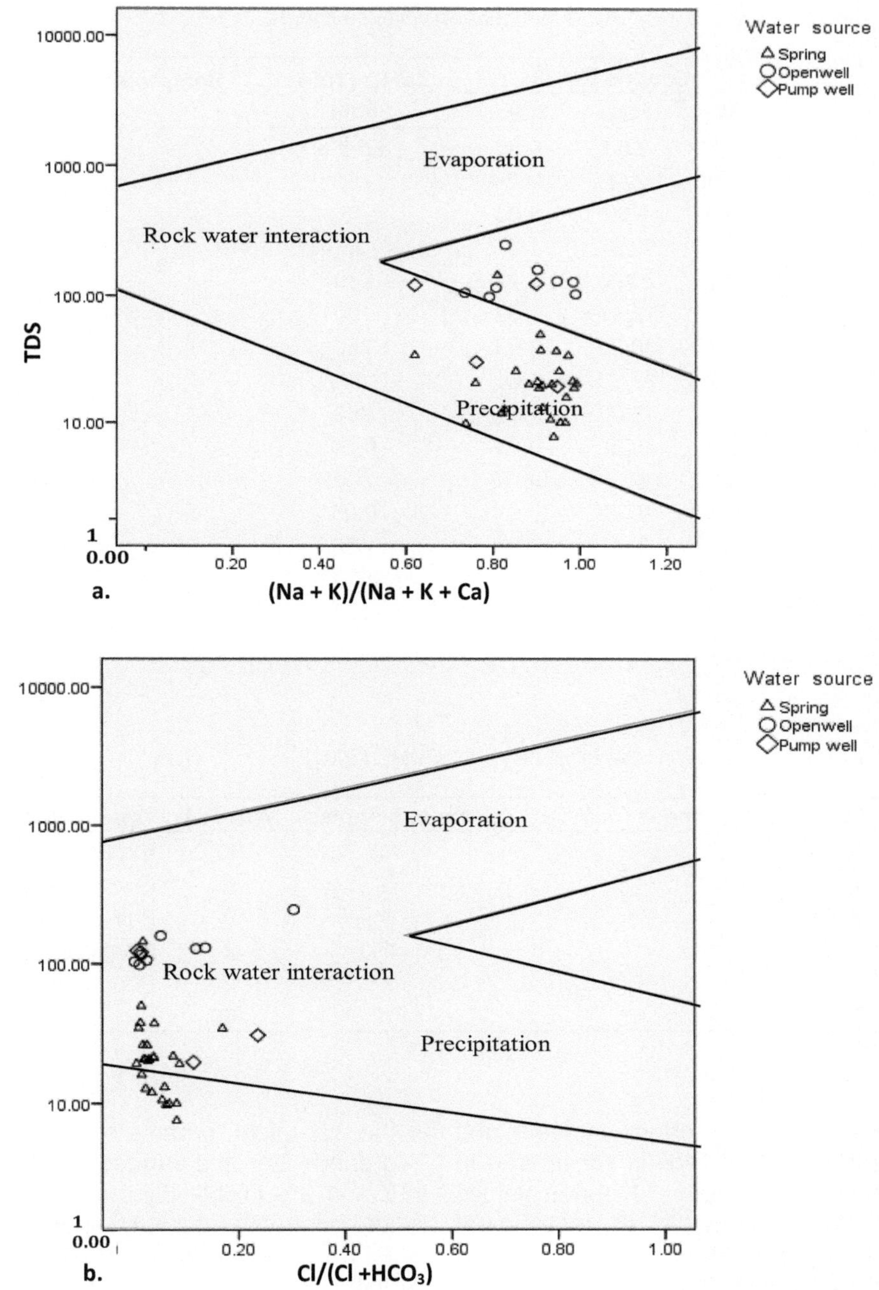

Figure 4. Mechanisms controlling the chemistry of groundwater (after Gibbs, 1970) a. Na+K/Na+K+Ca and b. Cl/Cl+HCO_3 respectively.

samples plotted in the water-rock interaction zone confirmed that rock dissolution is apparently the dominant process influencing the groundwater hydrochemistry.

Suitability of the water sources for drinking and agricultural purposes

The suitability of groundwater for drinking purposes was based on WHO (2004) limits (Table 4). Sixty percent (60%) of the groundwater samples had pH values different from the WHO (2004) guideline limit (Table 4) suitable for drinking water. The EC, TDS, cations, bicarbonate and sulphate of the groundwater sources are within the desirable limits of WHO (2004). Concentrations of Br^- (16%) and NO_3^- (10%) (Table 4) of the groundwater sources exceeded the WHO (2004) guidelines for drinking water.

The present study indicates low flouride values (0-0.33 mg/l) (Table 1) similar to those of the Ndop plain (0-0.39 mg/l) investigated by Wirmvem et al. (2013). Flouride

Table 4. Groundwater quality and compliance with WHO (2004) drinking water standards.

Parameter	Ground (N=38)			WHO (2004) limit	Sample above (or out of) WHO guideline limit (%)
	Min	Max	Mean		
pH	5.50	8.00	6.60	6.5-8.5	60.00
EC	3.00	274	41.54	750	0
Ca^{2+} (mg/l)	0.02	8.60	0.83	75	0
Mg^{2+} (mg/l)	0.04	12.87	1.34	30	0
Na^{+} (mg/l)	0.33	20.80	3.17	200	0
K^{+} (mg/l)	0.13	12.47	2.02	100	0
Cl^{-} (mg/l)	0.13	38.38	2.65	250	0
NO_3^{-} (mg/l)	0	77.28	4.85	10	10.00
SO_4^{2-} (mg/l)	0	11.00	1.31	250	0
PO_4^{2-} (mg/l)	0	0.39	0.03	0.30	0.50
HCO_3^{-} (mg/l)	4.88	118.34	38.08	200	0
Br^{-} (mg/l)	0	0.08	0.01	0.01	16.00
F^{-} (mg/l)	0	0.33	0.03	1.50	0
TDS (mg/l)	7.55	248.30	54.32	500	0
Total hardness (TH)	0.22	57.00	7.61	100	0
Total Coliform Counts (cfu/100 ml)	0	2400	1316.16	0	97.00

Table 5. Classification of water quality after CCME (2001).

CCME (2001) ranges for WQI (%)	No of samples	WQI (%)	Type of water
95-100	46	90	Excellent quality
80-88	3	6	Good quality
65-79	0	0	Fair quality
45-64	1	2	Marginal quality
0-44	1	2	Poor quality
Total	51	100	

is a health concern because of its effect on teeth and gums. Flouride concentration of ~1 mg/l is considered to be beneficial in preventing tooth decay. Flouride values >4 mg/l may cause dental fluorosis or mottling of permanent teeth and bone damage in children between the ages from 0 (birth) to 13 years (WHO, 1989). At low concentrations (< 0.2 mg/l), F^{-} in drinking water may result in dental caries (Edmunds and Smedley, 1996). Fantong et al. (2009) reported higher fluoride concentrations (>1.5 mg/l) in groundwater samples occurring in granitic terrain alongside the plain alluvium and the Limani-Yagoa ridge regions in north Cameroon which have caused severe cases of fluorosis as diagnosed by medical doctors (Fantong et al., 2013). Table 5 present the WQI range, water type and the calculated WQI for the analysed data. The water quality index is a rating technique that provides the composite influence of individual water quality parameters on the overall quality of water for human consumption (Mitra and ASABE Member, 1998). The WQI calculated using the adjusted form of CCME (2001) based on the major and the chemical parameters (TDS, pH, F^{-}, Br^{-}) for the groundwater and surface water data indicated that 90, 6, 2 and 2% (Table 5), provided excellent, good, marginal and poor water quality respectively for drinking purposes.

The excellent water quality (Table 5) implies that the water is very close to pristine conditions (CCME, 2001). Good quality water (Table 5) reflects water with minor incidences of impairment. Marginal water in the study area (open well at Bekora) had a high NO_3^{-} concentration characterised by a similar composition as marginal groundwater sources studied in the Keta Basin (Ghana) by Yidana et al. (2010). Elevated concentrations (> 5 mg/l) of nitrate in waters are an indication that the waters are at the risk of pollution (Edet et al., 2011) due to anthropogenic sources of fertilizers used in surrounding plantations and domestic waste discharge that penetrate into the subsurface.

High nitrate levels in water have been associated with methemoglobinemia, gastric ulcer (Canter and Knox, 1985), cancer and urinary tract diseases (Edet et al., 2011). In Douala, higher NO_3 (49.8-53.8 mg/l) concentra-

Table 6. Groundwater classification based on total hardness (TH) after Driscoll (2009).

Total hardness (TH) as CaCo3 (mg/l)	Water type	No of groundwater samples	Percentage of groundwater samples
<9	Soft	30	82
9-60	Slightly hard	8	18
60-120	Moderately hard	-	-
120-180	Hard	-	-
>180	Very hard	-	-
Total		38	100

tion was obtained in groundwater sources associated with organic nitrogen leached from pit latrines and solid waste dumps (Takem et al., 2010).

Based on the classification scheme by Driscoll (2009), 81% of the groundwater sources were classified as soft water, while 18% of groundwater sources were categorised as slightly hard water (Table 6). Hardness of the water is attributable to the presence of alkaline earths, that is, Ca^{2+} and Mg^{2+}. These ions react with soap to form precipitates (Jalali, 2006). Hardness of water supply intended for human consumption is between 80 and 100 mg/l (Discoll, 2009). The hardness of the groundwater sources were within the permissible limit of the WHO guideline of 100 mg/l (Ako et al., 2011).

Endeley et al. (2001) reported that 26 of the 31 (84%) water samples analyzed from sources around Mount Cameroon were soft waters. There is a link between water hardness and cardiovascular diseases. Dissanayake et al. (1992) found a negative correlation between water hardness and various forms of cardiovascular diseases and leukemia in Siri Lanka. The British committee on medical aspects of food policy (COMA, 1994) also found a weak inverse relationship between water hardness and cardiovascular disease death. This implies that people consuming soft waters have the likelihood to suffer from cardiovascular diseases than those consuming hard waters. Eighty-two (82) percent of the groundwater sources were soft waters meaning that people living in the Rio del Rey basin whose only source of drinking water are the groundwater sources have a high risk of contracting cardiovascular diseases.

The total coliform bacteria ranged from 0 to 2400 MPN/100 ml for groundwater sources (Table 4). From the results, all the water sources were polluted with the exception of the reservoir source in Mundemba town and a spring at Ekombe Liongo with values of 0 MPN/100 ml. According to the WHO (2004) guideline, 97% of groundwater sources were polluted water sources (Table 4). Based on the MPN count of total coliform bacteria following the classification by Cheesbrough (1992), the majority of the water sources were grossly polluted (75-2400 MPN/100 ml). Bacterial contamination in these water sources may possibly cause water-borne diseases such as cholera, typhoid fever and diarrhoea (WHO, 2004). The high total coliform counts were indicative of poor sanitary handling and/or environmental conditions of most of the water sources because they are not specific faecal matter indicators (Ejechi et al., 2007).

Sodium absorption ratio (SAR) values >10 are inacceptable for agricultural purposes (Yidana et al., 2010). EC values of the groundwater sources were <250 µs/cm and SAR <6 (Table 1) thus classifying the waters into low salinity hazard class (CI) and low sodium hazard class (SI; USSL, 1954) (Figure 5a). On the plot of SAR against EC, 99% of the water sources plotted in the low category (Figure 5a). The samples that plotted on the excellent to good curve (Figure 5b) had low salinity and sodium content and were the best water types for irrigation (Yidana et al., 2010). Five percent (5%) plotted on the good to permissible, 45% permissible to doubtful and 5% doubtful to unsuitable (Figure 5b). Based on residual sodium carbonate (RSC) values (Table 1), all the groundwater sources are suitable for irrigation with values that ranged from 0.05 to 1.07 with the exception of the open well at Bekora (Table 1) classified as marginal due to high values of 1.60.

High Na^+ waters are not suited for irrigation activities because Na^+ ion adsorbs onto cation exchange sites reducing soil permeability (Yidana et al., 2010). This implies that the studied water sources are suitable for irrigational practices due to their low salinities and Na %. Low salinity water (C1 class with EC < 250 µS/cm) (Figure 5a) can be used for most crops on most soils with little likelihood of soil salinity development (Richards, 1954). The relatively low SAR values imply that these groundwater sources would pose no Na^+ effect when used for irrigation (Yidana et al., 2010). Low salinity water (SAR < 6) can also be used for irrigation on most soils with little danger of developing harmful levels of exchangeable sodium (Richards, 1954).

Conclusions

A baseline study of the hydrogeochemical parameters of natural and anthropogenic influences on groundwater quality for drinking and agricultural uses in the Rio del Rey Basin within the remote Bakassi Peninsular has been generated for the first time. Sixty percent (60%) of the groundwater samples were slightly acidic (5.5-6.4) and out of the WHO (2004) standard for drinking water.

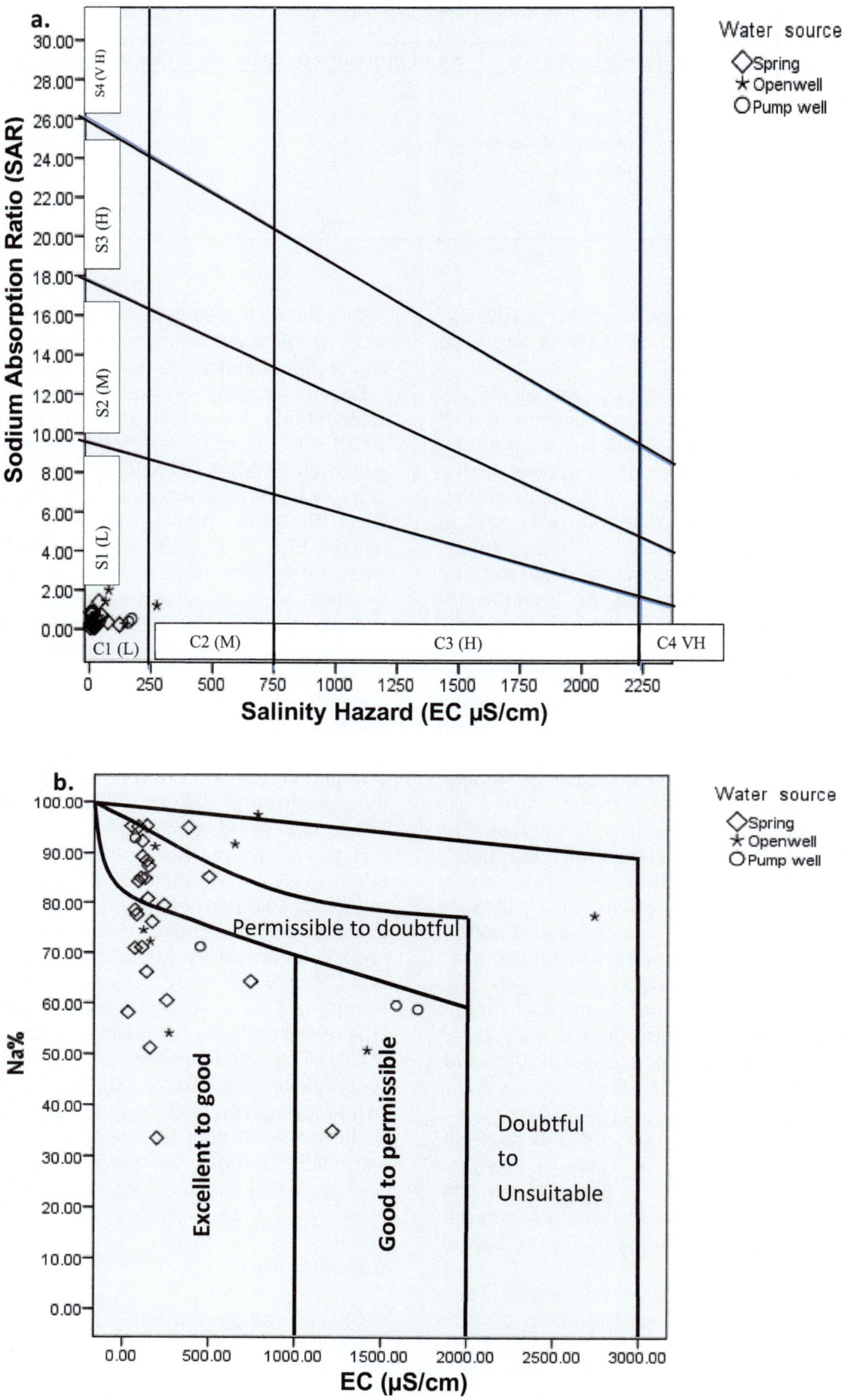

Figure 5. Classification of groundwater for irrigation. a. United States Salinity Laboratory (USSL) diagram after Richard (1954) and b. Wilcox diagram after Wilcox (1955) showing irrigational suitability in the study area.

The major ions of the groundwater sources were within the WHO (2004) guideline values for drinking water except for high concentrations of NO_3^- (77.28 mg/l) and Br^- (0.08) ions in a water source in the south east (Bekora) of the study area which increases the risk of the population here to diseases such as methemoglobinemia, gastric ulcer, cancer and urinary tract diseases. Sodium (Na^+) was the dominant cation, while HCO_3 was the dominant anion for all the groundwater sources. The relative abundance of cations and anions in the groundwater sources were as follows: Na^+>K^+>Mg^{2+}>Ca^{2+} and HCO_3^->NO_3^->Cl^->SO_4^{2-}, respectively.

The groundwater sources were fresh water with relatively low mineralisation of 7.6-248.mg/l TDS. The main sources of ions were from mineral dissolution of gneisses, basalts, sediments and clay units which constitute the geology of the area.

The water types in the area were: Na-HCO_3 (45%), Ca-HCO_3 (35%), Na-Ca-HCO_3 (15%), Ca-Mg-Cl (2%) and Na-Cl (2%). The main hydrogeochemical processes that influenced the chemical composition of the water sources were incongruent silicate dissolution and cation exchange. Mineral saturation indices suggest that the dissolution of carbonates, sulphates, halites and fluorides were insignificant in the groundwater.

The slightly acidic groundwater sources might lead to the dissolution of heavy metals in rocks in the study area. The majority of the groundwater sources were soft waters. Water quality index calculated using the adjusted form of CCME (2001) based on the major and minor ions (F^-, Br^-) suggested that 90, 6, 2 and 2% provided excellent, good, marginal and poor water quality respectively for drinking purposes.

However, based on the total coliform counts, 97% of these water sources were polluted and classified as polluted water sources due to poor hygienic practises of the population which predisposes these water sources to bacterial contamination thereby exposing the population to water-borne diseases such as typhoid, cholera, dysentery, diarrhoea and scabies. Based on EC, Na %, SAR, USSL classification and RSC, 90% of the ground-water sources were considered suitable for agricultural purposes.

Hydrochemical evaluation of potable water sources during the rainy season is necessary to check NO_3^- contamination which may arise from domestic waste discharges and agrochemicals into shallow groundwater. Due to the microbial contamination of the water sources it will be necessary to treat the waters before consumption.

ACKNOWLEDGEMENTS

We would like to acknowledge the contributions of the staff of the Institute of Agronomic Research for Development (IRAD) Ekona Water and Soils Laboratory for the analysis of bicarbonate, the Microbiology unit of the University of Buea for microbial analysis and members of the Takeshi Ohba Laboratory, Tokai University, Japan for the major ions analysis. The authors are grateful to the anonymous reviewers for their comments that improved the manuscript.

REFERENCES

Abderamane H, Razack M, Vassolo S (2013). Hydrogeochemical and isotopic characterization of the groundwater in the Chari-Baguirmi depression, Republic of Chad. Environ. Earth Sci. 69:2337-2350.

Agyingi CM, Foba JF, Epanty AF, Zisuh FA, Ongbwa AZ (2006). Carbonate resources of Cameroon and potential applications. Int. J. Natl. Appl. Sci. 2(2):57-66.

Ako AA, Shimada J, Hosono T, Ichiyanagi K, Nkeng GE, Fantong WY, Eyong GET, Roger NN (2011). Evaluation of groundwater quality and its suitability for dinking, domestic, and agricultural uses in the Banana Plain (Mbanga, Njombe, Penja) of the Cameroon Volcanic Line. Environ.Geochem. Health 33:559-575.

Al-mikhlafi AS, Das BK, Kaur P (2003).Water chemistry of Mansar Lake (India): an indication of source area weathering and seasonal variability. Environ. Geol. 44(6):645-653.

APHA (1992). Standard methods for the examination of water and waste water, 18thedn. Washington, D.C.: American Public Health Association.

Apha/Awwa (1998). Standard methods for the examination of water and wastewater. 20th edition, American Public Health Association, American Water Works Association and Water Environment Federation Publication, Washington D.C, USA.

Appelo CAJ, Postma D (1993). Geochemistry, groundwater, and pollution.Balkema, Netherlands. p. 536.

Atteia O (2005). Chemistry and pollution of underground waters. Tec and Do Ed., Lavoisier. p. 398.

Banton O, Bangoy LM (1997). Hydrogeologie. Multiscience environnnementale des eaux souterraines. Presse de l'univ. Du Quebec, AUPELF. ff. Quebec, Canada. p. 460.

Canter LW, Knox RC (1985). Septic tank system effects on groundwater quality. Lewis Publishers, Chelsea, MI, USA.

CCME (2001). Canadian water quality guidelines for the protection of aquatic life: Canadian Water Quality Index 1.0. Technical Report. Canadian Environmental Quality Guidelines, 1999. Winnipeg.

Chapman DV (1996). Water quality assessments: a guide to the use of biota, sediments and water in environmental monitoring 2nd edition Taylor and Francis. p. 625.

Chebotarev II (1955). Metamorphism of natural waters in the crust of weathering-1 Geochim. Cosmochim. Acta 8:22-48.

Cheesbrough M (1992). Water and Sanitation Decade Bacteriological Testing of Water Supplies. In Laboratory manual for tropical countries. Butterworth Heinemann Ltd. Jordan Hill Oxford 2: 206-244.

Chenini I, Khemiri S (2009). Evaluation of groundwater quality using multiple linear regression and structural equation modeling. Int. J. Environ. Sci. Technol. 6(3):509 -519.

COMA (1994). Nutritional aspects of cardiovascular disease. Committee on Medical Aspect of Food Policy No. 46 (186 pp). London: HMSO.

Das BS, Wrath JM, Inskeep WP (1999). Nitrate concentrations in the root zone estimated using time domain reflectometry. Soil Sci. Soc. Amer. J. 63:1561-1570.

De Neve S, Van de Steene J, Hartmann R, Hofman G (2000). Using

Dissanayake CB, Senaratne A, Weerassoriya VR (1992). Geochemistry of well water and cardiovascular diseases in Sri Lanka. Int. J. Environ. Stud. 19:195-203.

Domenico PA, Schwartz FW (1990). Physical and chemical Hydrogeology. John Wiley and Sons. 411:485-497.

Drever JI (1997). The geochemistry of natural waters. Surface and time domain reflectometry for monitoring mineralization of nitrogen from soil organic matter. Euro. J. Soil Sci. 51:295-304.

groundwater environments. New Jersey: Prentice Hall. p. 436.
Driscoll (2009). Water hardness based on concentration of calcium and magnesium. In: Poells, D.J., Smith, G.J. (Eds.). Encyclopedic Dictionary of Hydrogeology. Academic press, 30 Corporate drive, suite 400, Burlington, MA 01803, USA.
Dumort JC (1968). Carte Géologique de reconnaissance avec notice explicative feuile Douala-Ouest. Bureau des Recherché Géologique et minière, Yaoundé, 131.
Edet AE (2008). The hydro geochemical characteristics of coastal aquifers in the West Coast of Africa: A review. Trans boundary Aquifer Conference, Tripoli, Libya.
Edet AE (2009). Characterization of groundwater contamination using factor analysis in the Niger Delta (Nigeria). J. Hydrol. Eng. 14(11):1255-1261.
Edet A, Nganje TN, Ukpong AJ, Ekwere AS (2011). Groundwater chemistry and quality of Nigeria: A status review. Afr. J. Environ. Sci. and Tech. 5(13):1152-1169.
Edmunds WM, Smedley PL (1996). Groundwater geochemistry and health: An overview. In: Appleton JD, Fuge R, McCall GJH (eds.) Environmental geochemistry and health with special reference to developing countries. Geological Society Special Publication No.113, pp. 91-105.
Ejechi BO, Olobanyi SB, Ogban FE, Ugbe EC (2007). Physical and sanitary quality of hand-dug well water from oil producing area of Nigeria. Environ Monit. Assess. 128:495-501.
Elango L, Kannan R, Senthil, Kumar M (2003). Major ion chemistry and identification of hydrogeochemical processes of groundwater in a part of Kancheepuram district, Tamil Nadu, India. J. Environ. Geo. 10(4):157-166.
Ellis PA, Rivett MO (2007). Assessing the impact of VOC-contaminated groundwater on surface water at the city scale. J. Contam. Hydrol. 91:107-127.
Endeley RE, Ayonghe SN, Tchuenteu F (2001). A preliminary hydrogeochemical Baseline study of water sources around mount Cameroon. J. Cam. Acad. Sci. 1(3):161-167.
Eneke GT, Ayonghe SN, Chandrasekharam D, Ntchancho R, Ako AA, Mouncherou OF, Thambidurai P (2011). Controls on groundwater chemistry in a highly urbanised coastal area. Int. J. Environ. Res. 5(2):475-490.
Etia PM (1980). Climate. In: Atlas of the United republic of Cameroon. Groupe J.A. -51, Avenue des Ternes - 75017 Paris, pp 1-72 or to be specific. pp. 16-19.
Fantong WY, Satake H, Ayonghe SN, Aka FT, Asai K (2009). Hydrogeochemical controls and usability of groundwater in semi-arid Mayo Tsanaga River Basin, Far-North Cameroon. Environ. Geol. 58 (6):1281-1293.
Fantong WY, Takounjou AF, Fantong EB, Banseka HS, Gwanfogbe CD, Ayonghe SN, Tanyileke GZ (2013). Increased risk of fluorosis and methemoglobinemia diseases from climate change: evidence from groundwater quality in Mayo Tsanaga River Basin, Cameroon. J. Cam. Acad. Sci. 11(1):55-60.
Feng Q, Wei L, Su YH, Zhang YW, Si JH (2005). Distribution and evolution of water chemistry in Heihe River Basin. Environ. Geol. 45, 947-956.
Freeze RA, Cherry JA (1979). Groundwater. 2nd edn. Prentice Hall, Eaglewood, Cliff, New Jersey, USA. p. 604.
Gabche CE, Smith VS (2002).Water, salt and nutrients budgets of two estuaries in the coastal zone of Cameroon.W. Afr. J. Appl. Ecol. 3:69-89.
Ganyaglo, SY, Osae, S, Dampare, SB, Fianko, JR, Bhuiyam, MAH, Gibrilla, A, Bam, E, Ahialey, E, Osei, J (2012). Preliminary groundwater quality assessment in the central region of Ghana. Environ. Earth Sci. 66:573-587.
Gartland JS (1986). The biological importance of the Korup forest in Gartland, J.S.&H MACLEOD (1986) proceedings of the workshop on Korup National park, Mundemba, Ndian Division, Province, Republic of Cameroon.WWF/IUCN project 3206.
Gavaud M, Muller JP (1980). Soils. In: Laclavere, G., Loung, J.-F. (eds). Atlas of the United Republic of Cameroon. Editions JeuneAfrique 51, avenues des Terres 75017 Paris. ISBN: 2-85258-125-126.
Del-Rey Basin, South-West Cameroon. Afr. Geo. Review 01&02: 51-63.
Gibbs RJ (1970). Mechanisms controlling world water chemistry. Sc. 17: 1088-1090.
Gnazou MDT, Bawa LM, Banton O, Djanéyé-boundjou G (2011). Hydrogeochemical characterization of the coastal Paleocene aquifer of Togo (West Africa).Int. J. Water Res. Environ. Eng. 3(1): 10-29.
Hamer DH, Simon J, Thea D, Keusch GT (1998). Childhood diarrhea in Sub-Saharan Africa. Child Health Research Project Report 2 (1).
Jalali M (2005). Major ion chemistry of groundwaters in the Bahar area, Hamadan, western Iran. Environ. Geol. 47:763-772.
Jankowski J, Acworth RI, Shekarforoush S (1998). Reverse ion-exchange in deeply weathered porphyritic dacite fractured aquifers system, Yass, New South Wales, Austria, In: G.B., Arehart, J.R., Hulston (Ed). Proceedings of 9 "International symposium on water rock interaction Taupo, New Zealand Rotterdam: Balkema. pp. 243-246.
Katte VY, Fonteh MF, Guemuh GN (2003). Domestic water quality in urban center in Cameroon: a case study of Dschang in the West Province. Afr. Water. J. pp. 43-54.
Kuitcha D, Kamgang KBV, Sigha NL, Lienou G, Ekodeck GE (2008). Water supply, sanitation and health risks in Yaoundé, Cameroon. Afr. J. Environ. Sci. Technol. 2(11):379-386.
Kuitcha D, Ndjama J, Tita AM, Lienou G, Kamgang KBV, Ateba BH, Ekodeck, GE (2010). Bacterial contamination of water points of the upper Mfoundi watershed, Yaounde, Cameroon. Afr. J. Micro. Res. 4 (7):568-574.
Kumar M, Kumari K, Ramanathan A, Saxena R (2007). A Comparative evaluation of groundwater suitability for irrigation and drinking purpose in two intensively cultivated districts of Punjab, India. J. Environ. Geol. 53:553-574.
Lawrence AR, Goddy DC, Kanatharana P, Meesilp W, Ramnarong V (2000). Groundwater evolution beneath Hat Yai, a rapidly developing city in Thailand. Hydrol. J. 8:564-575.
Lawson DP (1993).The Reptiles and Amphibians of the Korup National Park Project, Cameroon.
Lerner DN, Barrett MH (1996). Urban groundwater issues in the United Kingdom. Hydrol. J. 4:80-89.
Ma J, He J, Qi S, Zhu G, Zhao W, Edmunds WM, Zhao Y (2013). Groundwater recharge and evolution in the Dunhuang Basin, northwestern China. Appl. Geochem. 28:19-31.
Meybeck M (1987). Global chemical weathering of surficial rocks estimated from river dissolved loads. Am. J. Sci. 287:401-428.
Mitra BK, ASABE Member (1998). Spatial and temporal variation of groundwater quality in sand dune area of Aomori Prefecture in Japan.
Mpakam HG (2009). Vulnérabilité à la pollution des ressources en eaux à Bafoussam et incidences socio-économiques et sanitaires : modalités d'assainissement. Th. Doct. PhD, Univ. Yaoundé I pp. 267.
Ndian Meteorology (2012).
Nduka JK, Orisakwe OE (2011). Water-quality issues in the Niger Delta of Nigeria: a look at heavy metal levels and some physicochemical properties. Environ. Sci. Pollut. Res.18:237-246.
Njoh OA, Petters SW (2008). Upper cretaceous foraminifera of the Rio-
Nola M (1996). Peuplement bactérien des eaux souterraines de Yaoundé (Cameroun) bio indicateurs et pathogènes opportunistes. Doctorate de 3e cycle, Yaoundé Univ. Cameroon. p.175.
Nwankwoala HO, Udom GJ (2011). Hydrochemical facies and ionic ratios of groundwater in Port Harcourt, Southern Nigeria. Res. J. Chem. Sci. 1(3):87-101.
Obenesaw OO, Njung T, Fongwei G (1997). Geographical reconnaissance for and undergroundwater source at Pamol Estate Lobe, EkondoTiti, Cameroon.
Odoh BI, Utom AU, Egboka BCE, Okeke HC (2012). Geoelectric sounding for predicting shallow aquifer properties using modified Archie equations. Paper presented at the SEG-AGU hydrogeophysics Workshop, Boise State University, Idaho, July 8-11.
Olufemi AG, Utieyin OO, Adebayo OM (2010). Assessment of Groundwater Quality and Saline Intrusions in Coastal Aquifers of Lagos Metropolis, Nigeria. J. Water Res. Prot. 2:849-853.
Palmer CE, Cherry JA (1984). Geochemical evolution of groundwater in

sequences of sedimentary rocks. J. Hydrol. 75 :27-65.
Piper AM (1944). A graphic procedure in the geochemical interpretation of water analyses. Am. Geophys. Union Trans. 25: 914-923.
Regnoult JM (1986). Synthesegeologique du Cameroon. SODEXIC, Yaounde.
Richard LA (1954). Diagnosis and improvement of saline and alkaline soils.US Department of Agriculture handbook. p. 60.
Roche MA (1993). Water Quality Issue in the Humid Tropics (chapter 15) In: Hydrology and Water Management in the Humid Tropics: hydrological Research Issues and Strategies for Water Management; Edited by Bonell M, Husfschmidt MM, Gladwell JS; UNESCO; p.590.
Rodier J, Bazin C, Broutin J-P, Chambon P, Champsaur H, Rodi L (2005). Water analysis: natural waters, residual waters, sea waters. 8th edition Dunod, Paris, p.1383.
Rogers RJ (1989).Geochemical comparison of groundwater in areas of New England, New York, and Pennsylvania. Groundwater 27 (5):690-712.
Schoeller H (1965). Hydrodynamique dans le karst (Hydrodynamics of karst). In: Actes du Colloques de Doubronik. IAHS/UNESCO, Wallingford. pp. 3-20.
Schoeller H (1967). Geochemistry of groundwater. In: Brown, R.H. (Ed.), Groundwater Studies - An International Guide for Research and Practice. UNESCO, Paris. pp. 1-18.
Singh KP, Malik A, Singh VK, Mohana D, Sinha S (2005) Chemometric analysis of groundwater quality data of alluvial aquifer of Gangetic plain. North India Analytica Chimica Acta 550:82-91.
Takem EG, Chandrasekharam D, Ayonghe SN, Thambidurai P (2010). Pollution characteristics of alluvial groundwater from springs and bore wells in semi-urban informal settlement of Douala, Cameroon, western Africa. Environ. Earth Sci. 61:287-298.
Tanawa E, Djeuda THB, Ngnikam E, Temgoua E, Siakeu J (2002). Habitat and suburban areas in African cities. Build. Environ. 37:269-275.
Tay CK (2012). Hydrochemistry of groundwater in the Savelugu-Nanton District, Northern Ghana. Environ. Earth Sci. 67:2077-2087.
Utom AU, Odoh BI, Egboka BCE (2013). Assessment of hydrogecochemical characteristics of groundwater quality in the vicinity of Okpara coal and Obwetti fireclay mines, near Enugu town, Nigeria. Appl. Water Sci. 3:271-283.
Vairavamoorthy K, Gorantiwar SD, Pathirana A (2008). Managing urban water supplies in developing countries - climate change and water scarcity scenarios. Phys. Chem. Earth, Parts A/B/C 33:330-339.
Welz B, Becker-ross H, Florek S, Heitmann U (2006). Front Matter, in High-Resolution Continuum Source AAS: The better way to do Atomic Absorption Spectrometry, Wiley-VCH Verlag GmbH & Co. KGaA, Weinheim, FRG. doi: 10.1002/3527606513.fmatter.
WHO (1989) Health guidelines for the use of wastewater in agriculture and aquaculture. Report of WHO scientific group-Technical report series 778, WHO, Geneva, 74 pp.
WHO (2004). Guidelines for drinking-water quality. 3rd ed., Geneva, Switzerland. p. 514.
WHO Regional Office for Africa (2012).Child & Adolescent Health. http://www.afro.who.int/en/clustersa-programmes/frh/child-and-adolescenthealth/programme-components/child-health.html (Assessed on April 2013).
Wilcox LV (1955). Classification and use of irrigation waters.U.S. Department of Agriculture Circular 969, Washington, D.C., USA.
William AB, Benson NU (2010). Interseasonal hydrological characteristics and variabilities in surface water of tropical estuarine ecosystems within Niger Delta, Nigeria. Environ. Monit. Assess. 165: 399-406.
Wirmvem MJ, Fantong WY, Wotany ER, Ohba T, Ayonghe SN (2013). Sources of bacteriological contamination of shallow groundwater and health effects in Ndop plain, Northwest Cameroon. J. Environ. Sci. Water Res. 2(4):127-132.
Wirmvem MJ, Ohba T, Fantong WY, Ayonghe SN, Suila JY, Asaah ANE, Tanyileke G, Hell JV (2013). Hydrochemistry of shallow groundwater and surface water in the Ndop plain, North West Cameroon. Afr. J. Environ. Sci. Technol. 7(6):518-530.
Xu Y, Usher B (2006). Groundwater Pollution in Africa. Taylor and Francis/Balkema, 2300 AK Leiden, The Netherlands. p. 349
Yidana SM (2010). Groundwater classification using multivariate statistical methods: Southern Ghana. J. Afr. Earth Sci. 57:455-469.
Younger PL (1994). Low-cost Groundwater Quality Investigation methods an example from Bolivia Altiplano.in NASH, H. and McCALL, G.J.H. (editors), Groundwater quality, Chapman and Hall, ISBN 0 412 58620 7, Chapter 5:55-65.
Zimmermann L (2000). A comparative study of growth and mortality of tress in ceasalp dominated low town African rainforest at Korup, Cameroon.

Elimination of Cu (II) and Zn (II) ions in mono-element and the bi-element aqueous solutions by adsorption on natural clay of Bikougou (Gabon)

Ndong Nlo Joseph[1], Eba Francois[1], Njopwouo Daniel[2], Beh Sima Didier[1] and Ondo Jean Aubin[1]

[1]Laboratoire Pluridisciplinaire des Sciences de l'Ecole Normale Supérieure de Libreville BP: 17009 Libreville, Gabon.
[2]Laboratoire de Physico-chimie des Matériaux Minéraux, Département de Chimie Inorganique Université de Yaoundé I BP: 812 Yaoundé, Cameroun.

This work was undertaken to evaluate the potential of the clay of Bikougou to remove copper (II) and zinc (II) ions (Cu^{2+} and Zn^{2+}) in mono-element and binary aqueous solutions. The experiments were carried out by employing the batch technique through the variations of pH, clay mass, ionic strength, concentration in metal, temperature and contact time. The quantity of ions coppers (II) or zinc (II) adsorbed increases when the pH increases up to pH=5 and beyond, there is precipitation of the metal hydroxides; the quantity of metal ions adsorbed decreases when the clay mass or the ionic force increases. The average adsorption energies of copper (II) and zinc (II) ions are indicative of endothermic reactions. The modeling of the experimental results is better achieved by application of Freundlich adsorption isotherm and Langmuir adsorption isotherm concerning the adsorption of Cu (II) ions with correlation coefficients of 0.98 and 0.93, respectively. The Langmuir adsorption isotherm and Freundlich adsorption isotherm are all applicable for describing the experimental result of isotherm adsorption of Zn (II) ions with correlation coefficient respectively, 0.99 and 0.96. The kinetic which shows more clearly the observed phenomenon is that of the pseudo-second order with the coefficients of correlation close to 0.99. In the binary system copper (II) - zinc (II) the adsorption of copper (II) is favoured compared to that of zinc (II). All these results have shown efficiency as adsorbent of the clay from Bikougou to eliminate $Cu^{2+}$and zinc (II) ions in mono-element or bi-element aqueous solutions. In the mono-element solution, the adsorption rate varied from 98.59 to 94.41% for the Cu (II) ions and from 86.76 to 53.80% for Zn (II) ions.

Key words: System of bi-element, copper and zinc, adsorption, technique in batch, natural clay mixture.

INTRODUCTION

Some heavy metals ions such as copper (II) and zinc (II) ions (Cu^{2+} and Zn^{2+}) released to the environment from industrial effluents and anthropogenic activities cause pollution of soils and water. Copper is an essential trace element required for proper health in an appropriate limit. Its high uptake may be harmful for human health and causes a number of symptoms for example growth retardation, skin ailments, gastrointestinal disorders among others. It causes a slow dysfunction of the liver and damage to the pancreas on the one hand, and the appearance of the Wilson disease on the other hand (Moore 1991). Zinc is reported as a coenzyme for over 200 enzymes involved in immunity, new cells growth, acid base regulation among others. Data related to Zn toxicity are very rare (Barone et al., 1998; Gyorffy and Chan, 1992). Zn toxicity symptoms can be gastric disturbance, cardiac arythmia among others (Salgueiro et al., 2000). The accumulation of Zn (II) ions causes a decrease in the cellular production and disturbance of the growth of the plants while that of Cu(II) ions leads to a delay in growth of plants (Moore,1991). The injection and inhalation of ions Zn (II) and Cu (II) have negative consequences on human health, as previously reported. That is why the World Health Organization has recommended that the concentrations of these ions in drinking water do not overpass 0. 5 mg/l. The same recommendation is applicable to industrial effluents containing these heavy metals before they are release to the environment.

Many physico-chemical methods such as: ions exchange, precipitation, phytoextraction, ultrafiltration, reverse osmosis, electroplating and adsorption using the activated carbons or zeolites have been developed. These methods are very costly which limits their generalization. The current research is focused on the development of the cheap adsorbents from the industrial and agricultural wastes compared to that of the active carbon and zeolites found expensive. In the technique of adsorption, considered as materials of low cost, the clays could sometime show an adsorption capacity non-negligible, comparable or even exceeding that of activated carbons in the same conditions of temperature and pH. In fact the porous nature, and the physico-chemical properties of clays: Large mechanical and chemical stability, large specific surface area, acid-base status of Brönsted and Lewis, negative charge on their surface, large cation exchange capacity and their structural variability may be associated to a high adsorption capacity (Ijagbemi et al., 2009).

Several works on heavy metals adsorption from aqueous effluents on the surface of different clay families have been reported: for example Kaolinites (Babel and Kurniawan, 2003), Bentonites (Babel and Kurniawan, 2003; Naseem and Tahir, 2001), Montmorillonites, Smectites, Palygoskites (Babel and Kurniawan, 2003; Gupta and Bhattacharya, 2005, 2008) exist in the literature. The essential results show that the adsorption capacity depends on the nature of the adsorbent and the characteristics of the solution. They are often better described by application of the isotherms of Langmuir (Langmuir, 1918) and/ or Freundlich (Gupta and Bhattacharyya, 2005; Freundlich, 1906) and their kinetic is either of the pseudo-first order (Eba et al, 2010), or that of the pseudo second order (Tien and Huang, 1991; Ho and McKay, 1999a,b; Ho et al., 2001).

The waters of discharges sometimes contain more than a heavy metal (Doula, 2009; Mohan and Singh, 2002); but, few studies are interested in this type of solution. In effect, there is in literature, many references on the adsorption of metals in the mono-element systems as compared to systems with several heavy metals. The main difficulty is the acquisition of adsorption models which describe exactly the experimental data of adsorption related to multi-body dissolved solution. Thus, Huh et al. (2000) and Kwong et al. (1998) proposed a model that allows to evaluate the adsorption capacity of a cation in a multi-element solution by application of the Langmuir extended model.

In a multi-element solution, when the report of the quantity adsorbed of an ion (i) qe (mix)i in the mixture is greater than or less than the quantity adsorbed of the same ion (i) in the mono- element solution qe (mono)i, the two ions compete to attach on the available adsorptions sites on the surface of the adsorbent. The comparison of the report of each ion in multi-element and mono-element solutions allows deducting the species respectively benefiting of preferential and/or decreasing adsorption.

$$\frac{q_e(\text{mix})_i}{q_e(\text{mono})_i} > 1 \text{ and/or } \frac{q_e(\text{mix})_i}{q_e(\text{mono})_i}$$

The present study was undertaken to assess the capacity of the clay from Bikougou (Gabon 11° 37' East and 1 ° 58' North) to attach the Cu (II) and Zn (II) ions in mono and bi-element solutions using as parameters pH of the solutions, mass of adsorbent, ionic strength, initial concentration of the Cu (II) ions and Zn (II) ions, interaction time clay-ions, temperature of the process and initial proportions of ions in the binary solutions.

MATERIALS AND METHODS

Properties of the clay

The natural clay from Bikougou deposit used as adsorbent has been previously characterized. By combining the X-ray powder diffractogram results (Figures 1 and 2) and the whole-rock chemical analyses (Table 1), the modal compositions have been calculated according a constrained multi-linear calculation (Eba et al., 2011). It indicated that the crude clay of Bikougou deposit is constituted mainly of 29.5% of kaolinite, 20% of albite, 16% of montmorillonite and 8% of illite and its other characteristics are presented in Table 2 (Eba et al, 2010).

Ionic solutions

The solutions of Cu (II) ions and Zn (II) ions were prepared by dissolving the appropriate weights of copper chloride ($CuCl_2$; molar mass 170.48 g, purity 99% Prolabo analytical grade) and zinc chloride ($ZnCl_2$: molar mass 136.2 g, purity 98.5 % Prolabo analytical grade) in deionized water. The pH is adjusted by adding drops of solution of HCl or NaOH. The effect of the variation of the ionic strength of the solutions of Cu^{2+} and Zn^{2+} has been achieved by dissolution of potassium chloride (KCl 74.55 g/mol analytical grade) in deionized water.

Adsorption procedure

The adsorption tests were performed by using batch technique. It

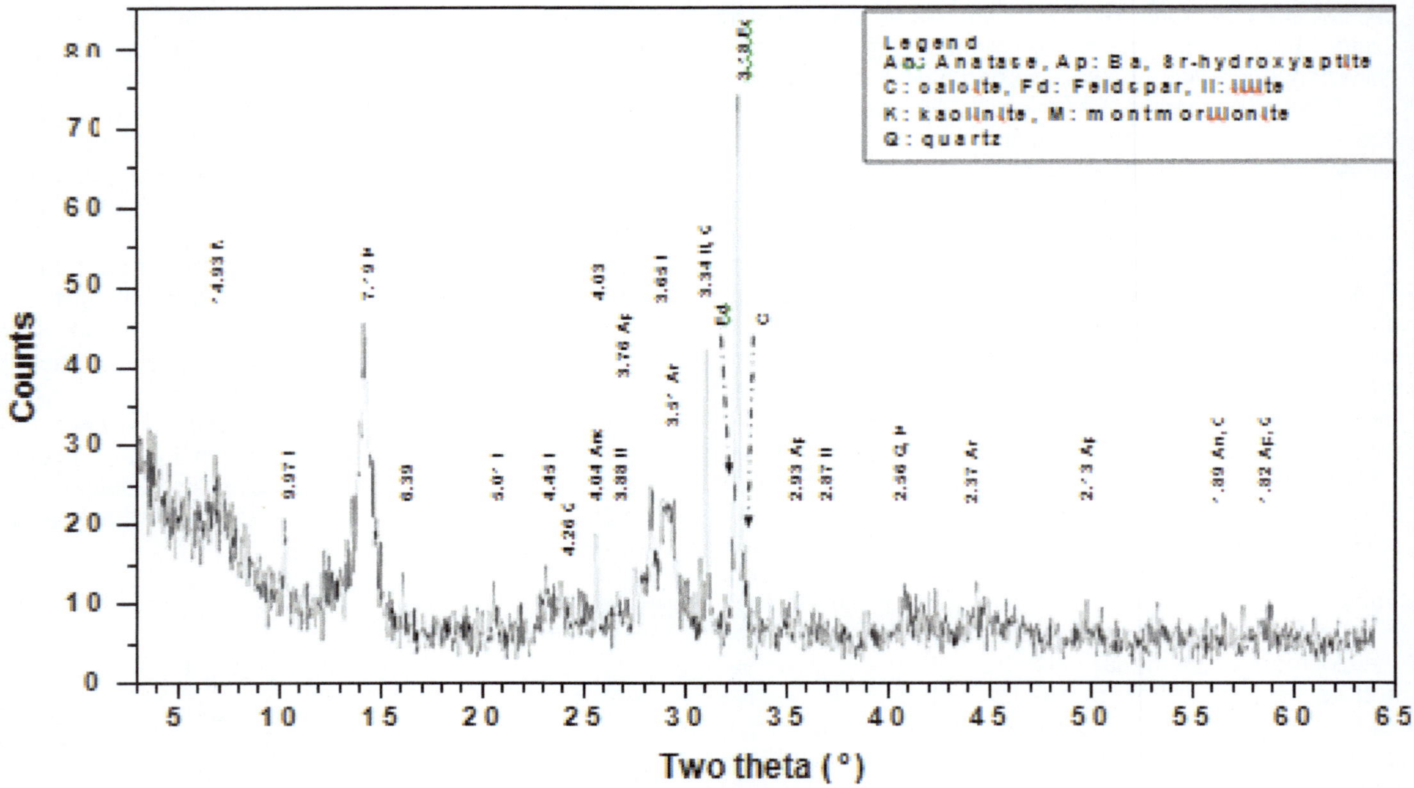

Figure 1. XRD-pattern of crude clay from Bikougou deposit.

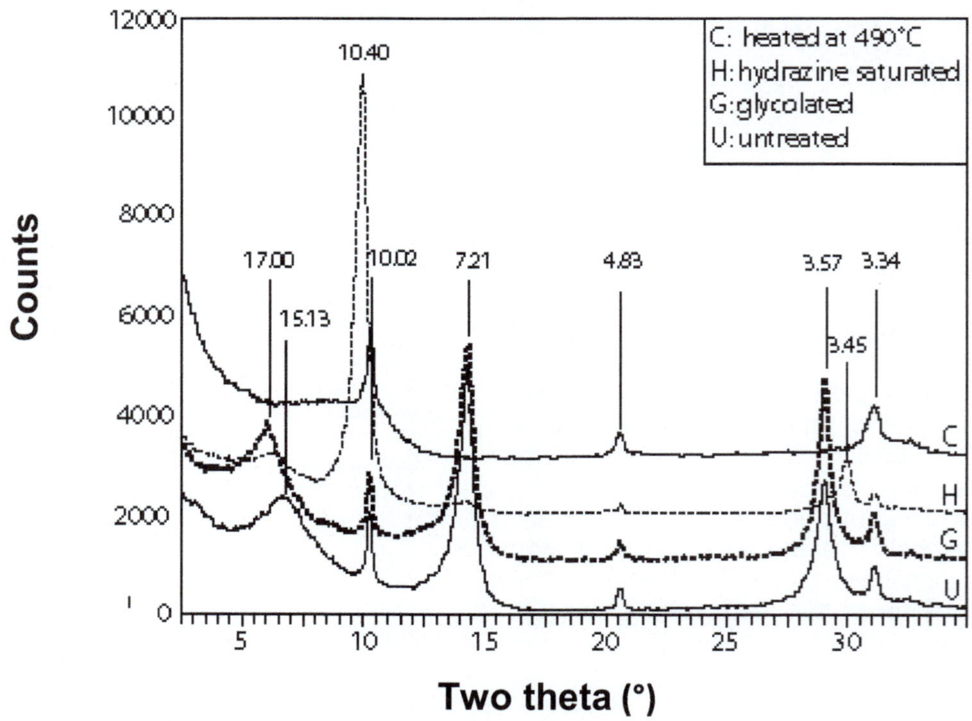

Figure 2. XRD-pattern of clay fraction from Bikougou deposit.

consists of adding an appropriate mass of clay (0.15 g except during the study of the effect of the mass of clay) to a solution (50 ml) containing the ion of the heavy metal (Cu^{2+} and/or Zn^{2+}) in a reactor maintained at constant temperature (except for the study of the influence of temperature) subjected to a constant stirring during a fixed time. After this, the mixture is filtered and the

Table 1. Mineralogical characteristics of clay from Bikougou.

Mineral	Raw clay (%wt)	Clay fraction (% wt)
Kaolinite	29.1	53
Albite	20.3	0.0
Montmorillonite	15.2	40
Illite	10.9	7.0
Quartz	19.1	0.0
Carbonate	3.0	0.0
Maghemite	1.2	0.0
Anatase	0.9	0.0
Ba -Sr hydroxyapite	0.8	0.0
Zircon	Trace	0.0

Table 2. Physio-chemical characteristics of Bikougou clay.

pH_{ZPC}	CEC (meq/100g)	Acidity neutralizations (meq/g)			Surface area (m^2/g)
		NaOH	NaHCO	Na CO	
3	12.87	2.79	0.285	0	170

Table 3. Adsorption models.

Isotherm model	Equation	Linear expression	Plot	Parameters	Reference
Langmuir	$q_e = \frac{b \times q_m \times C_e}{1 + b \times C_e}$	$\frac{C_e}{q_e} = \frac{C_e}{q_m} + \frac{1}{bq_m}$ (3)	$\frac{C_e}{q_e}$ vs Ce	b and qm	Langmuir (1918)
Freundlich	$q_e = K_F \times C_e^{1/n}$	$\ln q_e = \frac{1}{n}\ln C_e + \ln K_F$ (4)	$\ln q_e$ vs $\ln C_e$	K_F and n	Gupta and Bhattacharyya (2005)

residual concentration in Cu^{2+} and Zn^{2+} in the liquor is measured by ICP-AES.

Calculation

The adsorption capacity: q_e (mg/g) of the metal ion adsorbed is given by the equation:

$$q_e = (C_0 - C_e) \times \frac{V}{m} \quad (1)$$

The adsorption rate is given by the equation:

$$\tau\% = \frac{C_0 - C_e}{C_0} \times 100 \quad (2)$$

Where, C_0 and C_e are respectively the initial and the equilibrium concentrations cf the metal ions.

Theoretical basis

Adsorption models

The adsorption models used in this study are presented in Table 3. Where: q_m and b are the Langmuir constants representing respectively, the maximum adsorption capacity and the adsorption equilibrium constant. b value serve to definite the level of strength of the bonding between adsorbat-adsorbent. b high value indicate great the relationship of the bonding adsorbat-adsorbent, probably has resulting tothe chemically controlled process. With b weak value correspond to the low relationship of the bonding adsorbat-adsorbent, certainly as resulting to a physical process (Eba et al., 2011b). K_F and n are the Freundlich constants related respectively, to the maximum adsorption capacity and adsorption intensity. The isotherm adsorption parameters are determined from slopes and intercepts of linear plots related to each model. The Langmuir dimensionless factor (R_L) given by the equation:

Table 4. Kinetic models and adsorption process mechanisms equations.

Kinetic equation	Kinetic equation	Parameter	Reference
Pseudo-first order	$\ln(q_e - q_t) = \ln q_e - k_1 t$ (9)	k_1 and qe	Achak et al. (2009)
Pseudo-second order	$\frac{t}{t} = \frac{1}{2 \;\; e} + \frac{t}{e} \; q \;\; k$ (10)	k_2 and qe	Tien and Huang, (1991); Ho et al. (2001)
Intra particular diffusion	$q_t = k_W \times \sqrt{\ } + C^{ste}$ (11)	k_W et C^{ste}	Weber et al. (1963)
Liquid film diffusion	$LnC_e = -k_x \times t + C^{ste}$ (12)	k_x et C^{ste}	Wan Ngah and Hanafiah (2004)

$$R_L = \frac{1}{(bC_o + 1)} \quad (5)$$

The Freundlich parameter n serve to appreciate the nature of the adsorption. For $0 < R_L < 1$, the adsorption is favorable, $R_L > 1$ represents the adsorption unfavorable, $R_L = 1$ indicates a linear adsorption and $R_L = 0$ corresponds to an irreversible process (Juang et al., 1997; Kadirvelu and Namasivayam, 2003). The n values higher than unity (n>1) indicate a favorable adsorption process, while the n values weaker than unity (n<1) correspond to unfavorable adsorption (Fifi et al., 2010). The adsorption mean energy (E) given by the equation below is calculated to give how the solute is settled physically (1 kJ/mol ≤E ≤ 8 kJ/mol) or chemically (E>8kJ/mol) on the available adsorption sites (Eba et al., 2012).

$$E = \frac{1}{(-2\beta)^{0.5}} \quad (6)$$

In a system containing more than one element, it has been proposed the Langmuir extended isotherm equation to calculate the adsorption capacity of the heavy metal present in the mixture qe (mix) by the expression:

$$q_e(mix) = \frac{q_{m,i} bC_{e,i}}{1 + \sum_{i=1}^{N} bC_{e,i}} \quad (7)$$

Where, $C_{e,i}$ is the concentration of the solution to the equilibrium of a solute i in the mixture, b, and $q_{m,i}$ are the parameters determined from the Langmuir model in case of a mono-element solution. The applicability of the previous equation is tested by standard deviation (D. S.), which is defined by the following equation:

$$D.S. = 100\sqrt{\frac{\sum \left[\left(q_{e,i}^{cal} - q_{e,i}^{exp} \right) / q_{e,i}^{exp} \right]^2}{n-1}}$$

8

Inetics studies

Kinetics models used in this study are listed in Table 4. Where, k_1 (min^{-1}), k_2 (g/mg.min), k_W (mg/g. min$^{-0.5}$)and k_X (mg/g.min^{-1}) are respectively, the rate constants related to the pseudo-first order kinetic and pseudo-second order kinetic models, intra-particular diffusion and liquid film diffusion models. The applicability of a kinetic model is attested by the obtainment of its parameters with high correlation coefficient and of a good agreement between experimental and calculated adsorption capacities.

Thermodynamic studies

The equilibrium distribution constant (K_d) of the heavy metal between the solid and liquid phases is given by the equation:

$$K_d = \frac{q_e}{C_e} \quad (13)$$

The distribution constant (K_d) serves to calculate Gibbs free energy (ΔG) according to the expression:

$$\Delta G° = -RT \, ln \, K_d \quad (14)$$

and the enthalpy and entropy changes of the adsorption process by using the Van't Hoff equation

$$\ln K_d = -\left(\frac{\Delta H°}{R}\right) \times \frac{1}{T} + \frac{\Delta S°}{R} \quad (15)$$

RESULTS AND DISCUSSION

The variations of the adsorption capacity of the clay of Bikougou for the Cu^{2+} and Zn^{2+} in the mono-element and binary systems, were measured as function of pH, mass of adsorbent, ionic strength, time of adsorption equilibration, initial concentration of the solute and temperature.

Mono-element system studies

Effect of pH

The impact of pH of adsorption of Cu^{2+} and Zn^{2+} on the

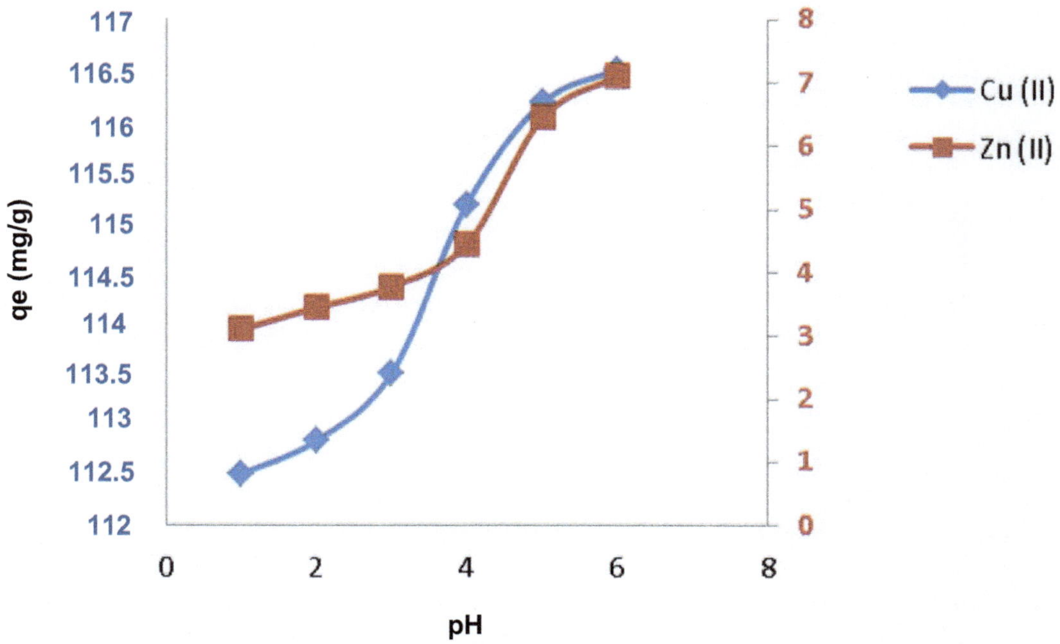

Figure 3. Variations of the amount of copper (II) ions (Co = 377.5 ppm) and zinc (II) ions (Co = 56.25 ppm) adsorbed as a function of pH (temperature= 25°C, mass of clay = 0.15g and contact time = 120min).

surface of the clay from Bikougou has been studied between pH=1 and pH=7, as presented in Figure 3. The quantities of metal ion adsorbed increase nearly linearly from pH=1 to pH=3 for Cu (II) ions and exponentially for pH higher than 3. A hydroxide precipitate appears for pH higher than 6. Similar results are reported in the literature with the adsorption of copper in aqueous solution by bentonite (Youcef and Ashour, 2006). The same behavior is observed for the Zn (II): low linear increase of the adsorption from pH=1 to pH=4 and strong increase beyond pH=4; appearance of a zinc (II) hydroxide to pH greater than 6. Similar results have been obtained by Kaya and Oren (2005) in the case of the adsorption of zinc by bentonite. To explain these results two sides of pH variation should be considered. In the sides of pH ranging from 1 to 3 for the copper (II) and 1 to 4 for the zinc (II), the relative low quantity of metal ions adsorbed is due to the repulsive competition between the metal ions in solution and the hydroniums ions also in the same solution but more mobile and in very large quantity to interact more easily on the adsorption sites (Gupta and Battacharyya, 2008). For the pH higher than 3 (for the copper (II) or 4 (for the zinc (II)), the quantity of hydronium ions is negligible that allows the adsorptions sites to be available for the coppers (II) and zinc (II) ions. When the pH is higher than 6 the precipitation of zinc and copper hydroxides occurs apparently as an increase of the adsorption (Yu et al, 2001; Bayramoglu et al., 2003). Because of the precipitation at pH > 6, adsorption experiments were achieved for pH ≤5 in this study.

Influence of clay dosage on the adsorption of Cu (II) and Zn (II) ions

The quantities of metal ions adsorbed per unit mass of clay as a function of the mass of clay diminish with the increase in clay mass from 0.05 to 0.5 g (Figure 4). Similar results were observed with the adsorption of copper (II) on the montmorillonite (Lin and Juang, 2002) and the adsorption of the zinc (II) on the bentonite (Youcef, 2006). This can be attributed to two reasons: (I) the decrease in the adsorption capacity observed in the two cases is explained by the fact that the concentration of metal ions in the solution being kept constant contrary of clay mass which grows, liberating more available adsorption sites and for that the amount of metal ions adsorbed per clay unit mass decreases (II) a large amount of adsorbent creates particles aggregates and, consequently, a decrease of the total reactive surface area. That contributes to the decrease of the adsorbed amount per mass unit (Shukla et al., 2002).

The effect of ionic strength on the adsorption of Cu ions (II) and Zn (II)

The results related to the study of the influence of the ionic strength on the evolution of the adsorption capacity of copper (II) or zinc (II) ions on the clay from Bikougou are presented respectively on Figure 5. These results show that the adsorption capacity of

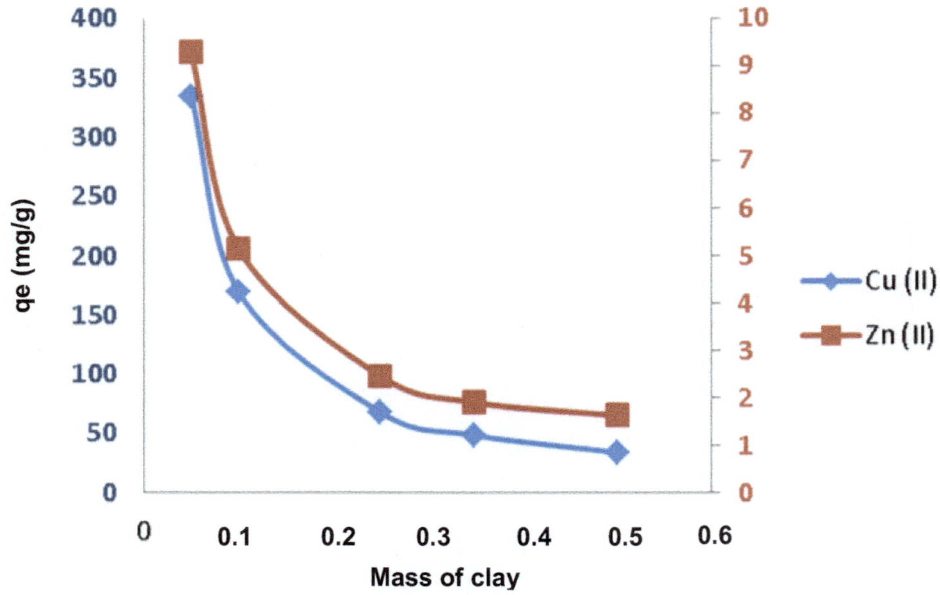

Figure 4. Variations of the adsorbed amounts of copper (II) ions (Co = 377.5 ppm) and zinc (II) ions (Co = 56.25 ppm) as a function of the mass of clay (temperature= 25°C, pH=5 and contact time =120 min).

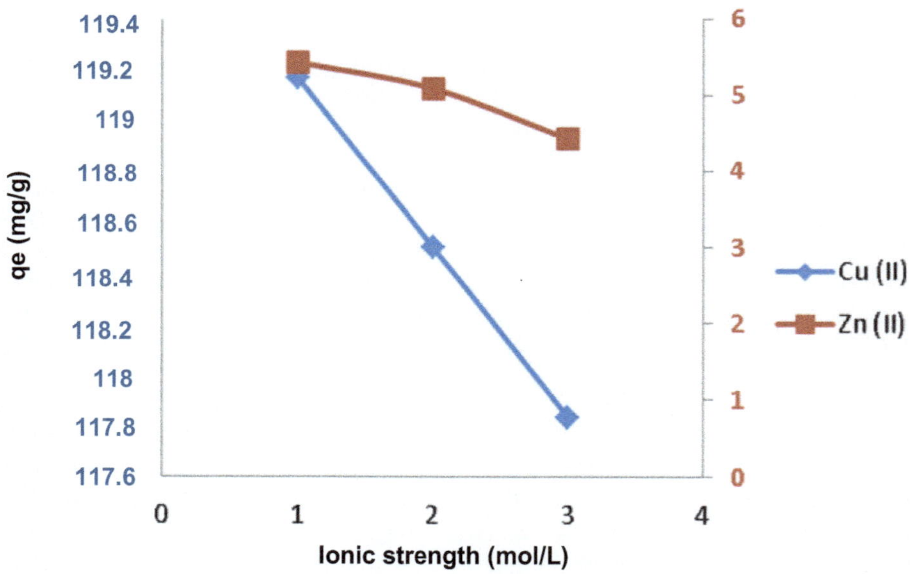

Figure 5. Variations of the amounts of copper (II) ions (Co = 377.5 ppm) and zinc (II) ions (Co =56.25 ppm) adsorbed as a function of the ionic strength (temperature= 25 °C, pH=5, earth clay =0.15 g and contact time =120 min).

both the metal ions decreases when the ionic strength increased from 1 to 3. Similar results were observed with the adsorption of Cu^{2+}on the vermiculite (El-Bayaa et al., 2009) and the adsorption of zinc (II) and Cu^{2+}on earthy sediment (Calvet and Msaky, 1990). These results may be explained by the repulsive competition between potassium ions and copper or zinc ions to attach on the adsorption sites of the clay (Chen and Lin, 2001).

Kinetic studies in the mono-element systems

The results on the evolution of the amounts of copper ions (II) and zinc (II) ions adsorbed per unit mass as a

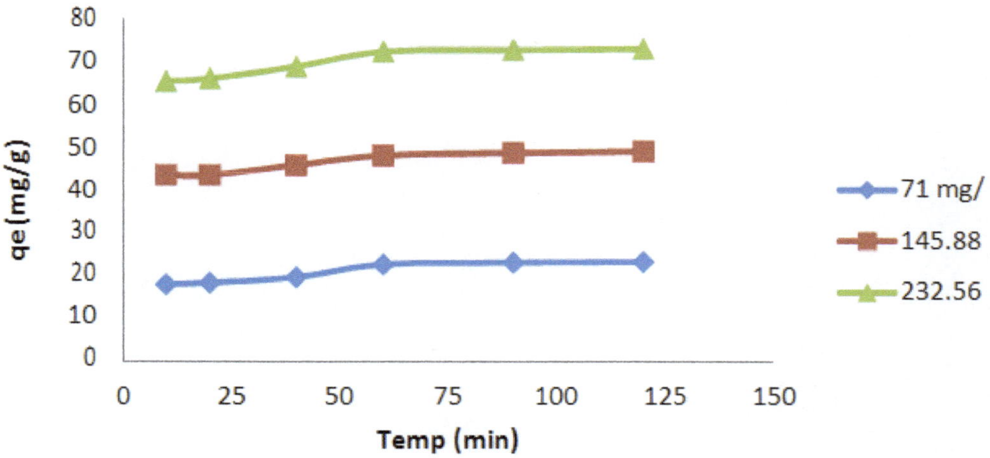

Figure 6a. The amount variations of Cu^{2+} ions adsorbed per unit of mass on the clay of Bikougou as a function of time (temperature= 25°C, pH=5 and mass = 0.15 g).

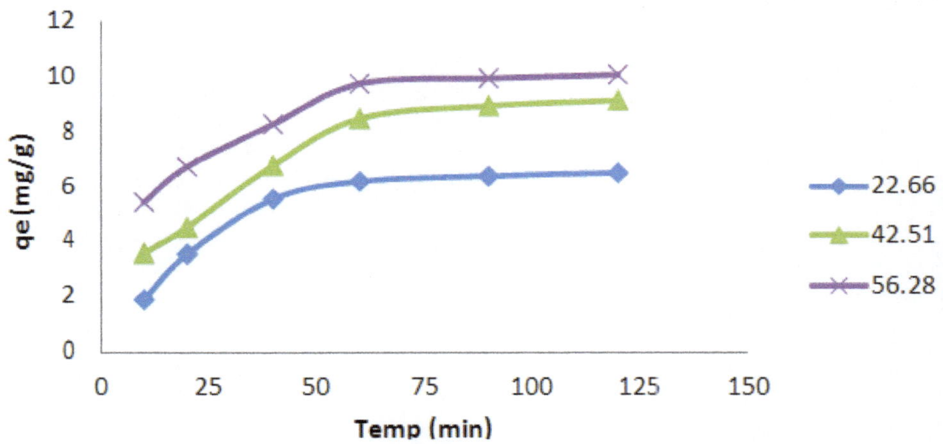

Figure 6b. The amount variations of Zn^{2+} ions adsorbed per unit of mass on the clay of Bikougou as a function of time (temperature= 25 °C, pH=5 and mass =0.15 g).

function of time are shown in the Figures 6a and 6b, respectively. The adsorption capacity increases rapidly as a function of time to achieve a maximum value of the adsorption equilibrium after 1 h of interaction. These results indicate that the kinetics of fixing copper or zinc ions on the clay surface are fast.

Similar results were observed with the adsorption of copper (II) on a clay of Nigeria (Njoku et al., 2011) and also with the adsorption of the zinc (II) on the bentonite (Mellah and Chegrouch, 1998). The kinetic results were modelized by the use of pseudo first model and pseudo-second order kinetics models.

The plots of pseudo-second order kinetic model for the adsorption of Cu (II) ions and Zn (II) ions on the clay surface are represented in Figures 7a and 7b. The kinetic constants related to the applicability of kinetic models to experimental data are reported in the Tables 5a and 5b for Cu (II) ions and Zn (II) ions, respectively.

The adhesion of the modeling tests to experimental results is better for each of the two ions through the use of the pseudo-second order kinetic model rather than of the pseudo-first order kinetic models. The pseudo-second order rate constant for Cu (II)- clay interaction ranged from $6.82*10^{-3}$ to $5.08*10^{-3}$ g/mg.min and the computed adsorptions capacities remained between 24.39 to 74.63 mg/g have been obtained with high correlation coefficients 0.99 when the initial concentration of Cu (II) solutions increase from 71 to 232.56 mg/L. By applying the pseudo-first order kinetic model, the adsorption rate constant ranged between $4.03*10^{-2}$ to $4.52*10^{-2}$ min^{-1} and calculated adsorption capacities remained between 10.49 and 17.3 mg/g have been obtained with correlation coefficients ranged between 0.708 to 0.824. Comparison between experimental

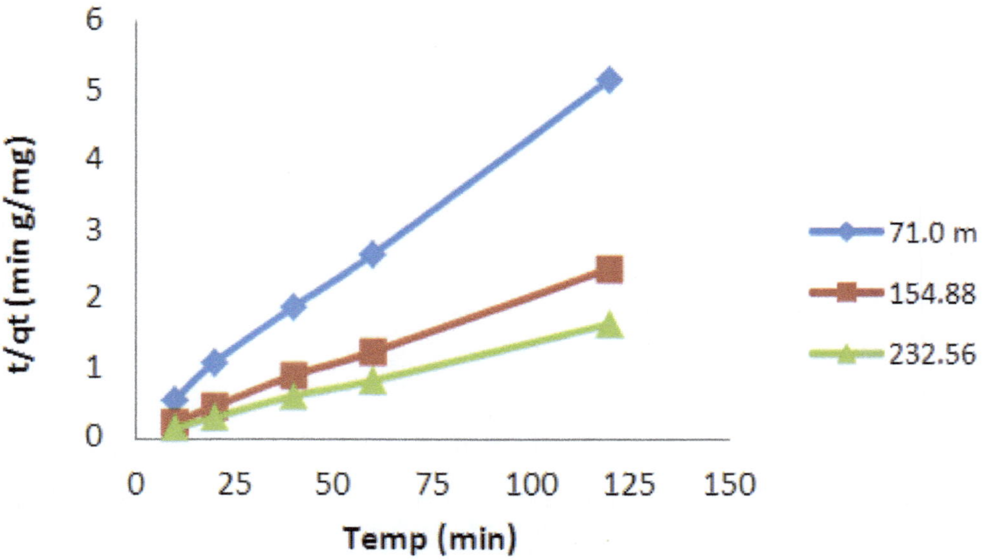

Figure 7a. Pseudo-second order kinetics plots for the adsorption of Cu^{2+} ions onto clay.

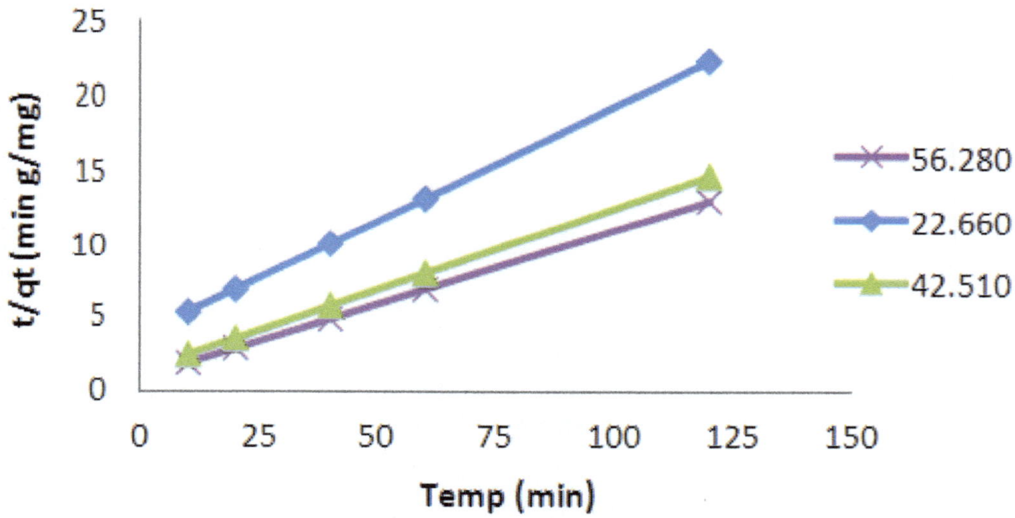

Figure 7b. Pseudo-second order kinetics plots for the adsorption of Zn^{2+} ions onto clay.

adsorption capacities (qe exp) and modelized adsorption capacities (qe th) shows that the qe (th) from pseudo second order model agrees very well with the experimental qe with a deviation varying from 0.96 to 1.44%. This small difference could be attributed to experimental uncertainties. The application of the pseudo-first order kinetic model gave the qe values which yield the experimental results with deviation varying from -12.84 to - 55.89%. In the case of zinc (II) ions, kinetic parameters have been obtained with correlation coefficient which are comparable between the kinetic model of the pseudo-second order (R^2 =0.99) and of the pseudo- first order (R^2 =0.99). In opposite, the adsorption capacities at the equilibrium outcome for the pseudo-first order kinetic model differ significantly from qe experimental values while those from pseudo-second order kinetic model are comparable to experimental data. The similar results were reported with the adsorption of ions Zn (II) by the montmorillonite (Lin and Juang, 2002).

Studies of reaction mechanisms in the mono-element

The adsorption process mechanisms listed in the Table

Table 5a. Constants and correlation coefficients of kinetic models for the adsorption of copper (II) ions on the clay surface.

Co (mg/l)	Pseudo-first order					Pseudo-second order				Intra-particular diffusion mechanism			Liquid film diffusion mechanism		
	Qe exp (mg/g)	K1 (min^{-1})	Qe théo (mg/g)	Δ	R^2	K2 (g/mg.min)	Qe théo (mg/g)	Δ	R^2	Kw ($mg/g.min^{-0.5}$)	Intercept	R^2	-kfd ($mg/g.min^{-1}$)	Intercept	R^2
71	23.33	4.03E-02	10.49	-12.84	0.824	6.82E-03	24.39	1.06	0.997	0.769	15.374	0.8928	0.0276	3.2189	0.937
145.88	49.29	3.30E-02	10.38	-38.91	0.742	6.50E-03	50.25	0.96	0.998	0.824	40.776	0.9154	0.012	3.2885	0.9187
232.56	73.187	4.52E-02	17.3	-55.89	0.708	5.08E-03	74.63	1.44	0.998	1.099	62.103	0.8946	0.0099	3.6177	0.847

Δ= qe théo- qe exp

Table 5b. Constants and correlation coefficients of kinetic models for the adsorption of zinc (II) ions on the clay surface.

Co (mg/l)	Pseudo-first order					Pseudo-second order				Intra-particular diffusion mechanism			Liquid film diffusion mechanism		
	Qe exp (mg/g)	K1 (min^{-1})	Qe théo (mg/g)	Δ	R^2	K2 (g/mg.min)	Qe théo (mg/g)	Δ	R^2	Kw ($mg/g.min^{-0.5}$)	Intercept	R^2	-kfd ($mg/g.min^{-1}$)	Intercept	R^2
22.66	6.55	2.98E-02	4.64	-1.90	0.998	3.09E-03	6.54	-0.0013	0.999	0.9611	-0.9127	0.9718	0.0293	3.0771	0.9849
42.51	9.17	3.89E-02	5.57	-3.59	0.993	4.43E-03	9.17	0.0043	0.999	0.9884	0.6315	0.9586	0.0118	3.5591	0.9817
56.28	10.09	4.86E-02	5.35	-4.73	0.991	5.78E-03	10.09	0	0.999	0.9288	0.7226	0.9982	0.0077	3.7527	0.9952

Δ= qe théo- qe exp

4 were applied to study mechanism of the adsorption of the Cu (II) and Zn (II) ions onto the clay surface. The plots related to liquid film diffusion model (Figures 8a and 8b) and these for intra-granular diffusion model (Figure 9a and 9b are linear and their main parameters have been obtained with high correlation coefficients ($R^2 \geq 0.91$ for Cu (II) ions and $R^2 \geq 0.98$ for Zn (II) ions).

For the zinc (II) ions, the low intercept values by applying intra-granular diffusion model and the high intercept values obtained using the liquid film diffusion model suggest that the adsorption in the case of the Zn (II) is not a phenomenon of surface. The intra- granular diffusion played a certain role in the adsorption mechanism beside other factors.

For the copper (II) the high values of the intercept from the intra-granular diffusion model and the related low values of the intercept for the liquid film diffusion model should be due to the fact that adsorption is not controlled by a surface phenomenon or by intra-granular phenol-menon but by other factors of the pseudo-second order kinetics.

Thermodynamic studies in the mono-element systems

The thermodynamic parameters of the adsorption of Cu (II) ions and Zn (II) ions may be linked to the distribution constant (k_d) by the Equation 13. The curves $\ln K_d = f(1/T)$ are linear and were obtained with correlation coefficients of 1 for the copper ion (II) (Figure 10) and 0.999 for the ion zinc (II) (not shown). We deduce from the slopes and intercept of these curves the thermodynamic parameters ΔS° ΔH° respectively (Table 6).

The positive values of the enthalpy for the Cu (II) ions on average of 26.694 kJ/mol and for the Zn (II) ions on average of 25.842 kJ/mol indicate that the adsorption of these ions is endothermic. Lin and Juan (2002) obtained the positive enthalpy values for the adsorption on montmorillonite for Zn (II) ions of 7.39 kJ/mol

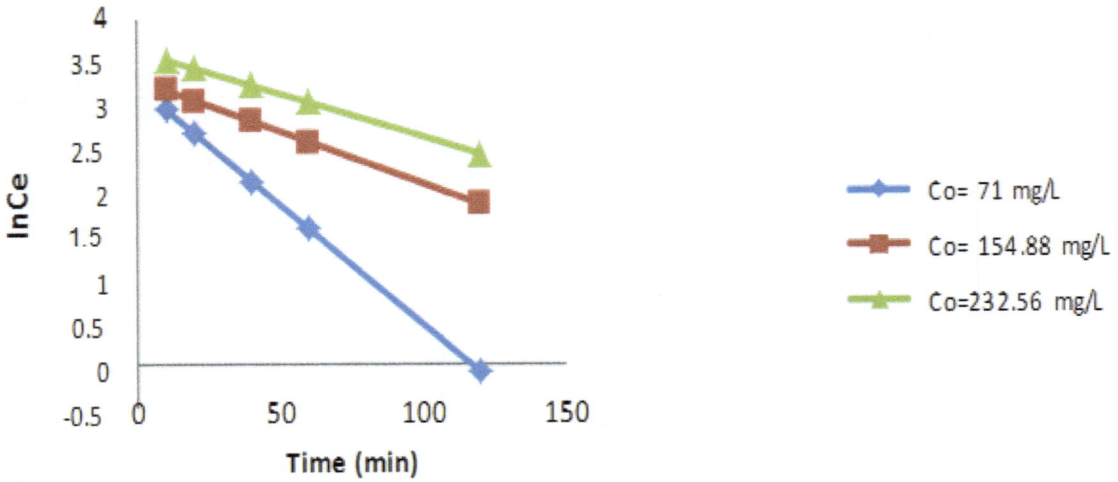

Figure 8a. Plots of liquid film diffusion model for the adsorption of Cu (II) ions onto clay surface.

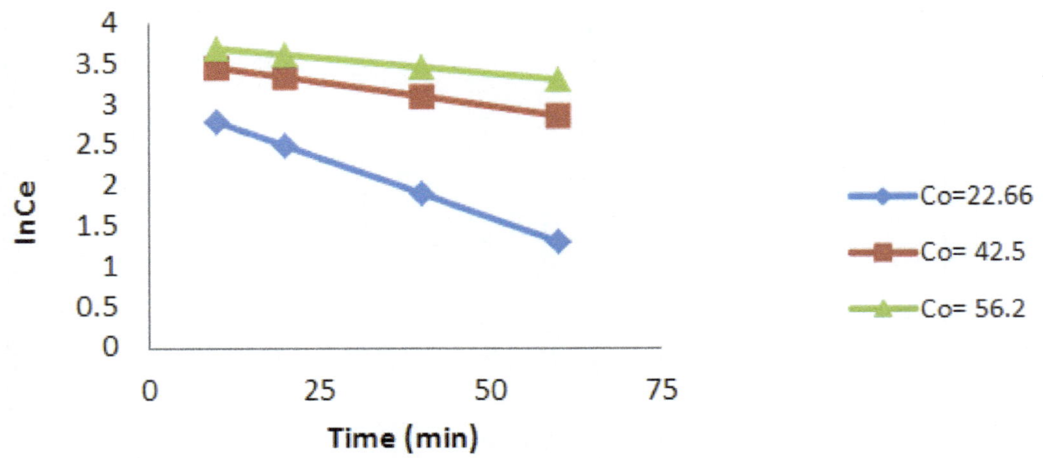

Figure 8b. Plots of liquid film diffusion model for the adsorption Zn (II) ions onto clay surface.

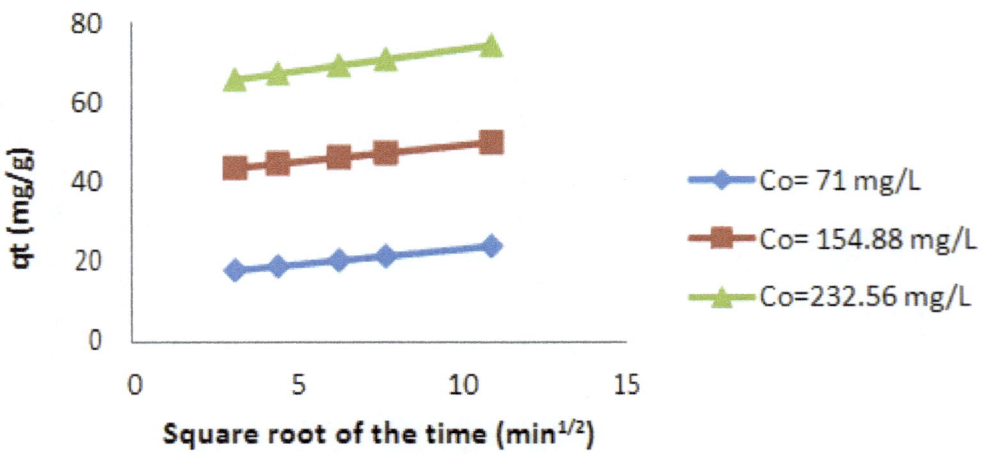

Figure 9a. Plots of change in adsorption capacity (qe) as a function of the square root of the time for copper (II) ions.

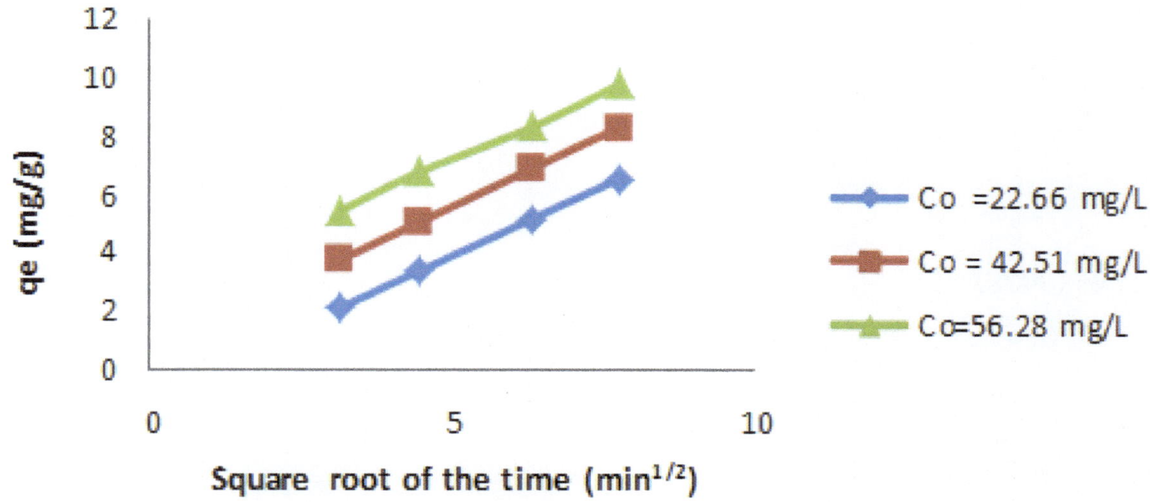

Figure 9b. Plots of change in adsorption capacity (qe) as a function of the square root of the time for the zinc (II) ions.

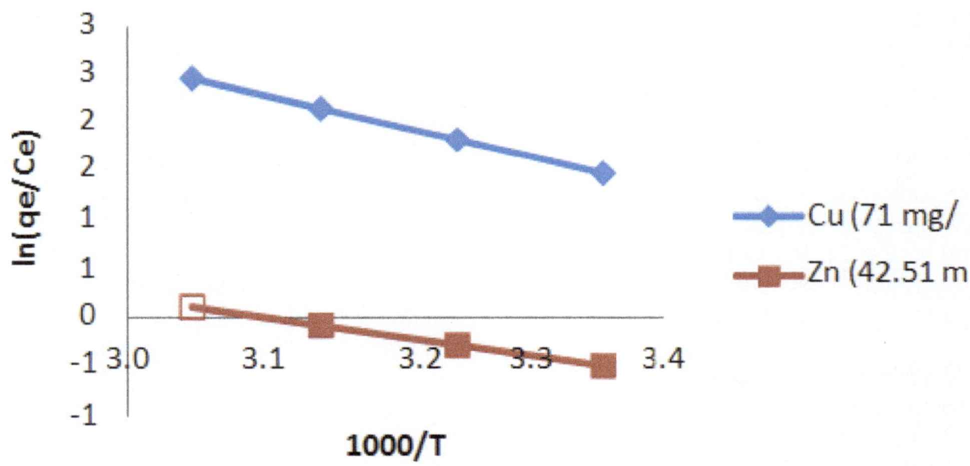

Figure 10. Variation of the logarithm of the constant distribution of copper (II) and Zn (II) ions adsorbed as a function of the temperature reverse.

Table 6. Thermodynamic parameters and statistical for the adsorption of Cu (II) and Zn (II) ions on the clay surface.

Ion	Initial concentration (mgl/L)	$\Delta_rG°$ (kJ/mol)				$\Delta_rH°$ (kJ/mol)	$\Delta_rS°$ (J/mol.K)	R^2
		298K	308K	318K	328K			
Cu^{2+}	154.88	-4.854	-5.912	-6.971	-8.029	26.693	105.79	1
	71	-3.668	-4.687	-5.706	-6.725	26.695	101.89	1
Zn^{2+}	42.51	1.217	0.391	-0.435	-1.262	25.842	82.633	0.999
	22.66	-1.094	-1.998	-2.902	-3.805	25.839	90.379	0.999

and for Cu (II) ions of 7.05 kJ/mol. In addition, Amer et al. (2010) obtained an enthalpy value for adsorption on

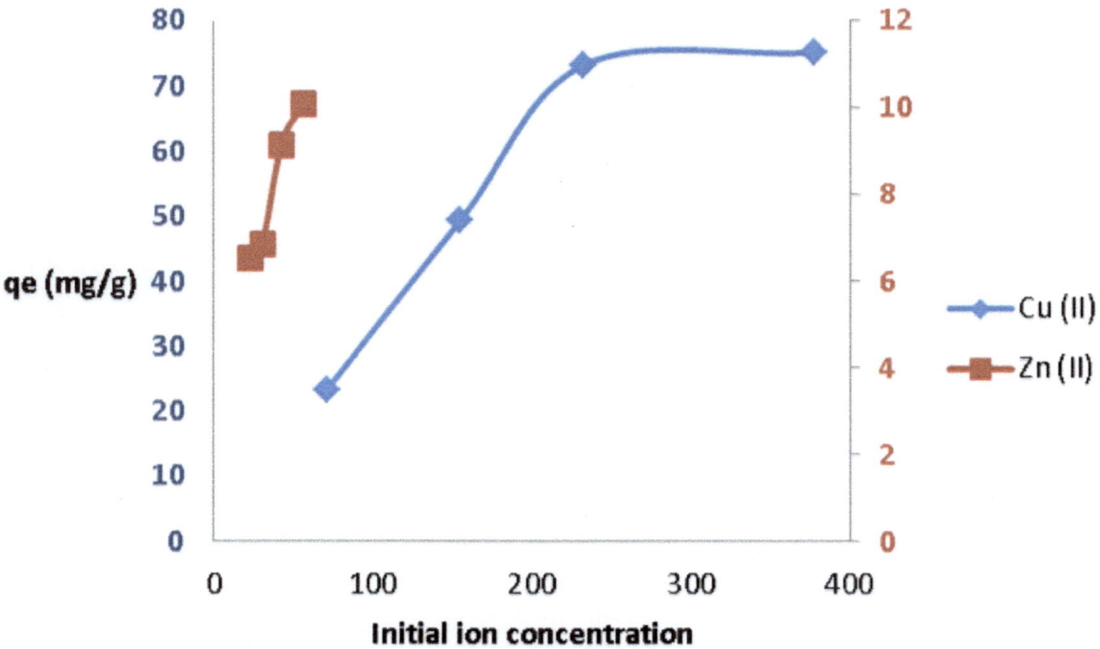

Figure 11. Experimental results of the effects of the ion concentration Cu (II) and Zn (II) on the adsorption by clay of Bikougou.

a kaolinite for Zn (II) of 24.49 kJ/mol, this value is comparable to that obtained in the present study. The positive values of the entropy for the Cu (II) ions of 103.84 J/mol.K and to the Zn (II) ions of 86.506 J/mol.K indicate that the distribution of ions at the surface of the clay is conducted more disorderly than in aqueous solution.

Lin and Juan (2002) have also obtained the values of the positive entropy for adsorption on montmorillonite for Zn (II) of 6.39 J/mol.K and to the Cu (II) of 9.09 J/mol.K. Amer et al. (2010) got a positive value of the entropy of 122.82 J/mol.K for the adsorption of Zn (II) ions on a kaolinite, a value near to that obtained in this study. The negative value of the free enthalpy explains the spontaneous character of the adsorption of each of these ions on the surface of the clay.

Adsorption isotherms studies

Mono-element system (at 25°C)

The experimental variations of the adsorption capacities at the equilibrium of Cu (II) ions and Zn (II) ions as a function of the initial concentration are represented in Figure 11. In each case, the adsorption capacity increases with the initial concentration before reaching the saturation equilibrium.

The adsorption of Cu (II) is that of a type I and that of Zn (II) of type V according to the IUPAC classification (Limousin et al., 2007). Similar results were observed for the adsorption of Cu (II) and Zn (II) on acidic soils (Arias et al., 2006). The results have been modeled by application of the Langmuir and Freundlich isotherms models. The adsorption isotherm model parameters have been determined from slope and intercept of each linear plot listed in Table 3.

The Langmuir maximum adsorption constant (q_m) 99.909 mg/g for the Cu(II) ions and 11.351 mg/g for Zn (II) ions (Figure 12b) and the energetical Langmuir constant (b) 0.262 L/g for Cu (II) ions and 0.258 L/g for Zn(II) ions have been obtained with correlation coefficients of 0.931 and 0.964, respectively. The separation factor R_L values of 0.010 with Cu (II) ions and 0.064 for Zn (II) ions indicate the favorable nature of the adsorption of these ions onto the clay surface. These b values may be associated to the physically controlled process.

The Freundlich maximum adsorption constant (K_F) of the order of 23.104 and 5.223 L/g and Freundlich intensity factor (n) of 2.392 and 4.950 with correlation coefficients (R^2) of 0.988 and 0.994, respectively, for the copper (II) and zinc (II) ions (Figure 12a). The Freundlich intensity factor (n) higher than unity indicates that the adsorption of copper ions (II) or zinc (II) ions is favorable on the surface of clay from Bikougou as reported by Arias et al. (2006) and Tellan and Owalude (2007).

The experimental adsorption for Cu (II) ions (Table 7a) results are best describe with Freundlich and Langmuir model and for Zn (II) ions (Table 7b) the two models are all applicable. The adsorption isotherm parameters observed in this study are compared with those of

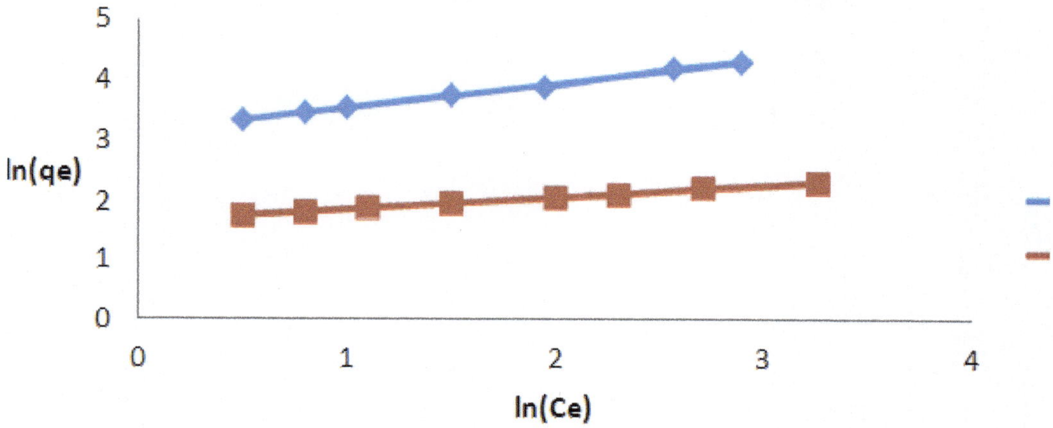

Figure 12a. adsorption isotherms plots of ions Cu^{2+} and Zn^{2+} or the clay obtained using the Freundlich model at pH =5.

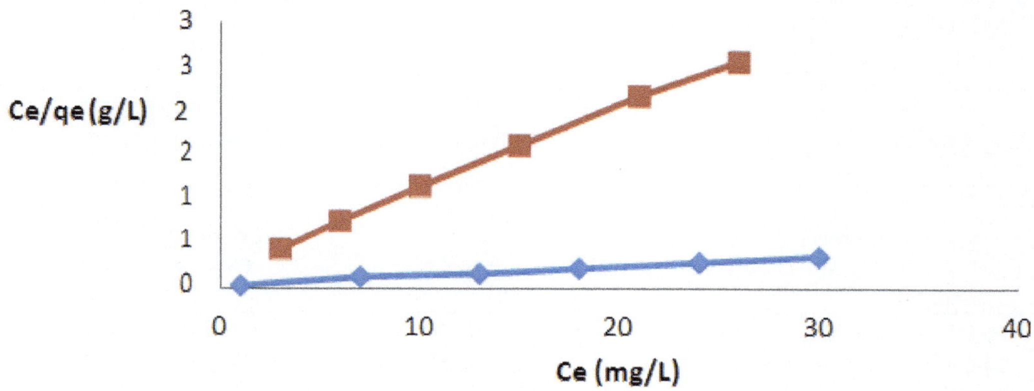

Figure 12b. Adsorption isotherms plots of ions Cu^{2+} and Zn^{2+} on the clay obtained using the Langmuir model at pH =5.

Table 7a. Langmuir and Freundlich isotherm model parameters and statisticals for the adsorption of Cu (II) ions on the clay surface.

Ions studied (Cu (II))	Langmuir parameter				Freundlich parameter			
	q_{max} (mg/g)	B (L/mg)	RL	R^2	n	1/n	KF (L/g)	R^2
Cu^{2+} this study	99.909	0.262	0.010	0.931	2.392	0.418	23.104	0.988
Njoku et al. (2011)	15.77	0.032		0.999	2.50	0.4	1.48	0.938
Lin and Juang (2002)	16.147	0.46	0.52	0.975				
Fifi et al. (2010)	0.70	1.81		0.90	1.27	0.78	0.45	0.92
Arias et al. (2006)				0.866-0.949				0.960-0994

Table 7b. Langmuir and Freundlich isotherm model parameters and statisticals for the adsorption Zn (II) ions on the clay surface.

Ion studied (Zn (II))	Langmuir parameter				Freundlich parameter			
	q_{max} mg/g)	B (L/mg)	RL	R^2	n	1/n	KF (L/g)	R^2
Zn^{2+} this study	11.351	0.258	0.064	0.964	4.950	0.202	5.223	0.994
Lin and Juang (2002)	13.267	0.47	0.52	0.986				
Arias et al. (2006)				0.894-0.961				0.939-0.994

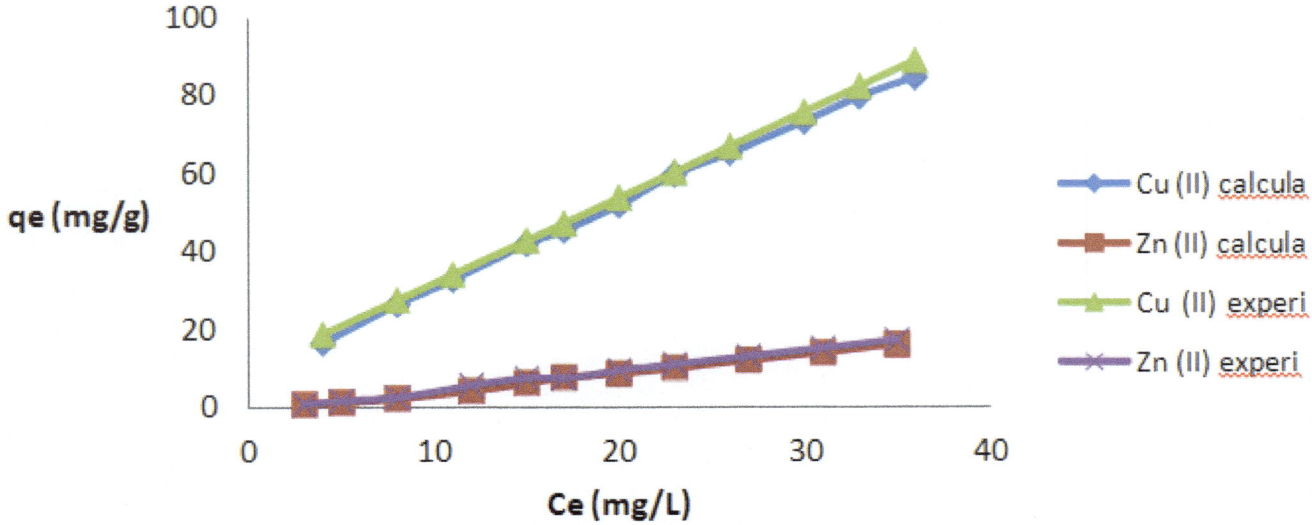

Figure 13. The calculated and experimental adsorption capacities variations of Cu (II) and Zn (II) ions as a function of adsorption equilibrium concentration in binary solutions.

Table 8. Theoretical and calculated adsorption capacities of copper (II) and zinc (II) ions in the mono and bi solutions element.

$\frac{[Cu]*100}{[Cu]+[Zn]}$ initial	$\frac{[Cu]*100}{[Cu]+[Zn]}$ at equilibrium	qe mix exp Cu	qe mix cal cu	qe mix exp Zn	qe mix cal Zn	qe mono exp Zn	qe mono exp Cu	$\frac{q_e mix}{q_e mono}Cu$	$\frac{q_e mix}{q_e mono}Zn$
96.40	96.98	0.58	88.67	84.44	0.89	0,766	1.651	70.694	1.25
93.39	94.62	1.23	82.11	79.56	1.6807	1.282	2.659	67.866	1.21
92	93.86	1.86	75.56	73.18	2.56	2.329	3.641	64.586	1.17
90.96	93.25	2.29	66.83	69.47	5.6	4.2	7.397	58.123	1.14
86.09	89.21	3.12	60.28	59.8	7.48	6.216	8.883	52.417	1.15
81.72	85.47	3.75	52.67	60.11	7,45	7.704	7.891	45.8	1.15
79	83.23	4.23	47.17	45.53	9,4687	8.683	9.832	42.119	1.12
74.6	80.62	6.02	42.81	42	11,0263	10.16	11.228	38.989	1.098
70.85	78.26	7.41	34.07	32.77	13,1031	12.13	13.155	31.870	1.069
66.1	74.74	8.64	27.51	26.39	15,1799	14.11	15.134	29.058	0.947
62.64	71.84	9.20	23.83	13.63	17,2567	16.08	16.592	33.005	0.722
60.28	69.84	9.56	18.78	16.63	0,89	0.766	0.7355	30.144	0.623

other authors. The clay from Bikougou is very efficient for Cu (II) ions adsorption has described by the Langmuir isotherm model.

Isotherm binary systems study

The evolution of the adsorption capacity of the ion Cu (II) and Zn (II) in binary solutions has been studied by varying the percentage of initial ions Cu (II) ([Cu] * 100/ [Cu] + [Zn]). Figure 13 shows calculated and experimental adsorption capacities of copper (II) and zinc (II) ions in binary mixtures. Calculated values of adsorption capacities are close to t experimental values of the two ions. The maximum standard deviation is 5.46 % for the copper (II) and the 5.30% for zinc (II) ions.

Lin and Juan (2002) obtain a standard deviation which is below 9% for the cooper and zinc by sorption in montmorillonite. The application of the Langmuir extended isotherm model makes good account in this study of the experimental results. It is shown in Table 8 that the experimental values of adsorption capacities of Cu (II) ions in the bi- element Cu (II)-Zn (II) solution are greater than those obtained for the same element in mono element Cu (II) solution. In opposite, the experimental adsorption capacities in the bi-solute Cu (II)-Zn (II) system are less than those of the solution containing only the Zn (II) ions. This indicates the manifestation of a competition between the two ions in a binary solution. Where, Cu (II) ions are selectively fixed on the adsorption available sites in comparison of Zn (II) ions as it is proved by the equilibrium distribution constant of each heavy metal from bi-element and mono-element solutions.

$$\frac{q_e(mix)}{q_e(mono)}$$

This constant is greater than the unity for Cu (II) ions and less than the unity Zn (II) ions adsorbed on the clay surface. When the ions compete to attach on the sites of adsorption on the surface of the adsorbent, the selectivity can be evaluated based on several assumptions: (i) of the standard deviation; (ii) report between initial fraction of a solute mixing and fraction to the surface of the solid.

A gap-positive type of a metal in a multiple system shows that the sorption of this metal is favored in the presence of other metals. If the standard deviation of a metal is equal to zero, the metals had no effect on one another. When the gap-negative type is obtained, the metal concerned is not preferentially adsorbed (Eba et al., 2012).

When the ratio is greater than 1, the ion is preferentially adsorbed and when the report is less than 1, the ion is less adsorbed on the adsorbent. These results show that the copper is selectively adsorbed in relation to the zinc. Table 8 presents the values obtained for the copper ion. A gap-type positive (0.60 to 12.80) and a report

$$\left[{q_e mix}/{q_e mono}\right]_{Cu} > 1$$

while for Zinc the standard deviation is negative and the report

$$\left[{q_e mix}/{q_e mono}\right]_{Zn} < 1$$

It is concluded therefore that the copper is preferentially adsorbed on the clay of Bikougou compared to the zinc.

According to Eba et al. (2012) the essential settings of the selectivity are: the valence, the electronegativity and the hydration enthalpy. If it is considered the electronegativity, the selectivity of an ion plays in favor of ions having a greater electronegativity. The ion with a greater electronegativity should be fixed more easily than the one that has a more low electronegativity. In this study, the electronegativity of Pauling of zinc (1.65) is lower than that of the copper (II) (Bernard and Busnot, 1996.), so the copper (II) ions would be more selectively adsorbed than zinc (II) ions. Similar results have been obtained in this study. If it is considered the enthalpy of heavy metal hydration reaction

$$M(II) \xrightarrow{H_2O} M(H_2O)_6^{2+}$$

the selectivity plays in favor of an enthalpy of hydration important. A high enthalpy of hydration is the origin of high binding. Copper which has an enthalpy of hydration (-2100kJ/mol) more important than the zinc (-2046 kJ/mol), should be more selectively fixed on the clay than Zn (II) ions as observed in the present study (Bernard and Busnot, 1996). Similar results have been reported by M'leyeh et al. (2002) about the adsorption of copper ions (II) and zinc (II) on clay in Tunisia or it has been proven that the proportion of copper adsorbed is always greater than that of the zinc regardless of its content in the original solution. We have similar results with the adsorption of copper (II) and zinc (II) on a montmorillonite (Lin and Juan, 2002).

Conclusion

The adsorption of Cu (II) and Zn (II) ions on the surface of the clay from Bikougou was made according to the technique in Batch: In the case of mono-element system the amounts of ion Cu (II) ions or Zn (II) ions adsorbed: Increase with the pH and decrease with the increase in ionic strength and in the mass of clay; this adsorption is endothermic for Cu (II) ions as well as for the Zn (II) ions and obeys the kinetics of the pseudo-second order. It is described by the isotherms of Langmuir and Freundlich concerning two ions; the adsorption of each heavy metal on the clay surface is spontaneous. In the binary systems, as this reflects the fact that the competition promotes the adsorption of Cu ions (II) and reduces that of the Zn (II) ions.

$$\left[{q_e mix}/{q_e mono}\right]_{Cu} > 1 \text{ and } \left[{q_e mix}/{q_e mono}\right]_{Zn} < 1$$

REFERENCES

Achak M, Hafidi A, Ouazzani N, Sayadi S, Mandi L (2009). Low cost biosorbent "banana peel" for the removal of phenolic compounds from olive mill wastewater: Kinetic and equilibrium studies. J. Hazard. Mater. 166:117-125.

Amer MW, Khalili Fawwaz I, Awwad, Akl M (2010). Adsorption of lead, zinc and cadmium ions on polyphosphate-modified kaolinite clay. J. Environ. Chem. Ecotoxicol. 2(1):001-008

Arias M, Pérec-Novov C, Lopez E, Soto B (2006). Competitive adsorption and desorption of copper and zinc in acids soils. Geoderma 133:151-159.

Babel S, Kurniawan TA (2003). Low-cost adsorbents for heavy metals uptake from contaminated water: A review. J. Hazard. Mater. 97:219-243.

Barone A, Ebesh O, Harper RG (1998). Placental copper transport in rats: Effects of elevated dietary zinc on fetal copper, iron and metallothionien. J. Nutr. 128:1037-1041.

Bayramoglu G, Bekta S, Arica MY (2003). Biosorption of heavy metal ions on immobilized white-rot fungus Trametes versicolor. J. Hazard. Mater. B 101:285-300.

Bernard M, Busnot F (1996). Usual of general and mineral chemistry; New re-examined and brought up to date edition. 3 édition dunod-Paris. pp. 90. ISSBN :2-10-003050-7.

Calvet R, Msaky JJ (1990). Factors influencing the experimental determination metal quantities of cations traces adsorbed by an earthy material. Science du sol. 28:1-14.

Chen JP, Lin M (2001). Surface charge and metal ion adsorption on an H-type activated carbon: experimental observation and modeling simulation by the surface complex formation approach. Carbon 39:1491-1504.

Doula MK (2009).Simultaneous removal of Cu, Mn and Zn from drinking water with use of clinoptilolite and its Fe-modified form. Water Res. 49:3659-3672.

Eba F, Ndong nlo J, Moussambi Membesti ZH, Ondo JA, Kouya Biboutou R, Edou Engonga P (2012). Isothermal and Kinetic studies of simultaneous removal Mn (II) and Pb (II) ions in aqueous solution by adsorption onto clay from Bikougou deposit (Gabon). Int. J. Chem. Technol. 4:1-16

Eba F, Gueu S, Mvongbote AE, Ondo JA, Yao BK, Ndong Nlo J, Kouya Biboutou R (2010). Evaluation of the adsorption capacity of the natural clay from Bikougou (Gabon) to remove Mn (II) from aqueous solution. Int. J. Eng. Sci. Technol. 2: 5001-5016.

Eba F, Nkoumbou C, Ondo JA, Omva-Zué J, Yvon J, Njopwouo D (2011b). Mineralogical and physico-chemical characterization of clay from Bikougou (Gabon). Annales de la Faculté des Sciences, Université de Yaoundé 1, série Chimie. 38(1):45-56.

Eba F, Ondo JA, Gueu S, Ndong Nlo J, Kouya Biboutou R, Yao BK (2011a). Treatment of lead content by using natural clay mixture of kaolinite-albite-montmorillonite-illite clay. J. Appl. Sci. 14:2536-2545.

El-Bayaa AA, Badawy NA, Abd AlKhalik E (2009).Effect of ionic strength on the adsorption of copper and chromium ions by vermiculite pure clay mineral. J. Hazard. Mater. 170(2-3):1204-1209.

Fifi U, Winiarski T, Emmanuel E (2010). Vulnerability of subterranean water with respect to the pollutants contained in urban rain water in the countries in development-Study of the adsorption of heavy metals on a ground representative of Port-au-Prince, Haiti. NOVATECH 2010 "7th International Conference on the techniques and strategies durable for the management of urban water in rainy weather", Graie, June 27th - July 1st, 2010, Lyon, France.

Freundlich HMF (1906). The adsorption in physical chemistry. Leipzig, 57A:385-470 (Original Article in German).

Gupta SS, Bhattacharyya KG (2005). Interaction of metal ions with clays: A case study with Pb (II). Appl. Clay Sci. 30:199-208.

Gupta SS, Bhattacharyya KG (2008). Immobilization of Pb (II), Cd (II) and Ni (II) ions on kaolinite and montmorillonite surfaces from aqueous medium. J. Environ. Manage. 87:46-58.

Gyorffy EJ, Chan H (1992). Copper deficiency and mycrocytic anemia resulting from prolonged ingestion of over-the-counter zinc. Am. J. Gastroenterol. 87:1054-1055.

Ho YS, McKay G (1999a). Pseudo-second order model for sorption processes. Process Biochem. 34:451-465.

Ho YS, McKay G (1999b). Competitive sorption of copper and nickel ions from aqueous solution using peat. Adsorption 5:405-417.

Ho YS, Ng JCY, McKay G (2001). Removal of Lead (II) from effluents by sorption on peat using second-order kinetics. Sep. Sci. Technol. 36:241-261.

Huh JI, Song DI, Jeon YW (2000). Sorption of phenol and alkylphenols from aqueous solution onto organically modified montmorillonite and applications of dual-model sorption model. Sep. Sci. Technol. 35:243

Ijagbemi CO, Baek MH, Kim DS (2009). Montmorillonite surface properties and sorption characteristics for heavy metal removal from aqueous solutions. J. Hazard. Mater. 166:538-546.Juang RS, Wu FC, Tseng RL (1997). Ability of activated clay for the adsorption of dyes from aqueous solutions. Environ. Technol. 18:535.

Kadirvelu A, Namasivayam C (2003). Activated carbon from cotonut coirpith as metal adsorbent: adsorption of Cd (II) from aqueous solution. Adv. Environ. Res. 7:471-478.

Kaya A, Oren HA (2005). Adsorption of zinc from aqueous solutions to bentonite. J Hazard Mater. 125(1-3):183-189.

Kwong SC, Song DI, Jeon YW (1998). Adsorption of phenol and nitrophenol isomers onto montmorillonite modified with hexadecyltrimethylammonium cation. Sep. Sci. Technol. 33:1981.

Langmuir I (1918). The adsorption of Bases on plane surfaces of glass, mica and platinum. J. Am. Chem. Soc. 40:1361-1403.

Limousin G, Gaudet JP, CHarlet L, Szenknect S, Barthes V, Krimissa M (2007). Sorption isotherm: A review on physical bases, modeling and measurement. Appl. Geochem. 22:249-275.

Lin S-H, Juang R-S (2002).Heavy metal removal from water by sorption using surfactant-modified montmorillonite. J. Hazard. Mater. 92(3):315-326.

M'leyeh A, Srasra E, Cheref A (2002). Fixing of heavy metals by clays of the garbage dump of Borj Chekir (SW of Tunis). Proceeding of Symposium on Environmental Pollution Control and Waste Management. pp. 533 -546.

Mellah A, Chegrouch S (1998). The removal of zinc from aqueous solutions by natural bentonite. Water Res. 31(3):621-629.

Mohan D, Singh KP (2002). Single and multi-component adsorption of cadmium and zinc using activated carbon derivated from bagasse and agricultural waste. Water Res. 36:2304-2318.

Moore JW (1991). Inorganic contaminants of surface water, research and monitoring priorities. Spinger -Verlag, New York. ISBN 10: 0387972811 / ISBN 13: 9780387972817.

Naseem R, Tahir SS (2001). Removal of Pb (II) from aqueous acidic solution by using bentonite as an adsorbent. Water Res. 35:3982-3986.

Njoku VO, Oguzie EE, Obi C, Bello OS, Ayuk AA (2011). Adsorption of Copper(II) and Lead(II) from Aqueous Solutions onto a Nigerian Natural Clay. Aust. J. Basic Appl. Sci. 5(5):346-353.

Salgueiro M, Zubillaga J, Lysionek M, Sarabia A, Caro MI, Paoli R (2000). Zinc as an essential micronutrient: A review. Nutr. Res. 20:737-755.

Shukla A, Zhang Y-H, Dubey P, Margrave JL, Shukla SS (2002). The role of sawdust in the removal of unwanted materials from water. J. Hazard. Mater. 95:137-152.

Tellan AC and Owalude SO (2007). Some Langmuir and Freundlich parameters of adsorption studies of chlorpheniramine maleate. Res. J. Appl. Sci. 2(8):875- 878.

Tien CT, Huang CP (1991). Kinetics of Heavy Metal Adsorption on Sludge Particulate. In: Heavy metal in the environment, Vernet, J.P. (Ed.). Elesvier Science Publishers, USA. pp. 313-328.

Wan Ngah WS, Hanafiah MAKM (2004). Adsorption of copper on rubber (Hevea brasiliensis) leaf powder: Kinetic, equilibrium and thermodynamic studies. Biochem. Eng. J. 39(3):521-530.

Weber JR, Morris JC (1963). Kinetics of adsorption on carbon from solution. American Society Civil Engineering. J. Sanit. Eng. Div. 89:31.

Youcef L (2006). Elimination of the mineral pollutants of water by processes of chemical precipitation and adsorption. Thesis of doctorate in Hydraulic Sciences, University of Bisrka, Algeria.

Youcef L, Achour S (2006). Elimination of the copper of water by processes of chemical precipitation and adsorption. Courrier du savoir. 7:59-65.

11

Ecohydrological characterization of the Nyando wetland, Lake Victoria, Kenya: A State of System (SoS) analysis

P. S. Khisa[1,2,3], S. Uhlenbrook[2,3], A. A. van Dam[2], J. Wenninger[2,3], A. van Griensven[2] and M. Abira[1]

[1]Water Resources Management Authority, Lake Victoria South Catchment Area, P.O Box 666-40100 Kisumu, Kenya.
[2]UNESCO-IHE Institute for Water Education, P.O. Box 3015, Delft, The Netherlands.
[3]Delft University of Technology, Water Resources Section, P.O. Box 5048, Delft, The Netherlands.

Lake Victoria floodplain wetlands have a complex hydrological setting characterized by transition from a terrestrial to an aquatic environment. A state-of-system (SoS) analysis was carried out in a papyrus dominated wetland in the Nyando River Delta, on the eastern shores of Lake Victoria, Kenya, to characterize and provide data for detailed ecohydrological studies. The objectives of the study were to: (1) determine the spatio- temporal changes in the wetland evolution and (2) analyze the main hydrological factors that have influenced wetland evolution. Multi-temporal dry-season Landsat MSS, Landsat TM and Landsat ETM+ imagery covering Nyando Wetland and its surrounding area were processed and analyzed to generate time series polygon and polyline maps of the wetland and river. Results show that the wetland increased in size from 5,925 ha in 1950 to 9,925 ha in 1973, and declined to 4,527 ha in 2008. In the last 60 years, Nyando River has migrated in a general eastward direction. Time series hydrological data (1950-2009) were statistically tested for homogeneity using the Spearman's rank test for linear trends, Pettit test and Standard Normal Homogeneity test (SNHT) for change point analysis, and split-record tests performed for variance (F-test) and mean (t-test). In addition, data were analyzed using descriptive statistics and frequency analyses. Statistical test results show that the hydrological data series were homogeneous. Results of change point analyses indicate that total annual rainfall in Nyando declined in 1979, while the mean annual discharge for Nyando River and Lake Victoria levels had significant upward shifts in 1961. The decadal mean discharges varied significantly over time and increased by 80% from 11.45 m^3/s observed in the 1950-1961 subset, reducing by 11.4 and 21.9% in the next two decadal sub-sets, before rising by 35.0% in 1990s and dropping by 24.0% in the last decade. The decadal mean annual lake levels increased from 1134.0 to 1135.43 m in the 1951-1961 and 1962-1972 and remained above the long-term mean of 1135.0 m for 43 years since 1962 before dropping drastically by 1.4 m to an average of 1134 m/year in 2005-2009. The highest recorded lake level at Kisumu Station was 1136.2 m in 1964 after increasing by 2.5 m from 1961. Discharge data exhibit trimodal seasonal patterns, while the lake levels had two peaks. The lake levels are more sensitive to direct lake rainfall. Changes in the Nyando wetland area are linked to the seasonal and episodic flood and drought events coupled with anthropogenic activities (regulation of lake levels, modification of river including cut-off meanders, river training and construction of dykes, drainage of wetland for cultivation, settlement and livestock grazing, abstraction of water for irrigation). A combination of these hydrological and human factors is the main cause of the Nyando Wetland evolution. If the land use trend continues unabated, then the increase in papyrus losses will pose a big challenge to the ecological functioning of the wetland and its support to sustaining community livelihoods.

Key words: Nyando Wetland, River, ecohydrology, Lake Victoria.

INTRODUCTION

Estuarine wetlands and floodplain wetlands near large lakes are characterized by complex interactions between vegetation types, surface water fluxes and pore water movements, and are very sensitive to small changes in topography and tidal fluxes (Hughes et al., 1998). The main factors affecting the hydrology of such wetlands are vegetation, rainfall, seasonal variations in evaporation and transpiration, extreme tidal or flood events, and variations in regional groundwater flow (Hughes, 1998). Hydrology is probably the single most important determinant of the establishment and maintenance of wetlands (Mitsch and Gosselink, 1993), and is essential in understanding and quantifying wetland functions and processes (Greeson et al., 1979). Anthropogenic hydrological modifications to rivers and their drainage basins, as well as, upland development have changed the physical, chemical and biological character of river-floodplain systems (Weilhoefer et al., 2008).

Floodplain wetlands are intimately linked to rivers and their watersheds through the pulsing of floodwaters (Weilhoefer et al., 2008), which is thought to be the major driver of biota in these environments (Junk et al., 1989). Floodplains are characterized by hydrological fluctuations, and the alternation of terrestrial and aquatic phases contribute to a high biodiversity (Gopal and Junk, 2000). The degree of tolerance of species to hydrological extremes (floods and drawdown events) determines the structure of macrophyte communities (van Geest et al., 2005a). Flooding events are important in the maintenance of macrophyte richness and succession (van Geest et al., 2005a), while drawdown events determine composition and abundance of aquatic plants (Wilcox and Meeker, 1991). Drawdown events can have catastrophic effects on aquatic vegetation, decreasing the aboveground biomass of most species (Richardson et al., 2002). Moreover, Van Geest et al. (2005a) observed macrophytes dominated by desiccation-tolerant species along shorelines of lakes disturbed by frequent drawdown events.

Analyses of remote sensing time series data show that more than 70% of swamps in the upper Nile and Congo basins have been lost in less than 20 years (Maclean, 2004), with about 50% of the papyrus wetlands in the Kenyan part of Lake Victoria being lost over the period 1969 to 2000 (Owino and Ryan, 2007). In common with the trend worldwide (Millennium Ecosystem Assessment, 2005; Mitsch, 2010), many researchers have attributed papyrus wetland losses around Lake Victoria to: (i) increase inhuman population pressures (Balirwa, 1995, van Dam et al., 2007), (ii) conversion to agriculture through drainage (Mafabi, 2000; Owino and Ryan, 2007; Schuyt, 2005), (iii) unsustainable levels of exploitation driven by perennial harvesting (Osumba et al., 2010), (iv) lake level fluctuations (Awange et al., 2008; Kiwango and Wolanski, 2008), and (vi) challenges presented by climate change (Odada et al., 2009). Over-exploitation and destruction of wetland vegetation curtail the filter function of wetlands as pollutants and nutrients are carried directly into the lake when the vegetation of wetland macrophytes is destroyed (Okeyo-Owuor, 1999).

Past studies on Lake Victoria papyrus wetlands (Jones and Muthuri, 1984; van Dam et al., 2007) have focused in detail, on some ecological component of the wetland, while socio-economic studies (Gichuki et al., 2001; Maclean, 2003) are few. Studies on wetland hydrology are quite limited, despite many works on catchment hydrology (Kizza et al., 2009; Mistry and Conway, 2003; Ogallo, 1988; Ongwenyi et al., 1993) in the Lake Victoria region. Because of the intimate relationship between shoreline wetland ecology, socio-economics and hydrology, knowledge of ecohydrology is critical if we want to predict and manage changes in wetland environments. Recent investigations by Kipkemboi (2006) and Kiwango and Wolanski (2008), have highlighted the importance of lake level fluctuations to the wetland's ecological functioning and impact on community livelihoods. Despite the many modifications on the hydrological flow regimes of major rivers flowing to Lake Victoria in the last 60 years, no studies have been undertaken to understand the ecohydrological functioning of the numerous floodplains and delta wetlands associated with such rivers.

Lake Victoria wetland ecosystems have been recognized to provide various functions and services (Gichuki et al., 2001; Kansiime and Nalubega, 1999; Kipkemboi, 2006), which include: groundwater recharge and discharge, flood control, water storage and purifycation, storm protection and windbreak, food chain support, freshwater fisheries, biodiversity, carbon storage and climate regulation. These functions and services provided by the wetlands have benefits for the wetland itself, surrounding ecosystems, and local and regional communities. The Nyando Wetland is one of the largest and economically important wetland ecosystems fringing the Lake Victoria shoreline and performs important hydrological, ecological and socio-economic functions (Kipkemboi, 2006; van Dam et al., 2011).

In this study, we determined the spatio-temporal changes (1950-2008) and analyzed the main influencing hydrological factors of a papyrus dominated floodplain wetland known as Nyando Wetland that is located on the eastern shores of Lake Victoria, Kenya. The study was carried out as a State of the System (SoS)

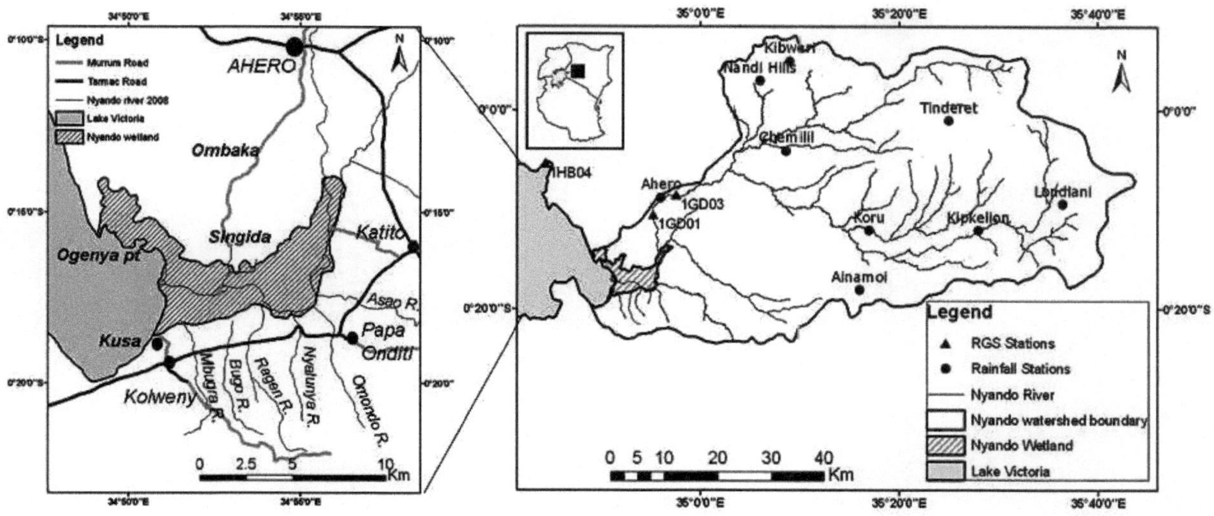

Figure 1. Location map of Nyando Wetland (left panel) and its Nyando Catchment (right panel) and hydrometric network and Nyando.

analysis to: (i) provide a basis for detailed experimental investigations and hydrological modeling, and (ii) integrate with the ecological and socio-economic studies under the ECOLIVE project (van Dam et al., 2011; http://www.unesco-ihe.org/ecolive). The study period was marked by extreme flood and drawdown events that have contributed to the current state of Nyando Wetland.

MATERIALS AND METHODS

Study area

This study was carried out in 2009-2010 at a 4,000 ha low-lying swamp, commonly referred to as Nyando Wetland, which is located between a terrestrial and aquatic environment on the eastern shores of Lake Victoria (Winam Gulf), Kisumu County, Kenya (Figure 1) (van Dam et al., 2011). The Nyando Catchment has an area of 3,600 km^2 while Lake Victoria, the second largest freshwater lake in the world is 68,800 km^2 in area. The extensive Kano floodplains and surrounding Nandi and Kendu Escarpments define the physiography of the wetland, which rises from 1134 m at the lake shoreline to 1145 m in the floodplains. Like most wetlands bordering Lake Victoria (Kansiime et al., 2007), Nyando Wetland vegetation is dominated by *Cyperus papyrus* L. (papyrus) and other less dominant species such as *Phragmities mauritianus* and *Vossia cuspidate*.

The most important sources of water supply to the wetland are the Nyando River, Lake Victoria and local rainfall. In general, the wetland experiences a bimodal rainfall pattern, with long rains occurring in March-May and short rains in October-December. The seasonal variability of mean temperature in Nyando Catchment is bimodal, with two peaks in March and October. Temperature ranges from 13 to 31°C in the wetland and 8 to 24°C in the highlands. Mean monthly temperatures are higher in the lowlands at 21°C and lower in the highlands at 16°C.

Geology and hydrogeological setting

The geology of the Nyando Wetland is characterized by the Pre-Cambrian system (grano-diorites), Tertiary and Quaternary volcanics (granites, rhyolites and phonolites) and Alluviums (Saggerson, 1947) (Figure 2). The metamorphic formations are the oldest rocks in the area dating back to Pre-Cambrian times (Shackleton, 1952). During Tertiary-Quaternary times, the region experienced the occurrence of several episodes of volcanic activity that were accompanied by widespread regional rifting and faulting.

The consolidated sediments were formed due to deformation and fracturing of the Kavirondo Rift Zone in Miocene-Pliocene times. In recent times, the unconsolidated sediments (sand, silt and clay) were deposited in Lake Victoria, and became interbedded over the Kano floodplains area with alluvium and colluvium brought down from the surrounding highlands (Millman, 1973). The last geological event is the migration of the Nyando River channel that has re-defined the geomorphology of Nyando Wetland environment.

The hydrogeological setting of the Nyando Wetland is defined by two separate groundwater systems (DHV Consulting Engineers, 1988). A deep groundwater system occurs in the Lower Pleistocene deposits, while a shallow groundwater system is found in the unconsolidated sediments. The two systems are separated by an aquiclude, which comprises mainly of clayey materials, for example, mudstone and siltstone (Figure 2). The shallow aquifers vary considerably in thickness (0 to 15 m) and are usually subject to seasonal fluctuations as compared to the thick (50 to 250 m) deep aquifers that store large volumes of groundwater. It is generally assumed that the shallow aquifers are recharged by rainwater percolation and the river system, and that the deep aquifers are recharged by shallow unconfined aquifers, shear zones and influent streams flowing from the escarpments. However, a more detailed understanding of the processes and their variability in space and time is subject to detailed and ongoing experimental and modeling investigations by the authors of this paper.

Soils

The Kano floodplains are characterized by a complex succession in the soil profiles. The plains are dominated by deep accumulations of colluvium that have been resorted by local flash

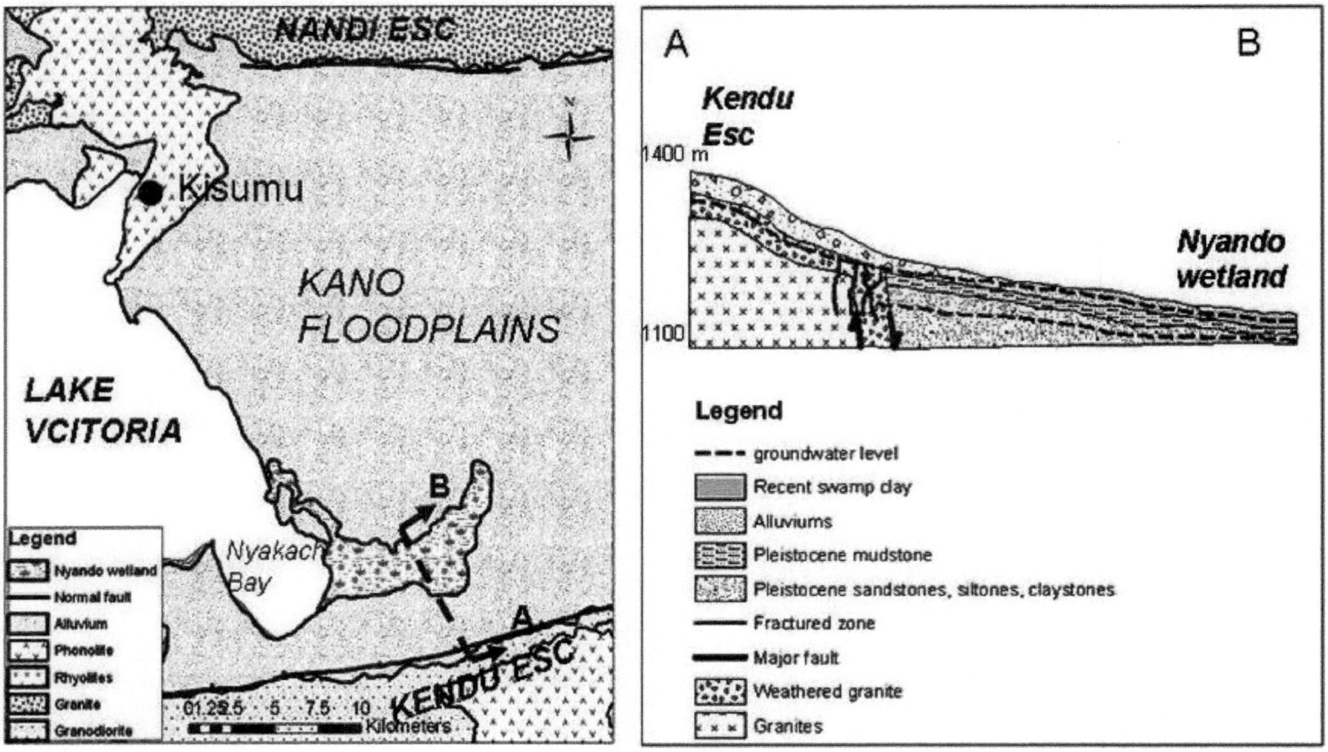

Figure 2. Geological map of Nyando Wetland (left panel) and hydrogeological cross-section (right panel) (source: Saggerson, 1950; DHV Consulting Engineers, 1988).

floods and become mixed with lenses of alluvium. Dark coloured clays and clay loam are the most widespread of the alluvial soil types; their colour vary from brown to black, which corresponds to a variation in clay content in the subsoil (Millman, 1973). When moistened, these soils become plastic, slightly sticky and loose any semblance of a structure, and crack on drying to form a weak, angular, blocky structure. These soils are usually termed 'black cotton soils' and represent considerable problems for agricultural development (Millman, 1973). The most common soil types include luvisols, greysols and fluvisols (Waruru et al., 2003).

Datasets and data properties

In this study, daily rainfall, discharge and lake level data collected at key monitoring stations in the Nyando catchment (Figure 1) for the period 1950-2009, were obtained from the Kenya Meteorological Department and the Water Resources Management Authority. The chronogram shown in Figure 3 shows a rise in data availability in 1970-1990 attributed to the WMO Hydromet Project (WMO, 1982). As observed by other researchers (Kizza et al., 2009), the expansion in the hydrometric network was regional, covering the Upper Nile, but declined in the 1990s when the project came to an end. In practice, the operational and maintenance costs of hydrometric networks are a big challenge, especially in the developing countries. This calls for further collaborative engagements between local and international players.

More than 50% of the stations had missing data ranging from a few days to several years. The data gaps were filled using linear regression and interpolation techniques. The most notable constraint for rainfall stations was that there was so much variability between nearby stations. Therefore, rainfall stations with only complete annual data series and having data for at least 30 years were used in the analysis. Double mass curves were used to test for the consistency of annual rainfall data. The river gauging stations were correlated with upstream stations to generate a long term time series. The lake levels had missing data for only two years, which was considered not to generally influence the long-term data series. Another constraint was limited recent rainfall data.

Statistical methods

Several statistical tests can be used to detect trends and homogeneity in time series hydrological data. These methods are broadly classified as parametric and non-parametric. Non-parametric (distribution free) tests are so called because the assumptions underlying their use are 'fewer and weaker' than those associated with parametric tests (Siegel and Castellan, 1988). For this reason, nonparametric methods are commonly used in hydrological investigations, especially when the data is considered skewed. However, parametric tests are much more flexible, and allow for testing a greater range of hypotheses. In this study, we used parametric and non- parametric methods.

We first applied the non-parametric Spearman's rank correlation method to evaluate the absence of linear trends in the data. Change-point analysis was then performed using both the non-parametric Pettitt (1979) test and the parametric Standard Normal Homogeneity (SNHT) test (Alexandersson, 1986; Alexandersson and Moberg, 1997). On identification of change-points in the data series, split-record tests on means (t-test) and variances (F-test) were carried out to quantify the magnitude of the changes and to test the significance of the differences in the subsets before and after the change point. All the statistical tests were carried out

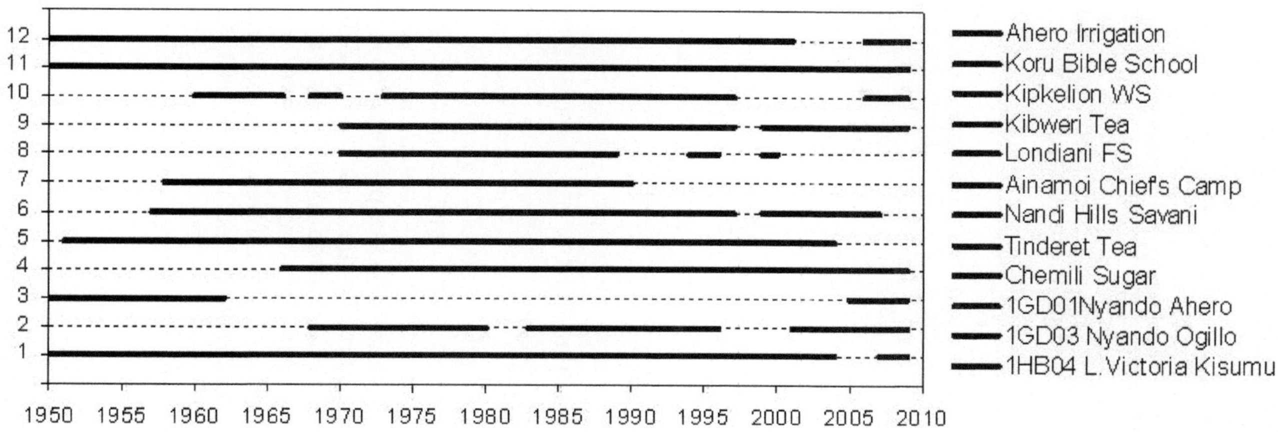

Figure 3. Chronograms showing years with complete total monthly rainfall data series.

Table 1. Spatial data used for characterization of Nyando wetland in this study.

Data type	Date	Scale/resolution	Sheet no./scene	Source
Geological map	01/07/1950	1: 125, 000	41	Mines and Geology, Kenya
Topographic maps	31/07/1967	1:50, 000	116/2 116/4	Survey of Kenya
Landsat 2 MSS	01/02/1973	30m	182/60	USGS
Landsat 4 TM	08/03/1986	30m	170/60	USGS
Landsat 7 ETM+	05/02/2001	30m	170/60	USGS
Landsat 5 ETM+	28/09/2008	30m	170/60	USGS

using the SPELL Stat package (Ashagrie et al., 2006). Critical values for the Spearman's rank correlation was less than or equal to 2.002 t-student distribution function at n-2 degrees of freedom at 5% significance level. The Pettit test had a critical level of 80% probability for significance of change points, while the SNHT test had 8.65 at 90% significance level. The year corresponding to the highest probability was considered the end of the first subset and the year following it as the first year of the next subset.

Image analysis

The methodology used for image analysis was a combination of optical image analysis and field survey as suggested by MacAlister and Mahaxay (2009). Spatial data were registered to the UTM World WGS84 coordinate system, zone 36 (south), using Arc GIS 9.3 program (Table 1). First, satellite images were pre-processed in the ERDAS Imagine 9.2 digital image processing program. Analyses of satellite data was limited to only single season (dry) imagery due to the lack of suitable (cloud free) wet season scenes. The dry season images were considered useful in assessing time series spatio-temporal changes in the wetland. We selected proven band combinations of the Landsat MSS (bands 2, 3, 4), Landsat TM (bands 2, 3, 4 and 5) and Landsat ETM+ (bands 3, 4, 5 and 7) images for wetland studies (Melack et al., 1994; Walsh et al., 1990). The images were ortho-rectified and layer stacked to generate false colour composites for wetland delineation using ERDAS Imagine 9.2 digital image processing system. Single panchromatic thermal bands (band 5) were selected for delineation of surface water bodies (river channel).

The pre-processed satellite images and geological map were registered to the 1:50, 000 topographic maps in Arc GIS 9.3.1, by utilizing the tie points (ground control points) technique but with an affine transformation. Eight ground control points (GCPs) were identified and selected in the four corners, at the east-west and north-south centers of the images. The maximum root mean square (RMSE) of rectification was <30 m. A standard histogram equalization image enhancement technique was applied for a better visibility of water-wetland-dry land boundaries (Perez et al., 2001). The swamp area (papyrus dominated wetland) and the main river course were visually interpreted on the satellite images and geological map, before on-screen digitization to generate polygon and polyline maps. From these maps, a number of numerical parameters (for example, total wetland area, perimeter length, lateral distance of river displacement) were calculated in order to evaluate the change in wetland size and river channel migration during the period 1950-2008.

Field survey was conducted in the Nyando Wetland and its surrounding areas in the period September 2009 to May 2010. The most common approach was to: (i) mark waypoints with Global Positioning System (GPS) along the main Nyando River channel and at boundaries of typical wetland habitats (ii) verify existing geological, topographic, soil, and land use and land cover maps of the wetland and its surrounding area. Geo-referencing of Nyando River channel was conducted aboard a canoe (local boat).

Field work was carried out along two main transects: lateral to the lake and across the river, taking into consideration the hydrological, ecological and socio-economic gradients. All geo-referenced data were plotted in Arc GIS. For accuracy assessment, field sample points (way points) were overlaid on the Landsat ETM+ 2008 imagery. Each waypoint was then reviewed

Table 2. Properties of total annual rainfall data series for Nyando Catchment.

Station name	WMO number	Annual mean (mm/year)	CV	SD (mm/year)	Skewness	Kurtosis	Max (mm/year)	Min (mm/year)	Start year	No. of years
Ahero Irrigation	9034086	1148	0.16	185	-0.35	2.24	1612	547	1950	56
Koru Bible School	9035148	1642	0.19	318	0.10	0.08	2499	680	1950	60
Kipkelion WS	9035258	1273	0.25	316	0.74	1.50	2238	650	1960	40
Kibweri Tea	8935161	1233	0.22	267	0.89	2.37	2150	684	1970	39
Londiani FS	9035002	1117	0.18	202	0.01	0.39	1499	677	1970	39
Ainamoi C. Camp	9035199	1605	0.26	418	-0.21	0.54	2708	748	1958	36
Nandi H. Savani	8935033	1587	0.21	338	-0.6	0.11	2124	739	1957	51
Tinderet Tea	9035263	1735	0.18	318	0.01	0.30	2401	926	1950	55
Chemilil Sugar	9035274	1452	0.14	200	-0.09	-0.75	1835	1106	1966	44

reviewed by comparing it with the corresponding river channel and wetland boundaries delineated on the image.

RESULTS AND DISCUSSION

Statistics

A summary of descriptive statistics for rainfall data series used in this study is presented in Table 2. The highest mean annual rainfall of 1735 mm/year was observed at Tinderet Tea station, with Londiani Forest station recording the lowest value of 1117 mm/year. In general, the highland stations (Tinderet Tea, Ainamoi Chief's Camp, Nandi Hills Savani, Koru Bible) receive more rainfall than stations in the plateau (Kipkelion Water Supply and Londiani Forest) and lowlands (Ahero Irrigation and Chemilil Sugar). Though located in the highlands close to Nandi Hills Savani, Kibwari Tea station has a mean annual rainfall of only 1233 mm/year. The location of the plateau stations on the leeward side of the surrounding highlands (Tinderet, North Tinderet, Mt Londiani and SW Mau Forest Complex) shields them from the south-easterly monsoon winds that bring about orographic rainfall. The lowland stations which are characterized by the Kano floodplains and Lake Victoria shoreline receive pre-dominantly convectional rainfall. The standard deviation varies from 185 mm/year at Ahero to 418 mm/y at Ainamoi. The coefficient of variation (CV) ranges from 14% at Chemilil Sugar to 26% at Ainamoi Chief's Camp, confirming that rainfall varies considerably at each station from year to year. This observation compares with the average 19% computed by Kizza et al (2009) for rainfall stations in Lake Victoria basin.

Statistical tests show that time series data from Ainamoi Chief 's Camp, Chemilil Sugar, Kibwari Tea Estate Nandi Hills, Kipkelion Water Supply, Londiani Forest Station, Tinderet Tea Station and Nandi Hills Savani rainfall stations, Nyando River RGS stations (1GD01 and 1GD03) and the lake level station at Kisumu bay (1HB04) are homogenous, since the null hypothesis for Spearman's rank correlation for absence of linear trend, was not rejected at 5% level of significance. Although the null hypothesis was rejected for Koru Bible School and Ahero Irrigation Research Stations, the data were still considered useful since lack of independence in the datasets was related to change of station locations.

Rainfall variability

The Pettit change point analysis indicates a shift in total annual rainfall in Nyando catchment in 1979, with a probability of 95%. The F-test and t-test parametric tests reveal a stable variance, and the presence of a significant downward jump in the mean from 1520 to 1403 mm/year as compared to the long-term mean of 1462 mm/year (Figure 5). These results are consistent with rainfall trends at Ainamoi Chief's Camp, Koru Bible School, Nandi Hills Savani and Tinderet Tea stations, but contrast with observations at Ahero and Kipkelion stations which show significant upward trends. The Chemilil Sugar, Kibwari Tea Estate and Londiani Forest stations did not show any significant trends. The decline in catchment rainfall is consistent with findings by Awange et al. (2008) for

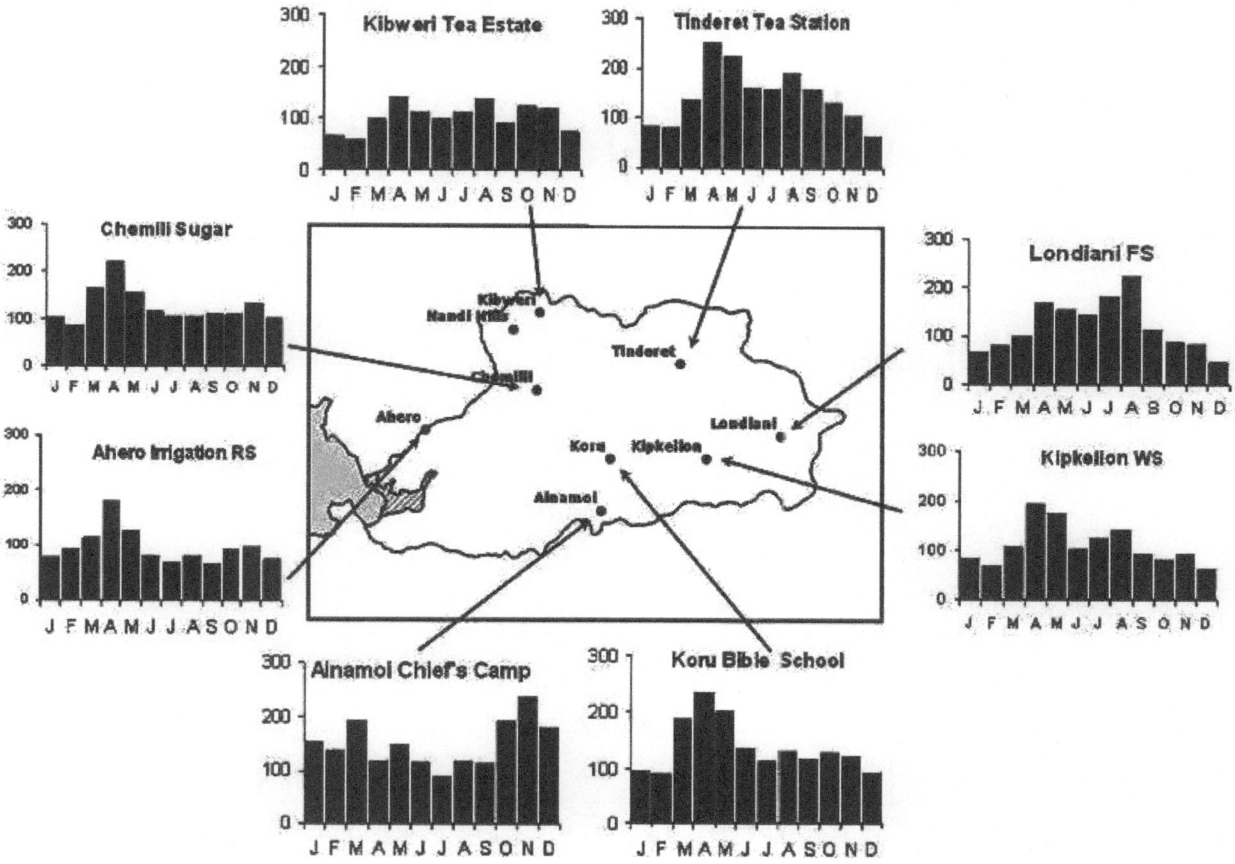

Figure 4. Long-term mean monthly rainfall variability at key selected stations in Nyando River basin.

the Lake Victoria basin. On a longer time scale, Kiiza et al. (2009), has shown that the Lake Victoria basin experienced a predominantly positive trend over the twentieth century with 1960s representing a significant upward jump in the basin rainfall. Mistry and Conway (2003) reported that rainfall in the Lake Victoria basin does not exhibit any marked trends across the twentieth century, although rainfall totals reveal an irregular harmonic pattern of quasi-periodic fluctuations on seasonal and annual time scales.

A visual inspection of the long-term mean monthly rainfall distribution from eight selected stations in the study area (Figure 4) shows a bimodal seasonal pattern of rainfall variability. The long rains occur in March-May (MAM) months at all stations. However, the occurrence of short rains varies between stations. In general, the lowland stations and some highlands stations receive short rains in the October-December (OND) months, while a few highland stations experience short rains in the July-September (JAS) months.

The Pettit and SNHT tests show a significant downward trend in the seasonal long rains (March-May) and no significant trends in the seasonal short rains (October-December) (Figure 5). Kizza et al. (2009) showed that, short rains tend to have more trends than the long rains in the Lake Victoria basin. The long rains season contribute 34%, while the short rains season account for 22% of the 1462 mm/year mean annual rainfall. The rains falling in July-September are higher than the short rainy season and contribute more than 24% of the annual total. In general, more than half (54%) of mean annual rainfall is received in the January-June months. Seasonal variations in rainfall respond to the north-south movement of inter-tropical convergence zone (ITCZ) (Millman, 1973; Mistry and Conway, 2003; Ogallo, 1988), and enhanced by a strong nocturnal land breeze from Lake Victoria (Flohn and Fraedrich, 1966) and westerly winds from the Congo basin.

A plot of mean annual rainfall anomalies (Figure 6), following the method suggested by Awange et al. (2007b) to define meteorological drought was used to determine the Drought Severity Index (DSI) (1st quartile) and the wet (3rd quartile) thresholds. Results show that the wet years in the Nyando River basin were observed in 1951, 1961-1964, 1968, 1972, 1977-1978, 1996-1998, 2002 and 2006. The episodic rainfall events of 1961-1964 (*'Uhuru'* rains) and 1996-1998 that exacerbated flooding problems in the entire Lake Victoria basin were reported by many authors (Awange et al.

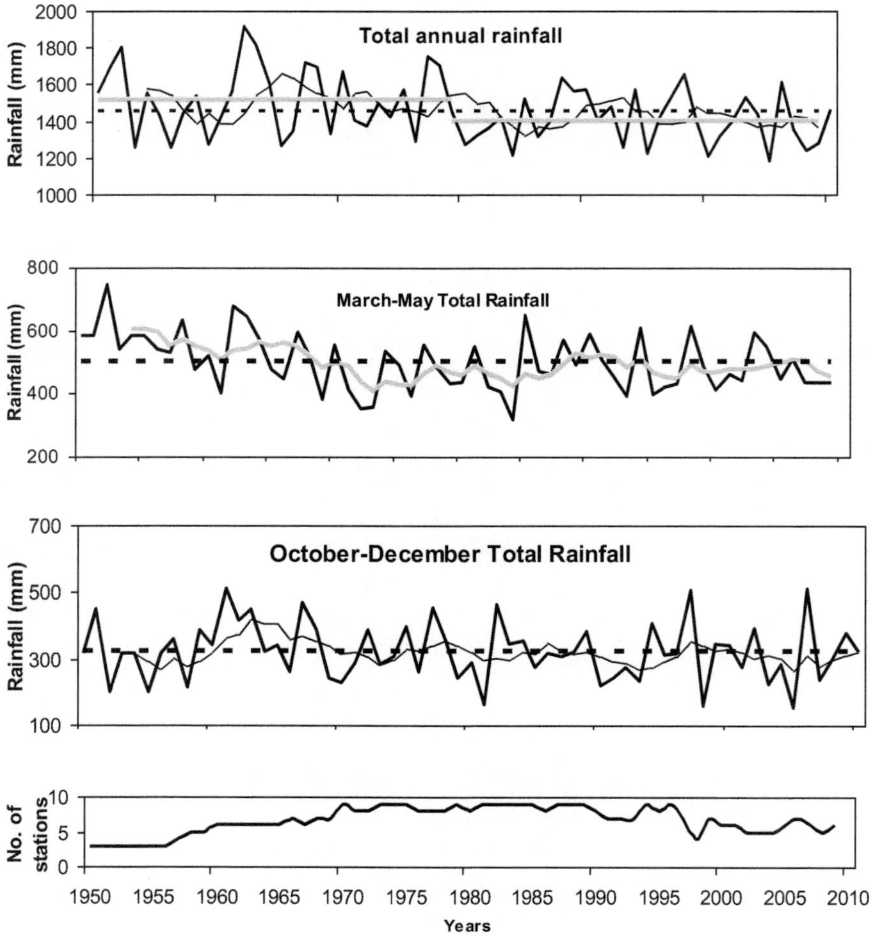

Figure 5. Total rainfall (annual, March-May and October-November) (dark thick line) for study stations with an average (dotted line), 5-year moving average (dark thin line) superimposed and change-point results (thick-grey line). The lower panel shows the number of stations used to compute the means.

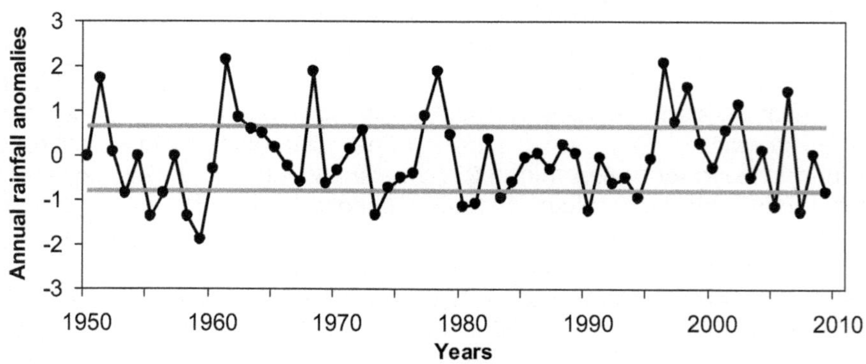

Figure 6. Time series of annual rainfall anomalies defining the wet, normal and drought years, determined by the drought severity index (DSI) and wet thresholds (horizontal lines).

2008; Kizza et al., 2009; Millman, 1973; Mistry and Conway, 2003; Ongwenyi et al., 1993). The drought years were observed in 1954, 1955-1956, 1958-1959, 1973, 1980-1981, 1984, 1990, 1994, 2005, 2007 and 2009, with the 1955-1956 period being the most devastating. The episodic flood and drought events have

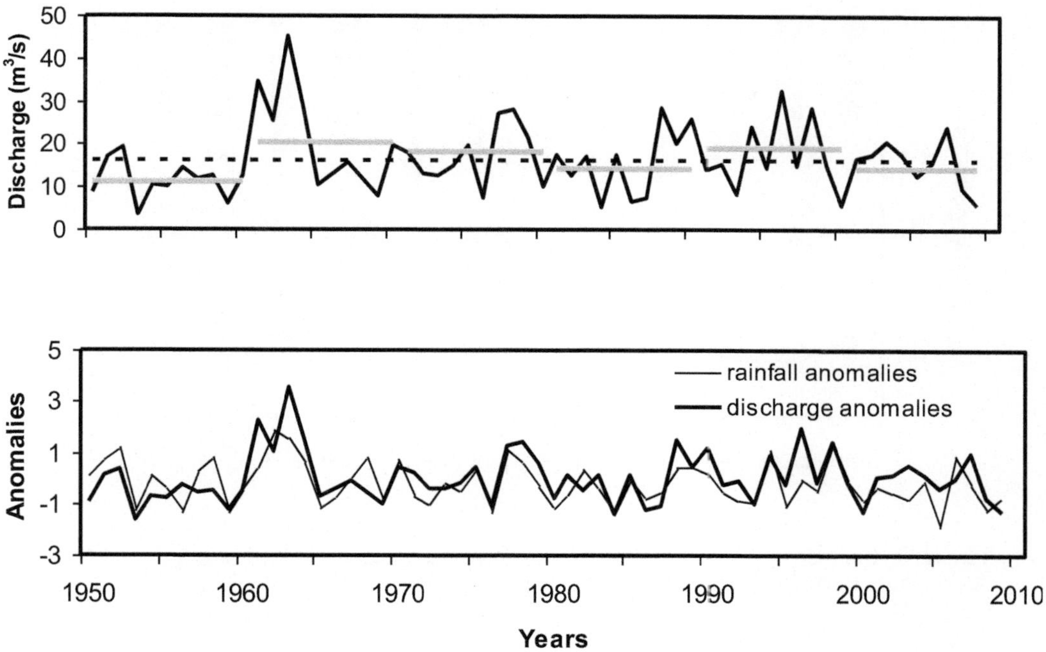

Figure 7. Mean annual discharges (top panel) (dark line) for Nyando River recorded at station 1GD03 showing changes in decadal means (grey horizontal lines) as compared to the 60-year long-term mean (dashed line) and comparison of total annual rainfall and mean annual discharge anomalies.

been linked to the El Niño Southern Oscillation (ENSO), Indian Ocean Dipole (IOD) and many other large-scale climate systems prevalent in the East African region (Ogallo, 1988; Mistry and Conway, 2003). Prominent rainfall cycles over Lake Victoria basin are 2 to 2.5 years, 2.7 to 3.3 years with less prominent cycles at 3.5 to 4.4, 5, 6 to 6.5, 7 to 8 and 10 to 11 years (Ogallo, 1993).

River flow variability

Statistical change-point analyses of river flow records show a shift in 1960 of mean annual discharge for Nyando River (Figure 7) with a stable variance. The decadal mean discharge increased by 80% in the 1961 to 1970 subset from the initial value of 11.45 m^3/s observed in the 1951-1960 subset. The river flows reduced by 11.4 and 21.9% in the next two decades, before rising by 35.0% in 1990s and dropping by 24.0% in the last decade. The decadal mean discharges were higher than the 60-year long-term mean of 16.4 m^3/s from 1960 to 1980 and 1990 to 2000, but remained lower in 1950s, 1980s and the last decade. The flow regime of Nyando River is characterized by high and low flows. The peak flows (based on maximum annual method) show an increasing trend with a change-point in 1987. The seven-day annual minimum discharge records indicate significant increases during and after prominent rainfall events of 1961-1964 and 1996-1998.

There is a positive correlation of 0.64 between mean annual river discharge and catchment mean annual rainfall (Figure 7). The upward shift in Nyando river discharge in the 1960s and 1990s coincided with El Nino events of 1961-64 and 1996-1998. The continued decline in catchment annual rainfall after 1979, may explain why the river has also exhibited declining trends, enhanced by an increase in water demand especially for irrigated agriculture. The highest mean annual discharge of 45.3 m^3/s was recorded in 1963 during the El Niño rains of 1961 to 1964 Fohn (1987) observed that the prominent rainfall event caused drastic rises in the White Nile discharges and widespread flooding. At seasonal scale, the river exhibits a tri-modal flow pattern with a primary peak occurring in May and secondary peaks in August and November in response to the seasonal rainfall pattern (Figure 8).

The mean monthly rainfall for the catchment has the highest value in April, August and November. The hydrograph of the river peaks after a one month time lag after the onset of the long rains, but peaks during the same months in the subsequent short rains periods. The long rains are usually preceded by the long dry spell, hence allowing the catchment to absorb the first rains before the river peaks. The high elevation areas of the catchment are generally wet on the onset of the short rains. Generally, discharges observed in the MAM, JAS and OND months contribute 30, 30 20% of total annual discharge, respectively. The mean monthly discharge is evenly distributed with 49%

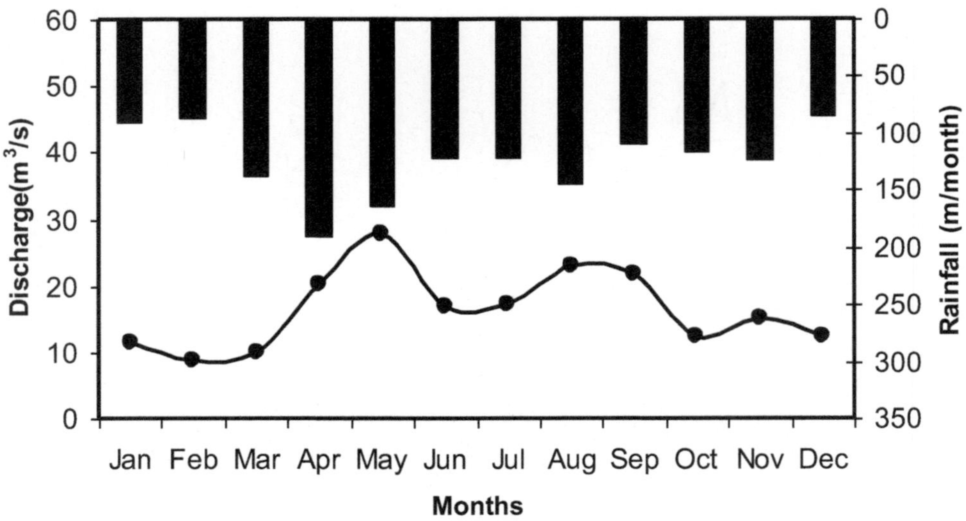

Figure 8. Comparisons of total mean monthly hyetograph (mean of 10 stations) and Nyando River hydrograph at station 1GD03.

occurring in the January-June months.

Flood frequency analysis using the Gumbel EV1 distribution function has indicated that a flood of 1.1 years return period is 41.79 m^3/s at 1% probability of non-exceedance. This may explain why Nyando River inundates the floodplain perennially, with a commencement-to-flow threshold of about 17-30 m^3/s along its reaches. The key factors that may be influencing the occurrence of floods in the wetland are climate variability and land use changes in the catchment.

As our results have shown, the relative large frequency and magnitude of floods in the basin may be explained by climate variability. The most prominent flood events have been observed during the OND short rains, which are influenced by the ENSO phenomena. This observation is consistent with findings by other researchers (Ashagrie et al., 2006) who suggested that the recent occurrence of frequent floods worldwide is likely due to climate variability where an increase in heavy precipitation might lead to an over-proportional increase in runoff due to non-linear processes. There are evidences that changes in land use have influenced the hydrological regime of various river basins (Tu et al., 2004, Ashagrie et al., 2006). It is, however, more challenging to quantify the impact of land use change on the rainfall-runoff relations for large basins where the interactions between land use, climatic characteristics and the underlying hydrological processes are often more complex and dynamic (Uhlenbrook, 2007, 2009).

Lake level fluctuations

The SNHT single test shows that Lake Victoria annual mean levels had a shift in 1961 (Figure 9). The F-test and t-test tests show significant upward jumps. The decadal mean annual lake levels jumped from 1134.0 to 1135.43 m in the 1951-1961 and 1962-1972 subsets, respectively. The level showed an insignificant downward trend in the next two decades before rising slightly in the 1995-2005 subset.

Consequently, the level remained above the long-term mean of 1135.0 m for 43 years since 1962, before dropping drastically by 1.4 m to an average of 1134 m in 2005-2009. The monthly lake levels fluctuations at Kisumu and Entebbe have similar trends with the Nile outflow in the period 1950-1999 (Figure 10). However, the Nile outflow increased steadily from 995.84 m^3/s to 1354.69 m^3/s in November 1999 to December 2004, as lake levels in Kisumu and Entebbe declined during the same period. It is most probable that this trend persisted up to 2009 as the lake level fluctuations in Kisumu bay have shown.

The highest recorded lake level at Kisumu station was 1136.2 m in 1964 after rising by 2.5 m from 1961. Similar observations were made by several authors (Millman, 1973; Kite, 1981; Nicholson, 1998; Awange et al., 2008; Conway, 2002; Flohn, 1987; Conway and Hulme, 1993; Birkett et al., 1999) who found that the lake level remained fairly stable throughout the first half of the twentieth century before rising abruptly in 1961 and continued until 1964, when the lake's level approached the high stands of the last century. According to Nicholson (1998), Lake Victoria was characterized by very low levels in the first half of the nineteenth century and a sharp rise of 4 to 5 m to remain high in the last decades of that century. The prolonged elevation of lake levels are attributed to the initial OND short rains in 1961 and 1997 followed by heavy seasons (Conway, 2002).

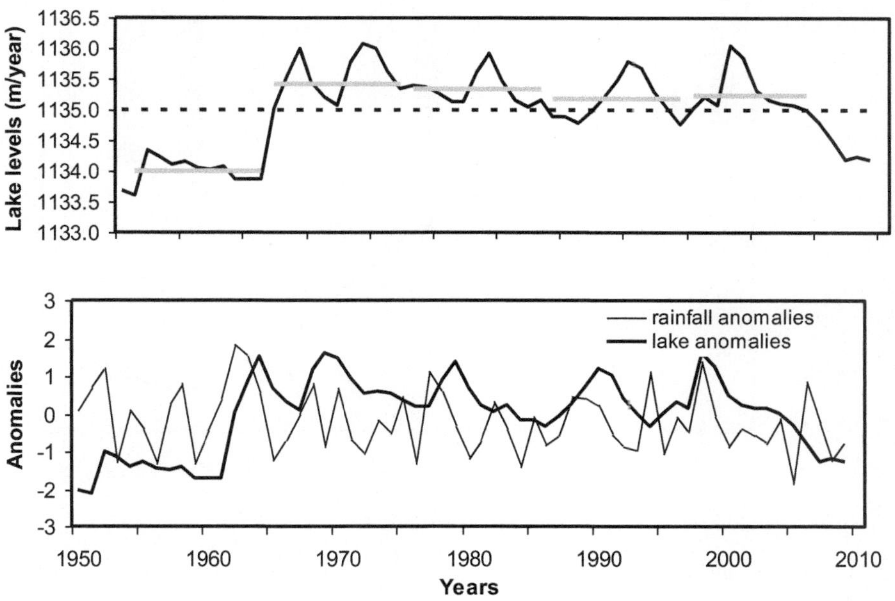

Figure 9. Comparison of mean annual lake level variability (thick dark line) with 60-year long-term mean (dotted line) and decadal means (grey horizontal lines) (top panel) and total annual rainfall and lake level anomalies (bottom panel).

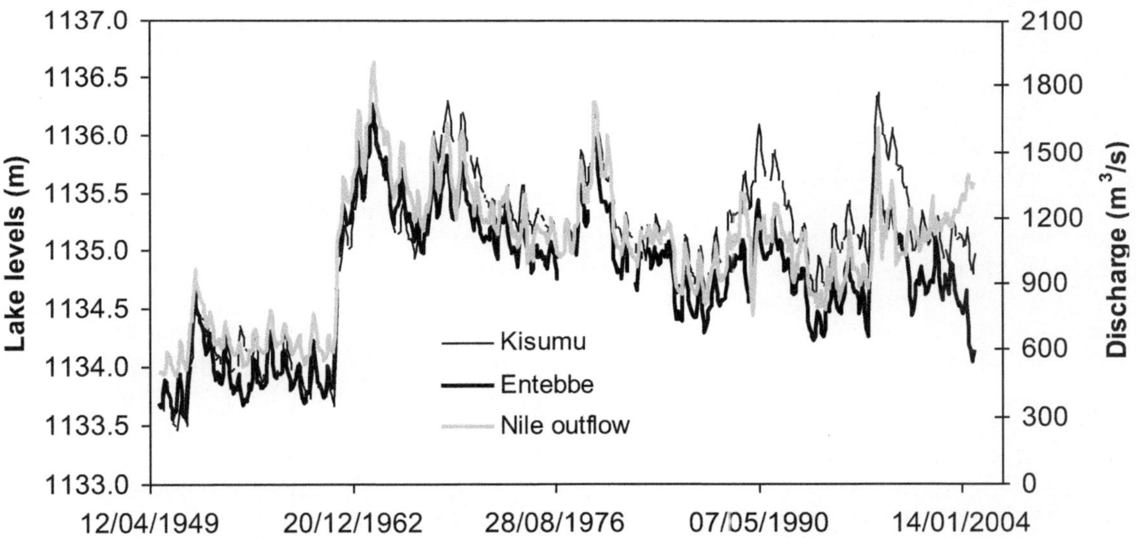

Figure 10. Long-term monthly lake level fluctuations at Kisumu and Entebbe stations and River Nile outflow discharge in the period 1950-2004.

The seasonal lake level pattern is bi-modal, with a primary peak in June (1135.14 m) and secondary peak in December (1134.94 m) (Figure 11). The monthly mean lake level is lowest in October and February at 1134.9 m. Difference between highest and lowest is 0.244 m. The lake levels respond to regional distribution of rainfall in the Lake Victoria basin, most of which occur in the MAM and OND months. The correlation coefficient of the total annual and seasonal rainfall in Nyando Catchment and the mean annual and seasonal lake level is 0.1 and 0.2, respectively. The main factor influencing lake level fluctuations can be explained by direct rainfall on the lake surface.

Mistry and Conway (2003) found a strong correlation between the lake rainfall series and the lake levels, with a time lag of 1 to 2 years. Flohn and Fraedrich (1966) estimated lake rainfall to be 25 to 3% higher than catchment rainfall using a combination of a convection model and rainfall observations around the lake and island stations. Similar differences between lake and

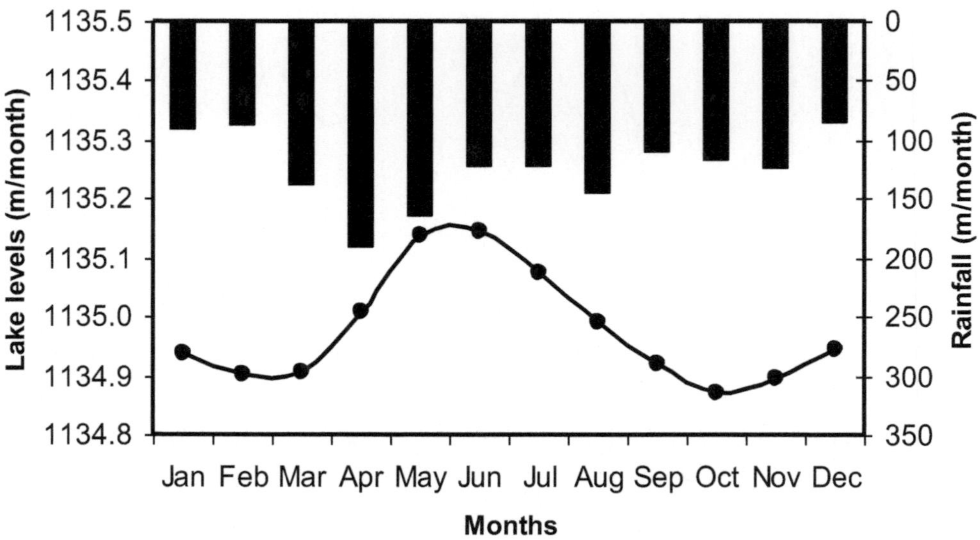

Figure 11. Comparison of Nyando catchment mean monthly rainfall and mean monthly lake levels from 1950-2009.

catchment rainfall were estimated by Ba and Nicholson (1998), using satellite data. Using their results, Yin and Nicholson (2002) determined that the water balance of the lake is dominated by rainfall over the lake (approximately 1791 mm/year) and evaporation (1532 mm/year). Tributary inflow and discharge averages are 338 mm/y and 524 mm/y, respectively. Lake rainfall is 82% of the total inflow; the remaining 18% is contributed by river inflow. River Kagera alone contributes 32% of the river inflow (COWI, 2002). Of the outflow, 76% is due to evaporation from the lake and the remaining 24% (745.2 m^3/s) is the outflow forming the White Nile River at Jinja (Scheren et al., 2000; COWI, 2002). Using the lake rainfall results of Yin and Nicholson (2002), this study has shown that lake rainfall is 23% higher than Nyando Catchment rainfall, which is comparable to findings by Flohn and Fraedrich (1966).

It is unlikely that the recent prominent decline in lake levels (Figure 9) can be explained by a decline in rainfall trends alone. Neither can it be wholly attributed to the prolonged long-term and short-term droughts and decrease in river flows to the lake as previously reported by Awange et al. (2008). The increase in the Nile outflow is perhaps the single most important phenomena that has impacted on the lake level fluctuations since 2000 (Figure 10). This finding is consistent with that of several authors (USDA, 2005; Kull, 2006; EAC, 2006; Kiwango and Wolanski, 2008) who attributed the recent episodic drop in lake levels to water withdrawals for generation of hydropower at the second Kiira Dam in Uganda. This would mean a deviation from the operating rule known as the Agreed Curve of 1954, which is a best estimate of the rating curve of the lake outlet under natural conditions.

River migration

The migration of the Nyando River has been described by Millman (1973) as the most recent geomorphological event to occur in the Kano floodplains. According to him, the river once reached Lake Victoria by a more direct course than at present, flowing westwards through a series of swamps to share a common outlet with Nyamasaria River at Nam Thoe swamp, southeast of Kisumu City. In the last 60 years, the river channel has been migrating laterally in an eastward direction and today debouches into Lake Victoria through the Miruka swamp (Nyando Wetland), having been joined by numerous tributaries from the surrounding southern highlands (Figure 12). The river mouth has been quite dynamic, moving further southwards and then northwards in Nyakach Bay, leading to the development of a delta (bird's foot type). In the period 1950-1973, the river channel migrated 4,900 m in an eastward direction representing a lateral displacement rate of 213 m/year. During this period, the river mouth was displaced 1,300 m southwards to Kusa. In the period 1973-2008, the river had a maximum lateral displacement of 1,500 m at a rate of 42 m/year in an easterly direction. The river mouth was displaced 3,000 m northwards and split into three channels to form the present delta.

The migration of Nyando River has been explained by natural processes and anthropogenic activities. Millman (1973) explained that the main reason why the river course changed its original course towards the Nam Thoe swamp in Kisumu, was because it was blocked by detritus material from small rivers flowing southwards from Nandi Escarpment to Winam Gulf of Lake Victoria. Similarly, changes in the lateral displace-

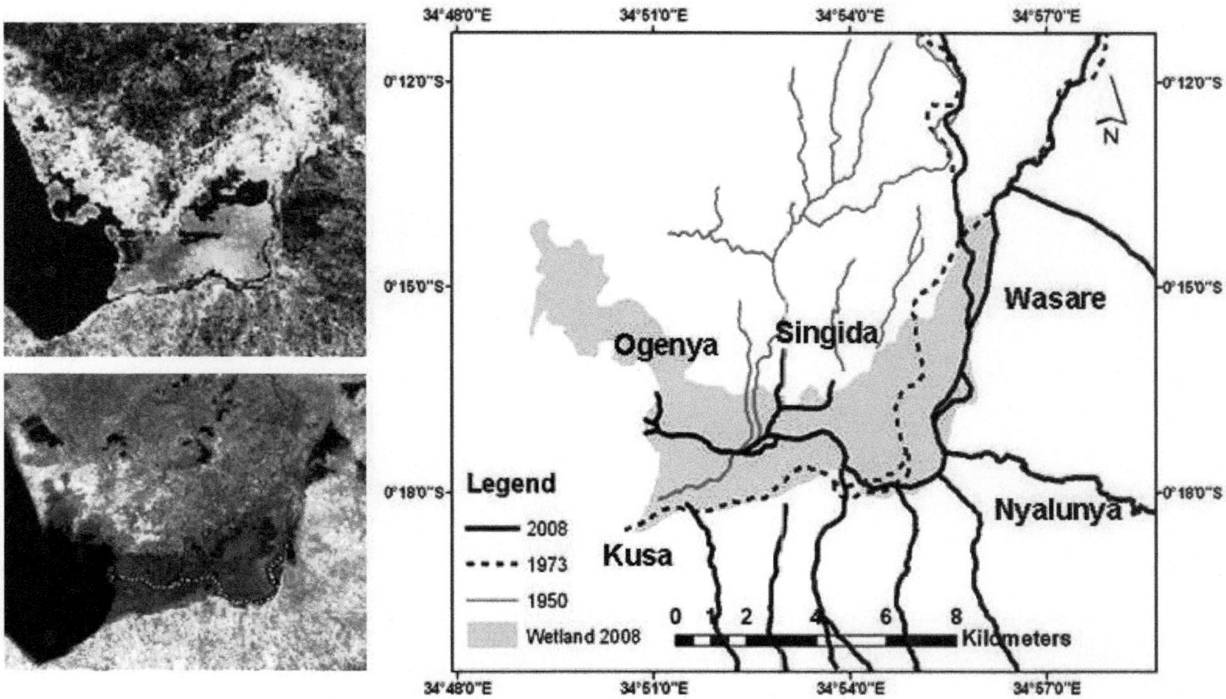

Figure 12. Landsat MSS imagery (Band 4) taken on 1st February 1973 (top left panel), Landsat 5 ETM+ dated 28th September 2008 (Band 5 with geo-referenced way points in small white dots) (bottom left panel) and polyline map showing the river migration patterns in the period 1950-2008 (right panel).

ment of the river course in the last 60 years is most probably due to deposition of excess sediment load carried by the river itself. The low gradient of the Kano floodplains and the river meanders have triggered deposition of sediments, leading to formation of several small islands, which block and force the river to split and finally change its course. The development of dykes and cut-off meanders to prevent persistent floods in 1975 (Ongwenyi et al., 1993) has limited lateral shift in the river course to within 4 km upstream of the delta. Instead the river channel has deepened 4 to 30 km from the delta as a result of increased conveyance capacity, leaving the wetland and adjacent floodplain decoupled from the river channel most of the time.

The most recent changes in river migration within the wetland area were related to changed land use activities and wildlife movements. According to the local community, the river exploited hippo trucks at Singida (Ligolo, personal comm.), and irrigation channels dug by Miruka at Wasare (Oliewo, personal comm.). The shift of the river mouth from Kusa Beach was linked to the blockage of the river channel by communities to prevent it from moving further south (Odhiambo, personal comm.). Historically, the migration of the Nyando River has been a major cause of conflicts between Nyakach and Kano communities, mainly because of loss or gain of land depending on which side of the bank is displaced. The Nyando River course and its main paleochannel, Miriu are still used today as administrative boundaries of Nyakach, Nyando and Kisumu East districts.

Wetland spatial and temporal changes

In the past 60 years, the Nyando Wetland has undergone drastic changes in its size (Figure 13). The most prominent spatial changes have taken place in a 5 to 10 km zone in the northern landward section (Singida-Kochogo areas) of the wetland. The littoral zone (lakeward section) of the wetland at Ogenya has experienced minimum spatial change of 0.5 to 3 km along the shoreline. The southern and southeastern landward sections were affected along a 1 km strip. The wetland area rose from 5,925 ha in 1950 to 9,925 ha in 1973, and dropped to 4527 ha in 2001 and 4,212 ha in 2008. The wetland increased by 67.5% growth in area in 1950-1973 and declined by 40% in 1973-1986, representing a negative growth rate of 305 ha/y. The wetland size reduced steadily by 24% from 1986- 2001 and 7% from 2001-2008. The flood and drawdown events prevalent in the Nyando Wetland present two scenarios of river-wetland and lake-wetland surface-water exchanges: (i) a connected phase where floodwaters link the river and lake to the wetland to make one hydrological unit, and (ii) an isolated phase during drawdown events when the river and lake are physically

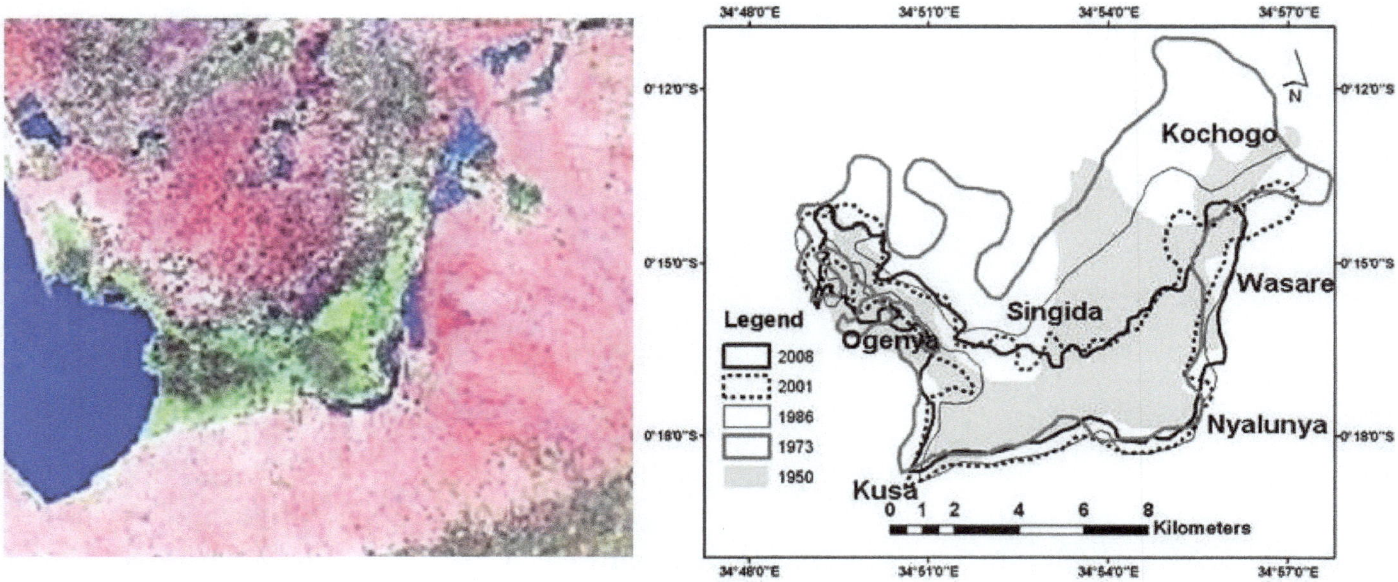

Figure 13. False colour composite of Landsat ETM+ taken on 28th September 2008 showing Nyando Wetland (green), Lake Victoria (deep blue) and other land cover types (pink) (left panel) and multi-temporal polygon maps of Nyando Wetland in the period 1950-2008 (right panel).

separated from the adjacent wetland with no surface water or minimum exchanges respectively. Hydrological connectivity between the wetland and the surface water bodies is important for water supply to the wetland ecosystem. The rise of river and lake levels result in drowning of papyrus wetlands and flooding on the landward section that can transform new lands into a wetland (Kiwango and Wolanski, 2008). However, the drying out during a fall in the lake level, is also a significant source of mortality of papyrus (Mnaya and Wolanski, 2002; Kiwango and Wolanski, 2008).

Flooding facilitates not only exchanges of water, but also sediments and nutrients between the wetland and river/lake systems (Jenkins and Boulton, 2003), considered to be essential for the functioning and integrity of these systems (Lake et al., 2006). Conversely prolonged lack of lateral connectivity can result in considerable habitat loss (Boulton, 2003). The changes in wetland size respond to prevailing hydrological factors and anthropogenic activities. The increase in wetland size in 1950-1973 is attributed to the episodic large flood events experienced in the Lake Victoria region in 1961-64. During this period, the entire Lake Victoria (Millman, 1973) and Nyando River (Ongwenyi et al., 1993) were connected with Kano floodplains, as a single hydrological unit. According to Ongwenyi et al. (1993), the frequent overtopping of Nyando River resulted in the development of swamps in Kano floodplains. It is likely that the wetland size was bigger during the floods than captured by satellites in 1973. Historically, the papyrus dominated swamps were viewed as non-productive nuisance areas, and were better converted to agricultural use (Juma, 1989; Lake Basin Development Authority, 1989; Muthuri, 1989). Consequently, the Nyando and Ombeyi Wetlands were drained for the large-scale West Kano and Ahero Rice Irrigation Schemes, through a government initiative.

The decline in wetland size in 1973-1986 was influenced by several key factors; (i) modification of the Nyando River (ii) prolonged drought in the 1980s and (iii) the establishment of small-scale rice irrigation schemes. In 1975, dykes measuring 8 km in length were constructed along the river banks from Ahero downstream to prevent persistent floods (Ongwenyi et al., 1993). In addition, the river was trained (cut-off meanders) to increase its conveyance capacity. While these human activities prevented flooding and reduced property and human losses in the area adjacent to the dykes, the frequency and depth of floods and flood inundation patterns in the Nyando Wetland were drastically altered. The prolonged drought of early 1980s meant that the lateral connectivity of the wetland and river was lacking for a longer period. The situation was further exacerbated by the declining water table as evidenced by the drying up of shallow wells, most commonly used for potable water by riparian communities (DHV Consulting Engineering, 1988).

In the 1986-2001 period, patterns of land use changes and migrations in the Nyando Wetland had emerged (Figure 14). The papyrus loss increased with further conversion of the wetland to rain-fed agriculture and expansion of small-scale irrigation schemes, livestock grazing and settlements. The drainage and

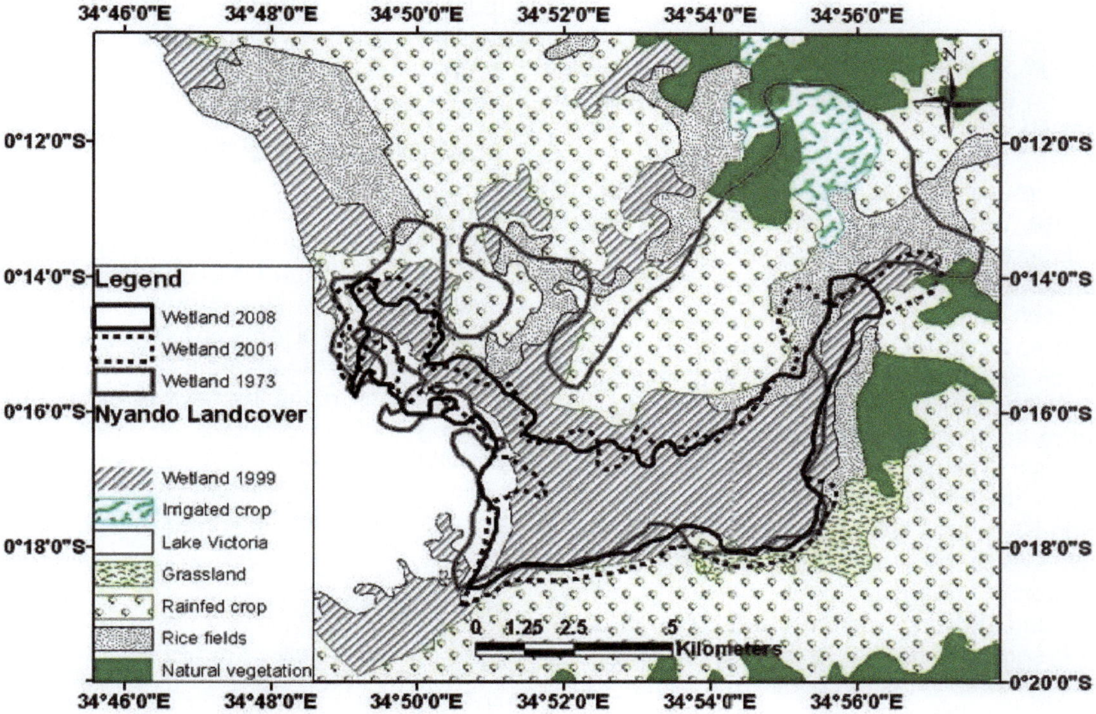

Figure 14. Land use and land cover (LULC) map of Nyando Wetland and its surrounding area in 1999, in relation to the wetland area coverage in 1973, 2001 and 2008 (source land use map: Afri-Cover 1999).

cultivation of papyrus swamps had intensified around the Kenyan part of Lake Victoria (Juma, 1989). The loss of papyrus in the landward and lakeward (littoral zone) sections, and riparian corridor occur quite sequentially. The process starts with the decline in lake and river levels making most parts of the wetland accessible for agricultural activities, livestock grazing and settlement. A combination of these factors decreases the overall rate of germination of papyrus and the opened fields become permanent and are expanded over time (Owino and Ryan, 2007). In this way, *C. papyrus* loses its key ecological functions (Harper and Mavuti, 2004) as they become more accessible for further clearance and animal trampling and livestock grazing. When episodic floods prevail, like the one that happened during the El Niño rains of 1996-1998, human activities in the wetland are abandoned and communities migrate to higher ground, giving room to papyrus to quickly re-generate. With persistent perennial floods, the seasonal calendar of activities for riparian communities in the wetland environment was clearly defined. These patterns of papyrus destruction were observed in other papyrus dominated sites in Kenya (Harper and Mavuti, 2004, Owino and Ryan, 2007) and around Lake Victoria (Lung'aya, 2001; Kansiime et al., 2007; Kiwango and Wolanski, 2008).

In general, there were minimal land use changes in the wetland area in the period 2001 to 2008. However, the wetland contracted on the southern landward section but extended on the lakeward section. The shift of the wetland towards the lake is likely due to the receding lake levels that result in exposure of the littoral zone. This may be related to the drastic drop in lake levels in recent times, mainly attributed to increased water withdrawal for generation of hydropower at the second Kiira Dam in Uganda (EAC, 2006; Kiwango and Wolanski, 2008). In addition, the expansion of the large-scale and small-scale rice irrigation schemes has increased water demand for irrigation and further altered the lateral connectivity of the wetland and river channel.

For example, the Ahero Pilot Irrigation Scheme pumps 544,081 m^3/day (6.30 m^3/s) directly from the river, which is more than the 60 year long-term 7-day minimum annual flow of 239, 328 m^3/day (2.77 m^3/s) and the reserve of 138,240 m^3/day (1.6 m^3/s). The reserve is defined as the amount of discharge required to meet environmental flows and basic human needs. With the planned expansion of the scheme and coupled with the numerous irrigation diversions downstream, there is clear need to provide storage facilities to meet domestic, public and irrigation water requirements. This may be the reason why a multi-purpose dam has been proposed to be developed at Koru at the foothills of the Nandi Escarpment and Kano floodplains. The operation of the multi-purpose dam is likely to regulate the river flow and perhaps alter the hydrological functioning of the wetland. In general, the environmental impacts assessment studies to be carried out during the feasibility phases of any proposed upstream dams should assess the Nyando Wetland water requirements.

Field surveys revealed that the shoreline has been exposed at Kusa Beach and turned into a cultivation zone, while supporting several fish ponds. Similarly, the Nyando delta has translated lake ward and the fringing papyrus has been replaced by hippo grass along the rim of Nyakach bay and the riverine riparian zone. The hippo grass is targeted for fishing and livestock grazing, especially during the dry season. Papyrus on the landward section in Ogenya and Singida Beaches is fast being replaced by sugarcane plantations, while the left bank and portions of the right bank of Nyando River at Wasare and Nyalunya are now covered by invasive species, most probably originating from the catchment. In this way, the chances of papyrus to re-generate and grow to maturity are quite limited, while the fish breeding grounds are severely destroyed. As observed by other researchers (Kansiime and Nalubega, 1999; Awange and Ong'ang'a, 2006; Owino and Ryan, 2007; Kiwango and Wolanski, 2008), the exposed shoreline and riparian zones are quickly encroached for settlements, livestock grazing and cultivation.

Conclusions

This study has shown that the evolution of the Nyando Wetland is influenced by hydrological and anthropogenic factors. The wetland spatial and temporal changes respond to seasonal wet-dry cycles and occasional episodic flood and drought events as observed in many tropical wetlands. The expansion in wetland size was caused by the El Niño floods of early 1960s that resulted in the papyrus translating on the landward section. The steady decline in wetland size is linked to a combination of key factors including the occurrence of prominent droughts (especially early 1980s) and seasonal dry spells and land use and land cover changes. The main human factors are regulation of the lake levels, modification of the river, irrigation water demand from Nyando River, encroachment of the wetland for agricultural activities, settlements and livestock grazing activities. Migration of the Nyando River is attributed to the blockage of the channel by natural sediment depositional processes, exploitation of hippo trucks and human drainage channels in the wetland. If the human activities continue unabated, the decline in size of the Nyando Wetland is likely to increase, with serious hydrological, ecological and socio-economic consequences. With increasing population pressure, coupled by climate variability, the wetland may be massively converted to irrigated and rain-fed agriculture. Clearly, there is need for trade-offs between conservation efforts and socio-economic activities.

ACKNOWLEDGEMENTS

The research described here was funded by the Netherlands Ministry of Development Cooperation (DGIS) through the UNESCO-IHE Partnership Research Fund (UPaRF). It was carried out jointly with UNESCO-IHE, University of Amsterdam, Egerton University, Water Resources Management Authority, and VIRED International in the framework of the Research Project 'Ecology of Livelihoods in East African Wetlands' (ECOLIVE). It has not been subjected to peer and/or policy review by DGIS, and, therefore, does not necessarily reflect the view of DGIS.

REFERENCES

Alexandersson H (1986). A Homogeneity Test Applied to Precipitation Test. 6:661-675.

Alexandersson H, Moberg A (1997). Homogenization of Swedish temperature data Part 1 Homogeneity tests for linear trends. Int. J. Climatol. 17:25-34.

Ashagrie AG, de Laat PJM, de Wit MJM, Tu Min, Uhlenbrook S (2006). Detecting the influence of land use changes on discharges and floods in the Meuse River Basin - the predictive power of a ninety-year rainfall-runoff relation? Hydrol. Earth System Sci. 10:1-11.

Awange E JL, Ogalo L, Bae KH, Were P, Omondi P, Omute P, Omullo M (2008). Falling Lake Victoria water levels: Is climate a contributing factor? Clim. Change 89:281-297.

Awange JL, Aluoch J, Ogallo L, Omulo M, Omondi P (2007b). An assessment of frequency and severity of drought in the Lake Victoria region (Kenya) and its impact on food security. Climate Res 33:135-142.

Awange JL, Ong'ang'a O (2006). Lake Victoria ecology, resources and environment. Springer, Berlin. pp. 243-257.

Ba MB, Nicholson SE (1998). Analysis of convective activity and its relationship to the rainfall over the Rift Valley lakes of East Africa during 1983-90 using the Metoesat infrared channel. J. Appl. Meteorol. 37: 1250-1264.

Balirwa JS (1995). The Lake Victoria environment: its fisheries and wetlands- a review. Wetlands Ecology and Management 3: 209-224.

Birkett C, Murtugudde R, Allan T (1999). Indian Ocean climate event brings floods to East Africa's lakes and the Sudd Marsh. Geophys. Res Lett. 26:1031-1034.

Boulton AJ (2003). Parallels and contrasts in the effects of drought on stream macroinvertebrate assemblages. Freshwater Biol. 48:1173-1185.

Conway D (2002). Extreme rainfall events and lake level changes in East Africa: recent events and historical precedents. In The East African Great Lakes: Limnology, Palaeolimnology and Biodiversity.

Conway D, Hulme M (1993). Recent fluctuations in precipitation and runoff over the Nile sub-basins and their impact on main Nile discharge. Climatic Change. 25:127-151.

COWI (2002). Integrated water quality/limnology study for Lake Victoria, Final Report Part II: LVEMP, Dar es Salaam.

DHV Consulting Engineers (1988). Water Resources Survey and Survey training Program. Rural Water Supply and Sanitation Programme. Report on Winam, Nyakach, Muhoroni and Nyando Divisions.

EAC (East African Community) (2006). Lake Victoria Basin Commission. Special report on the decline of water levels of Lake Victoria. Arusha, Tanzania.

Flohn H (1987). East African rains of 1961/62 and the abrupt change of the White Nile discharge. Paleoecology of Africa 18:3-18.
Flohn H, Fraedrich K (1966). The daily periodic circulation and distribution of rainfall over Lake Victoria. Meteorologische Rundschau 19:157-165.
Harper DM, Mavuti KM (2004). Lake Naivasha, Kenya: Ecohydrology to guide the management of a tropical protected area. Ecohydrol. Hydrobiol. 4:287-305.
Hughes CE, Binning GP, Willgoose GR (1998). Characterization of the hydrology of an estuarine wetland. J. Hydrol. 211:34-49.
Gichuki J, Dahdouh-Guebas F, Mugo J, Rabuor CO, Triest L, Dehairs F (2001). Species inventory and the local uses of the plants and fishes of Lower Sondu Miriu wetland of Lake Victoria, Kenya. Hydrobiolgia 458:99-106.
Gopal B, Junk WJ (2000). Biodiversity in wetlands: an introduction. In: Gopal B, Junk WJ, Davies JA (eds) Biodiversity in wetlands: assessment, function and conservation. Backhuys, Leiden, pp. 1-10.
Greeson PE., Clark JR, Clark JE (1979). Wetland Functions and Values: the State of our Understanding. American Water Resources Association. Minnepolis, MN, USA.
Jenkins K, Boulton AJ (2003). Connectivity in a Dryland River: Short-Term Aquatic Microinvertebrate Recruitment Following Floodplain Inundation Ecol. 84:2708- 2723.
Jones MB, Muthuri FM (1984). The diurnal course of plant water potential, stomtal conductance and transpiration in a papyrus (Cyperus papyrus L.) canopy. Oecologia 63:252-255.
Juma C (1989). Public policy and the Yala Swamp. Resources 1:21 -22.
Junk WJ, Bayley PB, Sparks RE (1989). The flood pulse concept in river-floodplain systems. Canadian Special Publication of Fisheries and Aquatic Sciences, Proceedings of the International Large River Symposium 106:110-27.
Kansiime F, Saunders MJ, LOISELLE SA (2007) Functioning and dynamics of wetland vegetation of Lake Victoria: an overview. Wetlands Ecol. Manage. 15:443- 445.
Kansiime F, Nalubega M (1999). Wastewater treatment by a natural wetland: the Nakivubo swamp, Uganda. Processes and implications. PhD Thesis, A.A. Balkema Publishers, Rotterdam, The Netherlands.
Kipkemboi J (2006). Finger ponds: seasonal integrated aquaculture In East African freshwater wetlands; Exploring their potential for wise use strategies. PhD Thesis UNESCO-IHE pp.1-4.
Kite GW (1981). Recent changes in level of Lake Victoria. Hydrol. Sci. Bull. 26:233-243.
Kiwango YA, Wolanski E (2008). Papyrus wetlands, nutrients balance, fisheries collapse, food security, and Lake Victoria level decline in 2000-2006. Wetlands Ecol. Manage. 16:89-96.
Kiiza M, Rodhe A, Chong-Yu Xu, Ntale EHK, Halldin S (2009). Temporal rainfall variability in the Lake Victoria Basin in East Africa during the twentieth century. Theor. Appl. Climatol. 98:119-135.
Kull D (2006). Connections between recent water level drops in Lake Victoria, Dam operations and drought. http://www.irn.org/programs/nile/pdf/060208vic.pdf.
Lake S, Bond N, Reich P (2006). Floods Down Rivers: From Damaging to Replenishing Forces. Advances in Ecological Research, 39: 41-62.
Lake Basin Development Authority (1989). Reclaiming the Yala Swamp. Resources 1:9-12.
Lung'aiya H, Sitoki L, Kenyanya M (2001). The nutrient enrichment of Lake Victoria (Kenyan waters). Hydrobiology 458:75-82.
Macalister C, Mahaxay M (2009). Mapping wetlands in the Lower Mekong Basin for wetland resource and conservation management using Landsat ETM images and field survey data. J. Environ. Manage. 90:2130-2137.
MacLean IMD (2004). An ecological and socio-economic of biodiversity conservation of East African wetlands. PhD Thesis, School of Environmental Sciences, University of East Anglia, Norwich, 210.
MacLean IMD, Tinch R, Hassal M, Boar R (2003). Towards optimal use of tropical wetlands: an economic valuation of goods derived from papyrus swamps in southwest Uganda. CSERGE Working Paper ECM 03-10. University of Anglia, Norwich.
Mafabi P (2000). The role of wetland policies in the conservation of water birds: the case of Uganda. Ostrich. 71:96-98.
Melack JM, Hess LL, Sippel S (1994). Remote sensing of lakes and floodplains in the Amazon Basin. Remote Sens. Rev. 10:127-142.
Millennium Ecosystem Assessment (2005). Ecosystems and human well-being: synthesis. Island Press, Washington. p. 155.
Millman R (1973). Problems of the Natural Environment of the Kano Plains of Western Kenya. Description orographique, hydrologique et climatolique d'une region naturelle a proximite du lac Victoria (golfe du Kavirondo). 13(50):182-192.
Mistry VV, Conway D (2003). Remote forcing of east African rainfall and relationships with fluctuations in levels of Lake Victoria. int. J. Climatol. 23:67-89.
Mitsch WJ (2010). Conservation, restoration and creation of wetlands: a global perspective. In Comin, F. (ed.), Ecological Restoration: A Global Challenge. Cambridge University Press, Cambridge. pp. 175-188.
Mitsch WJ, Gosselink JG (1993). Wetlands, 2nd ed. Van Nostrand Reinhold, New York.
Nicholson SE (1998). Historical fluctuations of Lake Victoria and other lakes in the Northern Rift valley of East Africa. In: Lehman JT (ed) Environmental Change and Response in East African Lakes. Kluwer, Dodrecht. pp.7-35 .
Mnaya B, Wolanski E, Kiwango Y (2006) Papyrus wetlands a lunar-modulated refuge for aquatic fauna. Wetlands Ecol. Manage. 14:359-363.
Muthuri FM, Jones MB, Mimbamba SK (1989). Primary productivity of papyrus Cyperus papyrus in a tropical swamp, Lake Naivasha, Kenya. Biomass 18:1-14.
Odada EO, Ochola WO, Olago DO (2009). Understanding future ecosystem changes in Lake Victoria basin using participatory local scenarios. Afr. J. Ecol. 47:147-153.
Ogallo LA (1993) Dynamics of the East African climate. Proc Indian Acad Sci (Earth Planet Sci.) 102:203-217
Ogallo L (1983) Relationships between seasonal rainfall in East Africa and the Southern Oscillation. J. Clim. 8:31-43.
Okeyo-Owuor JB (1999). A review of biodiversity and socioeconomics research in relation to fisheries in Lake Victoria. IUCN Eastern Africa Program, Nairobi. p. 77.
Ongwenyi GS, Denga FGO, Abwao P, Kitheka JU (1993). Impacts of floods and drought on the development of water resources of Kenya: case studies of Nyando and Tanacatchments. Hydrology of Warm Humid Regions (Proceedings of the Yokohama Symposium, July 1993). IAHS Publ. No. 216.
Osumba JJL, Okeyo-Owuor JB, Raburu PO (2010). Effect of harvesting on temporal papyrus (Cyperus papyrus) biomass regeneration potential among swamps in Winam Gulf wetlands of Lake Victoria Basin, Kenya. Wetlands Ecol. Manage. 18:333-341.
Owino AO and Ryan PG (2007). Recent papyrus swamp habitat loss and conservation implications in western Kenya. Wetlands Ecol. Manage. 15:1-12.
Perez AC, Damen MCJ, Geneletti D, Hobma TW (2001). Monitoring a recent delta formation in a tropical coastal wetland using remote sensing and GIS. Case Study: Guapo River Delta, Laguna de Tacarigua, Venezuela.
Pettit AN (1979). A non-parametric approach to the change-point detection. Appl. Stat. 28 126-135.
Richardson SM, Hanson JM, Locke A (2002). Effects of impoundment and water-level fl uctuations on macrophyte and macroinvertebrate communities of a dammed tidal river. Aquat. Ecol 36:493-510.
Saggerson EP (1947). The Geology of Kisumu District., Report no. 21, Nairobi.
Scheren PA, Zantingha, Lemmens AMC (2000). Estimation of water pollution sources in Lake Victoria, East Africa: application and elaboration of the rapid assessment methodology. J. Environ. Manage. 58:235-248.
Schuyt KD (2005). Economic consequences of wetland degradation for local populations in Africa. Ecol. Econ. 53:177-190.
Shakleton RM (1952). A contribution to the Geology of Kavirondo Rift Valley. Q. J. Meteorol. Soc. 196 (4).
Siegel S, Castellan NJ (1988). Non-parametric Statistics for the Behavioural Sciences. McGraw-Hill, New York, USA.
Uhlenbrook S (2009). Climate and man-made changes and their impacts on catchments. In: Kovar P., Maca P., Redinova J. (eds.).

Water Policy 2009, Water as a Vulnerable and Exhaustible Resource. Proceedings of the Joint Conference of APLU and ICA, 23-26 June 2009, Prague, Czech Republic. pp. 81-87.
Uhlenbrook S (2007). Biofuel and water cycle dynamics: what are the related challenges for hydrological processes research? HP Today. Hydrol. Proc. 21(26):3647-3650.
Tu Min., Hal MJ, de Laat PJM, de Wit MJM (2004). . Hydrology: Science and Practice for 21th century. 1:235-238.
USDA (United States Department of Agriculture) (2005). Low Water Levels Observed on Lake Victoria. Production Estimates and Crop Assessment Division, Foreign Agricultural Service, USDA.
van Dam AA, Kipkemboi J, Zaal F, Okeyo-Owuor JB (2011). The ecology of livelihoods in East African papyrus wetlands (ECOLIVE) DOI 10.1007/s11157-011-9255-6. Rev. Environ. Sci. Biotechnol.
van Dam A.A, Dardona P, Kelderman F, Kansiime F (2007). Simulation model for nitrogen retention in a papyrus wetland near Lake Victoria, Uganda (East Africa). Wetlands Ecol. Manage. 15:469-480.
van Geest GJ, Coops H, Roijackers RMM, Buijse AD, Schefer M (2005a). Succession of aquatic vegetation driven by reduced waterlevel fluctuations in floodplain lakes. J. Appl. Ecol. 42:251-260.
Walsh SJ, Cooper JW, von Essen IE, Gallager KR (1990). Image enhancement of Landsat Thematic Mapper data and GIS data integration for evaluation of resource characteristics. Photogram. Eng. Remote Sens. 56:1135-1141.
Waruru B, Njoroge C, Wanjogu S (2003). Biophysical baseline information for the Nyando catchment area: The soils of the Nyando catchment area Reconnaissance Soil Survey Report.
Weilhoeferi CL, Pan Y, Eppard S (2008). The effects of river floodwaters on floodplain wetland water quality and diatom assemblages. WETLANDS. 28(2):473-486.
Wilcox DA, Meeker JE (1991). Disturbance effects on aquatic vegetation in regulated and unregulated lakes in northern Minnesota. Can. J. Bot. 69:1542-1551.
WMO (1982). Hydrometeorological survey of the catchments of
Lake Victoria, Kyoga and Albert: project findings and recommendations. WMO/UNDP RAP/73/001, WHO, Geneva.
Yin X, Nicholson SE (2002). Interpreting annual rainfall from the levels of Lake Victoria. J. Hydrometeorol. 3:406-416.

12

Household-scale environmental health in the Ezulwini Valley, Swaziland

James D. Proctor, Ben Rathbun, Erin Scheibe and Samantha Shafer

Environmental Studies Program-MSC 62, Lewis & Clark College, 0615 SW Palatine Hill Road, Portland, OR 97219 USA.

An environmental health assessment of 210 households located in four communities in the Ezulwini Valley, Swaziland, is summarized. The assessment focused on household-scale environmental health in the context of four key resource sectors: drinking water, energy, solid waste and human waste, with availability and perceived adequacy considered for each sector. The survey was administered in the field by small teams of students alongside Swazi community members, utilizing a snowball sampling strategy with stratification by economic class. Electronic administration via mobile devices assisted in geolocating records, minimizing entry error and rapidly compiling results for daily review and analysis. Results indicate challenges in household access to basic resources and resource impacts, even in this relatively developed part of Swaziland; these results varied considerably by community and economic class, and were only somewhat comparable to previous national-scale assessments. In a larger context, international efforts toward improving household-scale environmental health conditions (e.g., via related UN Millennium Development Goals) are laudable, yet these results corroborate other research suggesting that progress can be difficult to measure, and is decidedly uneven by household location and socioeconomic status.

Key words: Swaziland, environmental health, water, sanitation, solid waste, fuelwood, Millennium Development Goals (MDGs).

INTRODUCTION

Situating global environmental health

The field of environmental health is as broad as our many needs for well-being and safe surroundings. According to the World Health Organization (WHO), environmental health "...addresses all the physical, chemical and biological factors external to a person, and all the related factors impacting behaviours. It encompasses the assessment and control of those environmental factors that can potentially affect health" (www.who.int/topics/environmental_health). The most recent WHO report (2013a) launches with a foreword citing the eight UN Millennium Development Goals (MDGs) for 2015, in which environment and health are central to at least four (www.un.org/millenniumgoals).

As with most global health statistics, however, environmental health is a key axis of differentiation between socioeconomic classes. For instance, WHO/UNICEF Joint Monitoring Programme for Water Supply and Sanitation (JMP) statistics for 2011 indicate that 67% of the population from World Bank low income group countries have access to improved drinking water sources, as compared to 99% in high income group countries, and only 37% from low income countries have access to improved sanitation as compared to 100% from high income countries (for further details, see WHO Global Health Observatory Data Repository, apps.who.int/gho/data; all statistics from 2011). One would expect that inequities in environmental health are present at the sub-country scale as well, given evidence suggesting marked disparities in other health indicators among the most and least wealthy members of a wide range of countries (WHO, 2013b, 143ff). Eradication of these disparities has been a major goal of the UN MDGs, yet the empirical question remains as to whether progress in global environmental health is moving quickly among poor people of the world; at least some published reviews of progress toward MDGs suggest otherwise (Poku and Whitman, 2011; Usua et al., 2012; Fehling et al., 2013).

Environmental health in Africa

Many people living in Africa face particular challenges relative to the rest of the world. The most recent World Health Organization statistical summary (2013b) launches by comparing regions of the world in progress toward MDG goals relative to 1990, and Africa's improvement in under-five mortality rates, as one indicator of health conditions, is higher than many other parts of the world, with an annual rate of decline close to 2% (p. 13). Yet absolute rates in maternal mortality, tuberculosis, HIV/AIDS and access to improved drinking water and sanitation are still considerably worse in Africa than in other parts of the world (pp. 14-17). These statistics suggest good progress over the last several decades, offset by continued struggles in securing adequate standards of general and environmental health (Gabay, 2011; Groenewald, 2011; Mukonka et al., 2014), especially in an equitable manner across socioeconomic class (Kangalawe et al., 2008; Yanda and William, 2010).

In the last five years, a great deal of attention has been paid to environmental health conditions in Africa. The proceedings of one major conference including 52 African country delegations and held in Libreville, Gabon in August 2008 launches by stating that "Africa continues to face the 'traditional' challenges of poor access to safe drinking water, hygiene and sanitation. Yet the continent must now also deal with new and emerging challenges, including the effects of climate change on health, accelerated urbanization and indoor and outdoor air pollution" (WHO, 2009a, 7). A frequently cited statistic in the proceedings is the estimate that nearly one-quarter of all deaths in Africa in 2002 were attributable to "environmental risk factors" (WHO, 2009a, 23; cf. Ogunseitan, 2007). The conference resulted in the Libreville Declaration on Health and Environment in Africa (WHO, 2009b), suggesting widespread commitment to improving environmental health conditions on the continent. The conference was followed by a smaller implementation meeting (WHO, 2009c). Overall, WHO pursued a wide range of public and environmental health initiatives in Africa in the latter part of the first decade of the 21st century (WHO, 2010).

The status of Swaziland

Swaziland is a small landlocked country in southern Africa, with a 2011 population of approximately 1.2 million inhabitants. According to World Bank statistics (data.worldbank.org), its general economic conditions are somewhat better than those of its peers: for instance, its 2011 gross national per capita income was estimated at US$2,830, over twice the average for developing countries in sub-Saharan Africa (US$1,248). Yet its health struggles are considerable: Swaziland has the highest HIV infection rate of any country in the world. According to the World Health Organization, in 2011 190,000 people or 15.8% of the overall population were HIV positive, including 26% of the population aged 15 to 49. This has contributed in large part to a life expectancy at birth of only 49 years, and at least 75,000 orphans due to parent death from AIDS (WHO Global Health Observatory Data Repository, apps.who.int/gho/data, all statistics from 2011).

The general environmental status of Swaziland has been summarized in a recent comprehensive assessment by the Swaziland Environment Authority (2012), focusing on five themes: land, water, atmosphere, biodiversity and human development. In the context of human health, the report concludes that "Swaziland faces many challenges in the human health sector among which limited capacity in terms of human and financial resources is one, and weak information systems particularly in relation to monitoring and evaluation of different priority health programmes is another" (p. 258). To compare this status with other countries in Africa, WHO health statistics estimate that 28% of Swaziland's population did not have access to improved drinking water sources in 2011, and 43% did not have improved sanitation (WHO, 2013b, 33-4), placing Swaziland in the 46th and 79th percentiles, respectively, among African countries (with higher percentiles being better status).

A more detailed health assessment was done via the 2010 Multiple Indicator Cluster Survey (MICS), administered in conjunction with UNICEF, and involving

Table 1. Selected Swaziland Multiple Indicator Cluster Survey (MICS) results, in percentage (n≈4800 households).

Item	Urban	Rural	Overall
Improved water	91	60	67
Water treatment	24	14	15
Improved sanitation	94	73	78

Table 2. Percent household access to improved water in Swaziland by wealth quintile (5 = richest; 1 = poorest).

Wealth index quintile	Improved water access
5	92
4	77
3	66
2	60
1	41

nearly 5000 Swazi households (Central Statistical Office and UNICEF, 2011). The UNICEF MICS instrument focuses on key MDGs, in particular those related to human health. As suggested in Table 1, the MICS found that 67% of the overall population had access to improved water sources and 78% had access to improved sanitation facilities. Though the MICS improved water access result is in rough agreement with the WHO figure provided above, the marked disparity between the MICS improved sanitation result of 78% and the WHO figure of 57% may suggest the need for greater methodological agreement. Table 1 also suggests that only 15% of the population had access to treated water sources, a much smaller proportion than those with access to "improved" sources, suggesting a key disparity to be discussed further below. A comparison between the overall population with access to improved water and those from rural vs. urban areas (Table 1) and poor vs. rich wealth quintiles (Table 2) demonstrates significant geographic and economic class variation masked in the overall percentage, as suggested in the JMP statistics above.

Further evidence on the state of Swaziland's rural environmental health can be gained from a series of annual assessments produced by the Swaziland Vulnerability Assessment Committee (2011, 2012). The SVAC reports include statistics on basic household resources such as water and energy necessary to secure livelihoods: for instance, in 2011 31% of rural respondents indicated an environmental hazard (solid or liquid waste, or liquid or sewage discharges) near their water source (SVAC, 2011, 48), and in 2012, 82% of rural households relied on fuelwood for cooking, vs. only 18% using electricity (SVAC, 2012, 13).

Overall, however, a more comprehensive picture of household access to safe resources, and impacts on these resources, has not been developed for Swaziland. By way of terminology, a household in Swaziland comprises an extended family unit dwelling on a homestead, which typically includes a number of adjacent dwelling and storage structures. Historically, Swazis generally dwelled in a more dispersed fashion on homesteads rather than in a nucleated fashion in villages, and some vestiges of this cultural past remain even in more densely settled parts of the country.

In light of the key role played by basic resources such as water and energy for household well-being, potential health effects of these resources on households (e.g., waterborne disease or indoor air pollution via fuelwood burning), and potential impacts of households on these resources (e.g., contamination by solid or human waste), this study focuses on household-scale environmental health in the context of four key resource sectors: drinking water, energy, solid waste and human waste. These resources are commonly considered in household-scale assessments in Africa (Joséphine et al., 2008; Oyelola and Babatunde, 2008; Ewodo et al., 2009; Awe et al., 2011; Mughogho and Kosamu, 2012), though they are less commonly brought together to present a more comprehensive view.

MATERIALS AND METHODS

Study site: Ezulwini Valley

The Ezulwini Valley (literally, Valley of Heaven) is classified in SVAC reports as part of a peri-urban corridor stretching between Mbabane and Manzini, the two largest urban areas in Swaziland (Figure 1). The Ezulwini Valley has great significance to Swazis, with the Parliament, National Archives, Somhlolo National Stadium, and major traditional sites located here. Additionally, Ezulwini has been a site of considerable commercial development, with numerous high-end resorts and a major shopping center. Based on this traditional and commercial significance, the Ezulwini Valley would appear to be superior to many other locations in Swaziland in terms of basic household resource needs. As a study site, then, Ezulwini is not "representative" of Swaziland (an impossibility at any rate, given geographic and economic class variability in basic resources as noted above); rather, it suggests how households are faring in a relatively developed part of Swaziland outside of Mbabane and Manzini.

Our connection to the Ezulwini Valley involved students from Lewis & Clark College in Portland, Oregon volunteering May through July 2013 in local community neighborhood care points, or NCPs, as part of an overseas program. Neighborhood care points were created in Swaziland in response to the HIV-AIDS crisis, as a means of providing basic food, health, and educational services to vulnerable populations, in particular orphans and vulnerable children (U.S. Fund for UNICEF, 2011). The original NCPs were created following a severe drought in 2003; there are now over 1500 distributed across Swaziland, providing a variety of services and exhibiting a wide range of structural improvements (NERCHA, 2011).

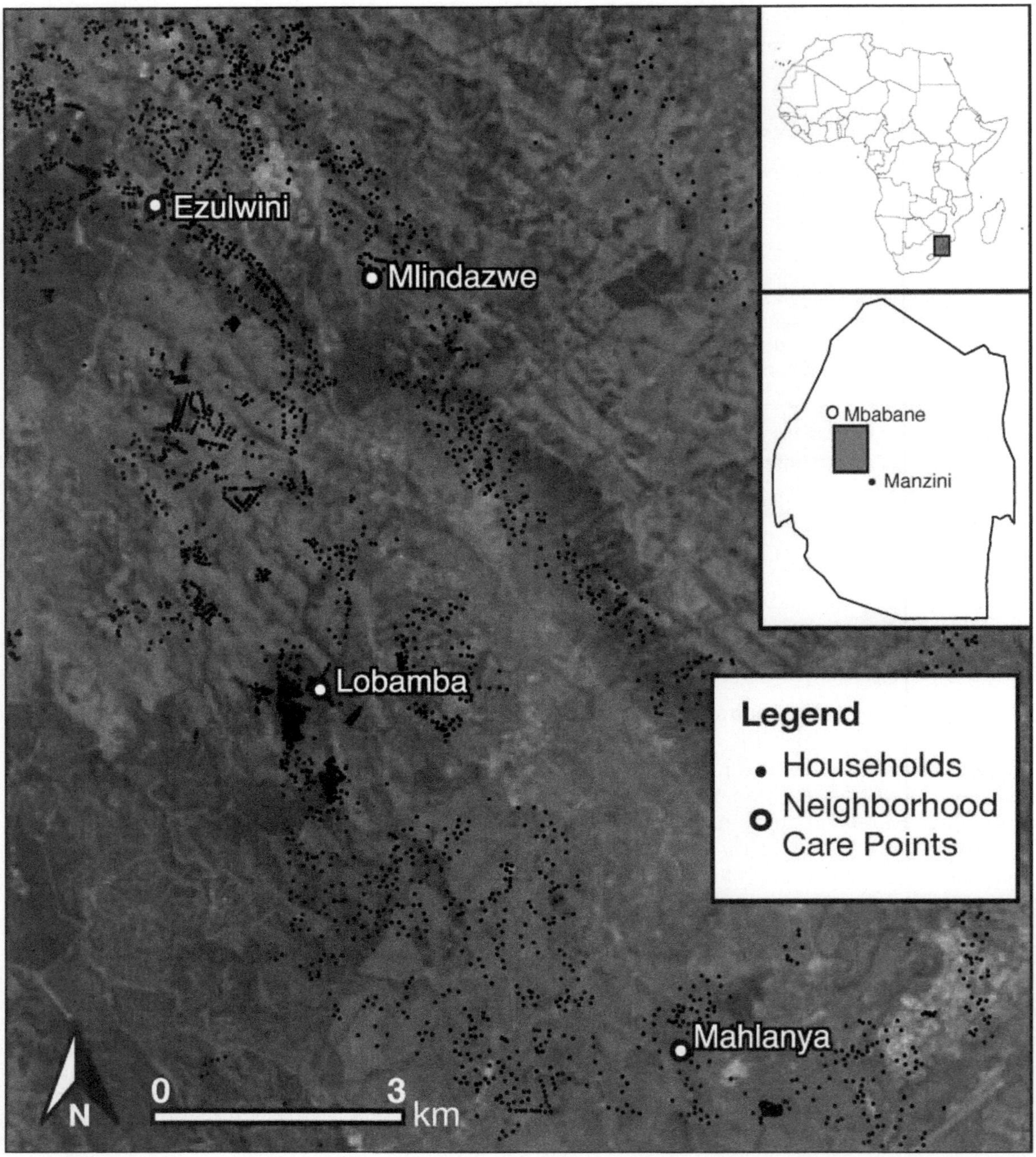

Figure 1. Location map of Ezulwini Valley study site in Swaziland, with four neighborhood care points and surrounding communities included in environmental health assessment.

Students worked in partnership with All Out Africa (www.alloutafrica.com), an organization based in the Ezulwini Valley that places volunteers in a number of local NCPs, which they also support via building assistance, educational materials, and other services. Students volunteered in four NCPs and conducted our environmental health assessment among households in the surrounding communities (Figure 1). Of these four NCPs, Ezulwini and Lobamba NCP are found in relatively densely settled peri-urban areas, whereas Mlindazwe and Mahlanya NCP are found in less densely settled areas with more rural characteristics. All four NCPs we selected are relatively close together, lying within a 12 km span of the Ezulwini Valley.

Research question

Our research question was "What is the condition and perception of water, energy and solid/human waste among households in communities surrounding four NCPs in the Ezulwini Valley?" In addition, we planned to compare results by community and economic level, and with country-scale Swaziland data. As is evident in these questions we were interested not only in the current state of these resources and resource impacts, but in how they were perceived and prioritized among the Swazis who lived in these communities, as any attempt to improve environmental health conditions must be grounded in knowledge of both. We also

Table 3. Environmental health assessment sections.

Section	Content
Pre- Survey Information	Survey location, reference number Economic class surrogates Respondent age/gender
Household General	Settlement duration Number of minor/adults Most important issues
Drinking Water	Sources Treatment Quantitative/qualitative adequacy Notes/recommendations
Household Energy	General energy sources Cooking energy sources Fuel wood cooking method/location Cooking energy adequacy Notes/recommendations
Solid Waste	Disposal locations Community adequacy Notes/recommendations
Human Waste	Sanitation facility locations Community adequacy Notes/recommendations

anticipated that differences would arise across communities and economic class, and in spite of the relatively unrepresentative nature of the Ezulwini Valley as a study location we wanted to compare our results with related data collected for Swaziland.

Survey design

The survey instrument was constructed in consultation with related assessments. Input and translation assistance to SiSwati was provided via key Swazi contacts (including Swazi students attending Lewis & Clark College). The survey was pretested in the field to ensure straightforward and consistent administration. We used actual photos of households in the Ezulwini Valley to train teams in calibrating the low/medium/high economic class categories, thus aiding consistency across community survey teams.

The instrument included six major sections, with primary survey items noted for each in Table 3. The overall flow of the survey was intended to begin with background information collected by survey administrators without prompting respondents, e.g., economic class surrogates and respondent approximate age and gender. It further solicited general household information, followed by each of the four priority resource sectors, starting with drinking water as a straightforward, recognized sector and ending with solid and human waste as more sensitive sectors. A general discussion of respondents' most important community issues was included prior to resource-specific sections to gauge the larger significance of these resources. For all sectors, respondent perception/prioritization was elicited via perceived adequacy items using a three-point scale (inadequate/somewhat adequate/adequate), where adequacy involved the household scale for drinking water (which included both qualitative and quantitative adequacy, that is, sufficient purity and volume) and energy, and the community scale for solid and human waste given the potential for cross-household impacts. The overall survey took between 10 and 30 min to administer per household.

Survey administration and analysis

Following student NCP volunteer work in the mornings, the environmental health assessment was conducted after the NCPs closed at midday, in teams of 1-3 students together with 1-4 Swazis (including Swazi Lewis & Clark students and NCP community members). Our goal was to survey approximately 50 households in each of the four communities, a decision based primarily on afternoon time available in each community, though we did oversample in Lobamba given its larger population, ending up with an overall sample of 210 households (Table 4). Due to technical and practical challenges in surveying a random sample of households, we adopted a snowball sampling strategy, starting with the households of contacts (e.g., cooks or participating youth) in the NCPs, then moving on to neighbors or other contacts. Given our interest in the effects of differing levels of wealth on household-scale environmental health, we informally stratified our community

Table 4. Neighborhood care point (NCP) communities and households.

Household	Ezulwini	Lobamba	Mahlanya	Mlindazwe	Total
Households within 0.5 km radius of NCP	201	401	66	72	740
Households surveyed	50	72	51	37	210

samples by economic class surrogates, primarily structural improvements and amenities (e.g., wall/roof materials, structure size, presence of satellite dish antenna). These sampling and quota decisions mean that our survey is not necessarily representative of these four communities in the Ezulwini Valley, but it does suggest some important characteristics of the households we surveyed, as well as similarities and differences by community and economic class.

The survey was administered using iPad Mini mobile devices running a survey app called Fulcrum (www.fulcrumapp.com). The iPads and Fulcrum app offered a number of advantages, including (a) concurrent administration via survey downloading to multiple devices, followed by uploading of data at the end of each day; (b) automatic geolocation of households, (c) minimization of entry error via response lists, branching logic and required questions; and (d) accumulation of data onto the Fulcrum site for ready monitoring and downloading. We were also able to provide daily updates and reminders directly on the app. The iPads provided community benefits in addition to survey administration: for instance, our students used them with NCP children during their volunteer time, and took (and printed for distribution) photos of NCP and community members upon request.

A small team of students reviewed data at the end of each day to check for entry error, geolocation error, and other potential problems as well as sample coverage. During the first few days of administering the final survey, these data review meetings proved invaluable in monitoring progress and making team-specific recommendations as well. Additional teams did follow-up reconnaissance of community water and fuelwood sources to determine quality. Water sources were tested for total and fecal coliform and inspected visually for probability of surface runoff contamination; fuelwood sources were assayed for extent, major species and reproduction rate.

Data from the household surveys were downloaded from the Fulcrum website in spreadsheet format, then recoded as necessary for descriptive statistical analysis; text responses were summarized to provide better interpretation of numerical results. Data were also uploaded in shapefile format to QGIS, an open-source GIS and mapping program, to discover spatial patterns. Analysis results were shared with Ezulwini Valley community members and Swaziland government officials in a variety of public events.

RESULTS

Drinking water

Selected drinking water results are shown in Figure 2 and Table 5. As noted above, the concept of "improved" drinking water, which has been central to MDG and Swaziland rural development goals, is arguably vague, as most assessments consider as "improved" any piped water irrespective of source protection or treatment. We thus adopted a more restrictive definition of "improved" in our survey. In brief, we distinguished in our survey between piped water (drinking water obtained via a piped distribution system, whether a community or private tap) and improved water (water considered safe to drink as a result of source protection or treatment); whereas virtually all improved water is piped, not all piped water is improved (safe to drink).

Table 5 thus lists the percentage of households in each community and overall with access to piped water, improved water and springs and streams (which are potentially unsafe due to surface runoff contamination), as well as those households who treated their drinking water (e.g., by boiling or adding bleach). Since some piped water was deemed improved (safe to drink), and some unimproved spring/stream sources were piped, the top three rows add up to more than 100%.

Overall, fully 70% of the 210 households we surveyed had access to piped water sources, which is understandable given the relatively developed nature of the Ezulwini Valley. Yet only 37% had access to what we determined as improved (safe to drink) sources, many (32%) regularly fetched drinking water from springs or streams, and only 9% did any form of household treatment. These results alone are highly important: nearly two out of three surveyed households in one of the most developed parts of the country regularly access unsafe drinking water sources, and less than one in ten regularly purifies their drinking water. Additionally, access time to water sources was significant even for this relatively developed area, with a mean roundtrip time of 50 min. Figure 2 and Table 9 suggest important differences between the four surveyed communities in their access to safe drinking water. The two more urban communities, Ezulwini and Lobamba, had relatively high proportions of households supplied by piped water, in contrast to the more rural community of Mahlanya, which relied heavily on springs and streams. Mlindazwe, the other community of a more rural nature, also had a high proportion of piped water, but in all cases these distribution systems relied on unimproved and untreated water sources. Access to improved water sources ranged from nil in Mlindazwe to 79% in Lobamba, the site of a former royal village and an important traditional and administrative center.

Perhaps surprisingly, access to piped water predictably was found to increase beyond the low economic class sector, but access to improved (safe to drink) water

Figure 2. Distribution of households using (a) piped (white) versus unpiped domestic water and (b) improved (white) versus unimproved domestic water sources.

Table 5. Water source type and water treatment practice (percent households surveyed) by community.

Item	Ezulwini	Lobamba	Mahlanya	Mlindazwe	Overall
Piped water	86	76	31	89	70
Improved water	20	79	20	0	37
Springs/streams	18	21	73	16	32
Any household treatment	14	3	10	13	9

actually decreased as economic class increased. How could this be? To be sure, a small number of wealthier households utilized their own private deep wells (boreholes), but most relied on piped systems shared with their communities, which as noted above were not consistently improved. One possible explanation for this anomalous result is that economic class may be a surrogate here for particular communities, given the vastly disparate community-scale results: in Lobamba, where improved water sources are commonly used, only one household in eight was rated upper economic class, whereas in Mlindazwe, for instance, where no surveyed households used improved water, almost seven out of eight were rated medium or high economic class.

Energy

Figure 3 and Table 6 present results of the household energy portion of our environmental health assessment. As could be expected in this relatively developed part of Swaziland, fully 71% of surveyed households utilized electricity as a general energy source (usually for lighting and outlets). In contrast, only 32% of households used electricity for cooking, a function of its perceived high cost relative to other options. The most common source of cooking energy was harvested wood, which roughly one-half of all households used; some respondents also purchased fuelwood from nearby vendors, who generally transport wood from other parts of the country to the

Figure 3. Distribution of households using (a) electricity (white) for general energy needs and (b) fuelwood (white) for cooking energy needs.

Table 6. Energy source type (percent households surveyed) by community.

Energy Source	Ezulwini	Lobamba	Mahlanya	Mlindazwe	Overall
Electricity (general)	76	72	59	81	71
Electricity (cooking)	50	28	14	43	32
Compressed gas	34	17	6	0	15
Fuelwood (harvested)	42	46	53	60	49
Fuelwood (purchased)	24	19	43	22	27

Ezulwini Valley given high demand relative to supply. Though fuelwood is traditionally used for cooking in Swaziland, this high percentage of harvested wood in the Ezulwini Valley is surprising given the relatively high settlement density as indicated in Figure 1. In all, 71% of households used harvested or purchased wood for cooking, exactly equal to the proportion that used electricity for general (non-cooking) energy needs.

A comparison of energy use by community suggests quite large differences in use of electricity for cooking (ranging from 14% in rural Mahlanya to 50% in peri-urban Ezulwini), presumably due in part to readily availability of fuelwood nearby for harvest or purchase. Trends in energy use by economic class (Table 9) are marked: high economic class households used electricity far more frequently for cooking than medium and low economic class households (e.g., roughly seven in ten among high vs. one in ten among low economic class households), and far less harvested wood. Evidently, decisions as to cooking energy are based on a tradeoff of cost verse labor,

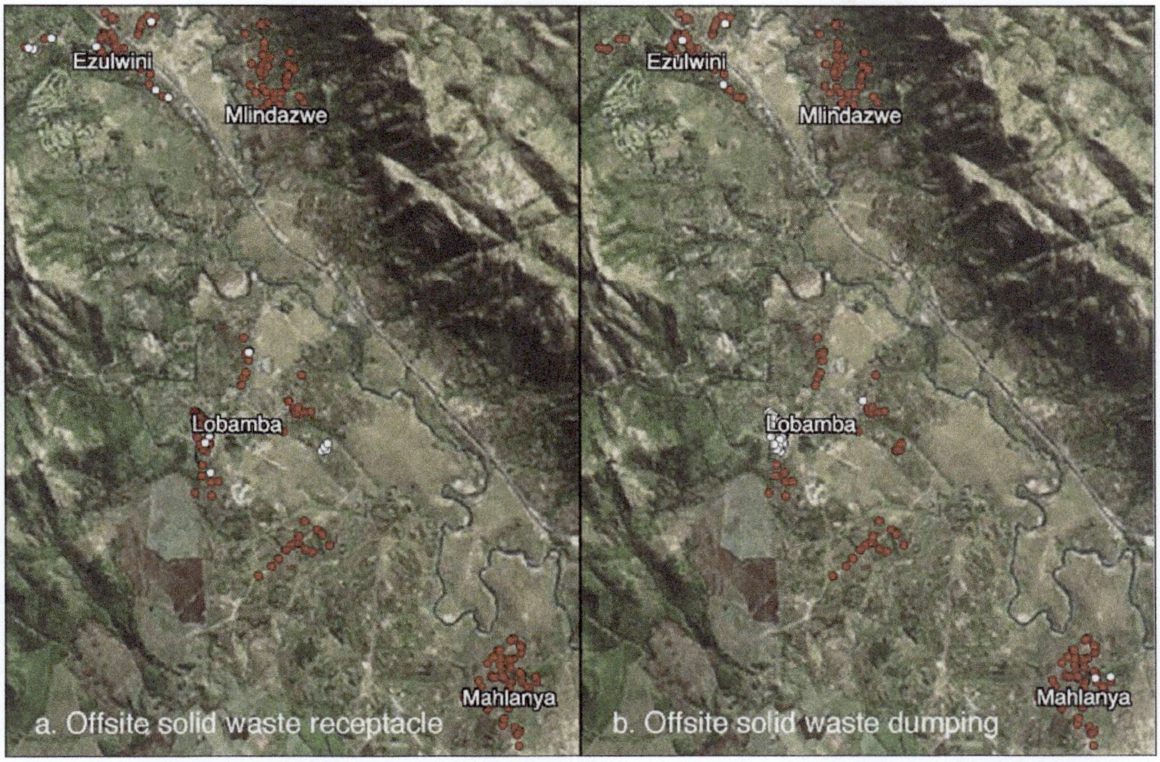

Figure 4. Distribution of households using (a) offsite solid waste receptacle (white) and (b) dumping solid waste offsite (white) in non-designated location.

Table 7. Solid waste disposal method (percent households surveyed) by community.

Disposal method	Ezulwini	Lobamba	Mahlanya	Mlindazwe	Overall
Burn onsite	62	51	94	95	72
Bury onsite	60	13	0	5	20
Offsite receptacle	18	11	0	0	8
Offsite dumping	8	26	4	3	12

as fetching fuelwood can take a great deal of time (2.8 h average roundtrip among reporting households) but communal fuelwood reserves are free.

This heavy reliance on fuelwood for cooking via open-air stoves, which are common in Swaziland, results in the well documented environmental health problem of indoor air pollution, for which sub-Saharan Africa is particularly afflicted (Legros et al., 2009). These results suggest that indoor air pollution is widespread even in relatively developed parts of Swaziland such as the Ezulwini Valley, and especially so among low economic class households, and among women who do almost all the cooking in a typical Swazi household. Similar to the relatively low proportion of households that purify their drinking water, however, the ubiquity of this indoor air pollution problem may render it difficult to successfully address.

Solid waste

The relation between households and their surrounding communities differs in the context of waste: though household drinking water and energy supply options depend on community-scale infrastructure in the Ezulwini Valley, solid and human waste decisions made by households generally have a cumulative impact on their surrounding communities. The household results for waste, then, should be considered at the scale of potential community as well as household impacts.

Figure 4 and Table 7 summarize selected solid waste survey results: overall, by far the most common method of dealing with solid waste was for households to burn or bury their waste onsite (in the immediate vicinity of their dwelling structures), but one in eight dumped their waste offsite (in a non-designated location vs. a landfill). The

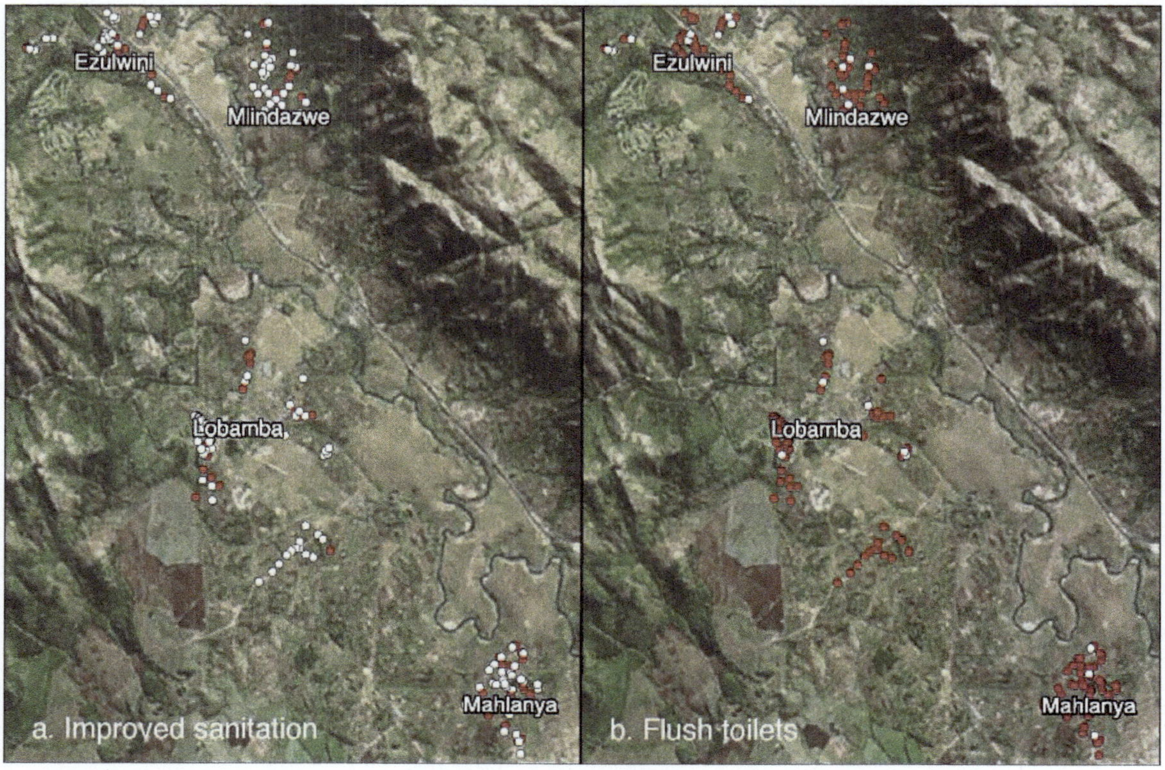

Figure 5. Distribution of households using (a) improved sanitation (white), and (b) flush toilets (white).

Table 8. Human waste disposal method (percent households surveyed) by community.

Disposal method	Ezulwini	Lobamba	Mahlanya	Mlindazwe	Overall
Overall improved sanitation	70	67	61	78	68
Flush toilet	22	8	6	14	12
Flush toilet	54	54	61	76	60
Improved latrine	28	19	35	19	25

Ezulwini Town Council maintains a set of solid waste collection receptacles in Ezulwini, with some receptacles in Lobamba as well; for the most part, however, these do not seem to be frequently used, and they are not available in the more rural communities of Mahlanya and Mlindazwe.

A comparison by community reveals, for instance, the stark conditions of peri-urban Lobamba, where crowded homesteads and a lack of collection receptacles have resulted in over one in four households dumping their waste offsite in non-designated locations. The accumulation of solid waste was clearly evident in these parts of Lobamba as well, increasing potential for disease transmission, especially among children who frequently walked or played near affected areas. Economic class seems to play a role as well (Table 9), with higher economic class households using offsite receptacles far more and dumping offsite far less.

Human waste

Globally, progress in access to sanitation has generally lagged behind provision of water supply, for a variety of political and other reasons (Rosemarin et al., 2008). This larger pattern may not be evident in the Ezulwini Valley: as shown in the human waste survey results in Figure 5 and Table 8. Overall, over two in three surveyed households (68%) used improved sanitation facilities for disposal of human waste, in contrast to only 37% having access to improved (safe to drink) water. These sanitation facilities are generally an improved pit latrine on the homestead (60 percent), though a small proportion (12 percent) of homesteads had flush toilets. Additionally, a full one-quarter of households used latrines that respondents or surveyors deemed unimproved (e.g., without a concrete slab or outdoors).

Comparison by community and economic level reveals

Table 9. Selected resource results (percent households surveyed) by economic level.

Item	Low economy	Medium economy	High economy	Overall
Piped water	52	78	78	70
Improved water	42	38	22	37
Electricity (general)	39	81	97	71
Wood (harvested)	61	49	28	49
Offsite receptacle	3	7	19	8
Offsite dumping	18	12	6	12
Overall improved sanitation	48	74	83	68
Flush toilet	2	7	44	12

some patterns in human waste disposal. Community results were generally consistent, with some community-specific differences: for instance, Ezulwini had a higher proportion of flush toilets (22%), in part connected to a community sewer scheme; and more Mahlanya households (35%) used unimproved latrines, perhaps in part due to the more rural nature of this community. Differences by economic level (Table 9) are more marked,with higher economic class households reporting much higher use of improved sanitation facilities (83% high vs. 48% low economic class). This difference is understandably even greater in the use of flush toilets (44% high vs. 2% low economic class). The generally strong performance of Ezulwini Valley homesteads in their use of improved sanitation, therefore, masks important differences by economic class.

Perceived adequacy

The results above reflect reported reality in the context of four key resources, all with environmental health implications, essential to household well-being in the Ezulwini Valley. To successfully address issues related to these four resources, empirical data on household perceptions of resource adequacy can help us understand the priorities of surveyed communities.

Figure 6 and Table 10 summarize these results, on a Likert-type scale, we devised ranging from 1 (low reported adequacy) to 3 (high reported adequacy). Overall, respondents rated water quality the most adequate (2.6 average), and sources of cooking energy the least (1.9 average). In comparison to the results above, respondents seemed more concerned about adequate drinking water quantity (2.3) than quality (2.6), even though our results indicate relatively few households have access to safe drinking sources. This may suggest challenges in gaining community support for improving drinking water quality if quantitative shortages are not also prioritized. The overall results also suggest that initiatives to address cooking energy challenges may receive strong community support.

Community comparisons summarized in Table 10 suggest considerable variance in perceptions of solid waste disposal adequacy, with the more peri-urban areas of Ezulwini and Lobamba perhaps understandably reporting much poorer adequacy than the more rural areas of Mahlanya and Mlindazwe. A similar difference between peri-urban and more rural communities was found with respect to human waste, and some variance was found in perceived cooking energy adequacy, though likely as a function of fuelwood availability. Water quantity and quality displayed less variance between communities.

When comparing adequacy by economic class (Table 11), the general pattern is one of the lower perceived adequacy among lower economic groups; for instance, perceived cooking energy adequacy ranged from an average of 1.5 among low economic class households to 2.4 among high economic class households, a possible function of the differing energy utilization patterns summarized above. This pattern is particular evident in perceived solid and human waste disposal adequacy, suggesting that lower economic class households may be disproportionately impacted. Less difference is evident in perceived quantitative and qualitative water adequacy.

DISCUSSION

Summary of results

A synthesis of the survey results above suggests clear overall patterns, as well as important differences across the four surveyed communities and three economic classes. Given that the Ezulwini Valley is one of the most developed areas in Swaziland, it is unsurprising that over two out of three surveyed households utilize piped drinking water, yet far more surprising that only slightly more than one out of three utilize water from protected or treated sources- a result that proved highly community specific. Access to safe drinking water would be a major environmental health priority, then, in this location; yet the perceived adequacy results summarized above remind us that drinking water quality alone is not the most important priority in these communities.

Figure 6. Perceived resource adequacy (white = adequate; red = inadequate) for (a) drinking water quality, (b) drinking water quantity, (c) solid waste, (d) human waste, (e) cooking energy and (f) overall average.

Table 10. Resource adequacy comparison by community (1 = inadequate; 3 = adequate).

Resource sector	Ezulwini	Lobamba	Mahlanya	Mlindazwe	Overall
Water (quantity)	2	2.5	2.1	2.6	2.3
Water (quality)	2.6	2.8	2.5	2.6	2.6
Cooking energy	2.1	1.8	1.5	2.4	1.9
Solid waste	1.7	1.9	2.6	3	2.2
Human waste	2.2	2	3	3	2.4

Table 11. Resource adequacy comparison by economic level (1 = inadequate; 3 = adequate).

Resource sector	Low economy	Medium economy	High economy	Overall
Water (quantity)	2.0	2.4	2.3	2.3
Water (quality)	2.5	2.7	2.8	2.6
Cooking energy	1.5	2.0	2.4	1.9
Solid waste	2.0	2.2	2.6	2.2
Human waste	2.3	2.5	2.7	2.4

A similar disparity is evident in overall household energy results, where over two out of three surveyed households have access to electricity, yet less than one in three use electricity for cooking- in this case, a result that varied significantly according to household economic class. From an indoor air pollution standpoint, and given the increasingly dense settlement (thus fuelwood demand) in the Ezulwini Valley, results from this important resource sector are unacceptable as well. In contrast to drinking water quality, however, there appears to be strong community prioritization of the need to address cooking energy challenges.

Solid and human waste disposal in surveyed communities of the Ezulwini Valley differs in part as a function of community type, where relatively dense peri-urban areas understandably struggle more with waste than do the more rural communities. Overall, relatively little attention has been devoted to solid waste disposal, and the vast majority of households fend for themselves, though perceived adequacy results suggest strong potential support for improved solid waste disposal in more peri-urban communities. Much more attention has gone to human waste, an even greater environmental health concern: here, over two in three surveyed households utilize improved sanitation facilities, though results differ significantly by household economic class.

Comparison of Ezulwini Valley and Swaziland

In comparing results from the Ezulwini Valley with the rest of Swaziland, recall the country-scale WHO and UNICEF-sponsored results above, where 67-72% of the country was reported to have access to improved (piped) water, and 57-78% was reported to have access to improved sanitation facilities. Relative to these estimates, the Ezulwini Valley is not significantly ahead of nor behind the national average- a relatively surprising result, given the Ezulwini Valley is one of the most developed areas in Swaziland. Households in the Ezulwini Valley do, however, utilize relatively more electricity and less fuelwood for cooking than in the country as a whole. Yet geographic and economic class differences in the Ezulwini Valley appear to be reproduced in the national-scale picture as well; the overall statistics at both the scale of this area and Swaziland thus apparently mask important differences by location and wealth.

Larger implications

The prominent role of environmental health in our most important development agendas, such as the UN Millennium Development Goals, is heartening. Yet, the above results from one of the most developed regions in Swaziland are sobering, as the status of drinking water, cooking energy and solid and human waste disposal reveals continued challenges for many households in the area. Indeed, simply assessing progress toward these important MDGs is methodologically challenging: in the key case of drinking water, our study reveals that relatively few "improved" sources may be safe for drinking. More accurately ascertaining progress toward ensuring safe drinking water for all is thus a far more demanding task, and may reveal even poorer overall performance, than efforts to date have suggested. Globally, this disparity between "safe" and "improved" drinking water sources has been recognized by the WHO/UNICEF Joint Monitoring Programme for Water Supply and Sanitation, which is planning a post-2015 (post-MDG) initiative focusing on evidence-based drinking water, sanitation, and hygiene assessment (WHO 2013b, 19). Additional research tends to corroborate the inadequacy of the MDG definition to ensure provision of truly safe drinking water (Lenton et al., 2008, Dar and Khan, 2011; Clasen, 2012; Sambu and Tarhule, 2013).

Our survey also suggests that overall results mean relatively little in the context of specific communities and socioeconomic classes. We selected four communities located quite close to each other, yet overall household-scale patterns were in many ways unique to each (with the exception of differences between peri-urban and more rural communities). Some enjoy good treated water schemes; some have better fuelwood resources; some worry less about waste. Effectively improving environmental health at the scale of household resource access and impacts, therefore, can only be done by gathering information on, and paying close attention to, community-specific needs and desires.

The additional layer of economic class is a critical one, perhaps no more striking than in the Ezulwini Valley, where among the most wealthy and many poorer households in Swaziland dwell in relatively close proximity. As suggested at the outset, however, this reality is not limited to the Ezulwini Valley nor Swaziland: in the context of urbanizing areas, for instance, "...the health impacts of the most serious problems are largely confined to lower-income groups....It is common for the residential areas of middle- and upper-income groups...to receive good quality water supplies, sewers, drains, electricity supplies and regular services to remove solid waste while 30% or more of the city population on the poorer residential areas receive little or nothing" (Hardoy et al., 2001).

Household-scale environmental health thus continues to be a challenge in our world, certainly more so among the poor, and with specific needs and priorities among particular communities; though the MDGs have contributed toward improvement, there is a long way to go, and the post-MDG scenario following 2015 is anything but clear (Poku and Whitman, 2011; Vandemoortele, 2011; Usua et al., 2012; Van Norren, 2012; Nayyar, 2013; Kesavan and Swaminathan, 2014). This study of the Ezulwini Valley, the Valley of Heaven, should remind us that there are households located even in relatively developed parts of the world that still face these challenges on a daily basis. Continued attention to fine-scaled data gathering and sharing, as was our survey's intent, may help these areas better identify their needs and desires as they move closer toward creating environments that are safe and healthy for their inhabitants.

Conflict of Interests

The author(s) have not declared any conflict of interests.

ACKNOWLEDGEMENTS

We acknowledge the Swaziland Environment Authority and All Out Africa in supporting our Ezulwini Valley environmental health assessment, all Lewis & Clark students who participated in the assessment, Ezulwini Valley community members, and Swaziland government officials who provided key assistance. We also acknowledge the generosity of The Andrew Mellon Foundation in supporting development of the Swaziland overseas program and our broader situated environmental research approach. Finally, we acknowledge the input of anonymous reviewers, whose comments contributed to improvements of the manuscript.

REFERENCES

Awe F, Osadebe CO, Imoagene E, Fashina AY, Eniola TS, Adeleke EO (2011). Assessment of rural households' objectives for gathering non-timber forest products (NTFPs) in Kogi State, Nigeria. Afr. J. Environ. Sci. Technol. 5(2):143-148.

Central Statistical Office and UNICEF (2011). Multiple indicator cluster survey 2010. Central Statistical Office and UNICEF, Mbabane, Swaziland.

Clasen TF (2012). Millennium Development Goals water target claim exaggerates achievement: Editorial. Trop. Med. Int. Health 17(10):1178-1180. doi:10.1111/j.1365-3156.2012.03052.x

Dar OA, Khan MS (2011). Millennium development goals and the water target: details, definitions and debate: Millennium development goals and the water target. Trop. Med. Int. Health 16(5):540-544. doi:10.1111/j.1365-3156.2011.02736.x

Ewodo MG, Ekwelgen C, Ntep F, Ekodeck GE (2009). Impact of urbanisation on the Mingosso watershed in the Yaounde periurban zone. Afr. J. Environ. Sci. Technol. 3(10): 272-285. Retrieved from http://www.ajol.info/index.php/ajest/article/download/56255/44700

Fehling M, Nelson BD, Venkatapuram S (2013). Limitations of the Millennium Development Goals: a literature review. Glob. Public Health 8(10):1109-1122. doi:10.1080/17441692.2013.845676

Gabay C (2011). Consenting to "Heaven": The Millennium Development Goals, Neo-liberal Governance and Global Civil Society in Malawi. Globalizations 8(4):487-501. doi:10.1080/14747731.2011.585852

Groenewald L (2011). Progress towards Millennium Development Goals? Strategies for housing and informal settlement in Gauteng, South Africa. Dev. South. Afr. 28(5):641-651. doi:10.1080/0376835X.2011.623912

Hardoy JE, Mitlin D, Satterthwaite D (2001). Environmental problems in an urbanizing world : Finding solutions for cities in Africa, Asia, and Latin America. Earthscan Publications, London.

Joséphine N, Véronique KKB, Luc SN, Georges E, Awah TM (2008). Water supply, sanitation and health risks in Douala, Cameroon. Afr. J. Environ. Sci. Technol. 2(12):422-429.

Kangalawe RY, Liwenga ET, Kabumbuli R, Walingo MK (2008). Livelihood diversification and implications on poverty and environment in the Lake Victoria Basin. Afr. J. Environ. Sci. Technol. 2(10):272-281.

Kesavan PC, Swaminathan MS (2014). From Millennium Development Goals to sustainable development solutions. Curr. Sci. (00113891), 106(4): 495-496.

Legros G, Havet I, Bruce N, Bonjour S (2009). The energy access situation in developing countries. United Nations Development Programme, New York, N.Y.

Lenton R, Lewis K, Wright AM (2008). Water, Sanitation and the Millennium Development Goals. J. Int. Aff. 61(2):247-258.

Mughogho BUG, Kosamu IBM (2012). Water supply arrangements in developing countries: A case study of Blantyre City, Malawi. Afr. J. Environ. Sci. Technol. 6(2): 94-103.

Mukonka VM, Malumo S, Kalesha P, Nambao M, Mwale R, Mwinga K et al. (2014). Holding a country countdown to 2015 conference on Millennium Development Goals (MDGs) - the Zambian experience. BMC Public Health 14(1):1-7. doi:10.1186/1471-2458-14-60

Nayyar D (2013). The Millennium Development Goals Beyond 2015: Old Frameworks and New Constructs. J. Hum. Dev. Capabilities 14(3):371-392. doi:10.1080/19452829.2013.764853

NERCHA (2011). Mapping of KaGogo centers and neighborhood care points in Swaziland—final report. National Emergency Response Council on HIV and AIDS, Mbabane, July 21.

Ogunseitan O (2007). Top Ten Environmental Priorities for Africa. Afr. J. Environ. Sci. Technol. 1(2). Retrieved from http://www.academicjournals.org/article/article1380104409_Editorial%20(1).pdf

Oyelola OT, Babatunde AI (2008). Characterization of domestic and market solid wastes at source in Lagos metropolis, Lagos, Nigeria. Afr. J. Environ. Sci. Technol. 3(12): 430-437. Retrieved from http://www.academicjournals.org/article/article1380186107_Oyelola%20and%20Babatunde.pdf

Poku N, Whitman J (2011). The Millennium Development Goals: challenges, prospects and opportunities. Third World Q. 32(1):3-8. doi:10.1080/01436597.2011.543808

Rosemarin A, Ekane N, Caldwell I, Kvarnström E, McConville J, Ruben C, Fogde MI (2008). Pathways for sustainable sanitation: achieving the millennium development goals. IWA Publishing, Stockholm.

Sambu DK, Tarhule A (2013). Progress of water service providers in meeting millennium development goals in Kenya. Afr. Geogr. Rev. 32(2):105-124. doi:10.1080/19376812.2013.790643

Swaziland Environment Authority (2012). Swaziland's state of environment report 2012. Swaziland Environment Authority, Mbabane, October 23.

Swaziland Vulnerability Assessment Committee (2011). Annual vulnerability assessment & analysis report 2011. Swaziland Vulnerability Assessment Committee, Mbabane, Swaziland.

Swaziland Vulnerability Assessment Committee (2012). Annual vulnerability assessment & analysis report 2012. Swaziland Vulnerability Assessment Committee, Mbabane, Swaziland.

U.S. Fund for UNICEF (2011). Caring for orphaned and vulnerable children in Swaziland. Every Child 3:4-5.

Usua N, Sado A, Nwachukwu F (2012). Rethinking the millennium development goals: ensuring development through communication and participation. Int. J. Acad. Res. 4(6):176-181. doi:10.7813/2075-4124.2012/4-6/B.27

Van Norren DE (2012). The Wheel of Development: the Millennium Development Goals as a communication and development tool. Third World Q. 33(5):825-836. doi:10.1080/01436597.2012.684499

Vandemoortele J (2011). If not the Millennium Development Goals, then what? Third World Q. 32(1):9-25. doi:10.1080/01436597.2011.543809

World Health Organization (2009a). Healthy security through healthy environments: Proceedings of the first interministerial conference on health and environment in Africa, Libreville, Gabon 26-29 August 2008. World Health Organization, Libreville, Gabon.

World Health Organization (2009b). Libreville declaration on health and environment in Africa, Libreville, 29 August 2008. World Health Organization, Libreville, Gabon.

World Health Organization (2009c). Meeting of partners on the implementation of the Libreville declaration on health and environment in Africa: The roadmap to support the implementation of the Libreville Declaration on Health and Environment in Africa (2009-2010), Windhoek, Namibia, 25-27 February 2009. World Health Organization and United Nations Environment Programme, Windhoek, Namibia.

World Health Organization (WHO) (2010). Public health and environment in the African region: Report on the work of WHO (2008-2009). World Health Organization, Brazzaville, Congo.

World Health Organization (WHO) (2013a). The world health report 2013: Research for universal health coverage. World Health Organization, Geneva, Switzerland.

World Health Organization (2013b). World health statistics 2013. World Health Organization, Geneva, Switzerland.

Yanda PZ, William C (2010). Livelihoods diversifications and implications on food security and poverty levels in the Maasai plains: The case of Simanjiro district, Northern Tanzania. Afr. J. Environ. Sci. Technol. 4(3): 154-166. Retrieved from http://www.ajo .info/index.php/ajest/article/view/56347

Assessment of environmental responses to land use/land cover dynamics in the Lower Ogun River Basin, Southwestern Nigeria

Awoniran D. R.[1], Adewole M. B.[2], Adegboyega S. A.[1] and Anifowose A. Y. B.[1]

[1]Department of Remote Sensing and Geoscience Information System, Federal University of Technology, Akure, Ondo State, SW, Nigeria.
[2]Institute of Ecology and Environmental Studies, Obafemi Awolowo University, Ile-Ife, Osun state, SW, Nigeria.

This study investigates the pattern of land use/land cover change in the Lower Ogun River Basin between 1984 and 2012. Two sets of topographical maps, a Landsat-5 TM image of 1984, Landsat-7 ETM+ of 2000 and a Google Earth image of 2012 were used for the study. The topographical maps and satellite images were digitally processed using ILWIS 3.2™ software and exported to ArcGIS 9.3™ for further processing and analysis. The processed images were subsequently classified using the maximum likelihood classification algorithm, resulting in the identification of seven land use classes. Furthermore, change detection analysis was carried out using Cross Module in ILWIS™. The result of the change detection analysis indicated that between 1984 and 2000, 80.08% of the land cover in the study area has been converted to other land uses while only 19.92% remained unchanged. Also, within the same period, light forests, non-forested wetlands and forested wetlands decreased at average annual rates of 8.26, 4.66 and 2.81%, respectively, while water bodies also decreased at an annual rate of 0.17%. On the other hand, farmland, shrubs and urban/built-up areas expanded at average annual rates of 7.23, 6.74 and 4.65%, respectively. The result further indicates that between 2000 and 2012, 49.86% of the land cover has been converted to other land uses, while 50.14% remained unchanged, and that farmlands, shrubs, urban/built-up areas and forested wetlands expanded at average annual rates of 6.01, 1.95, 0.89 and 0.17%, respectively, just as light forests, non-forested wetlands and water bodies decreased annually by 8.26, 2.70 and 1.40%, respectively. Five randomly collected soil samples were analyzed for their physicochemical properties. Findings show the growing impact of urban agriculture on wetland ecosystem within the study area, manifesting in soil degradation and biodiversity loss. The implication of these findings is that the area is susceptible to devastating flooding which can culminate in the loss of lives and properties. This study recommends the development of effective land management information system and policies that will ensure sustainable management of fragile ecosystem.

Key words: Change detection, river basin, urban agriculture, land use/land cover, wetlands.

INTRODUCTION

FAO (1995) stated that land use concerns the function or purpose for which the land is used by the local human population and can be defined as the human activities which are directly related to land, making use of its resources or having an impact on them. Land use and land cover change has emerged as one of the key independent themes in the global change, climate change, earth systems and sustainability research programs (Gutman et al., 2004). The environmental impacts of land use change have been documented in urban, suburban, rural and open space areas. For example, Awoniran (2012) observes that land use changes (land conversion) occur at the periphery of large urban concentration where urbanization and industrialization pressures frequently result in loss of prime agricultural lands and tree cover. These often result in unprecedented changes in the hydrological balance of the area, increase in the risk of floods and landslides, air and water pollution among others. Other local impacts are soil erosion, sedimentation, soil and groundwater contamination and salinization, extinction of indigenous species, marine and aquatic pollution of local water bodies, coastal erosion and land pollution.

Though humans have been modifying land to obtain food and other essentials for long, current rates, extent and intensities of landuse/landcover change (LULCC) are far greater than ever in history; driving unprecedented changes in ecosystem and environmental processes at local, regional and global scales (Lambin et al., 2007). Various estimates indicate that 50% of the ice-free land surfaces has been affected or modified in some ways by human activities (Vitousek et al., 1997), while 10 to 55% of the net primary productivity has been captured by human land use activities (Rojstaczer et al., 2001). The intergovernmental panel on climate change (IPCC) estimates that the cutting down of forests contributes close to 20% of the overall greenhouse gases that are entering the atmosphere, making the goal of reducing deforestation an urgent and immediate one (United Nations Development Programme, 2009). These anthropogenic-cally induced land use change has, to a greater or lesser extent, resulted in major environmental problems such as desertification, eutrophication, acidification, climate change, flooding, greenhouse effect and biodiversity loss.

The Lower Ogun River Basin, because of its proximity to the urban periphery of Lagos, is presently under intense pressure occasioned by increasing urbanization rate. Increasing population pressure, increasing demand for food, high cost of land and urban agricultural activities are having serious environmental impact on the basin (Tejuoso, 2006). The consequence of this is: land degra-dation, flooding and threatened food security.

The use of remotely sensed data with the integration of Geographic Information System technology provides a strong and analytical framework for assessing land use/land cover inventory, annual rate of change and evaluating the emerging environmental response at the periphery of a fast growing city. The importance of geospatial information to be generated from such an endeavor cannot be over emphasized as Adeniyi and Omojola (1999) submitted that information based on urban land use changes can shed more light on the growth process, since physical changes in the distribution of urban land uses are direct indications of social and economic changes.

This study assesses land use/land cover change, evaluate the proportion of changes and analyze the environmental impacts of the change on the Lower Ogun River Basin within 28 years using Remote Sensing and GIS technology. This is considered germane in addressing emerging environmental problems and ensuring sustain-able land use management which may be anchored on accurate and up to date land use data and map that serves as bedrock for evolving a sound land use planning and policy for the area under study.

The study area

The study area is located partly in Ifo South Local Government Area of Ogun State and Kosofe Local Government Area of Lagos State, and lies between Latitudes 06° 35'N and 06° 45'N and Longitudes 3°17'E and 3°25'E (Figure 1). It is approximately 161.4 km^2. It is located within the sub-equatorial zone, which is charac-terized by rainfall throughout the year with two maxima (May to July and September to October). December and January have very little rain, and the annual rainfall is between 1500 to 2000 mm. The effective temperature (ET) is between 32 and 36°C. However, the highest diurnal range of temperature in the dry season (mid November-mid March) is 20°C while the mean range is about 10°C during the warm and wet season (May to October). The highest air temperature occurs in April/May and the lowest occurs in December through February. (The mean annual temperature is about 27°C while the annual range does not exceed 6°C (Ekanade, 1985). It lies within the rainforest belt (dry lowland rainforest). The vegetation of the region is swamp and marsh forest, part of which had given way to the construction of houses, markets and other infrastructures. Tree species here consist of typical colonizer or invaded species. These are plants with numerous and easily dispersed seeds and capacity for fast and vigorous establishment in cleared or open location. The river channels are characterized by vegetation of the wet southern segment of the rainforest belt. The characteristic vegetation include tall trees like *Tarriefa utilis*, *Geophila* sp., epiphytic ferns (*placycerina* sp.), *Tuchomanes* sp. *Nephrolepis* sp. Mosses and Lierworts (Ogunbajo, 2005).

Generally, the relief of the area may be described as belonging to the belt of coastal plains. The land rises from the sandy beaches along the Atlantic Ocean to a belt of fresh water swamps with an intricate network of lagoons and creeks. The coastal belt is about 10 km wide and is generally less than 20 m in height. Further inlands, it gives way to a sandy plain which in turn leads up to a plain of eroded sandstones standing less than 20 m above

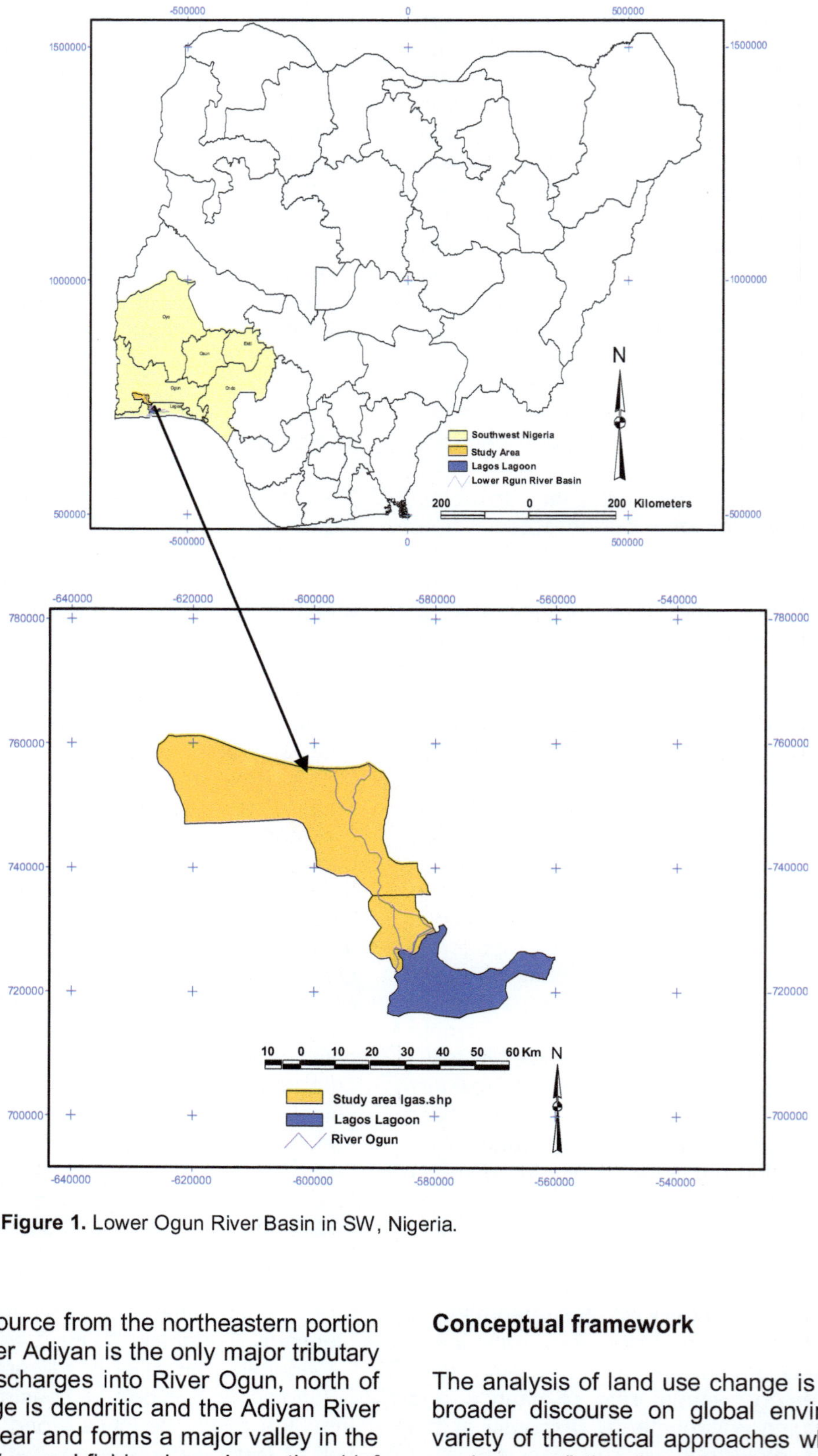

Figure 1. Lower Ogun River Basin in SW, Nigeria.

sea level. Taking its source from the northeastern portion of the study area, River Adiyan is the only major tributary to River Ogun and discharges into River Ogun, north of Iseri-Oke. The drainage is dendritic and the Adiyan River flows throughout the year and forms a major valley in the vicinity. Farming, hunting and fishing have been the chief occupations of the local people for many decades. Crops grown in the area include cocoa, vegetables, sugar cane and maize. However, sand mining along the channel of River Ogun that traverses the study area is a major human activity.

Conceptual framework

The analysis of land use change is embedded within the broader discourse on global environmental change, a variety of theoretical approaches which are called "man-environment" theories. More balanced theoretical frameworks for the study of nature-society relationship are offered by ecologically sensitive approaches known as "human ecology" or "cultural ecology" (Briassoulis, 2000). These approaches draw upon ecology and systems theories to provide comprehensive descriptions of the complex interactions between people and their bio-physical environment (Sack, 1990, Butzer, 1990).

Land use change research has evolved out of global efforts to identify, predict and manage ecologically damaging land use changes such as deforestation. Indeed, the international global environmental change research community has chosen land use/cover change as a major area of research not only because it provides broad scale data on changing carbon storage and sequestration by plants, but also because it provides an entry into understanding the human dimensions of environmental change (Turner et al., 1995; Lambin et al., 1999; Sherbinin, 2002). Political and cultural ecology, intensification theory, economic theories and other concepts have informed LULC research associated with the International Geosphere-Biosphere Progamme (IGBP).

The early interests of social scientists in explanation and ecologists in prediction are evidently convergent in the spatially explicit model. Ecologists have provided strong motivation with their focus on the landscape as a biogeographic unit of analysis subject to fragmentation processes. On their part, social scientists have filled the explanation gaps on the human side of the equation. Now, social scientists and ecologists are even found to publish jointly on the topic (Turner et al., 1996; Wear and Bolstad, 1998). The spatially explicit model is increasingly used to predict landscape change and is often evaluated both in terms of conventional inference on variable coefficients and goodness-of-fit and with respect to ability to predict actual landscape change (Nelson and Hellerstein, 1997; Wear and Bolstad, 1998; Mertens and Lambin, 2000).

MATERIALS AND METHODS

Research locale

The main data used for this study included a Landsat TM satellite image of December 1984, and ETM+ of February 2000 obtained from the Global Land Cover Facility and 2012 Google Earth® image. Two sets of topographical maps on a scale of 1:50000, Lagos NE Sheet 279 and Lagos SE Sheet 279, published in 1964 were obtained from the Federal Survey, and used as base map and supportive ground-truth information required for classification and accuracy estimation of the classified TM and ETM+ images was collected through a field survey carried out in 2012.

Digital image processing

Preprocessing operations in the form of linear contrast stretch and spatial filtering were performed on the images which were also geo-rectified to Universal Transverse Mercator (UTM) coordinate system. Subsets of the satellite images and topographical maps were geo-referenced using georeferenced tie points and affine transformation method. The supervised Maximum Likelihood Classification method was used for all the images. Training areas corresponding to each classification item (land use class) were chosen from among the training samples collected from the field and the topographical map of the study area.

Thereafter, seven land use/land cover classes were identified on the two images for change detection analysis. These include urban, light forest, forested wetland, non-forested wetland, farmland, shrub and water body. Change detection analysis was performed using the cross algorithm of the ILWIS software. Choice of the emerging land use classes was guided by i) the objective of the research ii) expectation of certain degree of accuracy in image classification and iii) the ease of identifying classes on the topographical map. The classified images were further subjected to majority filtering operation to smoothen and eliminate noise (salt and pepper) from the images. The evolving land use/land cover types were quantified using cross tabulation statistics to carry out land use/land cover change. The classified images were subsequently vectorized and exported to ArcGIS 9.3™ for the graphical illustration of land use/land cover change in the study area.

Soil sampling and analysis

Two study sites were selected on the classified images for soil quality analysis. That is, forested wetland which was relatively less disturbed and non-forested wetland which was under cultivation. Five composite soil samples, replicated thrice, each to the depth of 0-20 cm to give a total of 30 soil samples were randomly collected using soil auger and analyzed for their physicochemical properties to determine the impact of land use change on soil quality in the selected forested and non-forested wetlands. The coordinate of each soil samples were also taken using GPS.

The soil samples were air-dried for seven days, crushed and sieved through 2 mm opening. The hydrometer method (Bouyoucos, 1962) was used for particle size analysis. Soil pH was determined potentiometrically in H_2O at a ratio of 1:1 (soil to water) (Mclean, 1982). The Kjeldahl method was used to determine total nitrogen (Bremner and Mulvaney, 1982). The determination of soil organic carbon was based on the Walkley-Black chromic acid wet oxidation method (Nelson and Sommers, 1982), while available phosphorus was determined using Bray P1 method (Olsen and Sommers, 1982). Exchangeable cations (Ca^{2+}, Mg^{2+}, K^+ and Na^+) were determined using 1 M NH_4OAc (Ammonium acetate) buffered at pH 7.0 as extractant (Thomas, 1982). The K^+ and Na^+ concentrations in soil extracts were read on Gallenkamp Flame photometer while Ca^{2+} and Mg^{2+} concentrations in soil extracts were read using Perkin-Elmer Model 403 atomic absorption spectrophotometer. Descriptive statistics was used to detect the changes that have taken place in the study area from 1984 to 2000. Test of significance for differences in means of soil physico-chemical properties under forested and non-forested wetland was done using least square difference (LSD) method.

Vegetation sampling

In this study, four corners of each of the two sites were marked with pegs at the defined dimension (100 m^2) and were gridded into 10 quadrants having dimensions of 10 by 10 m. For each 10 m^2 grid observations of height of trees and herbs were recorded. During the study, a tree was taken to be a woody specie of erect posture with a minimum breast circumference of 7 cm and a minimum height of 2 m (Salami and Aladenola, 2003). Shrubs on the other hand are woody plants not having a main trunk but several main branches. In general, shrubs are smaller than trees. The vegetation data recorded in the field were entered into a database using Microsoft Excel® software and analyzed using the tools in the spreadsheet.

Focus group discussion (FGD)

As an additional source of ground-truth information, four focus discussion and the in-depth interviews were based on convenience

Table 1. Landsat TM 1984 land use/land covers accuracy assessment.

LULC	FML	FWT	LF	NFWL	SRB	URB	WB	TR
FML	487	14	429	0	66	60	0	1056
FWT	85	2321	301	200	0	25	2	2934
LF	1051	464	3087	239	176	202	1	5220
NFWL	134	242	185	2235	0	138	12	2946
SRB	917	43	762	24	730	21	3	2500
URB	19	46	38	16	5	4461	134	4719
WB	0	0	0	0	0	94	2182	2276
RELIABILITY	0.18	0.74	0.64	0.82	0.75	0.89	0.93	
LULC	FML	FWT	LF	NFWL	SRB	URB	WB	TR

Average accuracy = 68.55%, average reliability = 70.9%, overall accuracy = 71.6%.

Table 2. ETM+2000 land use/land cover accuracy assessment.

LULC	FML	FWT	LF	NFWL	SRB	URB	WB	TR
FML	1127	181	62	43	554	15	0	1982
FWT	208	1308	378	140	629	55	0	2718
LF	60	114	1913	97	67	73	0	2324
NFWL	33	42	26	1723	432	509	3	2768
SRB	114	77	33	932	2276	401	0	3833
URB	9	4	1	255	71	6531	53	6924
WB	0	0	0	17	0	59	2077	2153
RELIABILITY	0.73	0.76	0.79	0.54	0.56	0.85	0.97	
LULC	FML	FWT	LF	NFWL	SRB	URB	WB	TR

Average Accuracy = 71.4%, average reliability = 74.4%, overall accuracy = 74.7%.

and purposeful sampling (Miles and Huberman, 1984). The focus group had three themes for discussion, that is, the history of land use in the study area, the major types of land use, and the impact of land use.

RESULTS AND DISCUSSION

Land use/land cover accuracy assessment

The general acceptance of the error matrix as the standard descriptive reporting tool for accuracy assessment of remotely sensed data has significantly improved the use of such data. An error matrix is a square array of numbers organized in rows and columns which express the number of sample units (i.e. pixels and clusters of pixels) assigned to a particular category relative to the actual category as indicated by reference data (Coppin and Bauer, 1996). For instance, Table 1 shows the producer's accuracy for the image classification of Landsat TM 1984 into seven classes of land use/land cover types: Farmlands (Fm) = 46%; forested wetlands (FW) = 79%; light forest (LF) = 59%; non forested wetlands (NFW) = 76%; scrubs (SS) = 29%; urban (U) = 95%; water body (WB) = 96%.

Also, the user's accuracy as shown in the same table indicates the probability that a pixel classified into a given category are the true representation of that category on the ground: Farmlands (Fm) = 18%; forested wetlands (FW) = 74%; light forest (LF) = 64%; non forested wetlands (NFW) = 82%; scrubs (SS) = 75%; urban (U) = 89%; water body (WB) = 93%; Overall accuracy = TD (sum of major diagonal) divided by TR (row totals) = 15503/21651 x 100 = 71.6%. The accuracy is essentially a measure of how many ground truth pixels were classified correctly. Thus, for the 1984 land use/land cover classification an average accuracy of 66.23% and overall accuracy of 71.23% was generated. This indicates that error is considered to be consistent with limits of the available technology.

Table 2 shows the producer's accuracy for the image classification of Landsat ETM+ 2000 into seven classes of land use/land cover types: Farmlands (Fm) = 57%; forested wetlands (FW) = 48% light forest (LF) = 82%; non forested wetlands (NFW) =62%; scrubs (SS)= 59%; urban (U) = 94%; water body (WB) = 96%. Also, the user's accuracy as shown in the same table indicates the probability that a pixel classified into a given category are the true representation of that category on the ground:

Table 3. Google Earth Image 2012 land use/land cover accuracy assessment.

LULC	FML	FWT	LF	NFWL	SRB	URB	WB	TR
FML	1213.7	147.2	0	0	634.88	70.8	0	2066.72
FWT	294.76	1274	278.8	84.08	709.88	110.8	0	2752.68
LF	146.76	80.28	1813	41.08	147.88	128.8	0	2358.68
NFWL	119.76	8.28	0	1667.08	512.88	564.8	0.96	2873.76
SRB	200.76	43.28	0	876.08	2356.88	456.8	0	3933.8
URB	95.76	0	0	199.08	151.88	6586.8	50.96	7084.48
WB	86.76	0	0	0	80.88	114.8	2074.96	2357.4
RELIABILITY	0.84	0.64	0.8	0.48	0.52	0.92	0.98	
LULC	FML	FWT	LF	NFWL	SRB	URB	WB	TR

Average accuracy = 73.68%, average reliability = 73.75%, overall accuracy = 77.34%.

Farmlands (Fm) = 73%; forested wetlands (FW) = 76%; light forest (LF) = 79%; non forested wetlands (NFW) = 54%; scrubs (SS) = 56%; Urban (U) = 85%; water body (WB) = 97%; overall accuracy = TD (sum of major diagonal) divided by TR (row totals) = 16955/22702 x 100 = 74.9%. Thus, 2000 land use/land cover average accuracy is 70.40% and overall accuracy is 74.9%. This indicates that error is considered to be consistent with limits of the available technology.

Table 3 shows the producer's accuracy for the image classification of Google Earth Image 2012 into seven classes of land use/land cover types: Farmlands (Fm) = 80%; forested wetlands (FW) = 66%; light forest (LF) = 62%; non forested wetlands (NFW) = 55%; scrubs (SS) = 59%; urban (U) = 96%; water body (WB) = 99%. Also, the user's accuracy as shown in the same table indicates the probability that a pixel classified into a given category are the true representation of that category on the ground: farmlands (Fm) = 84%; forested wetlands (FW) = 64%; light forest (LF) = 80%; non forested wetlands (NFW) = 48%; scrubs (SS) = 52%; urban (U) = 92%; water body (WB) = 98%; overall accuracy is estimated to be 77.34%. Thus, 2012 land use/land cover average accuracy is 73.68%. This indicates that error is considered to be consistent with limits of the available technology.

Changes in land use/land cover between 1984 and 2012

Figures 2, 3 and 4 show the spatial distribution of the static land use/land cover of the study area in 1984, 2000 and 2012, respectively. Table 4a and b below showed the entire study area covering 16140 ha. In 1984, light forest constituted the most extensive land use/land cover occupying 5350 ha (33.2%) and urban/built up area was 2690 ha (16.7%). The non-forested wetland, forested wetland, shrub, water body and farmland occupied {ha, (%)} 2380 (14.5), 2170 (13.5), 1720 (10.7), 1180 (7.3) and 650 (4.0) respectively of the study area.

However, in 2000 urban or built up area expanded quite rapidly, increasing to 4690 ha (29.1%) of the study area. The shrub and farmland areas increased to {ha, (%)} 3580 (22.2) and 1400 (8.7) respectively. During the same period, non-forested wetland, forested wetland, light forest and water body decreased to {ha, (%)} 1960 (12.1), 1790 (11.1), 1570 (9.7) and 1150(7.1) respectively. By 2012 urban built up, shrub, farmland and forested wetland has increased to {ha, (%)} 5191 (32.16), 4417 (27.37), 2409 (14.92) and 1827 (11.32), respectively. On the other hand, light forest, non forested wetland and water body had decreased to {ha, (%)} 14 (0.09), 1325 (8.21) and 957 (5.93) respectively.

Also, the pattern of land use/land cover change within the period under consideration indicated that between 1984 and 2000 farmland, shrub and urban/built up were expanding at the rates of 7.2, 6.7 and 4.7% per annum. Light forest, non-forested wetland and forested wetland were decreasing at the average rates of 8.3, 4.7 and 2.8% per annum, respectively, and water body decreased marginally at an annual rate of 0.2% per annum. The table also indicated that between 2000 and 2012 farmland, shrub and urban/built up increased by 72.1, 23.4 and 10.7% per annum, and forested wetland increased marginally by 2.1%. On the other hand light forest, non-forested wetland and water body were decreasing at the average annual rates of 8.26, 2.70 and 1.40% respectively.

Land use/land covers conversion pattern between 1984 and 2000

Table 5a show the pattern of land use change between 1984 and 2000. In Table 5a, while 19.92% of the land use/land cover in the study area remained unchanged, 80.08% have been lost to other land uses. The table also reveals that 12.00% of farmlands, 20.74% of forested wetland, 9.16% of light forest and 23.99% of non-forested wetland, as well as 24.59% of shrub, 35.76% of urban and 20.42% of water body respectively remained unchanged between 1984 and 2000. It was also observed

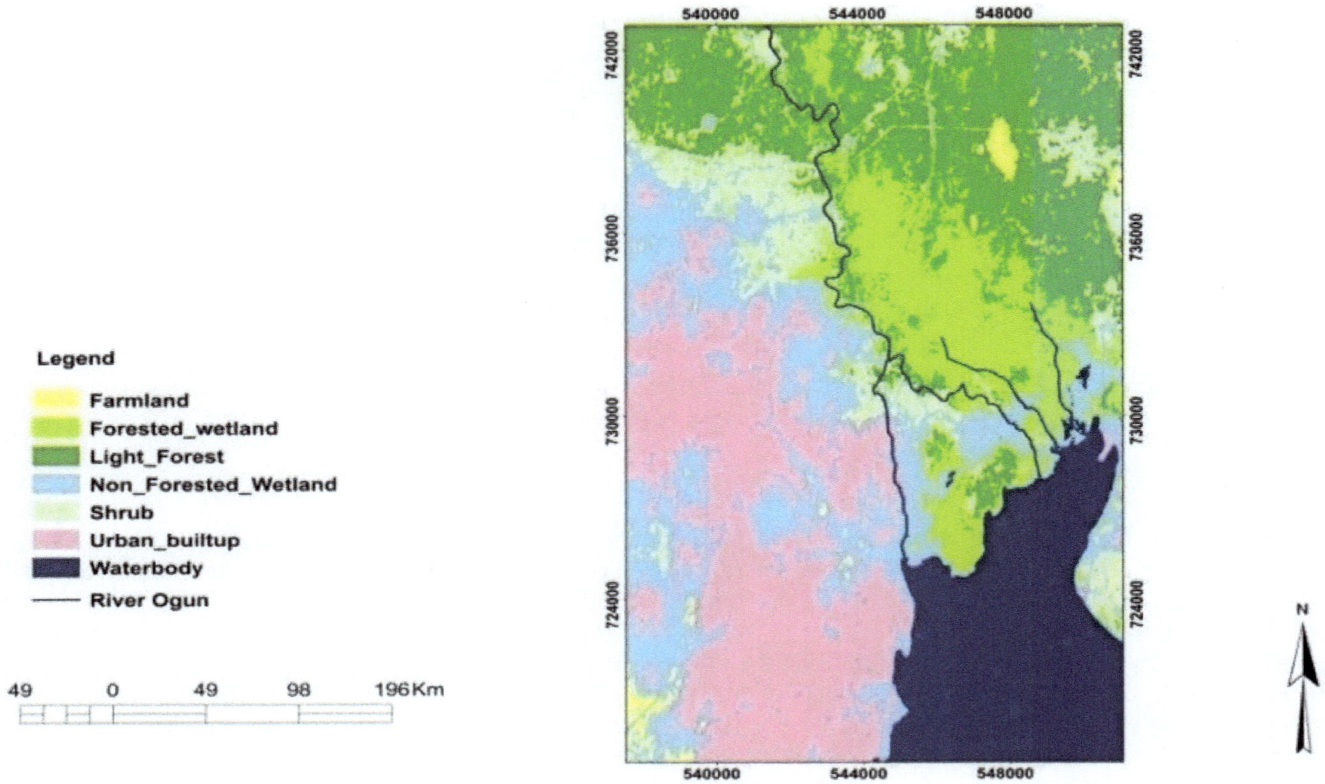

Figure 2. Land use/land cover map of Lower Ogun River Basin, SW Nigeria, in 1984.

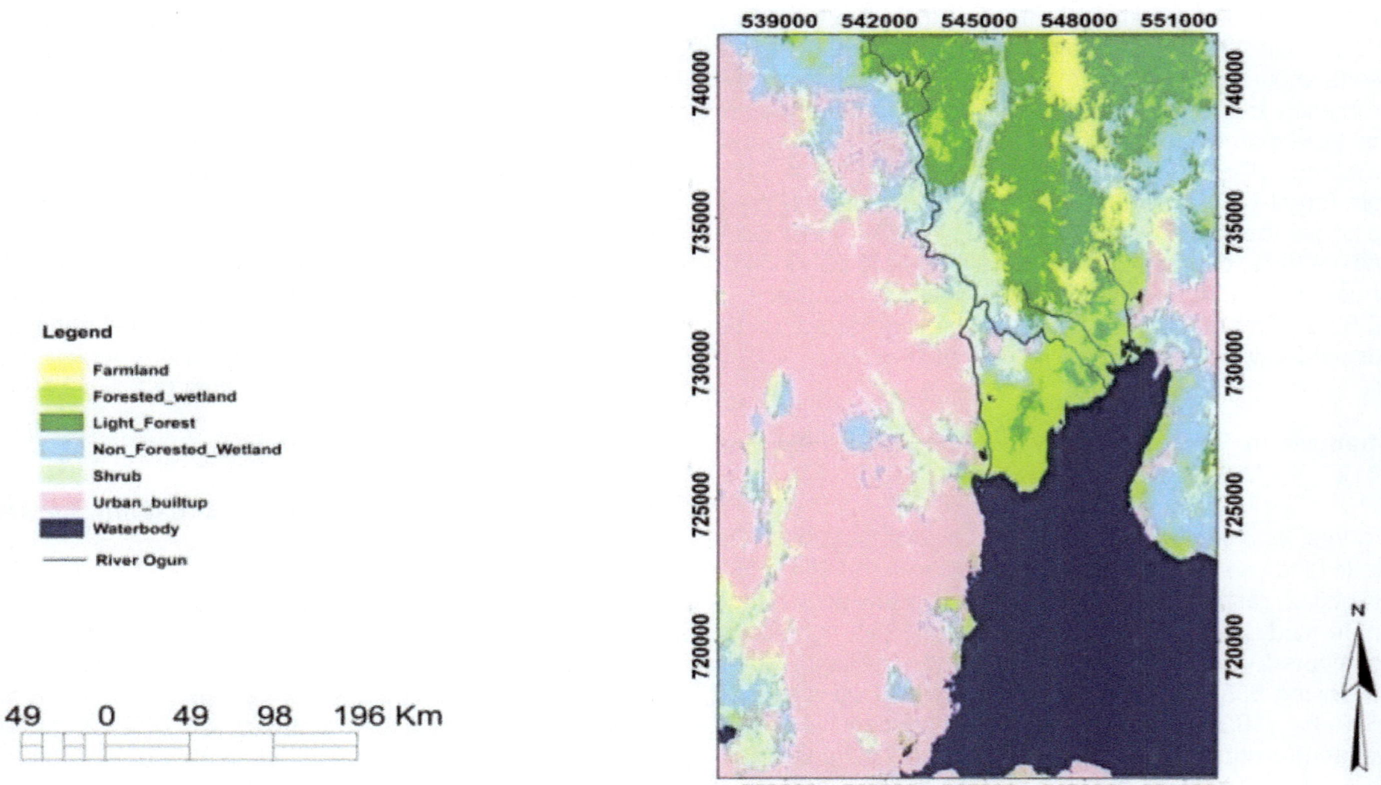

Figure 3. Land use/land cover map of Lower Ogun River Basin, SW Nigeria, in 2000.

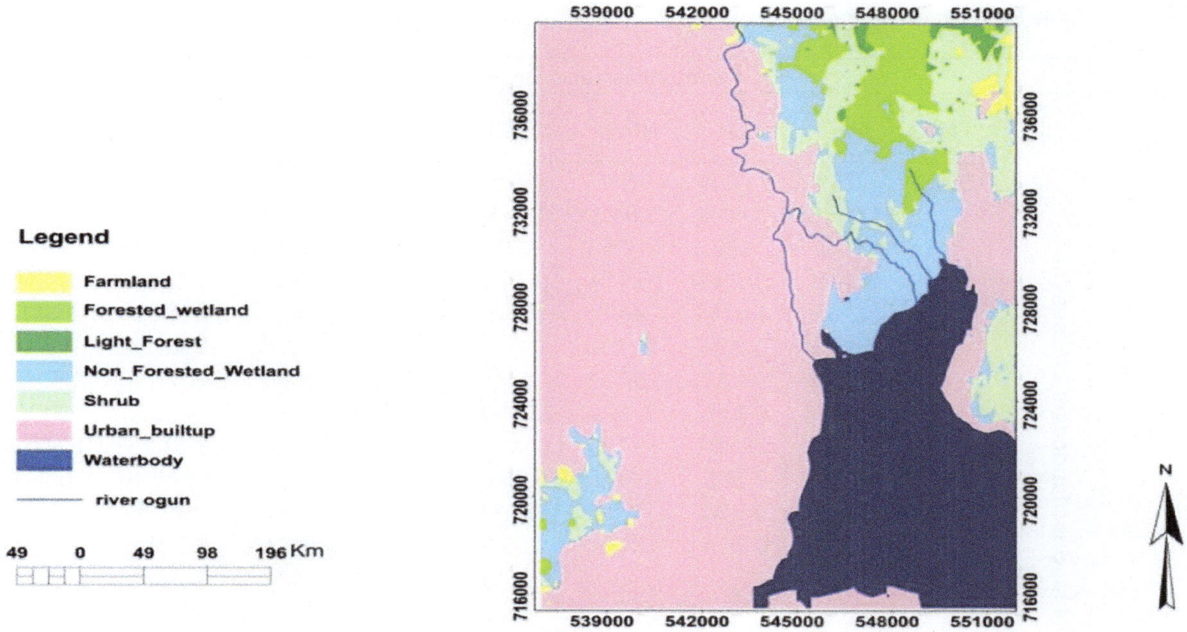

Figure 4. Land use/LAND COVER MAP of Lower Ogun River Basin, SW Nigeria, in 2012.

Table 4a. Extent and rate of change in land use/land cover from 1984 to 2000.

LULC	1984		2000		Change b/w 1984 & 2000		Av.Rate of Change	
	Area (ha)	Area (%)	Area (ha)	Area (%)	ha	%	ha	%
FML	650	4.0	1400	8.7	750	+115.4	47	+7.23
FWT	2170	13.5	1790	11.1	-980	-35.4	-61	-2.81
LF	5350	33.2	1570	9.7	-3780	-70.7	-442	-8.26
NFWL	2380	14.5	1960	12.1	-420	-17.7	-111	-4.66
SRB	1720	10.7	3580	22.2	+1860	+108.1	116	+6.74
URB	2690	16.7	4690	29.1	+2000	+74.3	125	+4.65
WB	1180	7.3	1150	7.1	-30	-2.5	-2.0	-0.17
Total	16140	100	16140	100	-	-	-	-

Authors' Image Analysis, 2013.

Table 4b. Extent and rate of change in land use/land cover from 2000 to 2012.

LULC	2000		2012		Change b/w2000 & 2012		Av.Rate of Change	
	Area (ha)	Area (%)	Area (ha)	Area (%)	ha	%	ha	%
FML	1400	8.7	2409	14.92	+1009	+72.1	+84.06	+6.01
FWT	1790	11.1	1827	11.32	+37	+2.1	+3.09	+0.17
LF	1570	9.7	14	0.09	-1556	-99.1	-129.68	-8.26
NFWL	1960	12.1	1325	8.21	-635	-32.4	-52.91	-2.70
SRB	3580	22.2	4417	27.37	+837	+23.4	+69.76	+1.95
URB	4690	29.1	5191	32.16	+501	+10.7	+41.75	+0.89
WB	1150	7.1	957	5.93	-193	-16.8	-16.08	-1.40
Total	16140	100	16140	100				

Authors' image analysis, 2013.

Table 5a. Land use/land cover change pattern between 1984 and 200

		Landuse/Landcover classes of 2000 (ha)							
	FML	**FWT**	**LF**	**NFWL**	**SRB**	**URB**	**WB**	**Total area**	**FML**
1984 landuse/landcover classes (ha)	FML	**78**	117.74	50.96	110.74	393.4	407.82	235.20	1400
		12.00%	8.41%	3.64%	7.91%	28.10%	29.13%	16.87%	100
	FWT	98.12	**450**	277.63	92.01	146.06	248.631	481.75	1790
		5.48%	**20.74%**	15.51%	5.14%	48.16%	13.89%	26.99%	100
	LF	93.62	144.44	**490**	154.80	346.50	346.19	311.017	1570
		5.96%	9.20%	**9.16%**	9.86%	22.07%	22.05%	19.81%	100
	NFWL	363.28	194.824	157.98	**571**	433.552	205.21	134.46	1960
		18.53%	9.94%	8.06%	**23.99%**	22.12%	10.47%	6.86%	100
	SRB	287.11	271.01	167.90	352.63	**423**	1113.74	157.16	3580
		8.02%	7.57%	4.69%	9.85%	**24.59%**	31.11%	4.39%	100
	URB	32.41	35.64	25.80	189.01	413.189	**962**	903.76	4690
		0.69%	0.76%	0.55%	4.03%	8.81%	**35.76%**	19.27%	100
	WB	0	0	0	0	0	202.75	**241**	1150
							17.63%	**20.41%**	10.0
	Total	952.329	1213.654	853.747	1291.577	2739.042	5614.572	3270.603	16140

Authors' image analysis, 2013.

that 88.00% of farmlands, 79.26% of forested wetland, 90.84% of light forest and 76.01% of non-forested wetland, as well as 75.41% of shrub, 64.24 and 79.58% of water body have been converted to other uses, within the same period.

Furthermore, the table also indicated that 5.48% of forested wetland, 5.96% of light forest, 18.53% of non-forested wetland, as well as 8.02% of shrub and 0.69% of urban have been changed to farmlands. While at the same time, 29.13% of farmland, 13.89% of forested wetland, 22.05% of light forest, as well as 10.47% of non-forested wetland, 31.11% of shrub and 17.63% of water body have been converted to urban land use. Thus, the conversion of light forest, forested and non-forested wetlands as well as shrub at the rates indicated above give an indication that farming and urbanization are the most important drivers of land use in the study area, and these also reflect the increasing scale of human activities in the study area.

Table 5b show the predominance of farmland and urban over other land use/land cover types with 80.7 and 53%, respectively. The table also reveal that 95.1 ha (10.5%) forested wetland, 35.8% light forest, 20.7% non-forested wetland, 15.2% shrub, 53% urban and 62.9% water body remain unchanged over the twelve years that constitute the latter part of the study period. According to the table, 13.7% of the farmland, 6.3% of forested wetland, 11% of the non-forested wetland and 1.9% of water body have been converted to urban. This suggests that environmentally fragile areas have been prone to urban encroachment without conscious effort to mitigate the devastating consequences such flooding, loss of biodiversity and destruction of ecosystem.

Table 6a and b present the results of the analyses on the proportions of LULC units gained and/or lost between 2000 and 2012 in the study area. From Table 6a, farmlands maintained 3360 ha equivalence of 41.5% but 805.5 ha (13.1%) of its total area over 12 years had been converted to other land use/land cover types. Nevertheless, farmlands had gained 2021.0 ha (20.3%) from other land use/land cover types, second highest to urban land use. This is further affirmed by the gained/lost analysis over the 12 years of the study which reveals that farmland record net gain of 1215.5 ha (60.1%). This suggests the growing interest of the urban dwellers in urban agriculture that constitutes one of the drivers of land use/land cover change in the study area. This is in line with the findings of Awoniran et al. (2011) who reported that urban agriculture and rapid urbanization are major drivers of land use/land cover change in the Lower Ogun River Basin.

This finding is further strengthened by the soil test carried out by the study which revealed the micro-variability of the soil, with significant ($p < 0.01$) differences in the properties compared. According to the results presented in Table 7, soils of the forested wetland areas are mainly clay while the non-forested areas are clay loam and sandy clay loam. The mean soil pH (1:1 soil-H_2O) ranged from 5.0 to 6.5, indicating slightly acidic soil condition. In the forested wetland areas, total nitrogen (TN; 0.98 to 2.63 g kg^{-1}), organic carbon (OC; 12.50 to 37.40 g kg^{-1}) and available phosphorus (24.05 to 44.80

Table 5b. Land use/land cover change pattern between 2000 and 2012.

2000 land use/land cover classes (ha)	2012 land use/land cover classes (ha) FML	FWT	LF	NFWL	SRB	URB	WB	Total
FML	**3360**	124.6	525.4	34.8	111.0	9.7	0	4165.5
	80.7%	3.0%	12.6%	0.8%	2.7%	0.2%	0	100
FWT	174.0	**95.1**	160.5	114.5	179.3	143.9	39.1	906.4
	19.2%	**10.5%**	17.7%	12.6%	19.8%	15.9%	4.3%	100
LF	517.6	262.2	**632.0**	92.7	225.0	33.1	3.2	1765.8
	29.3%	14.8%	**35.8%**	5.2%	12.7%	1.9%	0.2%	100
NFWL	252.3	183.5	349.0	**267**	185.0	50.8	3.8	1291.4
	19.5%	14.2%	27.0%	**20.7%**	14.3%	3.9%	0.3%	100
SRB	317.2	371.3	232.4	87.6	**241.0**	325.5	8.1	1583.1
	20.0%	23.5%	14.7%	5.5%	**15.2%**	20.6%	0.5%	100
URB	752.0	346.8	258.5	604.2	516.2	**2909.1**	105.8	5492.6
	13.7%	6.3%	4.7%	11%	9.4%	**53.0%**	1.9%	100
	7.9	3.7	0	63.8	161.1	110.9	**587.8**	935.2
WB	0.8%	0.4%	0	6.8%	17.2%	11.9%	**62.9%**	100
Total	5381	1387.2	2157.8	12 64.6	1618.6	3583	747.8	16140

Authors' Image Analysis, 2013.

Table 6a. Proportion of LULC unit gained and/or lost between 1984 and 2000.

Land use classes	Proportion of LULC in 2000 and unchanged in 2012		Proportion of LULC in 2000 lost to other LULC by 2012		Proportion of LULC in 2000 gained from other LULC type by 2012		LULC in 2012 (unchanged + gained)		Difference of (2000-2012) LULC gained-lost	
	ha	%	ha	%	ha	%	ha	%	ha	%
FML	3360	41.5	805.5	13.1	2021.0	20.3	5381	100.00	+1215.5	+60.1
FWT	95.1	1.2	811.3	13.2	1292.1	13.0	1384.2	100.00	+480.8	+37.2
LF	632.0	7.8	1133.8	18.5	1525.8	15.3	2157.8	100.00	+392	+25.7
NFWL	267	3.3	1024.4	16.7	997.6	10.0	1264.6	100.00	-26.8	-2.7
SRB	241.0	3.0	1342.1	21.9	1377.6	13.8	1618.6	100.00	+35.5	+2.6
URB	2909.1	36.0	673.9	11.0	2583.5	25.9	3583	100.00	+1909.6	+73.9
WB	587.8	7.3	347.4	5.7	160.0	1.6	747.8	100.00	-187.4	-117.1
Total	8092	50.14	8048	49.86	8048	49.86	16140	100	-	-

Authors' image analysis, 2013.

mg kg^{-1}) values were obtained. Also, the cation exchangeable capacity (CEC) in the forested areas ranged from 9.55 to 12.37 cmol kg^{-1}. However, in the non-forested areas, TN, OC, available phosphorus and CEC values were low. Within each of the study area, soil properties varied significantly from one sample location to another. These results are in agreement with those reported by Adepetu et al. (1979) under continuous soil manipulation as a result of human activities. The soil physical properties such as soil texture known to be relatively 'stable' over-time (Mbagwu, 2008) changed from clay in the forested wetland to sandy clay loam in nearby non-forested wetland where anthropogenic activities are on-going.

Though, urban land use retained 2909.1 ha (36.0%) in 2012 and lost 673.9 ha (11.0%) to other LULC but still maintained its predominance by gaining 2583.5 (25.9%) from other land use/land cover types. The net gained analysis shows that this land use has gained 1909.6 ha, an equivalence of 73.9% over the twelve years of the latter study period. This revelation provides a strong signal about possible lost of farmlands that constitutes second largest land use in the area to urban land use in the nearest future. This may be attributed to the fact that

Table 6b. Proportion of LULC unit gained and/or lost between 2000 and 2012.

Land use classes	Proportion of LULC in 1984 and unchanged in 2000		Proportion of LULC in1984 lost to other LULC by 2000		Proportion of LULC in 1984 gained from other LULC type by 2000		LULC in 2000 (unchanged+gained)		Difference of (1984-2000) LULC gained-lost	
	ha	%	ha	%	ha	%	ha	%	ha	%
FML	78	12	572	88.0	1322	94.43	1400	100.00	750	53.57
FWL	450	20.74	1720	79.26	1340	74.86	1790	100.00	-380	21.23
LF	490	9.16	486	90.84	108	68.79	1570	100.00	-3780	-40.76
NF W	571	23.99	1809	76.01	1389	70.87	1960	100.00	-420	-21.43
SHR	423	24.59	1297	75.41	3157	88.18	3580	100.00	1867	52.15
URB	962	35.72	1728	64.24	3728	79.49	4690	100.00	2200	46.91
W B	241	20.42	939	79.58	909	79.04	1150	100.00	-30	-2.61
Total area	3215	19.92	12925	80.08	12925	80.08	16140	100.00	-	-

Authors' Image Analysis, 2013.

Table 7. Mean values of physico-chemical properties of soils of forested and non-forested wetland.

	pH	Sand	Silt	Clay	P	OC	TN	Mg	Ca	K	Na	Exchangeable acidity	Textural Class
			$(g\ kg^{-1})$		$(mg\ kg^{-1})$		$(g\ kg^{-1})$			$(cmol\ kg^{-1})$			
Non forested wetland													
1	5.1c	320	300	380	21.66d	20.30^{a}	1.93^{a}	1.23d	5.50d	0.17^{b}	0.50^{b}	2.00^{b}	CL
2	5.4b	440	250	310	29.57^{b}	15.60d	1.30^{c}	1.22d	6.52^{a}	0.23^{ab}	0.48^{b}	1.80^{b}	CL
3	5.3b	360	210	430	23.15^{c}	19.90^{b}	1.67^{b}	2.23^{b}	5.33d	0.27^{a}	0.48^{b}	1.22^{b}	C
4	5.0c	400	240	360	31.36^{a}	13.70e	1.07d	2.03^{c}	5.95^{c}	0.20^{b}	0.46^{b}	1.40^{a}	CL
5	5.9a	500	200	300	29.42^{b}	16.80^{c}	1.38^{c}	3.25^{a}	6.28^{b}	0.33^{a}	0.59^{a}	1.30^{b}	SCL
Forested wetland													
1	5.4c	120	270	610	44.80^{a}	22.50^{c}	1.72^{b}	2.22^{a}	8.70^{a}	0.93^{a}	0.52^{b}	1.00^{bc}	HC
2	6.2b	100	370	530	33.25^{b}	37.40^{a}	2.63^{a}	1.83^{b}	8.20^{b}	0.67^{a}	0.59^{a}	1.10^{b}	C
3	5.3c	120	380	500	31.63^{c}	18.70e	1.57^{c}	0.81^{c}	8.30^{b}	0.65^{a}	0.59^{a}	1.30^{a}	C
4	6.5a	40	400	560	33.15^{b}	19.50d	0.98e	1.83^{b}	7.45^{c}	0.56^{ab}	0.57^{a}	1.40^{a}	C
5	5.2c	100	320	580	24.05d	24.20^{b}	1.24d	1.83^{b}	6.90d	0.30^{b}	0.52^{b}	0.90^{c}	C

Means with the same alphabet(s) are not significantly different at $p< 0.01$. CL = Clay loam, C = clay, SCL = sandy clay loam, HC = heavy clay.

farmlands possess some characteristics that favour urban development and hence increases its vulnerability to urban encroachment. This implies a devastating decline in urban food production if the trend is not controlled to a sustainable level.

It is equally important to note from the table that a substantial loss in areas occupied by water body and non-forested wetland estimated at 117.1 and 2.7%, respectively, occurred over the study period. Keeping this trend going will not only amount to loss of vital parts of ecosystem that may trigger the conditions encouraging local climate modification but increase the vulnerability of the study area to destructive and pernicious effects of urban flooding that may culminate into loss of lives and properties magnitude of which may be horrified to imagine. Meanwhile, forested wetland, light forest and shrub, according to the table have net gain of 37.2, 25.7 and 2.6% over the study period but Table 5b reveals that these land covers are prone to conversion to other land uses particularly farmland and urban. This suggests that it requires conscious effort to stem down the inevitable conversion of these land covers to the sustainable path. The modification of these land covers through a programme of Lagos State Government; greenness and beautification of environment will go a long way to sustainably conserve the threatened land covers. The study has been able to identify two major processes of land use/land cover change in the study area, namely conversion and modification.

Table 6b shows that 19.3% of the area occupied by

farmland over the study period has been converted to other LULC, in which light forest recorded 12.6% which happens to be the highest in the study period, followed by forested wetland 3.0%, shrub (2.7%), non-forested wetland (0.84%) and urban (0.23%). In spite of this loss, farmland has continued to increase in spatial extent by gaining 19.2% from forested wetland, light forest (29.3%), non-forested wetland (19.5%), shrub (20.0%), urban, that is undeveloped plots of land within urban (13.7%) and water body through reclamation of land (0.8%). Urban land use obviously is competing with farmland for available space in the process 33.1% of light forest, 20.6% of shrub, 11.9% of water body, 3.9% of non-forested wetland and 0.23% of farmlands that are previously non-forested wetland or forested wetland or shrub or light forest have been converted to urban use within the study period. This may partly provide clue to contemporary issue of building collapse in the Lagos mega city where large hectares of land have been reclaimed from water, non-forested wetland and shrub as revealed by this study and utilized for urban development without appropriate geotechnical analysis. The modification noticed is in the area of tree planting and beautification programme targeted at promoting green environment which, to a large extent, deserves encouragement.

Drivers of landuse/landcover change in the study area

Land use/land cover changes reflect the dynamics observed in the socio-economic condition of a given area. Similarly, changes in the socio-economic situations cause land use/land cover changes through their influence on land management techniques used and other various aspects of farming systems, institutional settings, environmental policy and others (Mengistu and Salami, 2007). As many researchers indicated in Nigeria, several factors have been modifying the original form of land cover. These include human activities such as agricultural colonization (Adejuwon, 1971); spread of rural settlements (Osunade, 1991); evolution of rural road networks and government policy (Ekanade et al., 1996) noted in Salami (2001).

The results obtained from the change detection analysis gave a clear indication that most of the light forests have disappeared, having been cleared for property development, road construction and provision of utilities such as the two waterworks in the study area.

Information obtained from field work also confirmed that the forested wetlands were being cleared for the cultivation of vegetables and sugarcane. Andreas and Hugh (2009) in a study of land use/land cover change in sub-Saharan Africa over a 25 year period concluded that land cover change in Africa is mainly driven by the expansion of croplands. The expansion of croplands may lead to a growth in agricultural output such as food and fiber production and impact positively on the socio-economic condition of the people. However, at the same time the land left available for future agricultural expansion is decreasing, and with population increase the agricultural zone itself is more crowded exacerbating potential friction amongst and in between agriculturalists and pastoralists alike. Such changes require rapid adjustments to land management so as to avoid crises in food security and conflict over dwindling access to natural resources, which are becoming more evident. The continued expansion of urban built up and farmlands have serious implications for the hydrology, microclimatology and soil health of the river basin, which would only be revealed by further studies.

Environmental consequences of land use/land cover change

From the results of the analyses carried out the dominant land use/land cover (drivers of land use/land cover) in the study area are urban and farmlands. Therefore, the analysis of land use impact that follows will examine the impacts of these two land use classes.

The results obtained from the various ecological approaches adopted in this study, that is, change detection analysis, focus group discussion, soil and vegetation sampling, were used to assess the biophysical status of the Lower Ogun River Basin ecosystem. This assessment focused primarily on: (a) soil quality (b) biodiversity (c) socio-economic impacts and (d) hydrological impact.

Soil quality has been viewed as the capacity of a soil to function within its ecosystem boundaries and interact positively with the environment external to that ecosystem (Olayinka, 2009). Soil quality has emerged as a unifying concept to address the larger issue of sustainability of ecosystem in general and agriculture in particular (Eswaran et al., 2005). From the point of view of land use and land management decision making soil quality is a measurable component of the environment which provides a quantitative basis for evaluating different land use options and impacts of technology. As indicated in Table 7, the impact of LULC is quite noticeable on the soil quality in the study area. Conversion of native forests to cultivation is usually accompanied by a decline in soil organic carbon and nutrients, and deterioration of soil structure. The result obtained from the soil analysis carried out indicated that the growth of urban agriculture in the study area engendered by rapid urbanization is having negative impact on the physical, chemical and organic properties of soil in the study area. The result of the vegetation profile conducted in the study area indicated that the combination of urban growth and the concomitant increase in population has resulted in the loss of biodiversity as manifested in almost complete degradation of the non-forested wetland.

Table 8. FGD findings on the impact of landuse change.

Impacts of land use	Maidan (forested wetland)			Isasi (non-forested wetland)		
	Sand Miner	Farmer	Fishermen	Sand miner	Farmer	Fishermen
Increased flooding	+	+	+	+	+	+
Pollution	+	+	+	+	+	-
Reduction in plant and animal species	+	+	+	+	+	+
Global Warming	-	-	-	-	+	+
Income Generation	+	+	+	+	+	+

+, Where the opinions were expressed by the respondents; -, where the opinions were not expressed at all.

Thus the original vegetation cover has given way to shrub and herbs while trees have completely disappeared, in contrast with the forested wetland. This development also has a serious implication on the microclimate of the study area and carbon emission. Information gathered from field work and focus group discussion also revealed that the dominant life form in the non-forested wetland was herb which covered 83% of the study site, while the remaining 17% was covered by shrub, there were no trees, as plant height ranged between 0.1 to 5 m. This result also indicated that this site has been seriously degraded. On the other hand, life forms in the forested wetland showed degradation from low shrub, medium shrub, tall herb and tall trees, which were the dominant life form, at 43%. This result also demonstrated the characteristic feature of tropical rainforest in which tree canopies occur in tree layers. Thus heights of plants ranged from <2 to >25 m.

A major source of hydrological impact of land use/land cover change in the study area are the two water works (Adiyan and Iju water works) sited within the study area. Ogunbajo (2005), in a study carried out on the impact of the two water works reported that, flooding of the immediate environment, change in the flow regime of river Ogun, silting of the river bed and pollution from sludge biomass were the observable impacts of the two waterworks. The waste water and high amount of silt contained therein are discharged into the downstream sector of river Adiyan resulting in the alteration of the morphology of the area which induces large scale erosion when it rains and consequently large flooding of the area.

Urban agriculture as with other urban activities has both positive and negative social, environmental, and economic impacts and externalities. Urban agriculture is an important source of income for some of the inhabitant of the communities within the study area. Findings from FGD, according to Table 8, confirmed the fact that in addition to income from sales of surpluses, farming households save on household expenditure by growing their own food.

Conclusion

Changes in the patterns of land use/land cover as demonstrated in this study are a reflection of increasing anthropogenic pressure on landscape which needs to be seriously addressed. Rapid population growth and its twin brother, urbanization remain the major causes of environmental degradation. More and more people, especially the urban poor, are getting engaged in urban agriculture using the mostly unprotected and uncontrolled wetlands in the Lower Ogun River Basin. Although this could lead to increased food production and food security there is need for further research into sustainable management and utilization of fragile ecosystem such as wetlands. There is a need for the control and planning of activities around the existing wetlands by government to avoid unsustainable encroachment of the fragile wetlands.

REFERENCES

Adejuwon JO (1971). Agricultural Colonization in twentieth century in Western Nigeria: Distribution and Vegetation Characteristics. J. Trop. Geogr. 39:1-10.

Adeniyi OP, Omojola AS (1999). Landuse/Landcover change evaluation in Sokoto-Rima Basin of N.W. Nigeria based on archival remote sensing and GIS techniques. geoinformation technology applications for resource and environmental management in Africa (Edited by Adeniyi, O.P.) African Association of Remote Sensing of the Environment.

Adepetu JA, Obi AO, Aduayi EA (1979). Changes in soil fertility under continuous cultivation and fertilization in southwestern Nigeria. Niger. J. Agric. Sci. 1:15-20

Andreas BB, Hugh DE (2009). Monitoring 25 years of land cover change dynamics in Africa: A sample based remote sensing approach. J. Appl. Geogr. 29: 501-512.

Awoniran DR (2012). Geo-Spatial Analysis of Land Use and Land cover Change in the Lower Ogun River Basin. Unpublished M. Sc. Thesis Obafemi Awolowo University, Ile-Ife.

Awoniran DR, Adewole MB, Salami AT (2011) Wetland Conversion and Fragmentation Pattern with its Impacts on Soil in the Lower Ogun River Basin. Ife Research Publication in Geography, Vol.10, Issue 1, pp. 125-133.

Bouyoucos GJ (1962). Hydrometer Method improved for Making Particle Size Analysis of Soils. Agron. J. 54:464-465.

Bremner JM, Mulvaney CS (1982). Nitrogen-Total. In: Methods of soil analysis. 2nd ed., Part 2 ed., A. L. Page, R. H. Miller, and D. R. Keeney. eds. Agronomy Monograph No. 9. Madison, WI: American Society of Agronomy. Pp. 595-624.

Briassoulis H (2000). Analysis of landuse change: Theoretical and modeling approaches. The Web Book of Regional Science, Regional Research Institute West Virginia University, Virginia, total number of pages.

Butzer, KW (1990). The realm of cultural-human ecology: Adaptation and change in historical perspective. In: The earth as transformed by human action, Eds., B.L. Turner, II, WC Clark, RW Kates, JE Richards, JT Mathews, and WB Meyer, pp. 685-701. Cambridge: Cambridge University Press.

Coppin PB, Bauer ME (1996). Change detection in forest ecosystems with remote sensing imagery. Remote Sensing Rev. 13:207-234.

De Sherbinin A (2002). Land-Use and land-cover change. In: A CIESIN Thematic Guide. Palisades, NY: Center for International Earth Science Information Network of Columbia University.

Ekanade RO (1985). The Impact of Cocoa Cultivation on Soil Characteristics in Southwestern Nigeria. Unpublished PhD Thesis, University of Ife. p. 268.

Ekanade RO, Salami AT, Aborode M (1996). Flouristic Changes in the Tropical Rainforest of South Western Nigeria. Malays. J. Trop. Geogr. 27:8-13.

Eswaran H, Almaraz R, Reich P, Zdruli P (2005). Soil Quality and Soil Productivity in Africa, US Department of Agriculture, Natural Resources Conservation Service.

Food and Agriculture Organization (FAO). 1995. Planning for sustainable use of land resources. FAO Land and Water Bulletin 2. Rome: Food and Agriculture Organization of the United Nations.

Gutman G, Janetos AC, Justice CO, Moran EF, Mustard JF, Rindfuss RR, Skole D, Turner BL II, Cochrane MA (eds) (2004) Land change science: Observing, monitoring and understanding trajectories of change on the Earth's surface. Remote Sensing and Digital Image Processing Series 6, Kluwer Academic, Dordrecht Boston London

Lambin E, Helmut JG, Erle E (2007). Causes of land-use and land-cover change: Encyclopedia of Earth. Eds. Cutler J. Cleveland (Washington, D.C.: Environmental Information Coalition, National Council for Science and the Environment). 2007; http://www.eoearth.org/article/Causes of land-use_and_land-cover_change

Lambin EF, Baulies X, Bockstael N, Fischer G, Krug T, Leemans R, Moran EF, Rindfuss RR, Sato Y, Skole D, Turner BL II, Vogel C (1999). Land-use and land-cover change (LUCC): Implementation strategy. IGBP Report 48, IHDP Report 10, International Geosphere-Biosphere Programme, International Human Dimensions on Global Environmental Change Programme, Stockholm Bonn. p. 125.

Mbagwu JSC (2008). Environmental sustainability: Reversing the degradation trends in sub-saharan Africa. In: Joseph A., Nwigwe C., Agbicodo E., Opoku A., Obidiegwu J., and Acha I. Eds. Sustainable agriculture in sub-Saharan Africa: A critical look into the constrains and prospects. Proceedings of the 13th Annual Symposium of the International Association of Research Scholars and Fellows, 13th February, 2008, International Institute on Tropical Agriculture, Ibadan Nigeria. pp. 72-83.

Mclean EO (1982). Soil pH and lime requirement. In Methods of Soil Analysis. 2nd ed., Part 2 ed., A. L. Page, R. H. Miller, and D. R. Keeney. eds. Agronomy Monograph No. 9. Madison, WI: American Society of Agronomy. pp. 595-624.

Mengistu DA, Salami AT (2007). Application of remote sensing and GIS in Land use/Land cover mapping and change detection in a part of south western Nigeria. Afr. J. Environ. Sci. Technol. 1(5):99-109.

Mertens B, Lambin E (2000). Land-cover-change trajectories in southern Cameroon. Ann. Assoc. Am. Geogr. 90 (3): 467-94.

Miles MB, Huberman AM (1984). Qualitative Data Analysis: A Source Book of New Methods. Beverly Hills, California. 263 p.

Nelson DW, Sommers LE (1982). Total carbon, organic carbon and organic matter. In: Methods of soil analysis. 2nd ed., Part 2 ed., AL Page, RH Miller, DR Keeney. eds.. Agronomy Monograph No. 9. Madison, WI: American Society of Agronomy. pp. 539-579.

Nelson GC, Hellerstein D (1997). Do roads cause deforestation? Using satellite images in econometric analysis of land use. Am. J. Agric. Econ. 79:80-88.

Ogunbajo MI (2005). Environmental Assessment Study for the Lagos State Water Corporation under the World Bank Assisted Project Preparation Facility for the National Water Rehabilitation Project.

Olayinka A (2009). Soil microorganism, wastes and national food security. Inaugural Lecture delivered at Obafemi Awolowo University, Ile-Ife Nigeria. 26th May, 2009. p.40.

Olsen SR, Sommers LS (1982). Phosphorus. In Methods of Soil Analysis. 2nd ed., Part 2 ed., A. L. Page, R. H. Miller, and D. R. Keeney. eds. Agronomy Monograph No. 9. Madison, WI: American Society of Agronomy. pp. 403-430.

Osunade MAA (1991). Agriculture Change by Supplanting Process in a Traditional Farming System. Int. J. Ecol. Environ. Sci. 17:201-209.

Rojstaczer S, Sterling SM, Moore NJ (2001). Human Appropriation of Photosynthesis Products. Science 294:2549-2552.

Sack, R.D. (1990). The Realm of Meaning: The Inadequacy of Human-Nature Theory and the View of Mass Consumption. In: *The Earth as Transformed by Human Action*, BL Turner, II, WC Clark, RW Kates, JE Richards, JT Mathews, WB Meyer Eds.. pp. 659-671. Cambridge: Cambridge University Press.

Salami AT (2001). Agricultural colonization and floristic degradation in Nigeria's rainforest ecosystem. Environmentalist 21: 221-229.

Salami AT, Aladenola VO (2003). Management practices and ecological status of a humid tropical rainforest. Int. J. Environ. Stud. 60:215-228.

Tejuoso OJ (2006). Wetland uses/Dynamics For agricultural purposes and its health implications in Lower Ogun River Basin, Lagos, Nigeria. Technical Report submitted to International Development Research Centre (IDRC) Ecosystem approach to human health team, August, 2006.

Thomas GW (1982). Exchangeable cations. In: Methods of Soil Analysis. 2nd ed., Part 2, A. L. Page, R. H. Miller, and D. R. Keeney. Eds, Agronomy Monograph No. 9. Madison, WI: American Society of Agronomy. pp. 159-165.

Turner BL II, Skole D, Sanderson S, Fischer G, Fresco L, Leemans R (1995). Land-Use and Land-Cover Change. Science/Research Plan: International Geosphere-Biosphere Programme.

Turner MG, Wear DN, Flamm RO (1996). Land ownership and land-cover change in the Southern Appalachian highlands and the Olympic peninsula. Ecol. Appl. 6 (4): 1150-72.

United Nations Development Programme Annual Report (2009). Environment and sustainable development: Harnessing the green economy UNDP, May 2009. p. 27.

Vitousek PM, Mooney HA, Lubchenco J, Melillo JM (1997). Human domination of Earth's ecosystems. Science 277:494-499.

14

Geophysical study of saline water intrusion in Lagos municipality

OLADAPO, Michael Ilesanmi[1], ILORI, Olugbenga Bashiru[2] and ADEOYE-OLADAPO, Oluwakemi Olanike[3]

[1]Department of Applied Geophysics, Federal University of Technology, Akure, Ondo State, Nigeria.
[2]Lagos State Water Corporation, Lagos, Lagos State, Nigeria.
[3]Department of Physics, Adeyemi College of Education, Ondo, Ondo State, Nigeria.

Saline water intrusion presently constitutes serious concerns in the Lagos municipality just like many other coastal cities, thus necessitating its intervallic study. The present study involving 52 borehole logs (consisting of natural gamma and electrical resistivity components) was aimed at delineating intruded and vulnerable zones. Saline water columns defined by low resistivity values in the range of 0.1 and 20 Ωm as compared to fresh water (≥ 100 Ωm) were delineated on 22 logs. Four of the geo-sections generated in this study indicate saline water intrusion at depths varying from surface in Satellite Town, Kirikiri, Ijora, Iganmu, Apapa, Lagos Island, Ikoyi, Victoria Island and Lekki to depths ranging from 40 m at Iganmu to 158 m at Lekki. Intrusions of 47 m (143 - 190 m) and 60 m (56 - 116 m) were delineated at Ajah; 50 m (265 - 315 and 258 - 308) at Lakowe; 57 and 112 m (51 - 108 m and 198 - 308) at Ibeju Lekki, Akodo and 122 m at (233 - 355m) at Awoyaya. The hydrogeologic importance of the Coastal Plain Sand aquifer unit in Lagos is under severe threat of continued sea water intrusion on its southern flank. This study illustrates the current extension of the sea water intrusion. It highlights the depreciation of the water resource due to over pumping at higher rate than the natural recharge and slow sea level rise.

Key words: Borehole logs, saline water intrusion/incursion, natural gamma, resistivity, freshwater sand.

INTRODUCTION

Seawater intrusion is an inevitable problem of coastal fresh water aquifer associated with urban area (Hwang et al., 2004). Coastal aquifers constitute a vital source of fresh water in these regions, and are increasingly used to meet the water supply needs (Pareek et al., 2006). There is vital need to monitor the feasible risk of saline water intrusion of the coastal aquifers because, once saline intrusion into coastal aquifer has occurred, it is extremely difficult to overcome and improve the management of the water resources based on long term strategy. Less than 2% of seawater intrusion in the fresh water can diminish the water's potability (Custodio, 1987). Frequently, bore-holes have to be abandoned and other water sources sought, often at high cost. The challenge of saline water contamination in coastal aquifers is driven by a violation of the delicate hydrogeological balance that exists between freshwater and seawater in coastal aquifers (Goldman and Kafri, 2004) due to large-scale ground-water abstraction occasioned by rapid urbanization (Pareek et al., 2006).

Due to the proximity of Lagos to the Atlantic Ocean (Figure 1), the general population is faced with problems of freshwater abstraction from the subsurface. It is becoming harder for groundwater developers to construct

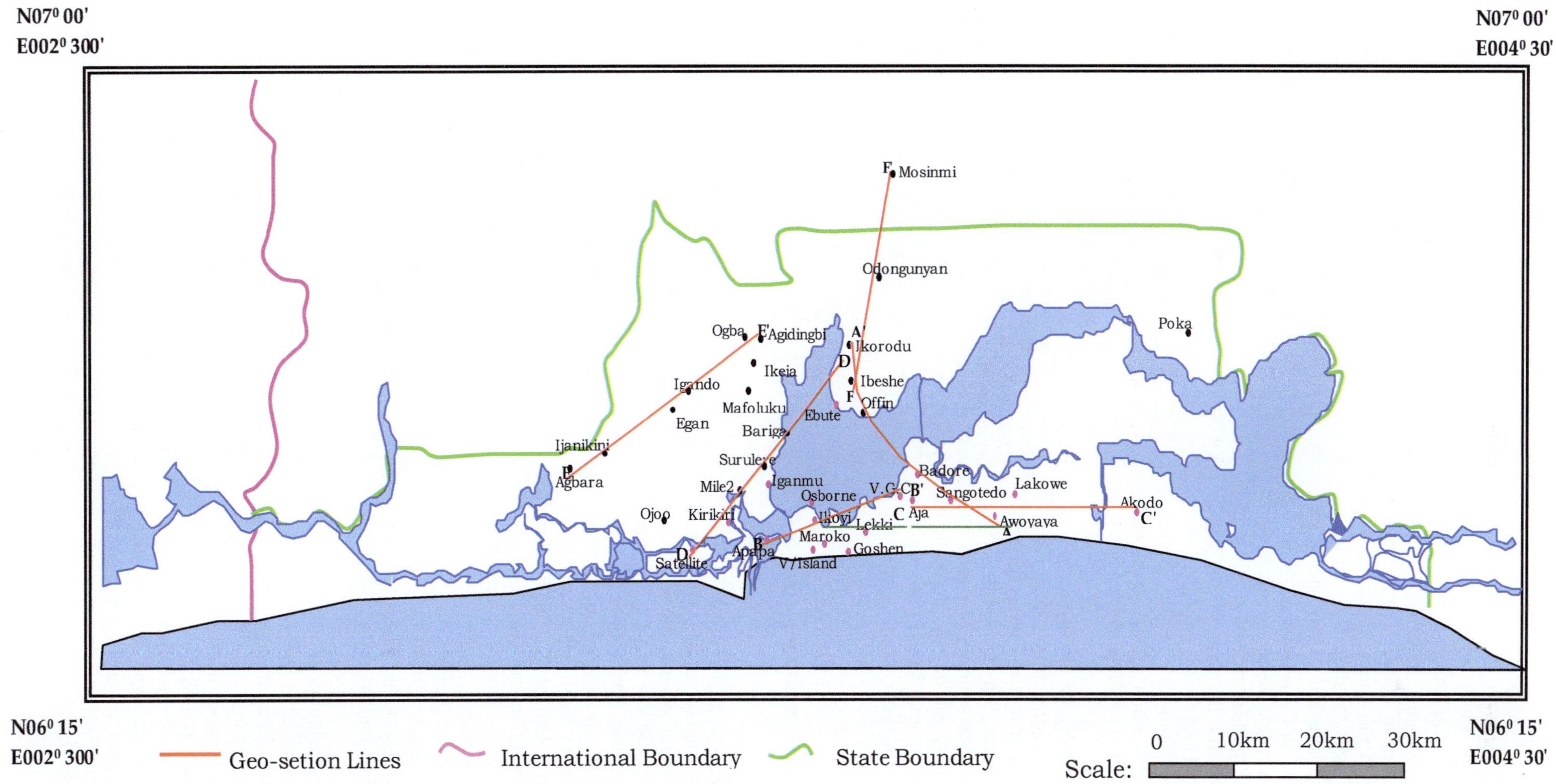

Figure 1. Map of Lagos State showing location of boreholes utilized for study and geo-section lines.

boreholes in areas adjoining the sea without encountering salt water. Some localities deemed to be very problematic within Lagos metropolis have been previously investigated using the electrical resistivity method (Ayolabi et al., 2003; Adepelumi et al., 2008).

Saline water intrusions vary in character and thickness. Accordingly, identification, distribution and nature of the intrusion can be used to identify areas where aquifer(s) may be more vulnerable to contamination to mitigate the degradation of the resources. The approximate thickness of the intrusion can be used to identify potentially sensitive areas.

The large resistivity contrast between salt water-saturated formation and the fresh water-saturated formation has been used for studying the salt water intrusion in coastal areas (Bates and Robinson, 2000; Hwang et al,, 2004). Fitterman et al. (1999), Nowroozi et al. (1999) and Paillet et al. (1999) have reported the applicability of geophysical well logging and surface geophysical surveys for the evaluation of the seawater intrusion characteristics. Numerous measurements and studies have established correlation between resistivity values and groundwater salinities (Zarroca et al., 2011). In overall majority of cases, the portions of aquifers saturated with seawater present resistivity values that are generally below 3 Ωm (Goldman and Kafri, 2004). Portions saturated with brackish water exhibit resistivity values between 3 and 10 Ωm (Yechieli et al., 2001; Repsold, 1989).

This study is aimed at identifying the lithological units, saline water and fresh water horizons of the aquifer and thus enabling the determination of the interface between the saline water and freshwater aquifer sands. The hydrogeological setting of the coastal areas of Lagos as may be observed from this study will reveal information on the geological stratigraphy and the fresh water extension so that fresh groundwater aquifer can be effectively protected for future use.

Description of the geology, hydrogeology and geomorphology of Lagos

Lagos is underlain by the Benin Basin (Figure 2). The rocks of the Benin Basin are mainly sands and shales with some limestone which thicken towards the west and the coast as well as down dips to the coast (Oteri and Atolagbe, 2003). The stratigraphic description of sediments in the basin has been provided by various authors (Elueze and Nton, 2004; Nton, 2001; Okosun, 1990; Omatsola and Adegoke, 1981; Ako et al., 1980). Five lithostratigraphic formations covering the cretaceous to Tertiary ages have been described. The formations from the oldest to the youngest include: Abeokuta Group (Cretaceous), Ewekoro Formation (Paleocene), Akinbo Formation (Late Paleocene - Early Eocene), Oshosun Formation (Eocene) and Ilaro Formation (Eocene). The Abeokuta Group present an unconformity with the basement complex.

The Abeokuta Formation constitutes a deep aquifer only in the northern parts of Lagos city (Ikeja area) where boreholes to the aquifer are about 750 m deep. The Ilaro and Ewekoro Formations are not key aquifers in Lagos as they are predominantly composed of shale/clay. The only source of hydraulic information on the Ilaro formation was obtained at Lakowe where no fresh water horizon was intercepted. It has not been possible to differentiate the Ewekoro as a target aquifer in any boreholes or existing wells in the metropolis. The formation apparently represents a minor groundwater resource in Lagos.

The Coastal Plains Sands is the main aquifer in Lagos that is exploited through hand-dug wells and boreholes. It forms a multi-aquifer system consisting of three aquifer horizons separated by silty or clayey layers (Longe et al., 1987; Kampsax and Sshwed, 1977). The aquifer thickens from its outcrop area in the north of the city to the coast in the south and the sand percentage in the formation also changes from north to south (Longe et al., 1987).

MATERIALS AND METHODS

Fifty-two (52) borehole logs were used for the study. The logs consist of natural gamma ray, single point resistance, short and long normal resistivity (Figure 3). The log analysis involved the identification of lithology, fluid content, saline water zones, establishing of saline water horizon thickness and delineating of freshwater aquifer.

Resistivity is measured by means of the introduction of a known voltage through a probe consisting of R16, R32 and R64 units in a water filled hole. The natural gamma radiation of the logged rock encountered in the hole is also measured with the same sonde but with a different probe. The natural gamma content and the resistivity values are measured with ELGG^{TM} sonde. The logging tool utilized is the RG PC Logger II^{TM} (Plate 1) with PCL II^{TM} software. The natural gamma ray log component is interpreted qualitatively to describe the lithological sequence of the borehole formation while the resistivity component is interpreted to describe the water quality (Figure 4).

Coode et al. (1996) has correlated the natural gamma ray log obtained from different borehole locations in Lagos and their work forms the basis on which different sand aquifers were identified and classified. The resistivity data presented on the logs were interpreted qualitatively in terms of resistivity values of formation at various zones. Interpretations of the logs were based on the classifications of Zohdy and Martin (1993) and modified by Ibrahim (2008) (Table 1).

RESULTS AND DISCUSSION

The results of the study are presented as geo-sections and maps. The geo-sections enabled the identification of the various lithological units within the study area. The maps were utilized to illustrate the fresh/saline water lateral interface boundaries. Six sections were generated from the study for the following axes:

A-A' Ikorodu - Badore - Sangotedo - Awoyaya; B-B' Apapa - Ikoyi - Victoria Island - Lekki - Ajah; C-C' Victoria Garden City - Sangotedo - Awoyaya - Lakowe - Akodo; D-D' Satellite Town - Kirikiri - Iganmu - Bariga - Ikorodu; E-E' Agbara - Egan - Idimu - Agidingbi; F-F' Mosinmi - Odongunyan - Ikorodu - Ibeshe.

Twenty-two (22) of the 52 logs interpreted intercepted saline water at various depths. The shallowest saline water interface was obtained at Victoria Island area (southern flank of the metropolis) while the deepest interface was delineated at depth of 308 m in Lakowe area (southeast of the metropolis).

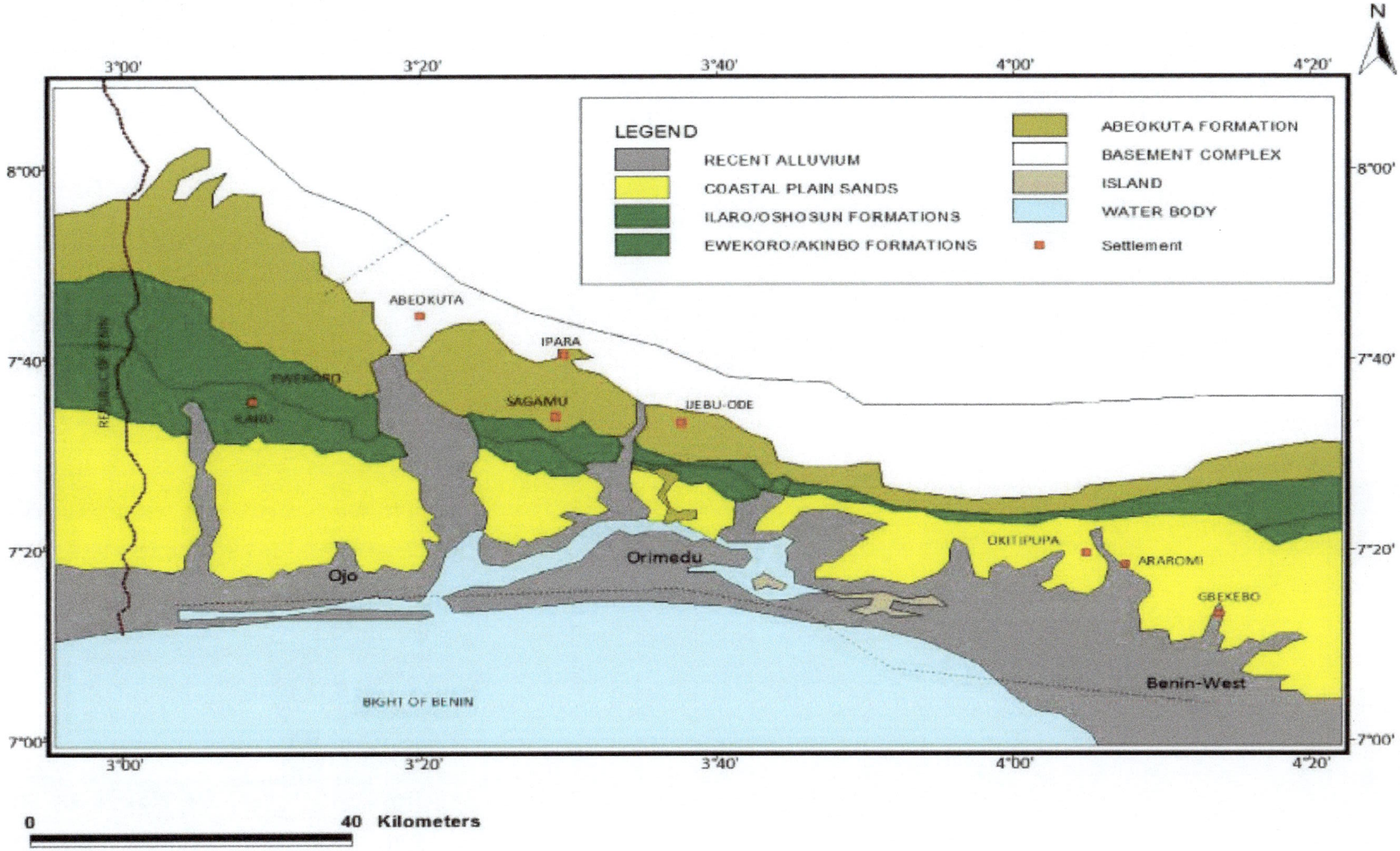

Figure 2. Geological map of Lagos.

Geo-section A-A': (Ikorodu - Badore - Sangotedo - Awoyaya)

The geo-section for profile A-A' is presented in Figure 5 while the interpretation summary is presented in Table 2. The section shows a general lithological dip in the south laterally in the NW-SE direction. Seven borehole logs covering four areas - Ikorodu, Badore, Sangotedo and Awoyaya were utilized for generating the section. The section shows a significantly thick clay/shale column from depth of 111 to 454 m (thickness of

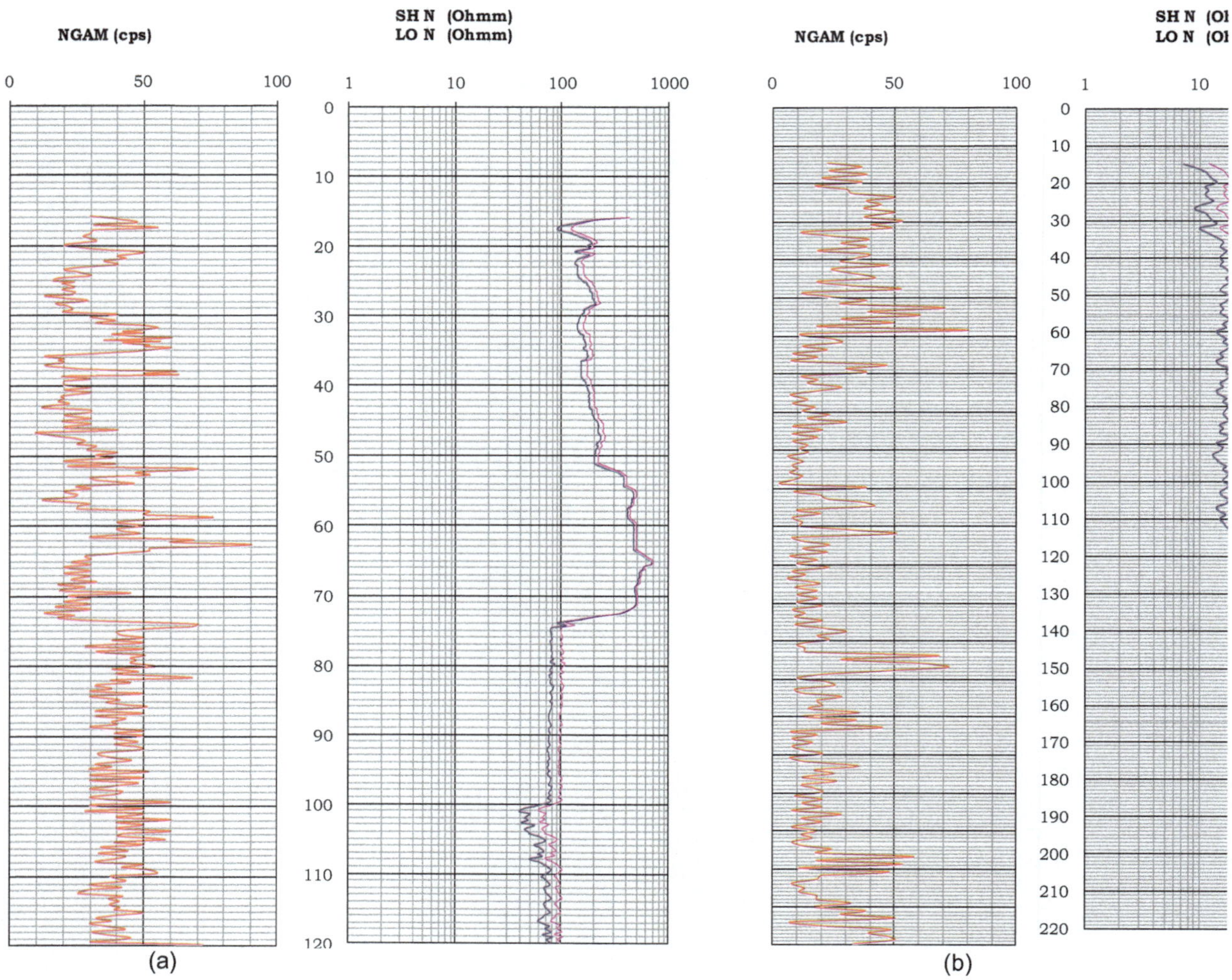

Figure 3. Typical borehole logs with (a) Odogunyan and (b) Osborne Estate Ikoyi logs (as examples) utilized for study.

Plate 1. The RG PC Logger II^{TM} Unit mounted on sports utility vehicle used for borehole loggings utilized for the study.

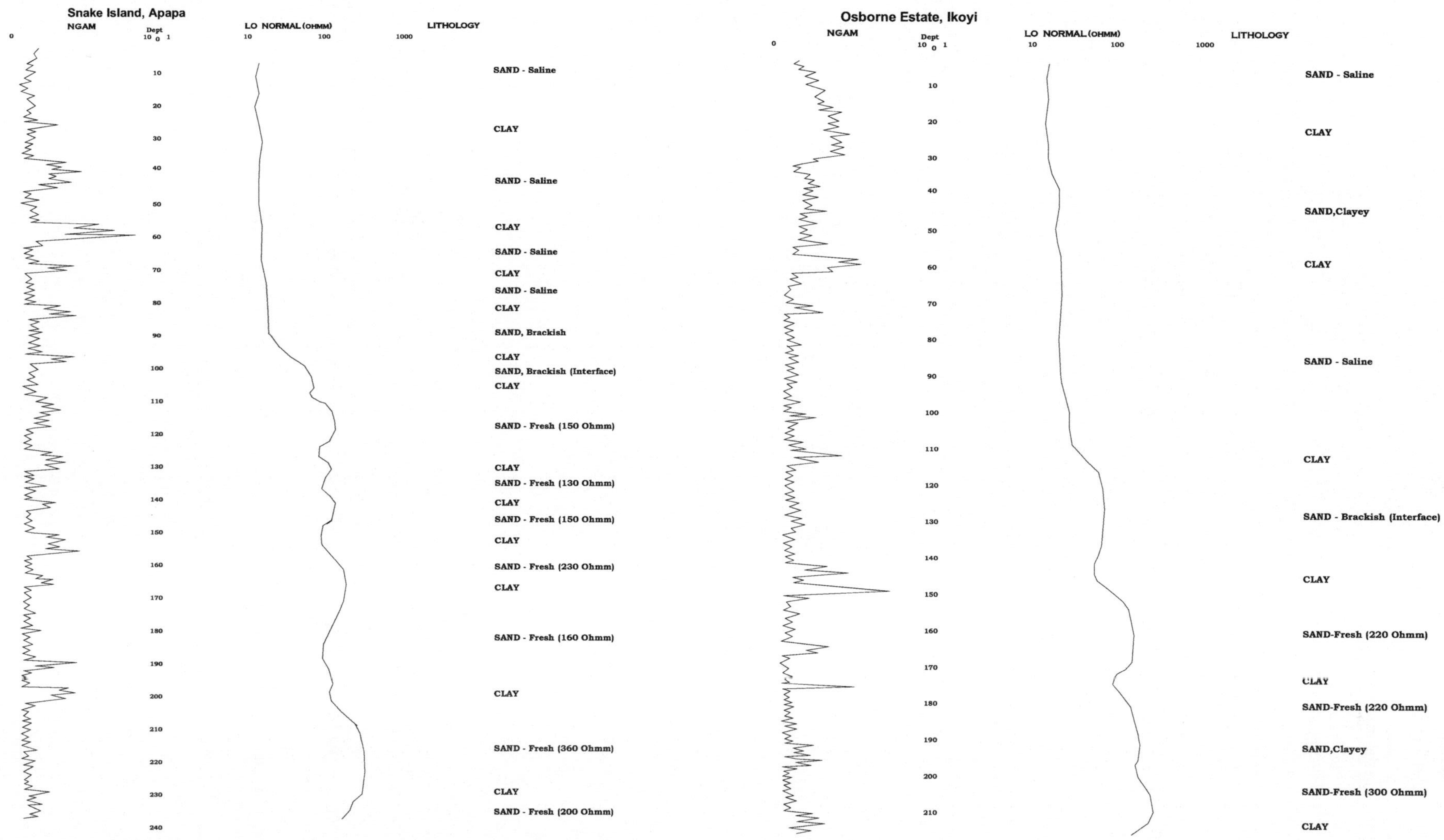

Figure 4. Typical logs and the lithology-water quality interpretation from the study area.

of 343 m) at Ikorodu. Thus, domestic boreholes may be terminated at limited depth of 110m within Ikorodu and environ while industrial boreholes should be drilled to depths in excess of 455 m.

Geo-section B-B': (Apapa - Ikoyi - Victoria Island - Lekki - Ajah)

This axis constitutes the western half of west-east flank of the State and it is proximal to the Atlantic Ocean shore line. The axis connects the south-west edge of the metropolis to the south edge. Fourteen (14) borehole logs studied within the axis

Table 1. Resistivity values for water and sediment rock.

Resistivity (Ω-m)	Sediment rock	Interpretation
0.5 - 2.0	Very porous sand, or saturated clay	Seawater, very saline water TDS = 20.000 mg/l
2.0 - 4.5	Porous sand, or saturated clay	Saline water, TDS= 10.000 mg/l
4.5 - 10.0	Sandy saturated or Sandy clay	Salty Brackish water, TDS = 10.000-1500 mg/l
10.0 -15.0	Sandy clay, sandy gravel	Brackish water, TDS = 5000-1500 mg/l
15.0 -30.0	Sand, gravel. Some clay	Poor quality fresh water, TDS = 1500 - 700 mg/l
30.0 - 70.0	Sand, gravel, minor clay	Intermediate quality fresh water, TDS ~ 100 mg/l
70.0 - 100.0	Sand, gravel, no clay	Good quality fresh water, TDS small
More than 100.0	Coarse sand, gravel, no clay	Very good quality fresh water, TDS very small

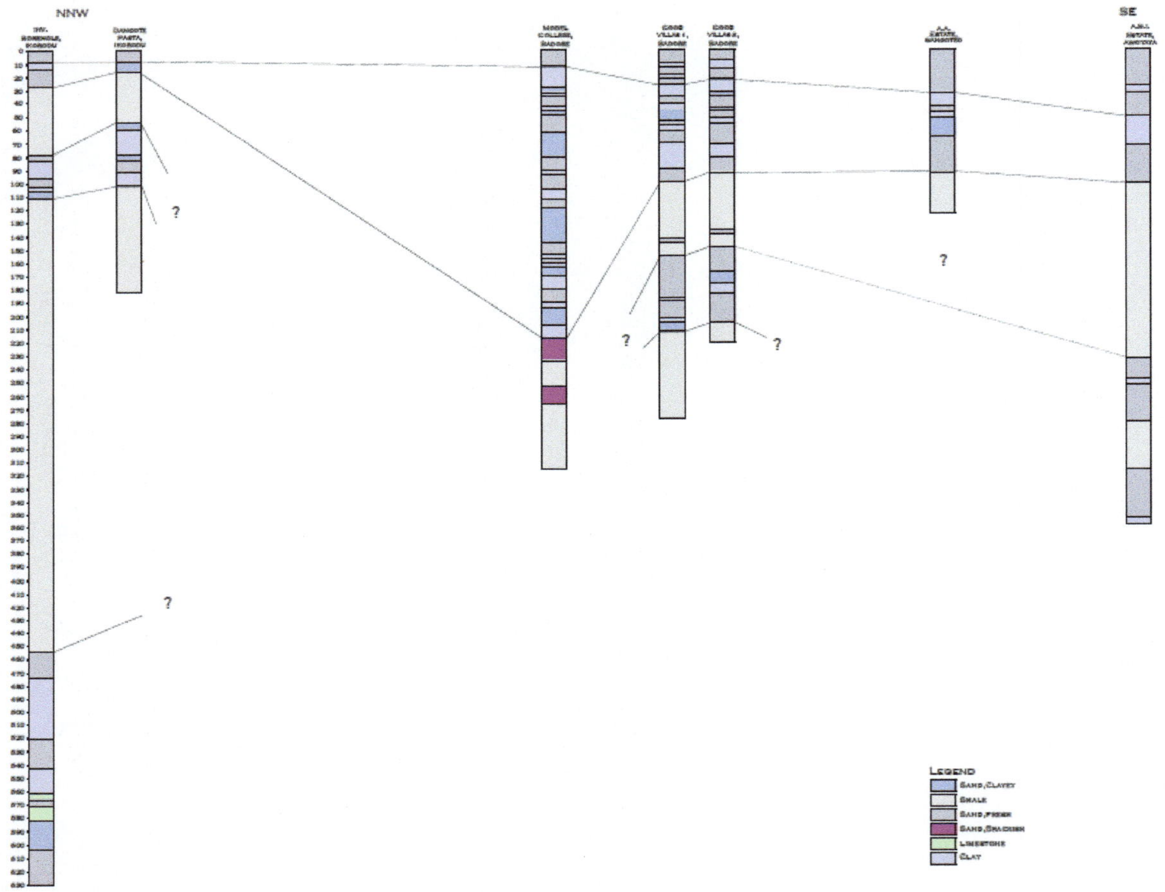

Figure 5. Geo-section A-A' showing lithology correlation across Ikorodu - Badore - Sangotedo - Awoyaya axis.

Table 2. Interpretation summary of borehole logs within profile A-A'.

Location	Elevation (m)	Fresh water zone depth (m)	Saline water zone depth (m)
Lagos Road waterworks, Ikorodu	114	15-27, 78-83, 96-102,105- 111,455-474, 520-544, 566-571, 605-627	None
Dangote Spata, Ebute	18	80-90	None
Model College, Badore	31	40-48,54-64,76-112,118-124,144-154, 180-196	None
Coop Villa Bh1, Badore	28	12-17, 20-24.8, 155-166, 173-186, 188-202,	None
Coop Villa Bh2, Badore	27	13-20, 62-70, 150-164, 176-188, 196-202,	None
Ajayi Apata, Sangotedo	22	16-33, 43-46	None
Eko-Akete, Awoyaya	20	18-24, 30-44	233 - 355

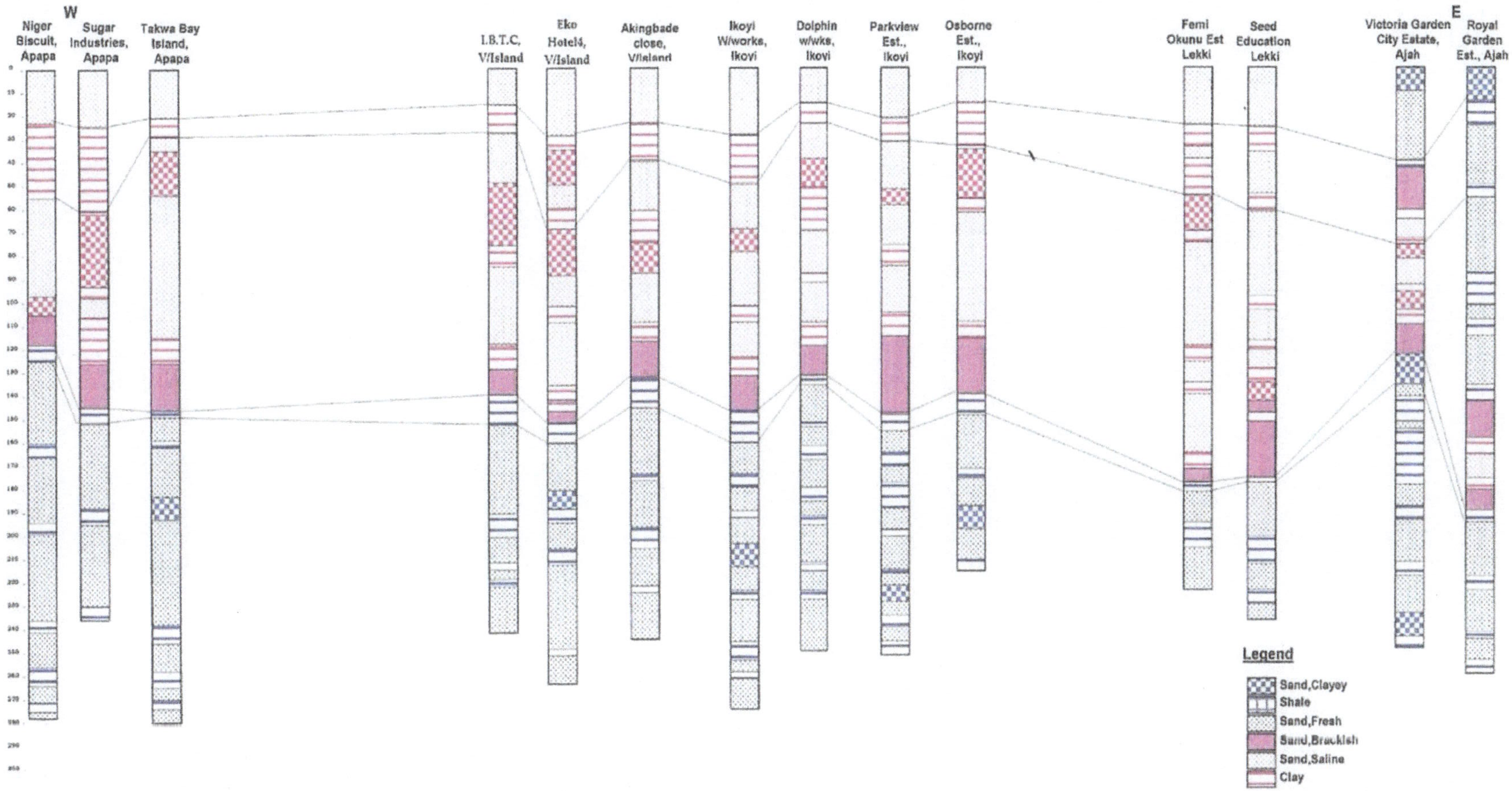

Figure 6. Geo-section B-B' showing lithology correlation across Apapa - Ikoyi - Victoria Island - Lekki - Ajah axis.

were obtained at Ikoyi, Dolphin, Victoria Island and Lekki. The geo-section generated for the axis is presented in Figure 6 while the interpretation summary is presented in Table 3. Within the axis, the delineated sand units are saline water saturated to depths varying between 118 and 196 m in Apapa and Ajah, respectively. At Ajah however, lenses of fresh water were delineated within the thick saline water sand. Net thickness of saline water sand units varies between 21 and 48 m in Ajah.

Geo-section C-C': VGC (Ajah) - Sangotedo - Awoyaya - Lakowe - Akodo Axis

This is the eastern half of the west-east axis of the study area which connects the southern limit to the south-eastern edge. Eight borehole logs were obtained for this study in the axis. The section generated from the logs is presented in Figure 7 while the interpretation summary is presented in Table 4. The net thickness of saline water sand increases eastwards (from Sangotedo) within the axis thus making freshwater abstraction more difficult on the eastern flank. However, at Akodo two lenses of freshwater were delineated.

Table 3. Interpretation summary of borehole logs within profile B-B'.

Location	Elevation (m)	Fresh water zone depth (m)	Saline water zone depth (m)
Niger Biscuit, Apapa	4	148-159, 162-183, 192-238, 246-258, 264-270, 274-280	0 - 125
Sugar Industries 1, Apapa	6	149-159, 162-184, 208-226, 232-236, 239-248	0 - 136
Tarkwa Bay Island, Apapa	10	148-155, 158-168, 171-179, 182-186, 188-193, 206-226	0 - 130
Ikoyi Bh1 W/works, Ikoyi	15	161-174, 179-190, 193-204, 215-224, 228-246, 253-259, 261-276	0 - 138
Dolphin W/wks, Ikoyi	5	137-152,154-162,169-180,187-192,195-214, 218-224, 229-250	0 - 131
Parkview Estate, Ikoyi	8	156-165, 171-178, 189-198, 201-215, 217-222, 230-235, 240-245	0 - 138
Osborne Estate, Ikoyi	5	164-172,176-193,202-208	0 - 148
IBTC, V/Island	20	154-191, 203-213, 215-219, 222-242	0 - 144
Eko Hotel Bh4, V/Island	19	153-193,194-208, 214-237	0 - 144
Akingbade Close, V/Island	18	148-174, 176-198, 208-220, 224-242	0 - 140
Femi Okunnu Estate, Lekki	27	182-196, 206-224	0 - 140
Seed Education, Lekki	15	166-188,192-196,198-232	0 - 158
Victoria Garden City Estate., Ajah	17	30-40, 136-141,152-155, 179-188, 193-212, 217-234	56 - 116
Royal Garden Estate, Ajah	17	46-50, 57-88, 102-108, 114-136, 190-218, 223-242, 245-252	None

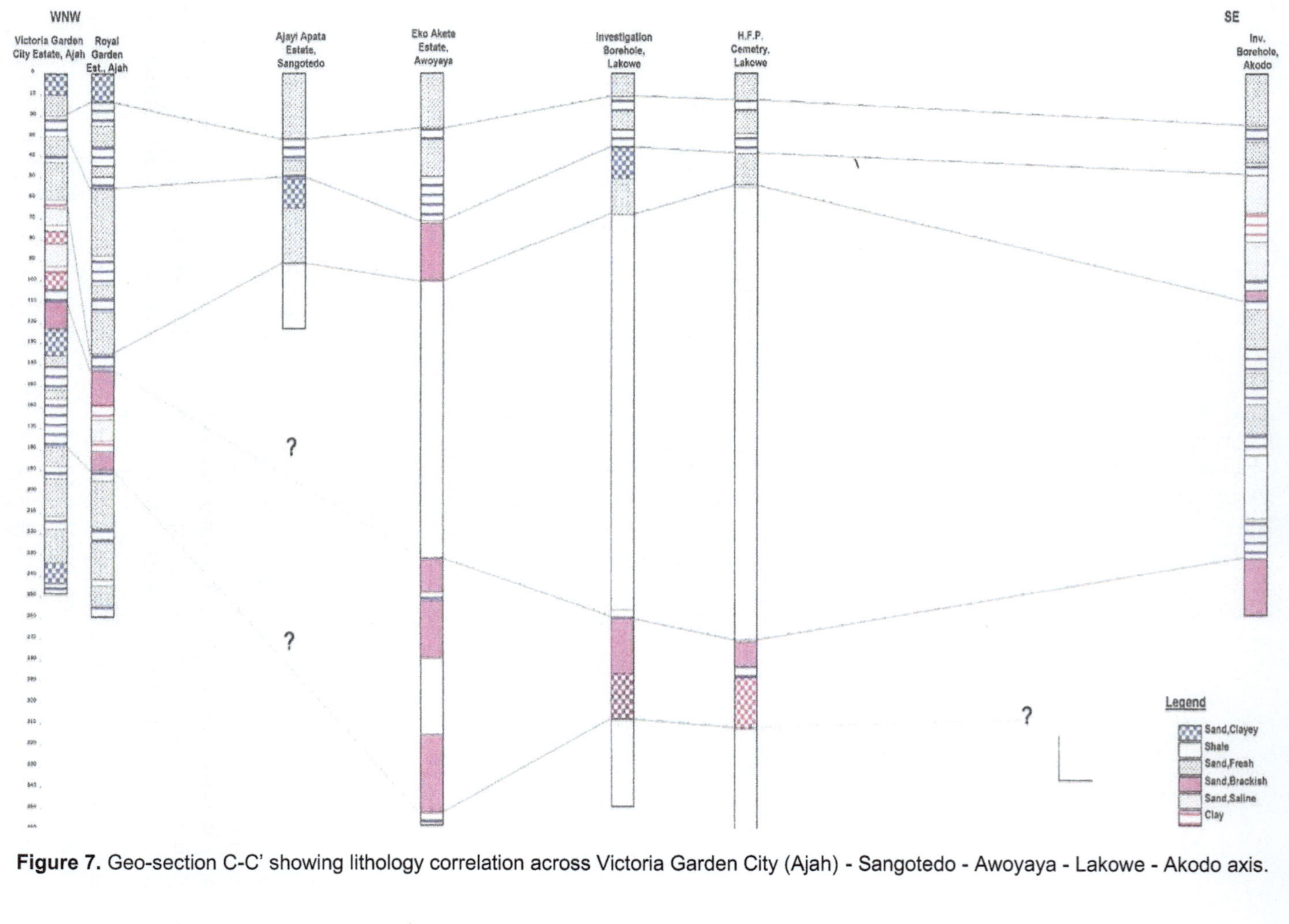

Figure 7. Geo-section C-C' showing lithology correlation across Victoria Garden City (Ajah) - Sangotedo - Awoyaya - Lakowe - Akodo axis.

Table 4. Interpretation summary of borehole logs within profile C-C'.

Location	Elevation (m)	Fresh water zone depth (m)	Saline water zone depth (m)
Victoria Garden City Est., Ajah	17	30-40,136-141,152-155, 179-188, 193-212,217-234	56 - 116
Royal Garden Est., Ajah	17	26-35, 46-50, 57-88, 102-108, 114-136, 190-218, 223-242, 245-252	143 - 190
Golden Park Est., Sangotedo	22	20-36, 44-57, 60-68, 82-100, 112-124	None
Ajayi Apata, Sangotedo	22	16-33, 43-56	None
Eko-Akete, Awoyaya	20	18-24, 30-50	233 - 355
HFP Cemetry, Lakowe	29	15-54	265 - 315
Lakowe Gram.Sch. Lakowe	26	21-27	258 - 308
Ibeju-Lekki Secretariat, Akodo	10	12-18, 36-46,108-132, 144-150, 159-168	51 - 108, 196 - 308

The hydrogeological setting on the axis is fairly complex. Saline water intrusion within the coastal plain sand within the axis is interspersed with lenses of freshwater sand units on the western flank around Ajah. At Victoria Garden City, the saline water sand is dominant from 62 to 124 m. At Sangotedo, Awoyaya and Lakowe, a significantly thick clay/shale column predominates with thickness varying from 133 m to 214 m at Awoyaya and HFP Cemetry in Lakowe, respectively, thus truncating the coastal plain sands (Figure 7). At Akodo, the coastal plain sand is again intercepted with a sequence of consisting of alternation of fresh and saline sand units.

Geo-section D-D': Satellite Town - Kirikiri - Iganmu - Bariga - Ikorodu Axis

The geo-section (Figure 8) which transverses southwest to northeast of the metropolis, was generated using seven logs obtained from Satellite town, Amuwo-Odofin, Iganmu, Bariga and Ikorodu. Summary of the interprettation is presented in Table 5. Saline water intrusion is prevalent at shallow levels on the South-western and central areas of the section around Satellite Town, Kirikiri, Ijora Badia and Iganmu.

Geo-section E-E': E-E' Agbara - Egan - Idimu - Agidingbi axis:

The geo-section (Figure 9) transverses the north-western flank of the metropolis utilizing seven geophysical borehole logs. The geo-section covers Agbara, Igando and Ogba and spans about 39 km. According to the summary (Table 6), there is no saline water sand unit incursion within the flank. Thus, the area is generally underlain by the vital fresh water sand units. A significantly thick shale unit was penetrated at Agbara.

Geo-section F-F': Mosinmi - Odongunyan - Ikorodu - Ibeshe axis

Geo-section of profile F-F' is presented in Figure 10. The profile transverses the northern flank of the metropolis in SSW-NNE direction and crosses into nearby Ogun State in the north. Just like the Geo-section of profile E-E', there is no observable incursion of saline water in the northern flank of the metropolis from Ibeshe in the south to Mosinmi in the north at the time of study (see Table 7).

Saline water intrusion map of Figure 11 was generated from the twenty two geophysical borehole logs showing penetration into saline water incursion zones (Figure 11). The map highlights the lateral extent of saline water incursion within the city. We can observe that the saline water intrusion is severe on the southern coastal areas of the city. The areas mostly affected are Apapa, Kirikiri, Ijora, Satellite Town, Iganmu, Bariga, Lagos Island, Victoria Island, Lekki, Ajah, Badore, Sangotedo, Awoyaya, Lakowe and Akodo areas.

Conclusions

Four of the geo-sections generated in this study (profiles A-A', B-B', C-C' and D-D') indicate saline water incursion at depths varying between 0 m (surface) in Satellite Town, Kirikiri, Ijora, Iganmu, Apapa, Lagos Island, Ikoyi, Victoria Island and Lekki to depths ranging from 40 m at Iganmu to 158 m at Lekki. Intrusions of 47 (143 - 190 m) and 60 m (56 - 116 m) were delineated at Ajah; 50 m (265 - 315 and 258 - 308) at Lakowe; 57 and 112 m (51 - 108 and 198 - 308 m) at Ibeju Lekki, Akodo and 122 m (233 - 355 m) at Awoyaya. The hydrogeologic importance of the Coastal Plain Sand aquifer unit in Lagos is under severe threat of continued sea water incursion and intrusion on its southern coastal flank. Results presented in this paper show that the process of salt water intrusion is active on southern flank of Lagos metropolis and geophysical wireline logs have proved to be precious

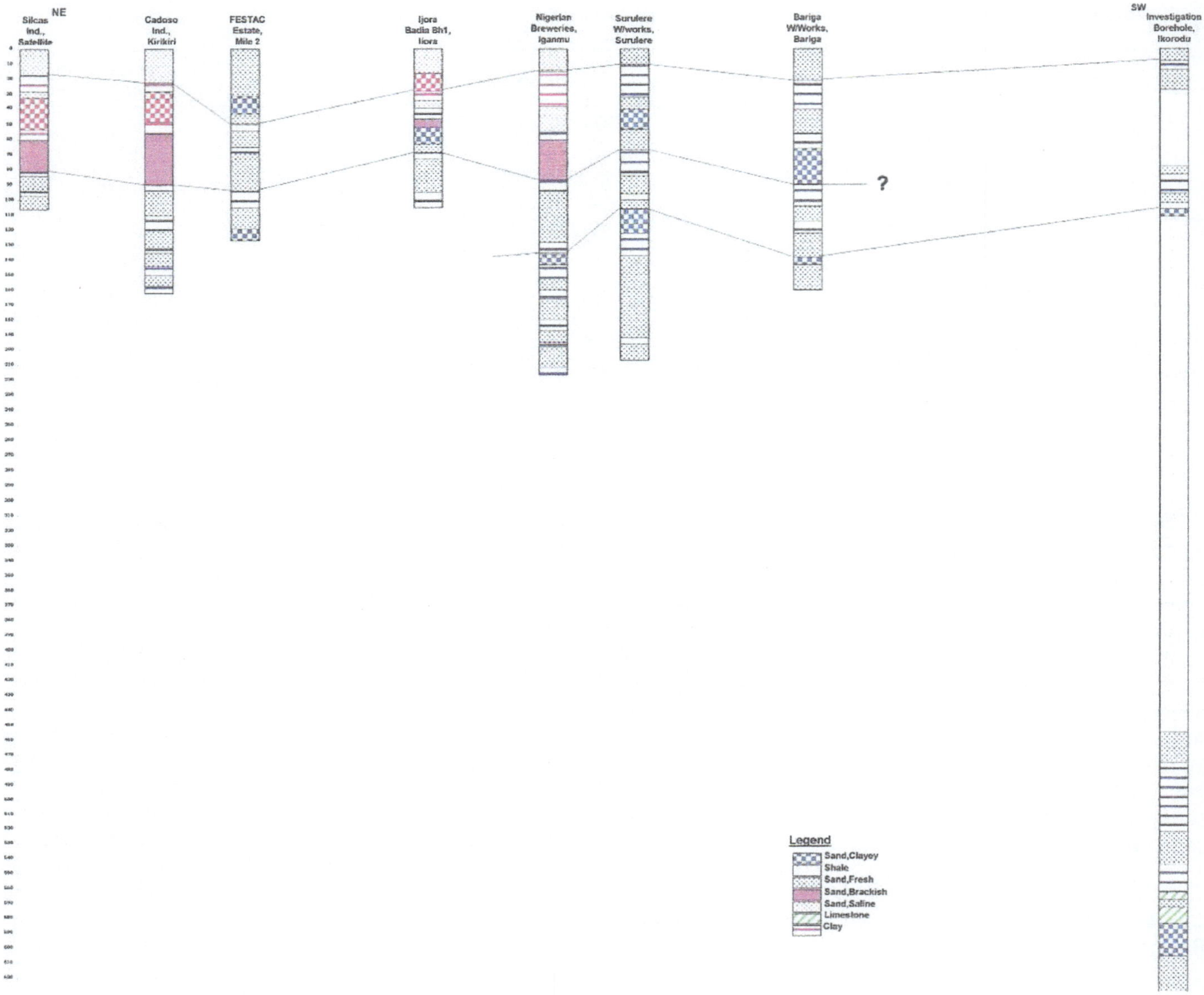

Figure 8. Geo-section D-D' showing lithology correlation across Satellite Town - Kirikiri - Iganmu - Bariga - Ikorodu axis.

Table 5. Interpretation summary of borehole logs within profile D-D'.

Location	Elevation (m)	Fresh water zone depth (m)	Saline water zone depth (m)
Silcas Ind., Satellite Town	23	82-94, 96-104	0 - 80
FESTAC Estate, Mile 2	31	18-32, 40-50, 56-64, 66-94, 106-118	None
Cadoso Ind., Kirikiri	27	100-110, 120-132, 134-144, 149-156	0 - 92
Ijora Badia Bh1, Ijora	17	62-68, 72-92	0 - 54
Nigerian Breweries, Iganmu	23	58-88, 96-128, 152-160, 166-180, 188-194, 198-212	0 - 40
Surulere W/works, Surulere	49	32-40, 54-66, 84-95, 100-106, 138-198, 194-206	None
Bariga W/works, Bariga	37	14-20, 40-56,106-114,122-138, 144-160	None
Lagos Road waterworks, Ikorodu	115	15-27, 78-83, 96-102,105- 111, 455-474, 520-544, 566-571, 605-627	None

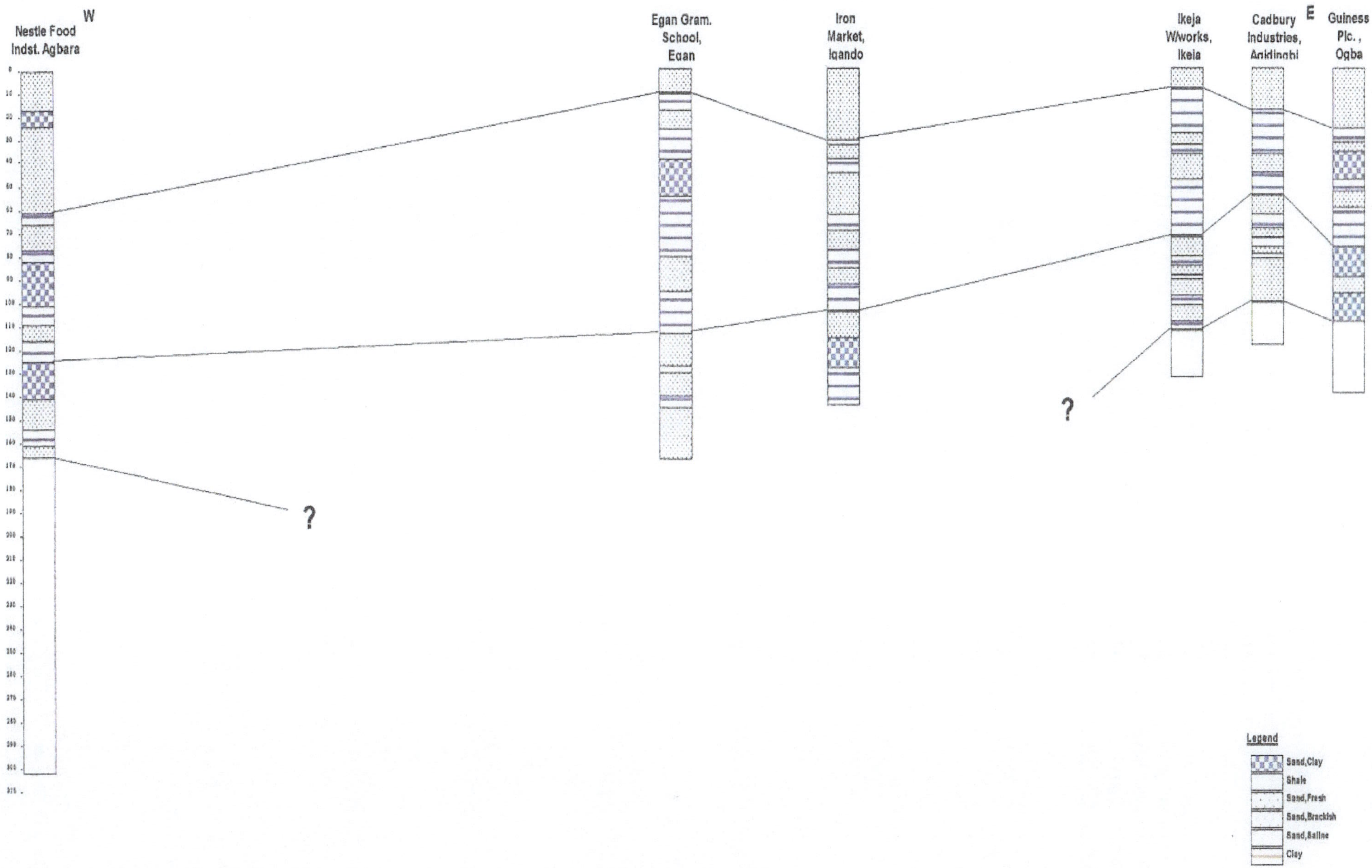

Figure 9. Geo-section E-E' showing lithology correlation across Agbara - Egan - Idimu - Ikeja - Ikorodu axis.

Table 6. Interpretation summary of borehole logs within profile E-E'.

Location	Elevation (m)	Fresh water zone depth (m)	Saline water zone depth (m)
Nestle, Agbara	29	24-54, 56-61, 82-101, 110-114, 142-155,	None
Egan Gram. Sch. Egan	19	81-96,114-129,132-141, 146-164	None
Iron Market, Igando	66	14-32, 34-39, 46-62, 70-76, 88-93, 106-116	None
Ikeja W/works, Ikeja	64	29-32, 40-48, 74-80, 86-88, 90-96, 104-108	None
Cadbury, Ogba	44	57-62, 65-72, 77-80, 82-100	None
Guiness, Agidingbi	65	14-25, 30-36, 52-60, 82-108	None

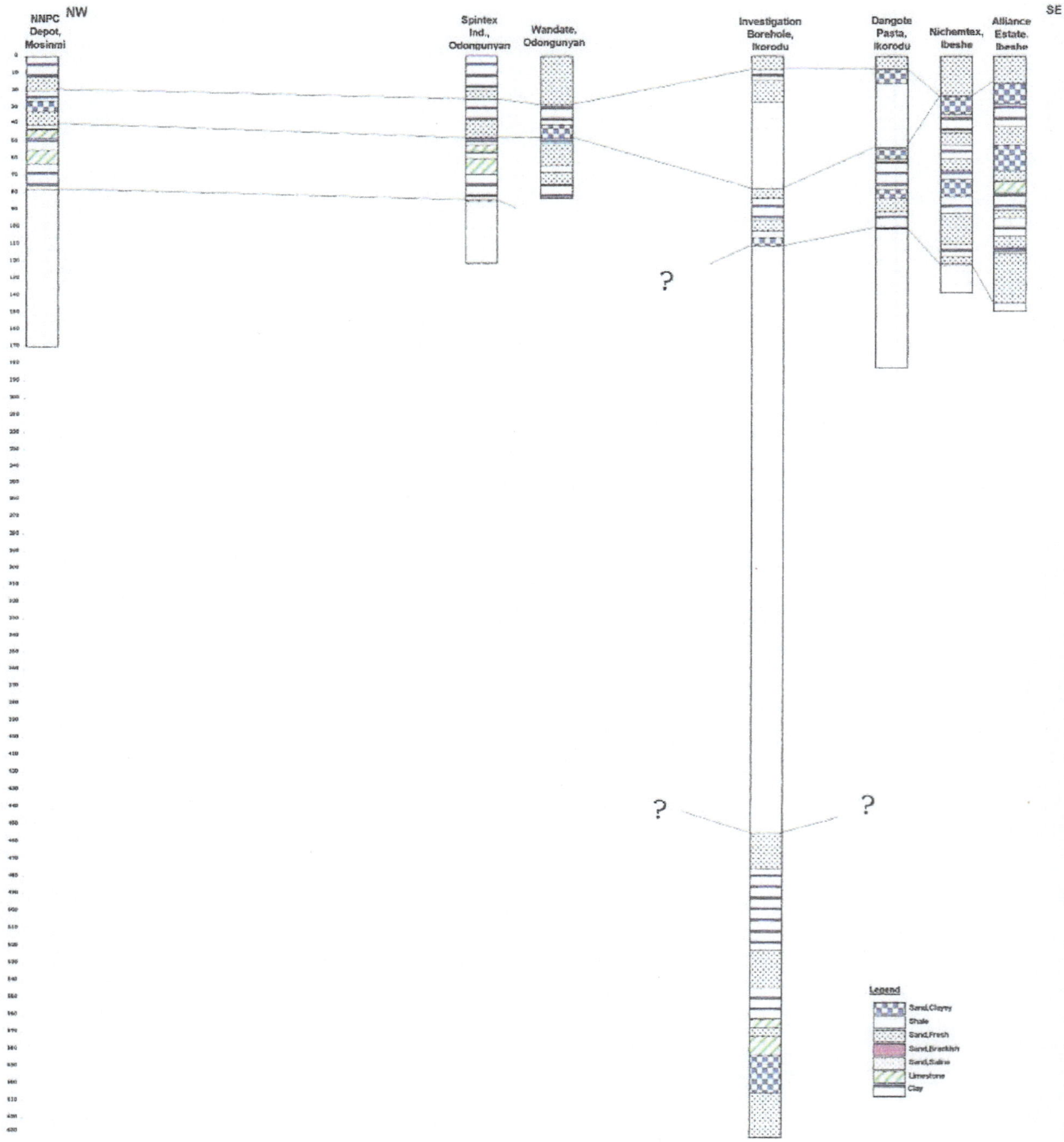

Figure 10. Geo-section F-F' showing lithology correlation across Mosinmi - Odongunyan - Ikorodu - Ibeshe axis.

Table 7. Interpretation summary of borehole logs within profile F-F'.

Location	Elevation (m)	Fresh water zone depth (m)	Saline water zone depth (m)
NNPC Depot, Mosinmi	94	13-20, 34-40, 44-47, 56-63	None
Spintex, Odongunyan	46	22-24, 42-48, 58-69	None
Wandate, Odongunyan	91	12-30, 50-64, 67-78	None
Lagos Road waterworks, Ikorodu	115	15-27, 78-83, 96-102,105- 111, 455-474, 520-544, 566-571, 605-627	None
Dangote Spata, Ebute	18	80-90	None
Nichemtex, Ibeshe	15	46-52, 60-68, 91-110, 115-122	None
Alliance Est. Ibeshe	18	42-52, 104-111, 114-144	None

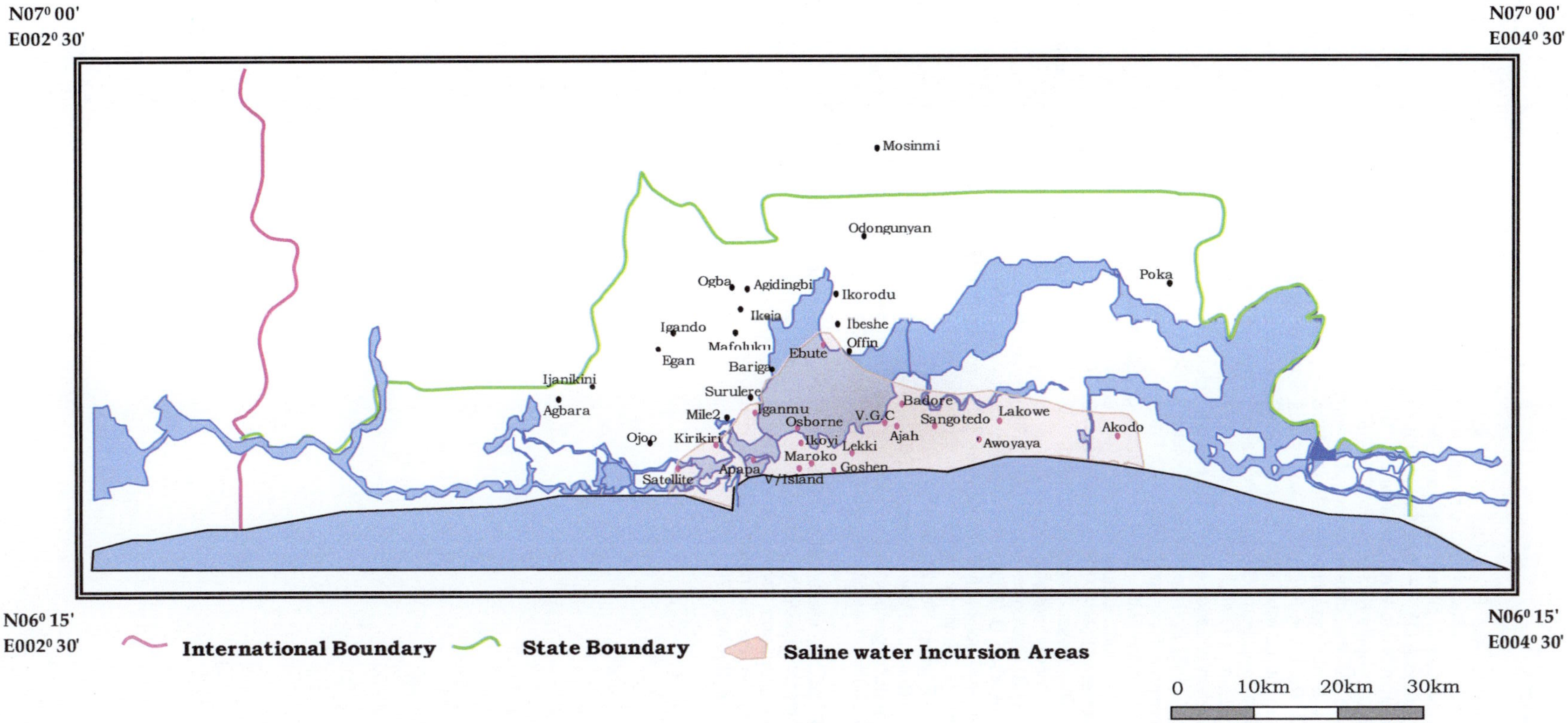

Figure 11. Map of Lagos showing the lateral extent of saline water intrusion within the metropolis.

tools for periodically monitoring the intrusion.

REFERENCES

Adepelumi AA, Ako BD, Ajayi TR, Afolabi O, Omotoso EJ (2008). Delineation of saltwater intrusion into the freshwater aquifer of Lekki Peninsula, Lagos, Niger. J. Environ. Geol. 56(5):927- 933.

Ako BD, Adegoke OS, Petters SW (1980). Stratigraphy of the Oshosun Formation in South-Western Nigeria. J. Min. Geol. 17:97-106.

Ayolabi EA, Oyedele KF, Oyeyemi EO, Ewiwile, EE (2003). Geoelectric sounding for the determination of groundwater conditions around Iwaya area of Lagos, Nigeria. Glob. J. Pure Appl. Sci. 9(4):539-545

Bates CR, Robinson R (2000). Geophysical Surveys for Groundwater Modeling of Coastal Golf Courses EAGE 62nd Conference and Technical Exhibition - Glasgow, Scotland, 29 May - 2 June. pp. 1- 5

Coode Blizard, Oteri, AU, Rofe KL, (1996). Hydrogeological Investigation of Lagos State. Final Report, Vols I & II,

Custodio E (1987). 'Groundwater Problems in Coastal Areas', Studies and Reports in Hydrology, UNESCO, Paris.

Elueze AA, Nton ME (2004). Organic geochemical appraisal of limestones and shales in part of eastern Dahomey basin, southwestern Nigeria. J. Min. Geol. 40(1):29-40.

Fitterman DV, Deszc-Pan M, Stoddard CE (1999). Results of time-domain electromagnetic soundings in Everglades National Park, Florida: U.S. Geological Survey Open-File Report, 99-426.

Goldman M, Kafri U (2004). Hydrogeophysical Applications in Coastal Aquifers. In: Applied Hydrogeophysics. Eds Vereecken, H. Binley, A., Cassiani, G., Revil, A. and Titov, K. Springer NATO Science Series IV: Earth Environ. Sci. 71:233-254.

Hwang S, Shin J, Park I, Lee S (2004). Assessment of seawater intrusion using geophysical well logging and electrical soundings in a coastal aquifer, Youngkwang-gun, Korea. Explor. Geophys. 35(1):99-104.

Ibrahim E (2008). Identification of Groundwater Spreading by Using Geoelectrical Resistivity Method. Int. J. Sci. Eng. Technol. 1(3):55-59.

Kampsax-Kruger and Sshwed Associates (1977). Underground water resources of the Metropolitan Lagos, Final Report to Lagos State Ministry of Works, p.170.

Longe EO, Malomo S, Olorunniwo MA (1987). Hydrogeology of Lagos Metropolis. Afr, J. Earth Sci. 6 (2):163 - 174.

Nowroozi AA, Horrocks SB, Henderson P (1999). Saltwater intrusion into the freshwater aquifer in the eastern shore of Virginia: a reconnaissance electrical resistivity survey: J. Appl. Geophys. 42:1-22.

Nton ME (2001). Sedimentological and Geochemical Studies of Rock Units in the Eastern Dahomey Basin, Southwestern Nigeria. Unpublished Ph.D thesis, University of Ibadan. p. 315.

Okosun EA (1990). A review of the Cretaceous stratigraphy of the Dahomey Embayment, West Africa. - Cretaceous Research, v. 11, p. 17-27.

Omatsola ME, Adegoke OS (1981). Tectonic Evolution of Cretaceous stratigraphy of the Dahomey Bas\in. J. Min. Geol. 18 (1):130-137.

Oteri AU, Atolagbe FP (2003). Saltwater Intrusion into Coastal Aquifers in Nigeria. Paper presented at The Second International Conference on Saltwater Intrusion and Coastal Aquifers - Monitoring, Modeling, and Management. Mérida, Yucatán, México, March 30 - April 2

Paillet FL, Hite L, Carlson M (1999). Integrating surface and borehole geophysics in groundwater studies - an example using electromagnetic soundings in south Florida: J. Environ. Eng. Geophys. 4:45-55.

Pareek N, Jat MK, Jain SK (2006). The utilization of brackish water, protecting the quality of the upper fresh water layer in coastal aquifers. Environmentalist 26:237-246.

Repsold H (1989). Well logging in groundwater development. Int. Assoc. Hydrogeol. 9:136.

Yechieli Y, Kafri U, Goldman M, Voss CI (2001). Factors controlling the configuration of the fresh-saline water interface in the Dead Sea coastal aquifers. Hydrogeol. J. 9:367-377.

Zarroca M, Bach J, Linares R, Pellicer XM (2011). Electrical methods (VES and ERT) for identifying, mapping and monitoring different saline domains in a coastal plain region (Alt Empordà, Northern Spain). J. Hydrol. 409:407-422.

Zohdy AAR, Martin RJ (1993). A Study of Seawater Intrusion Using Direct-Current Soundings in the Southern part of the Oxward Plain: California, Open-File Report 93-524 U.S Geological Survey 139.

15

Promoting a low cost energy future in Africa

Robert Kirchner and Adeleke Salami

African Development Bank Group, AfDB Temporary Relocation Agency (Tunis),15 Avenue du Ghana, P.O.Box 323-1002, Tunis-Belvedère, Tunisia.

With a large part of the population not having access to modern energy services in their daily life, energy poverty remains one of the most pressing development challenges on the African continent. Africa's fossil fuel resources as well as its renewable energy potential can serve as the means to achieve this. For Africa's social and economic development in the 21st century, however, the benchmark for these sources is to deliver energy that is affordable, reliable and sustainable. The following study offers a comparison between the two energy sources according to economic, social and environmental indicators. As the analysis shows, renewable energy technologies increasingly become the preferred option for Africa's energy challenge. The study then concludes with a description of policies for African countries to realize the up-scaling of these technologies.

Key words: Renewable energy, green growth, energy transition.

INTRODUCTION

Energy poverty remains a pressing development challenge in the African continent. A large part of the African population does not have access to power in their daily life. Improving this situation can be achieved through the use of fossil fuel technologies. Indeed, Africa is endowed with immense resources of oil, gas and coal making this an appealing solution. According to estimates, 45 out of the 54 African countries possess proven and/or probable oil and/or gas reserves, and most of these resources are untapped. Nevertheless, Africa is also well-placed to use renewable energy technologies [1](RETs) to address energy poverty due to its abundant renewable energy potential. The hydro potential in Africa alone equals three times Africa's current electricity production (IRENA, 2011a).

Whether Africa relies on fossil fuels or renewable energy technologies for its energy future, is subject of a detailed comparison between these two sources of energy provision. Important criteria are their levelized costs of energy[2] as well as their impacts on energy independence and sustainability. After all, Africa's social

[1]Throughout the text renewable energy is defined as any energy resource that is naturally regenerated over a short time scale and derived directly from the sun (such as thermal, photochemical, and photoelectric), indirectly from the sun (such as wind, hydropower, and photosynthetic energy stored in biomass), or from other natural movements and mechanisms of the environment (such as geothermal and tidal energy) (EREC, 2004)

[2]Levelized cost of energy (LCOE) is the ratio of lifetime costs to lifetime electricity generation, both of which are discounted back to a common year using a discount rate that reflects the average cost of capital. As an economic assessment of the cost of the energy-generating system including all the costs over its lifetime, it is a useful indicator to compare the electricity costs from different sources.

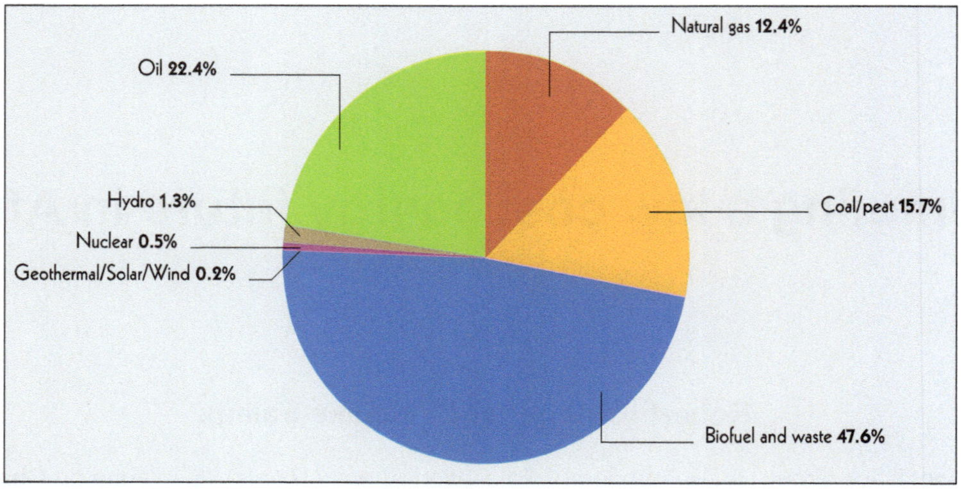

Figure 1. Share of Total Primary Energy Supply in Africa in 2009. Source: IEA (2012).

and economic development in the 21st century depends on affordable, reliable, and sustainable energy. Current trends show that RETs become the more suitable solution over time to reach these objectives.

This study offers a comparison of the two different models for Africa's energy future according to the criteria mentioned above. It summarizes policy options for African policymakers for the large-scale diffusion of the more suitable model, namely RETs, to reach a low cost energy future in Africa. First, a short overview of the energy sector in Africa is given. Then, the study discusses the potential of RETs as compared to fossil fuel technologies. In the end, the study recommends policy options that will contribute to the diffusion of RETs within African countries.

THE ENERGY SECTOR IN AFRICA- FOSSILS FUELS AND RURAL ENERGY POVERTY

Fossil fuels, biofuels and waste currently take up a pivotal role in Africa's energy mix as the most important sources of energy. Fossil fuels represent about 54% of total primary energy supply. As shown in Figure 1, oil, coal and natural gas contributed respectively 22, 16 and 12% of the continent's total primary energy supply in 2009. In 2010, about 80% of the continent's electricity was generated from fossil fuels.

Biofuels and waste, in comparison, amounts to 48% of energy supply (Figure 1). The reliance of many Africans on traditional biomass shows that Africa is still lagging behind the rest of the world in terms of access to modern energy, energy infrastructure, and institutional and technical capacity. Yet, access to modern energy is vital for improving the social and economic conditions for the African population. For instance, electrification enhances lighting, gives access to communication tools, allows the mechanization of production, and enables refrigeration, which in turn helps in improving food security and healthcare conditions. Figure 2 shows that the majority of the African population (58%) lacks access to modern energy. However, out of this 58%, 47% are located in rural areas compared to only 11% in urban ones. This trend will not undergo an alteration at least until the near future (Figure 2). As a result, efforts to address energy poverty will have to focus on rural areas.

POTENTIAL OF RENEWABLE ENERGY TECHNOLOGIES COMPARED TO THEIR FOSSIL-FUEL COUNTERPARTS

The most economical solution (in rural areas)

Renewable power generation now represents close to half of new annual capacity additions globally (IRENA, 2013). This massive up-scaling is the main driver behind the trend of falling costs of RETs through economies of scale, learning-by-doing mechanisms and increased competition in the various RETs markets. At the same time, it also shows how increasingly cost competitive RETs are becoming towards fossil fuel technologies.

In fact, renewable energy solutions such as hydropower, wind power, biomass and solar photovoltaic (PV) are already the most economical solution for off-grid and mini-grid electrification in remote areas in Africa, as well as in some cases of centralized grid supply too (IRENA, 2013). The cost advantage of RETs with regards to off-grid and min-grid electrification can be explained on the basis that only 15% of the rural population in Africa lives within 10 km of a substation (or within 5 km of the medium-voltage line) so that only a small proportion of the rural population can be added to the electricity grid at relatively low cost (Mafalda et al., 2010). But even for grid-connected projects, RETs are increasingly becoming the most economical solution compared to fossil fuel

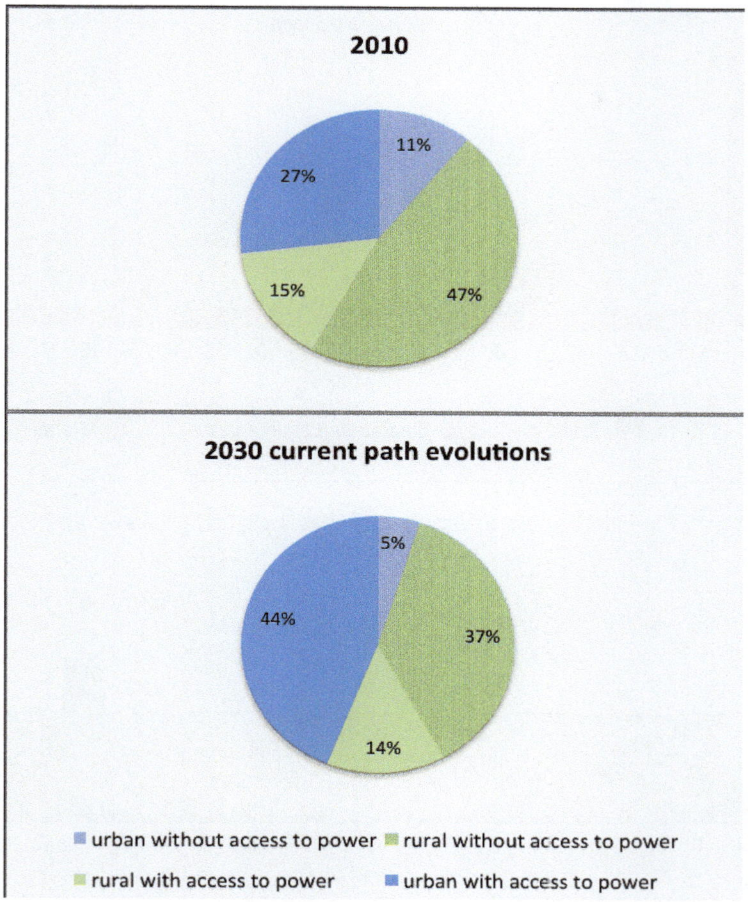

Figure 2. Comparison of Rural and Urban Electricity Access in 2010 and 2030, if current trends in Africa continue. Source: IRENA (2013).

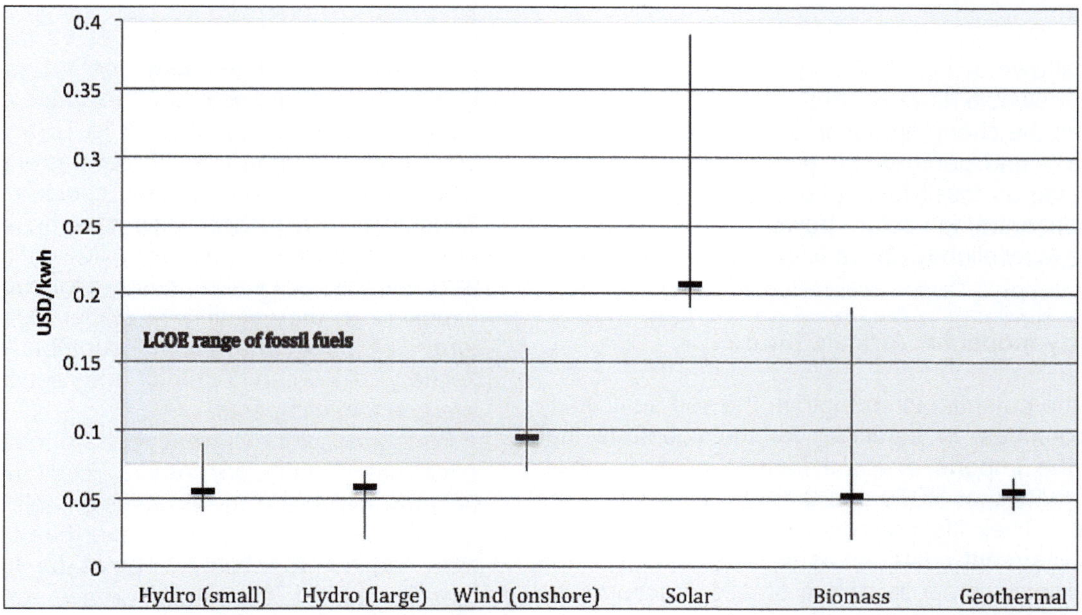

Figure 3. LCOE Ranges of Renewable Energy Technologies in Africa as compared to Fossil Fuels (for selected grid-connected projects). Source: IRENA 2013; IRENA 2012. Note: The horizontal black bars are the capacity weighted average value. Values for LCOE range of fossil fuels summarize ranges from South and East Africa as in IRENA 2013.

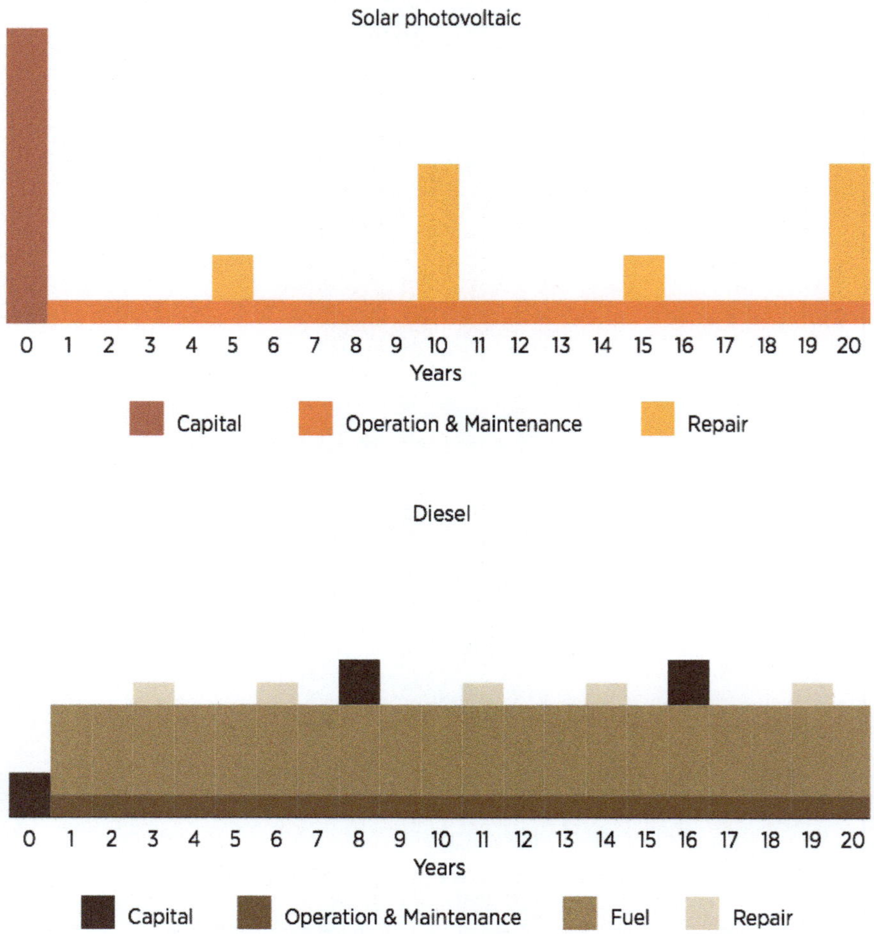

Figure 4. Comparison of Lifetime Costs of PV Project with Diesel Generator (indication). Source: IRENA 2013.

options. As shown in Figure 3 that compares the LCOE ranges of renewable energy and fossil fuels in Africa, hydropower is the cheapest option to generate electricity. Wind, biomass and geothermal remain below or on the same cost level as fossil fuel options. Only solar energy including concentrated solar power (CSP) and PV remains on a level slightly above fossil fuel technologies.

A safe energy model for Africa's future

RETs have the potential to strengthen the self-sufficiency of African countries, in particular for the countries that depend on the imports of fossil fuels. Figure 4 shows exemplary for PV that RETs see their on-going costs fall once installed. Their lifetime costs are mainly composed of upfront investment costs; costs for operation, maintenance, and repair compose only a smaller fraction. The costs of fossil-fuel technologies represented by the diesel generator in Figure 4, in contrast, are more evenly distributed throughout their lifetime. Lower upfront capital costs are offset by higher ongoing costs, mainly because fossil fuels are continuously needed as input to produce electricity. The oil price as an example of these fossil fuel inputs more than quadrupled to US$ 112 per barrel in 2012 from less than US$ 20 per barrel in 1999 becoming less and less affordable for customers (AfDB, 2013). Thus, fossil fuel technologies have higher ongoing uncertainties in their costs, due to fluctuations in fuel prices whereas all renewable costs are known. Limiting the exposure to the volatility of global fossil fuel markets in terms of price and supply contributes to the energy security[3] of African countries and reduces possible negative economic impacts.

Furthermore, the current prices of oil pose a burden on government budgets in Africa. Fossil fuel subsidies have become more and more unsustainable for many African governments. In 2010-2011, over half of all African countries had some subsidy in place for fuel products, and these subsidies consumed, on average, 1.4% of GDP in

[3]Defined as the availability of sufficient supplies at affordable prices following Yergin, 2006

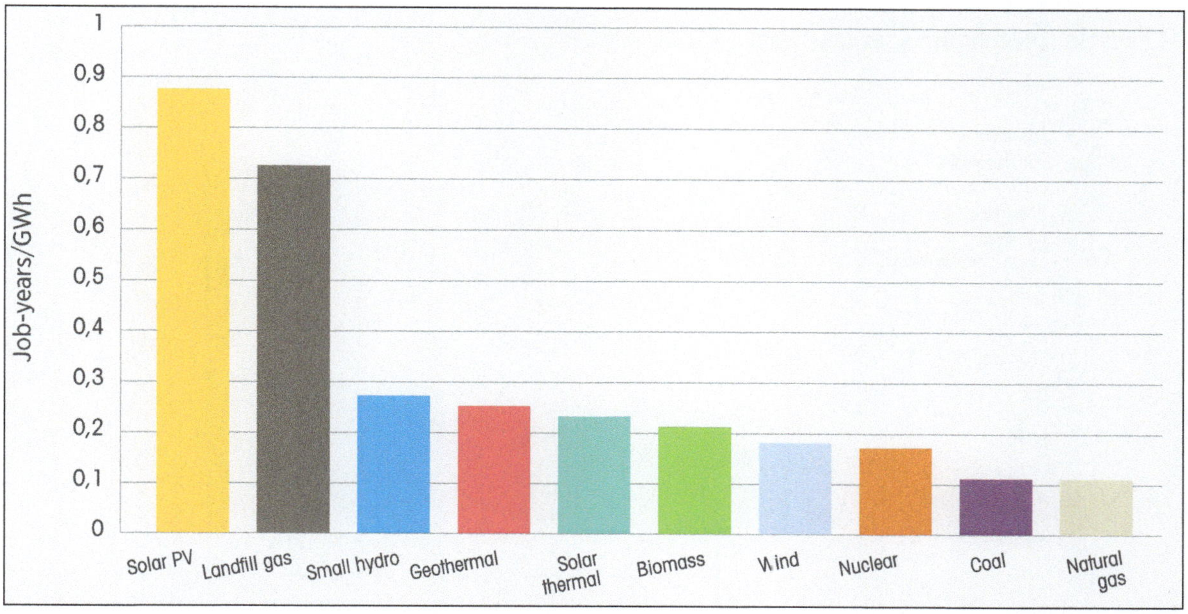

Figure 5. Comparison of Job-Years across different Energy Technologies (Job-Years/GWh). Source: Wei et al. 2010 as illustrated in IRENA (2011b).

public resources. Of the 25 countries with fuel subsidies, the fiscal cost of subsidies in six countries—primarily oil exporters—was at or above 2% of GDP in 2011. The fiscal cost for oil exporters was almost two-and-a-half times the levels observed for oil importers (World Bank, 2012).

The engines of job creation

The promotion of RETs is an efficient way of addressing poverty by creating additional income opportunities and new enterprises. There are important details about job creation with respect to RETs, particularly small scale RETs and their use in rural areas. Many of the jobs are in the service end of the supply chain and include distribution and sales, installation, maintenance, and so on. For some RETs (biogas plants and improved cook stoves) there are also opportunities for manufacturing or construction jobs at the beginning of the supply chain, which are unlikely in the case of PV modules because of the high level of skills required. These job opportunities could potentially benefit women and thus promote gender equality in access to labor markets.

Compared to fossil fuel technologies, the promotion of renewable energy will imply the creation of more job opportunities. Wei et al. (2010) averaged estimates about the labor intensity of different energy technologies over a range of studies. Their findings presented in Figure 5 show that all RETs have higher labor intensity than fossil fuel technologies. For example, the labor intensity of solar PV is more than eight-times that of natural gas. Within RETs, the above-mentioned solar PV has the highest labor intensity whereas wind has the lowest.

The environmentally sustainable energy generation

The deployment of RETs puts Africa on a more environmentally sustainable development path. Firstly, it leads to reduction in GHG emissions. Figure 6 gives a summary overview of lifecycle GHG emissions from a selection of technology groups. The figure reports this number within a range, particularly with regards to bioenergy. This has partly to do with the way lifecycle assessments are conducted. Some will be more comprehensive than others. But it is also related to the range of technologies within each group. Nevertheless, there is little overlap between the worst performing biofuels and the fossil energy sources, all of which produce much higher levels of GHG emissions than the "other renewables", except when carbon capture and storage (CCS) is used.

Secondly, RETs also result in the reduction of local pollutants, especially particulates. According to Johansson et al. (2012), this could mean "a saving of 20 million disability adjusted life years (DALYs) from outdoor air pollution and more than 24 million DALYs from household air pollution" compared with just the introduction of air-quality legislation that is currently planned. The authors also suggest that these positive health impacts could help to persuade individuals to adopt RETs, more so than asking them to make changes to achieve global benefits such as the mitigation of climate change.

Lastly, non-bioenergy RETs can significantly reduce the rate of deforestation within Africa[4]. The clearing of forests serves as an important source of energy and food

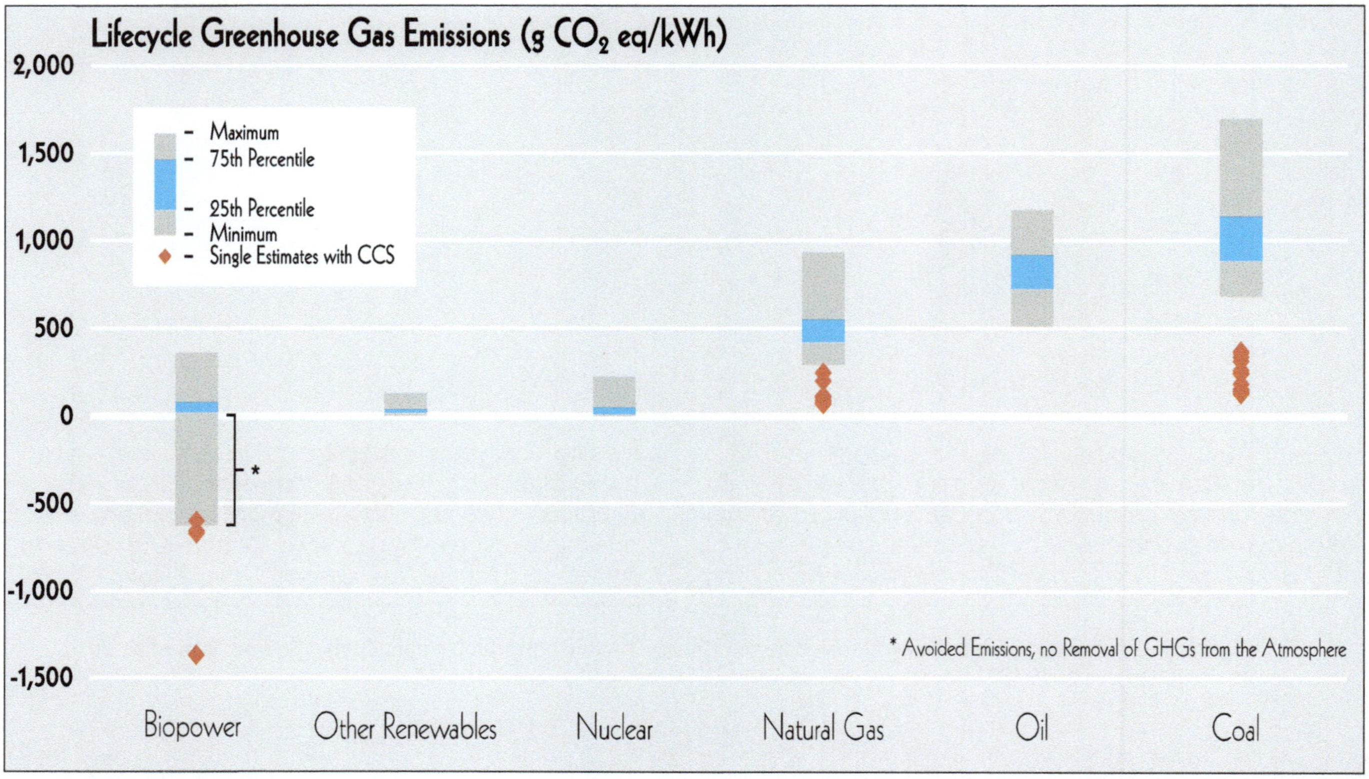

Figure 6. Lifecycle GHG Emissions of Renewable Energy, Nuclear Energy and Fossil Fuels. Source: Moomaw et al. (2012) as illustrated in AfDB (2013).

security, especially for Africa's rural poor. This trend is likely to continue as the total number of households in absolute terms relying on this traditional biomass is predicted to increase from 657 million to 922 million (OECD/IEA, 2010). This indicates, on the one hand, that increases in population outweighs shifts to "modern" energy services but, on the other hand, stresses the need to up scale the deployment of RETs to stop this trend.

POLICY OPTIONS

Policies play a crucial component in the large-scale deployment of RETs. Investments in the energy sector last for decades so that investors and the industry need the right policy environment ensuring reasonable returns over the lifetime of projects. This can be achieved through a mix of policies that send reassuring signals, provide relevant information, and offer long-term guarantees. Furthermore, governments need to ensure financing is adequate to set the right incentives for customers to uptake renewable energy.

[4]In contrast, bioenergy could potentially lead to deforestation (or negatively affect food production). Policies to promote bioenergy need to take these negative effects into consideration for their design.

Announcing a long-term target for the share of renewable energy

Announcing a long-term target for the share of renewable energy in total energy supply can signal investment security in RETs. This can leverage private investment for the diffusion of RETs because having such a target as part of national strategies ensures the investors of the long-term commitment on the side of the corresponding government. As of 2011, 16 African countries had already announced a long-term renewable energy target (UNEP, 2012). Morocco, for example, aims at an ambitious target of 20% of power supply coming from renewable energy in 2020.

Providing detailed information on the country's renewable energy potential

Providing detailed information on the country's renewable energy potential can contribute to strengthen the confidence of financial institutions and investors in RETs. The publication of a national solar and wind atlas, for example, informs potential investors about suitable areas and reduces the costs for feasibility studies (Renewable Energy Ventures, 2012). Due to a lack of knowledge and project experience with RETs, obtaining financing for RETs is currently more difficult than for fossil-fuel plants despite a comparably better economic case (IRENA, 2013). Banks are often reluctant to finance projects, or agree to finance but only at premium rates. Mapping renewable energy potentials including their techno-economic feasibility can contribute to overcoming financial burdens faced by investors and in helping financial institutions making profitable investments.

Offering long-term guarantees to renewable energy producers

Offering long-term guarantees to renewable energy producers through power purchase agreements such as feed-in-tariffs is the key policy for the diffusion of RETs. Market liberalization is an important prerequisite for renewable energy producers to be able to enter energy markets currently dominated by national utilities in a lot of African countries. Long-term guarantees along with market decentralization policies can ensure the easy access for new renewable energy producers. At the same time, these guarantees mean to obtain necessary financing for projects because producers are offered a price guarantee for a fixed period of time. In 2011, 7 African countries had already used feed-in tariff policies (UNEP, 2012). This group of African countries includes a small-island state dependent on fuel imports (Mauritius), the continent's biggest carbon polluter (South Africa), countries with less than 3% rural electrification (Tanzania), and others with almost universal access to electricity (Egypt and Algeria) (Renewable Energy Ventures, 2012). The design of feed-in tariffs can be aligned towards achieving a country's specific development objective whether this is the building of large-scale renewable energy plants or an increase of energy access in rural areas. In the first case bidding processes are the preferred options whereas for the latter ones differentiated tariffs ensuring that smaller installations are included are better (Renewable Energy Ventures, 2012). For the design of the tariff costs of generation, return on investment, impact on electricity prices and the costs of the support need to be taken into account; the choice of a tariff structure should be based on a holistic and long-term analysis based on the lifetime costs of RETs (IRENA, 2013; Renewable Energy Ventures, 2012)

A financing strategy for the diffusion of RETs

A financing strategy for the diffusion of RETs should be in place that ideally sets incentives for the poorer population to use renewable energy sources. In Africa, energy subsidies benefit industrial user and richer households. For example, an estimated 44.2% of fossil fuel subsidies go to the richest 20%, while the poorest 20% benefit from only 7.8% of these subsidies (AfDB, 2013). Social-transfer mechanisms and the cross-subsidization of low-income households through higher tariffs to rich customer could potentially change the current status-quo in favor of the poorer population. As an example for the former, South Africa already provides a monthly quota of free electricity to low-income households. For the latter, tariff structures in Kenya and Ethiopia are based on a pricing scheme with low prices for low consumption. Minimally, a financing strategy needs to ensure that feed-in tariffs will not incur an additional cost burden for households, especially in poor rural areas. This could potentially have negative consequences to achieve the objective of improving energy access. Additional funding will thus become necessary to protect poorer households from higher prices. For African countries, there are several strategies to finance feed-in tariffs by focusing on inter alia harnessing fiscal and environmental policy tools and leveraging global financing options (AfDB, 2013). Algeria and Mauritius have both taxed fossil fuels in order to fund renewable energy. Meanwhile Ghana and Uganda are looking to international climate finance such as through Nationally Appropriate Mitigation Actions (NAMAs) under the UNFCCC and the new Green Climate Fund (Renewable Energy Ventures, 2012). The African Development Bank and other organizations can help African countries here by facilitating awareness, knowledge sharing and upstream technical support or through their role as manager or host of a range of innovative financing instruments. For example, throughthe Climate Investment Funds (CIFs) and the Sustainable Energy Fund for Africa (SEFA), the Bank has several funding instruments that help promote scaling up of clean energy solutions at different levels. Robert et al. (2013) reported about the Ecological Modernization of the German Economy by a modern environmental policy. Dessau-Roßlau: German Federal Environment Agency.

CONCLUSION

The study has shown that RETs provide the better economic case for Africa's energy future compared to fossil-fuel technologies. They are the lowest cost option for different kinds of electrification projects and will even improve this status if current trends of rising fossil-fuel prices will continue. More income earning opportunities can be created through the deployment of RETs and customers can expect lower energy costs in the middle-to long-term. Also, in terms of energy independence and environmental sustainability, RETs make a more compelling case than their fossil-fuel counterparts.

For the large-scale deployment of RETs, the study recommends policy options to address two critical issues for a low cost energy future in Africa: ensure investment security and set incentives for low income households. For the former, a mix of policies is necessary that send signals of the government's long-term commitment, provides relevant information for profitable investments and offers guarantees for reasonable returns over the lifetime of projects. For the latter, a financing strategy needs to be put in place that ideally sets incentives, but minimally ensures that guarantees will not result in additional costs for low-income households. For African countries, options include the harnessing of fiscal and environmental policy tools and leveraging global financing options.

REFERENCES

African Development Bank (AfDB) (2013). The African Development Report 2012- Towards Green Growth in Africa. Tunis: The African Development Bank Group.

European Renewable Energy Council (EREC) (2004). Policy paper for the international conference on renewable energy. Available at:http://www.stakeholderforum.org/practice/renewables/Position%20Papers/RE%20Manufacturers%20&%20Suppliers.pdf (Accessed 28 November 2013).

International Energy Agency (IEA) (2012). IEA Energy Statistics. Available at: http://www. iea.org/stats (Accessed 05 July 2013).

International Renewable Energy Agency (IRENA) (2011a). "Scenarios and Strategies for Africa, Working Paper Presentation." IRENA-Africa High-Level Consultations, Abu Dhabi, 8th and 9th July.

IRENA (2011b). Renewable Energy Jobs: Status, Prospects & Policies. Abu Dhabi, United Arab Emirates: IRENA.

IRENA (2012). Renewable Power Generation Costs: An Overview. Available at: http://irena.org/DocumentDownloads/Publications/Overview_Renewable%20Power%20Generation%20Costs%20in%202012.pdf (Accessed 29 August 2013).

IRENA (2013). Africa's Renewable Future- The Path to Sustainable Growth. Abu Dhabi, United Arab Emirates: IRENA.

Johansson T, Nakicenovic N, Patwardhan A, Gomez-Echeverri L, Arent D, Banerjee R (2012). "Technical Summary." In The GEA Writing Team (eds.) Global Energy Assessment – Toward a Sustainable Future. Cambridge, UK, New York, USA and Laxenburg, Austria: Cambridge University Press and the International Institute for Applied Systems Analysis.

Mafalda D, Nagarajan S, Brixiova Z (2010). "Financing of Sustainable Energy Solutions". Committee of Ten Policy Brief, No. 3. Tunis: The African Development Bank Group.

Moomaw W, Yamba F, Kamimoto M, Maurice L, Nyboer J, Urama K, Weir T (2011). "Introduction." In Edenhofer, O., R. Pichs-Madruga, Y. Sokona, K. Seyboth, P. Matschoss, S. Kadner, T. Zwickel, P. Eickemeier, G. Hansen, S. Schlömer and C. von Stechow (eds.). IPCC Special Report on Renewable Energy Sources and Climate Change Mitigation. Cambridge, Cambridge University Press.

OECD and International Energy Agency (2010). Energy Poverty. How to Make Modern Energy Access Universal? Special Early Excerpt of the World Energy Outlook 2010 for the UN General Assembly on the Millennium Development Goals (September). Available at: http://content.undp.org/go/cmsservice/stream/asset/?asset_id=2822269 (Accessed May 2013).

Renewable Energy Ventures (2012). "Powering Africa through feed-in tariffs". Study for the World Future Council (WFC), the Heinrich Böll Foundation (HBF) and Friends of the Earth England, Wales & Northern Ireland (FoE).

Robert K, Jutta K, Ingmar M, Walter K, Jürgen B, Dietmar E, Wolf-Peter S, Christian S, Rainer W (2013). Ecological Modernization of the German Economy by a modern environmental policy. Dessau-Roßlau: German Federal Environment Agency.

United Nations Environment Programme (UNEP) (2012). Feed-in Tariffs as a Policy Instrument for Promoting Renewable Energies and Green Economies in Developing Countries. Nairobi: UNEP.

Wei M, Patadia SS, Kammen D (2010). "Putting renewables and energy efficiency to work: How many jobs can the clean energy industry generate in the US?". Energy Policy 38: 919-931.

World Bank (2012). Africa's Pulse: An Analysis of Issues Shaping Africa's Economic Future, Volume 5. Washington, DC: World Bank.

Yergin D (2006). Ensuring Energy Security. Foreign Affairs. Issue March/April 2006.

16

Isolation of diatom *Navicula cryptocephala* and characterization of oil extracted for biodiesel production

Sanjay K. R.*, Nagendra Prasad M. N., Anupama S.[#], Yashaswi B. R.[#] and Deepak B.[#]

Department of Biotechnology, Sri Jayachamarajendra College of Engineering, Mysore- 570006, India.

Over the past decade diatoms are screened for high lipid content. Geologists claim that much of crude oil comes from diatoms. In this study diatom *Navicula cryptocephala,* isolated from fresh water source was grown on suitable media for extracting and characterizing the oil for biodiesel production. Three methods namely, shake flask, polythene bag and photobioreactor were adopted for culturing the diatom. The maximum yield of biomass was obtained in photobioreactor (5.95 g/ 100 ml of media). Compound microscope and electron microscopic studies were carried out to identify the diatom. Both soxhlet and sonication methods were used for extraction of oil from the diatom biomass. The yield of oil is higher in sonication method (0.364 g/g of dry biomass). Further characterization of oil using gas chromatography showed the presence of Palmitic acid, Oleic acid, Palmitolic acid and linoleic acid as major fatty acids. HPLC analysis of crude oil showed the presence of OLL (dilinoleoyloleoylglycerol) and OOL (linoleoyldioleoylglycerol) as major triacylglycerols (TAGs).

Key words: Diatom, biodiesel, triaclyglycerols, photobioreactor.

INTRODUCTION

Biodiesel is a domestic, renewable fuel derived from natural oils and fats. Biodiesel offers economic, environmental, fuel quality and energy security benefits vs. petroleum-based diesel (Knothe et al., 1997). Natural oils can be converted to biodiesel by a relatively simple refining process called transesterification. In this process vegetable oils or animal fats or microalgal oils are esterified using alcohol (ethanol or methanol) in the presence of a catalyst (sodium or potassium hydroxide) to form fatty esters (ethyl or methyl ester) (Vasudevan and Briggs, 2008).

Over the past few decades, more than thousand species of algae including diatoms have been screened in search of high lipid content. It was found that, on an average, polyunsaturated fatty acid constitutes approximately 25% of algal mass (Hu et al., 2006). This content may vary noticeably between species, and interestingly, the lipid content increases when cells are subjected to unfavorable culture conditions, such as photo-oxidative stress or nutrient starvation. This is due to the shift in lipid metabolism from membrane lipid synthesis to the storage of neutral lipids (Qiang et al., 2008).

Diatoms were shown under appropriate conditions to produce up to 60% of their cellular mass as triacylglycerols (TAGs) under certain growth conditions. These TAGs can be easily converted into biodiesel through a transesterification reaction (Sheehan et al., 1998). Diatoms produce oil intracellularly as reserve food material during the vegetative period of growth. Besides high lipid and fatty acid content, there is an abundance of eicosapentaenoic acid, a polyunsaturated fatty acid (PUFA)) in diatoms (Lebeau and Robert, 2003).

*Corresponding author. E-mail: sanjayrajesh@gmail.com.

[#]Equal contribution to the work.

Theoretical calculations based on the photosynthetic ability and growth potential of diatoms has shown thatthey can yield more than 30,000 liters of oil per hectare per annum. Diatoms can yield 100 to 200 times as much oil as soybean, 10 to 200 times more than oil seeds and 7 to 31 times more than oil palm, which is considered as the next best source of oil. Therefore, sustainable energy could be made from diatoms. This requires development of appropriate cultivation and extraction of oil, using advanced technologies that mimic the natural process while cutting down the time period involved in oil formation (Ramachandra et al., 2009). Despite many years of research on the relationship between diatoms and crude oil, we still know very little about the oil inside diatoms itself. To conduct a proper chemical analysis of the oil inside diatom oil droplets, a method for separating out oil droplets inside diatoms from the shell and cytoplasm must be developed. The mechanisms of crude oil formation by natural phenomena are now partially understood, and technology for crude oil synthesis is in the budding stage (Ramachandra et al., 2009). The manipulation of diatom lipid quantity using available biochemistry and quality could be very significant and help us in effectively using this renewable resource as energy. The present work explains, isolation, culturing and identification of diatom *Navicula cryptocephala* species from fresh water sources, and extraction and characterization of lipids from its biomass for biodiesel production.

MATERIALS AND METHODS

Culturing of diatoms

Water samples were collected for isolation of diatoms from seven different fresh water sources in and around Mysore, Southern India. The collected samples were transferred to Chu medium containing Calcium nitrate (40 mg/L), Dipotassium hydrogen phosphate (10 mg.L^{-1}), Magnesium sulphate heptahydrate (25 mg.L^{-1}), Sodium carbonate (20 mg.L^{-1}), Sodium silicate (25 mg.L^{-1}), Ferric chloride (8 mg.L^{-1}). These media, without vitamins, were autoclaved at 120°C for 20 min. 0.1 mg.L^{-1} Thiamine, 0.5 µg.L^{-1} Biotin and 0.5 µg.L^{-1} Cyanocobalamin were added in the same proportions to both media. Vitamin solutions and the oligo-element solution were filtered (0.2 µm) and frozen before use.

Three methods of culturing were followed for diatoms such as shake flask, polythene bag and photobioreactor culturing. In shake flask method, 1000 ml of Chu medium was prepared and transferred to 500 ml conical flasks (100 ml for each flask). The flasks were sterilized in autoclave at 121oC, 15 lbs pressure for 15 minutes. 10 ml of collected samples were transferred aseptically to the conical flasks. The flasks were kept in incubator shaker provided with illuminator (LT IL, Scigenics Biotec, Intencity 6 Watts) at 20°C and 120 rpm for 10 days. After 10 days diatoms was identified using stereomicroscope and isolated from other algal species using capillary tube method (Blanco et al., 2008; Debenest et al., 2009). The isolated species of Navicula were inoculated in to fresh medium for further multiplication and used as inoculum for further studies.

In polythene bag culturing method autoclaveable polythene bags (22 cm × 35 cm Breadth and Length respectively) were used. 200 ml of medium was transferred to each bags and sterilized at 121°C, 15 lbs pressure for 15 min. The bags were inoculated with the 20 ml of pure culture of *Navicula* aseptically and kept at 20°C below the light source (6 Watts). Aerators with the sterile filters were used for sufficient aeration for the growth of diatoms.

Vertical photobioreactor made up of glass chamber (80 cm × 40 cm × 50 cm) was designed for culturing of diatoms. The reactor was provided with a source of light of (12 W), aerator (with the sterile filters), 2.5 l/min, voltage AC 220/240- 50 Hz., Power-3 W), thermometer, inlet and outlets. 15 L of sterilized Chu medium was transferred to the reactor and 1.5 L of inoculum was added in aseptic condition. The temperature was maintained at 20°C, with sufficient aeration and light and incubated for 15 days.

The diatom biomass was harvested by centrifugation at 10,000 rpm for 10 min (Ronald et al., 2011). The biomass was transferred to pre-weighed clean Petri plate and dried at 50°C to determine the dry weight of the biomass and stored under refrigerated condition.

Identification of diatoms

Diatoms were identified in compound microscope according to the method of Karthick et al. (2010). The sample was mixed with concentrated HCl in 1:1 ratio and kept for 24 h at room temperature. After 24 h, two to three drops of hydrogen peroxide was added and samples were observed under microscope. For electron microscope studies dried samples were sprinkled over double-stick tape pasted on the copper stubs used for mounting specimens for scanning electron microscope, and coated with about 20 nm gold particles in a sputter coater (EMS-550). The sample were observed and photographed with a JEOL 100CX-II electron microscope fitted with a scanning attachment (ASID-4D) at 20 kV.

Extraction of lipids from diatom biomass

The dried biomass of diatom was ground to fine powder in a mortar pestle. The lipid was extracted from dried biomass by sonication (David and Stuart, 1998) and soxhlation methods (Ronald et al., 2011). Sonication was carried out in a probe sonicator (model). An aliquote of sample (2.5 g) was suspended in 5 ml of hexane and subjected to sonication for 5 min at 31 Amp. The diatom biomass (2.5 g) was subjected to soxhlation using hexane as solvent for 16 h.

Chemical and physical analysis of diatom oil

The weight of oil extracted per gram of biomass was measured to determine the lipid content. Results were expressed as the percentage of oil in the dry matter of diatom biomass. Further the oil was characterized for its density (ASTM D 4052-96), kinematic viscosity (ASTM D 445-06), acid value (AOAC Official Method Cd 3a- 63), saponification value (MPOB Official Test Method, 2004) and peroxide value (AOAC Official Method 965.33).

Determination of fatty acid composition

The free fatty acids were characterized using gas chromatography (Shimadzu 2010 plus, Japan) equipped with flame ionizing detector using RTX/1 bonded with Poly (Dimethyl siloxane) stain less steel capillary column (30 mm × 0.25 mm × 0.25 mm). Injector and detector temperatures were set as 220 and 230°C respectively, with

Table 1. Yield of diatom biomass and total lipid content.

Culturing techniques	Diatom biomass dry weight basis^A	Crude lipid obtained by sonication method	Crude lipid obtained by soxhlation method
Shake flask	3.210	0.320	0.251
Polythene bag	5.124	0.350	0.342
Photobioreactor	5.945	0.364	0.362

^A (g/ 100 ml media), *(g/g of dry biomass).

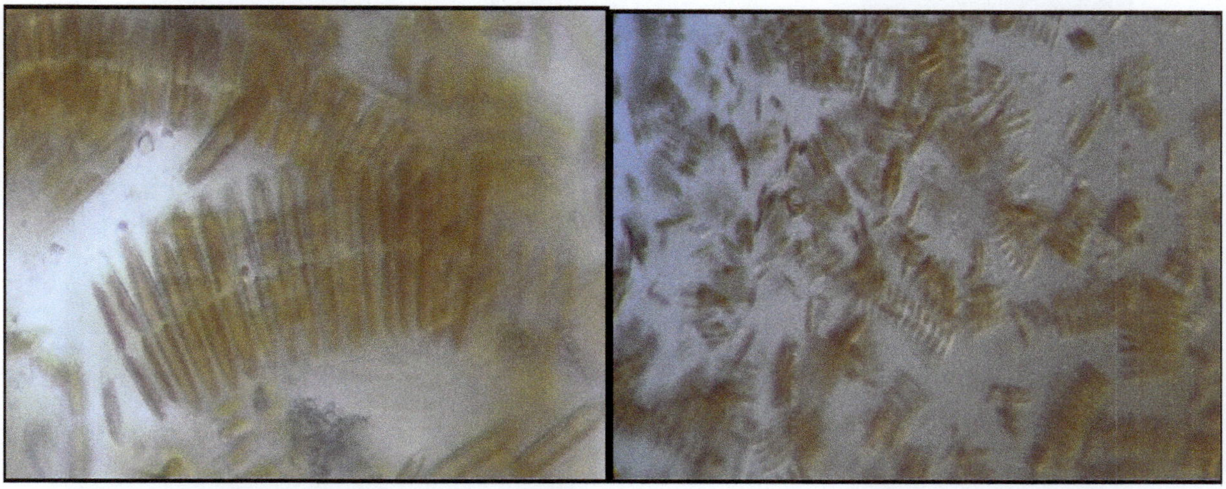

Figure 1. Structure of *N. crptocephala* under light microscope.

flow rate of 0.8 ml/min. The temperature programmed to 120°C for 5 min and 5°C increase/min up to 280°C.

Detection of TAGs composition

Thin layer chromatography was carried out to detect the TAGs using hexane: Ether: acetic acid (60:35:5) as a solvent system, along with the standard TAGs. Further determination of TAGs was carried out by HPLC (Shimadzu LC) equipped with ELSD 800 detector (altech). The TAGs was separated using commercial C-18 Column from waters (250 × 4.6 mm). The mobile phase was a mixture of acetonitril: dichlormethane (60:40) set at a flow rate of 1.0 ml/min, with pressure 2.3 bar. TAGs peaks were identified based on the retention time of available commercial TAGs (Sigma Aldrich, Germany).

RESULTS

Culturing of diatoms

For culturing of diatoms, initially, three media namely WC, Chu and Polytomella media were used. Out of the three, Chu medium was seen to facilitate the growth of diatoms very well. In the Chu medium, visible growth of diatom biomass was seen in 72 h after inoculation. It was observed that the growth of diatoms in three culturing techniques (shake flak, polythene bag and photobioreactor culture) was dependent upon factors such as surface area, amount of inoculum added, temperature and degree of aeration. Table 1 shows the yield of diatom biomass by different culturing techniques.

Identification of diatoms

The samples collected from different water sources were subjected to microscopic studies (Figure 1). In all the samples collected the major diatom species identified was *Navicula cryptocephala* (Kingdom-Chromista, Phyllum- Ochrophyta, Class -Bacillariophyceae, Family- *Naviculaceae*, Genus- *Navicula*). In this species the cells are either free or united into ribbon-like colonies. Frustules are generally linear in nature; lateral longitudinal ribs may be present in some forms. A characteristic feature of diatom cells is that they are encased within a unique cell wall made of silica (hydrated silicon dioxide) called frustules (Figure 2).

In pennates, 'sex' involves two adult cells lining up alongside each other, dividing and then swapping one daughter cell each. The new pairs of daughter cells fuse, giving two cells that are half of each 'parent'. They then

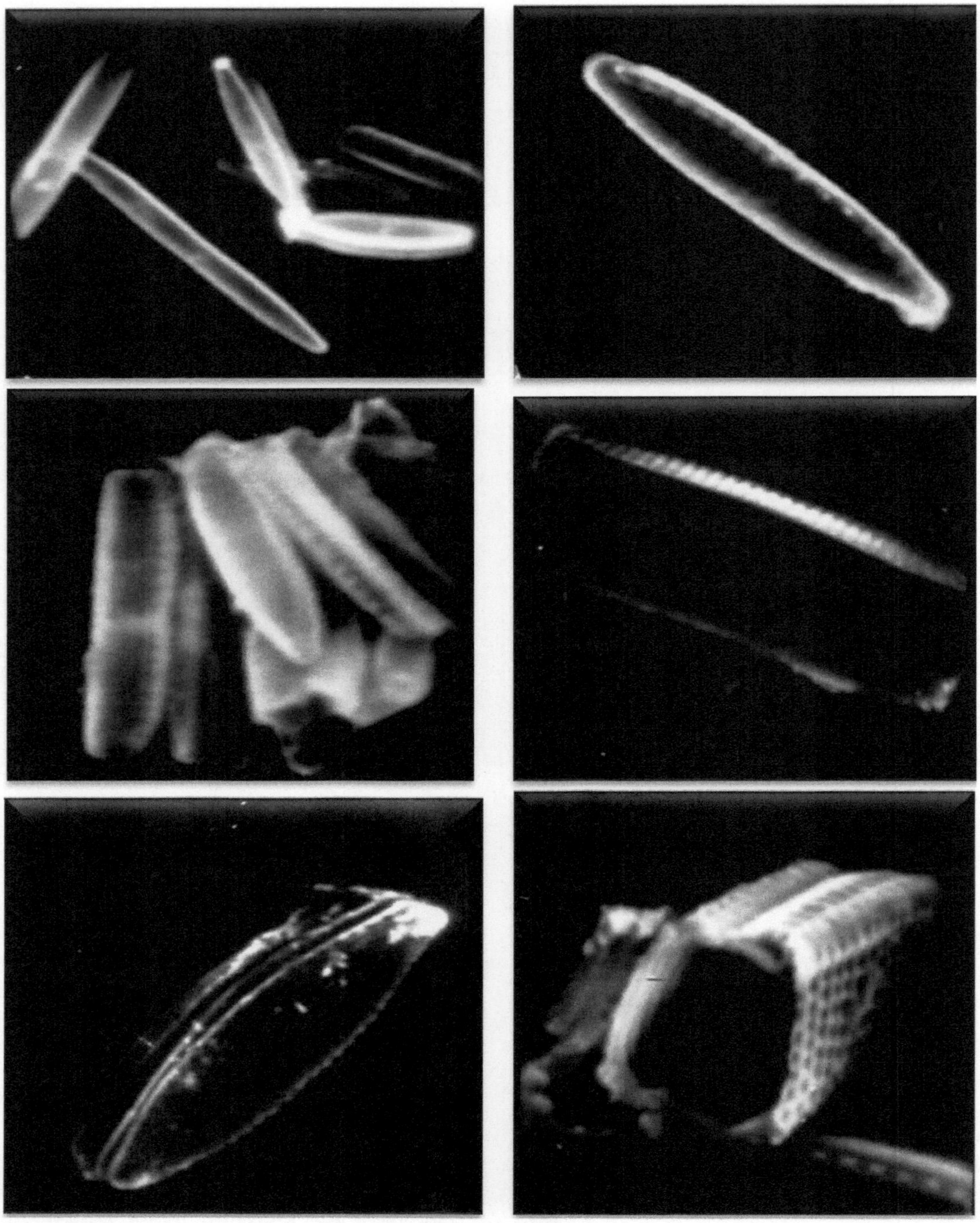

Figure 2. Scanning electron microscopic pictures of *N. cryptocephala*.

produce new, large frustules and glide away (Figure 2).

Lipids from diatoms biomass

The results of different culturing methods of diatoms showed that the yield of biomass is comparatively higher (5.954 g/100 ml of media used) in photobioreactor then the other two methods. This may be due to effect of availability of surface area and aeration. The lipids from diatom biomass was extracted by sonication and soxhlation method, the amount of crude lipid obtained from sonication and soxhlation method with different culturing method was shown in Table 1. Sonication method was found to be more effective compare to soxhlation method for extraction of lipids from diatom

Table 2. Physical and chemical analysis of crude oil.

S/N	Analytical parameter	Value
1	Density (g/cm³)	0.91
2	Viscosity (at 31°C), Centistokes	38.42
3	Iodine Value	89.35
4	Saponification value	168.44
5	Acid Value	35.64

Table 3. Fatty acids composition of crude oil.

S/N	Fatty acid	Retention time	Area	Concentration (%)
1	Myristic acid	18.48	63796	3.412
2	Palmitoleic acid	22.23	204084	10.916
3	Palmitic acid	22.73	635475	33.989
4	Linoleic acid	25.84	151128	8.083
5	Oleic acid	26.02	562564	30.089
6	Stearic acid	26.53	111163	5.946

biomass.

Characterization of crude oil obtained from diatom biomass

The results of physical and chemical analysis such as density, kinematic viscosity, acid value, saponification value and peroxide value of crude oil extracted from diatom biomass are shown in Table 2. Gas chromatogram results for free fatty acid detection showed that the major free fatty acids present in the oil extracted from diatom biomass were myristic acid, palmitoleic acid, palmitic acid, linoleic acid, oleic acid and stearic acid (Table 3). Out of six fatty acids, palmitic and oleic acid were found as major fatty acids (Figure 3). Results of HPLC analysis for estimation of TAGs content in the oil extracted from diatom biomass showed that the presence of OOL (19.82%), followed by OLL (Oleoyl-dilinoleoylglycerol) (15.62%), POL (Palmitoyloleoyllinoleoyl) (11.25%), SOO (Stearoyl-dioleoylglycerol) (2.48%). Monounsaturated that has been detected are MPP+OOO (12.25%), POO (palmitoyl-dioleoyl-glycerol) (8.72%), and PLP+MOP (1.05%) (Table 4).

DISCUSSION

The basic line of reasoning of geologists in attributing crude oil to diatoms is that they comprise the bulk of the ocean phytoplankton, so they must be a major source of the oil (Levorsen and Bery, 1967). The pattern of fatty acids in recent sediments matches that of diatoms (Volkman and Joans, 1977) and the 24-norcholestane biomarkers are indicators of diatom- formed oil (Holba et al., 1998; Rampsen et al., 2007).

In preliminary studies three media were used for culturing diatoms, however the Chu No. 10 medium was selected for its suitability. The comparison of modified freshwater "WC" medium (Guillard and Lorenzen, 1972) with Chu No. 10 medium indicated that freshwater "WC" medium was rapidly colonized by cyanobacteria and other algae (green algae) (Nichols, 1973). Such contamination may be related to the absence of inhibitor elements in the micronutrient solution, such as vanadium for cyanobacteria (Nalewajko et al., 1995). Although the diatom biomass growth rate was higher in freshwater "WC" medium, the contamination of the culture by other microorganisms made it unsuitable for further ecotoxicological experiments. These microorganisms produce EPS that can agglomerate diatoms and other algae and thus disturb diatom cellular observations under the microscope. In contrast, Chu No. 10 allowed sufficient production of diatom biomass (Debenest et al., 2009).

Diatoms have been regarded as useful neutral lipid sources of liquid-fuel precursors. Besides high lipid and fatty acid content, there is an abundance of eicosapentaenoic acid (polyunsaturated fatty acids in diatoms) (Renaud et al., 1994). *Nitzschia laevis* is a potential producer of eicosapentaenoic acid, as shown by extracting the lipid and analyzing it via thin layer chromatography (TLC) and gas chromatography (GC) (Chen et al., 2007). The lipids present are neutral lipids (accounting for, 75%), glycolipids, and phospholipids. Fatty acids that dominate the organisms include

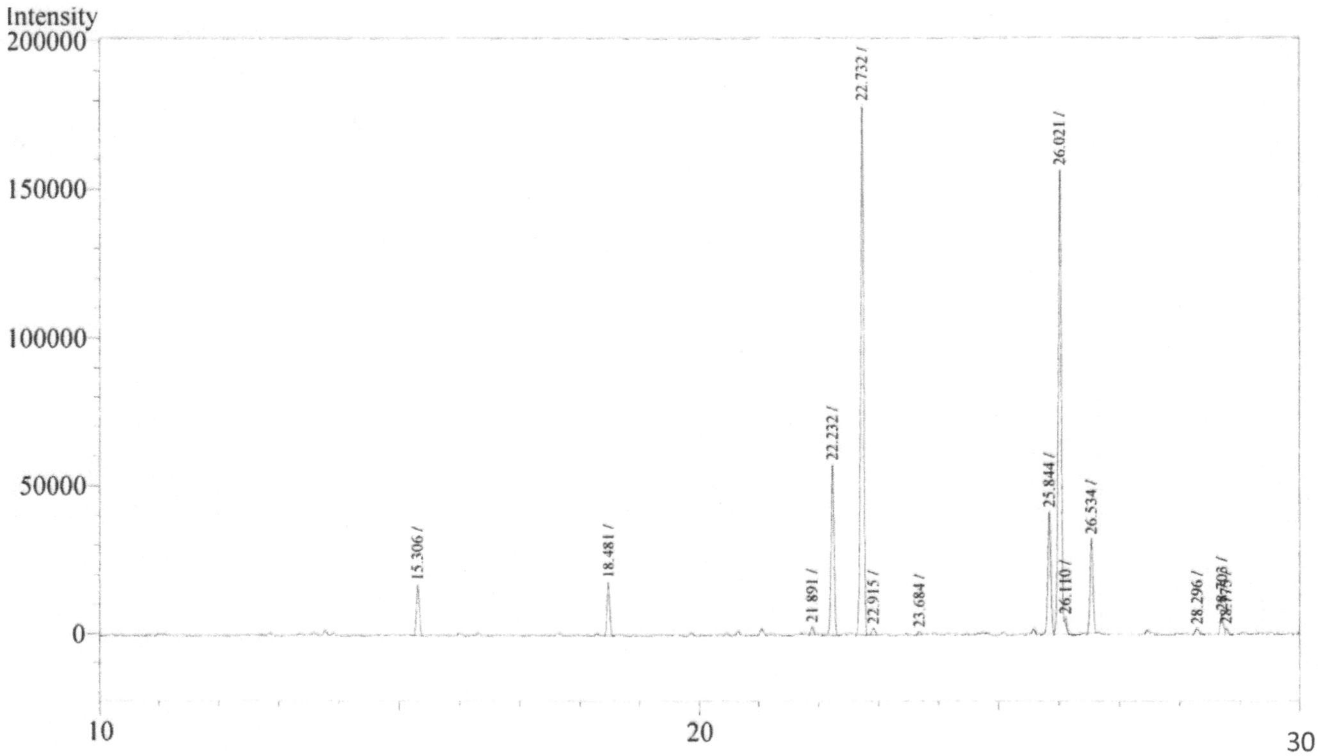

Figure 3. Gas chromatogram of diatom oil showing fatty acid composition.

Table 4. Triacylglycerol composition.

S/N	Triacylglycerol	Relative composition (%)
1	OOL	19.82
2	OLL	15.62
3	MPP+OOO	12.25
4	POL	11.25
5	POO	8.72
6	SOO	2.48
7	PLP+MOP	1.05

tetradecanoic acid, hexadecanoic acid and palmiloteleic acid. Many of the C25 (haslene) and C30 (rhizene) alkenes are biosynthesized by a restricted number of diatom genera, particularly some species of *Haslea*, *Rhizosolenia*, *Pleurosigma*, and *Navicula* (Grossi et al., 2004). In *Haslea ostrearia*, highly branched isoprenoid alkene (haslene) biosynthesis proceeds even under axenic conditions, indicating de novo biosynthesis (Wraige et al., 1999).

Lipid valorization as biodiesel using diatoms was reported with *Hantzschia* DI-6067 and *Chaetoceros muelleri* (McGinnis et al., 1997). The production of fuel (diesel, gasoline) through the transesterification and catalytic cracking of lipids accumulated in algal cells has been reported, including diatoms (Nagle and Lemke, 1990). The main raw material for diatom-based biodiesel is the enormous range of triglycerides (monoglycerides, diglycerides, and triglycerides), which are indeed compounds of fatty acids and glycerol. In the transesterification process, an alcohol (such as methanol) reacts with the triglyceride oils that are contained in diatom fats, forming fatty acid alkyl esters (biodiesel) and glycerin (Lebeau and Robert, 2003).

The fatty acid composition of oil plays an important role in the performance of biodiesel in diesel engines. Saturation fatty acid methyl esters increase the cloud point, octane number, and improve stability whereas more polyunsaturation reduces the cloud point, cetane number and stability (Kumar et al., 2003). The oil extracted from *N. cryptocephala* showed the presence of palmitic, oleic, palmitoleic and linoleic acids as main fatty acids detected which can be compared to Jetropha oil and rape seed oil used for biodiesel production. The free fatty acid content is one of the important factors for alkali-catalyzed transesterification. This is due to the free fatty acids reacting with the alkaline catalyst to produce soap, which inhibits the reaction and this result in the reduction of biodiesel yield (Meher et al., 2006).

Total lipid content on dry weight basis and total lipids made up of triglycerides were comparable with the other algal and plant sources used for biodiesel production. An additional algal characteristic for biodiesel production is the suitability of lipids for biodiesel in terms of the type and amount produced by an algal species, e.g. chain length, degree of saturation and proportion of total lipid made up by triglycerides. These influence the quality of biodiesel produced. The majority of lipid-producing algal species have a similar lipid profile, generally equivalent to vegetable oil from land plants suitable for biodiesel production (Xu et al., 2006). The proportion of various lipid classes (particularly triglycerides) varies widely with environmental conditions (Rodolfi et al., 2008), making it difficult to compare algal species across experimental conditions (Molina Grima et al., 1994). Lipid productivity can be calculated as the product of biomass productivity (grams dry weight per liter per day) and lipid content (%dw) to give an indicator of oil produced on a basis of both volume and time. High lipid content may, however, improve the efficiency of biomass processing (Rodolfi et al., 2008).

Conclusion

Diatoms would seem to fare, in average dry weight that they can synthesize as lipids, only a little better than green algae (24.5% vs. 17.1%), although their average dry weight is enhanced by a factor of 2 to 3 by nitrogen deprivation. However in spite of all these evidences there are only few reports on extraction and characterization of lipids from diatoms for biodiesel production. This study resulted in isolation of diatom *N. cryptocephala,* further the characterization of oil extracted showed presence of fatty acids and triglycerides content suitable for biodiesel production by transesterfication. Further studies on mass culturing of diatom, transesterification process of oil extracted and characteristics of biodiesel produced are in progress.

ACKNOWLEDGEMENTS

The authors acknowledge Department of Science and Technology, Govt. of India for their gracious funding of the project, Principal SJCE, Mysore and Head, Department of Biotechnology, SJ College of Engineering, Mysore, India for his constant support and guidance.

REFERENCES

Blanco S, Alvarez I, Cejudo C (2008). A test on different aspect of diatoms processing techniques. J. Appl. Phycol. 20:445-450.

Chen GQ, Jiang Y, Chen F (2007). Fatty acid and lipid class composition of the eicosapentaenoic acid-producing microalga *Nitzschia laevis*. Food Chem. 104(4):1580-1585.

David S, Stuart H (1998). Extraction and Quantification of chlorophyll A from Freshwater Green Algae. J. Sci. Technol. Charles Sturt Univ. 32:2220-2223.

Debenest T, Silvestre J, Coste M, Delmas F, Pinelli E (2009). A New Cell Primo-culture Method for Freshwater Benthic Diatoms communities. J. Appl. Phycol. 21:65-73.

Grossi V, Beker B, Geenevasen JAJ, Schouten S, Raphel D, Fontaine MF, Sinninghe Damste JS (2004). C25 highly branched isoprenoid alkenes from the marine benthic diatom *Pleurosigma strigosum*. Phytochemistry 65(22):3049-3055.

Guillard RRL, Lorenzen CJ (1972). Yellow-green algae with chlorophyllide. J. Phycol. 8:10-14.

Holba AG, Tegelaar EW, Huizinga BJ, Moldowan JM, Singletary MS, McCaffrey MA, Dzou LIP (1998). 24-norcholestanes as age-sensitive molecular fossils. Geology 26(9):783-786.

Hu Q, ZhangCW, Sommerfeld M (2006). Biodiesel from Algae: Lessons Learned Over the Past 60 Years and Future Perspectives. Juneau, Alaska: Annual Meeting of the Phycological Society of America, July 7-12:40-41 (Abstract).

Karthick B, Taylor JC, Mahesh MK, Ramachandra TV (2010). Protocol for Collection, preservation and Enumeration of Diatoms from Aquatic Habitats for Water Quality Monitoring in India. Icfai Univ. J. Soil Water Sci. 3:25-60.

Knothe G, Dunn RO, Bagby MO (1997). Biodiesel: The use of vegetable oils and their derivatives as alternative diesel fuels. ACS Symp. Ser. 666:172.

Kumar MS, Ramesh A, Nagalingam B (2003). An experimental comparison of methods to use methanol and Jatropha oil in a compression ignition engine. Biomass Bioenergy 25:309-318.

Lebeau T, Robert JM (2003). Diatom cultivation and biotechnologically relevant products. Part II: current and putative products. Appl. Microbiol. Biotechnol. 60(6):624-632.

Levorsen AI, Berry (1967). FAF Geology of Petroleum. In; Freeman WH and Co, 2nd Edition.: San Francisco, CA, 1967.

McGinnis KM, Dempster TA, Sommerfeld MR (1997). Characterization of the growth and lipid content of the diatom *Chaetoceros muelleri*. J. Appl. Phycol. 9(1):19-24.

Meher LC, Sagar DV, Naik SN (2006). Optimization of alkali-catalyzed transesterification of *Pongamia pinnata* oil for production of biodiesel. Bioresour. Technol. 97:1392-1397.

Molina Grima E, Robles Medina A, Gimenez Gimenez A, Sanchez Perez JA, Garcia Camacho F, Garcia Sanchez JL (1994). Comparison between extraction of lipids and fatty acids from micro algal biomass. J. Am. Oil Chem. Soc. 71:955-959.

Nagle N, Lemke P (1990). Production of methyl ester fuel from microalgae. Appl. Biochem. Biotechnol. 25(1):355-361.

Nalewajko C, Lee K, Jack TR (1995). Effects of vanadium on freshwater phytoplankton photosynthesis. Water Air Soil Pollut. 81(1-2):93-105.

Nichols HW (1973). Growth media-freshwater. In: Stein JR (ed) Handbook of phycological methods: culture methods and growth measurements. Cambridge University Press, New York.

Qiang H, Milton S, Eric J, Maria G, Matthew P, Michael S, Darzins A (2008). Microalgal triacylglycerols as feedstocks for biofuel production: perspectives and advances. Plant J. 54:621-639.

Ramachandra TV, Mahapatra DM, Karthick B (2009). Milking diatoms for sustainable energy: Biochemistry engineering versus gasoline-secreting diatoms solar panels. Ind. Eng. Chem. Res. 48(19):8769-8788.

Rampsen SW, Schouten S, Abbas B, Panoto FE, Muyzer G, Campbell

CN, Fehling JS, Damste JS (2007). On the origin of 24-norcholestanes and their use as age-diagnostic biomarkers. Geology 35(5):419-422.

Renaud SM, Parry DL, Thinh LV (1994). Microalgae for use in tropical aquaculture I: Gross chemical composition and fatty acid composition of twelve species of microalgae from the Northern Territory, Australia. J. Appl. Phycol. 6(3):337-345.

Rodolfi L, Zittelli GC, Bassi N, Padovani G, Biondi N, Bonini G, Tredici MR (2008). Microalgae for oil: strain selection, induction of lipid synthesis and outdoor mass cultivation in a low-cost photobioreactor. Biotechnol. Bioeng. 102(1):100-112.

Ronald H, Brendan G, Michael K D, Paul AW (2011). Oil extraction from microalgae for biodiesel production. Bioresour. Technol. 102:178-185.

Sheehan J, Dunahay T, Benemann J, Roessler PG (1998). US Department of Energy's Office of Fuels Development, July 1998. A Look Back at the US Department of Energy's Aquatic Species Program – Biodiesel from Algae, Close Out Report TP-580-24190. Golden, CO: National Renewable Energy Laboratory.

Vasudevan PT, Briggs M (2008). Biodiesel production-current state of the art and challenges. J. Indian Microbiol. Biotechnol. 35:421-430. PMID: 18205018.

Volkman JK, Johns RB (1977). The geochemical significance of positional isomers of unsaturated acids from an intertidal zone sediment. Nature 267:5613-693.

Wraige EJ, Johns L, Belt S T, Masse G, Robert JM, Rowland S (1999). Highly branched C25 isoprenoids in axenic cultures of *Hasleaostrearia*. Phytochemistry 51(1):69-73.

Xu H, Miao X, Wu Q (2006). Biodiesel production from heterotrophic microalgal oil. Bioresour. Technol. 97:841-846.

17

The next generation feedstock of biofuel: *Jatropha* or *Chlorella* as assessed by their life-cycle inventories

Pu Peng[1] and Wenguang Zhou[2]

[1]State Key Laboratory of Catalytic Material and Reaction Engineering, Research Institute of Petroleum Processing, SINOPEC, 18 Xue Yuan Road, Beijing 100083, China.
[2]Center for Biorefining, Bioproducts and Biosystems Engineering Department, University of Minnesota, 1390 Eckles Ave., Saint Paul, MN 55108, USA.

Promising energy crops such as *Jatropha curcas Linnaeus* (*JCL*), which are planted on marginal lands, or microalgae such as *Chlorella*, which are cultivated in ponds located on mudflats or deserts, have been regarded with high hopes to solve the shortage of food crops and increase the amount of biodiesel (fatty acid methyl ester, FAME) production. However, the annual yields of biomass and transport fuels (t/ha) of both are still unclear and often exaggerated in the literature. Large portions of *JCL* biomass, including tree trunks and leaves, can also be used to generate electricity along with FAME, which is produced from seed lipids. Meanwhile, lipid extracted algae (LEA) is composed of proteins, polysaccharides and lipids other than glycerides which are unable to be esterified to form FAME and much more abundant in the microalgae than oil cake in the oil crops. Therefore, it was strongly suggested that not only transesterification or esterification but also Fischer-Tropsch (FT) process and bio-electricity generation should be considered as routes to produce biofuels. Otherwise, the yield of biofuel would be extremely low using either *JCL* or *Chlorella* as feedstock. The life-cycle inventories (LCI) of the biofuel processes with whole biomass of *JCL* and *Chlorella* were compared based on their net energy ratio (NER) and CO_2 emission saving (CES). It was shown that the technological improvement of irrigation, cultivation and processing for either economic-crops or microalgae are all necessary to meet the requirements of commercial biofuel production.

Key words: Biodiesel, biofuel, *Jatropha*, microalgae, chlorella, LCA, LCI.

INTRODUCTION

The need for sustainable biofuels can be attributed to both an increase in energy consumption and the tighter restriction of greenhouse gas (GHG) emissions. It was believed that the use of biodiesel instead of fossil diesel results in a significant reduction in CO_2 emission. The development of biodiesel (fatty acid methyl ester, FAME) has met with large scale success in the EU and US with the use of rape seed and soybean, respectively, during the past 10 years. The EU hopes to radically cut GHG emissions and reduce dependency on fossil fuels through encouraging the production and use of sustainable biofuels. The arable land used for biodiesel production has been around 3 M ha (million hectares). Meanwhile, similar research has been conducted in other countries such as Brazil, Thailand, West Africa, and China. There have been more difficulties in China as there is much less arable and marginal land and a lower climate temperature than the countries in South Asia, Southeast Asia, and Africa.

Until 2008, US and EU biodiesel production has been up to over 2 million tonnes (Mt) and near 10 M t, respectively (Timilsina and Shrestha, 2011). However, 10 Mt is only 3% of diesel consumption in EU and far from the renewable energy directive (RED) 10% target by 2020 in 29 European countries (E27 plus Norway and Switzerland). In order to meet the gap between the 3 and 10% and find more sustainable feedstock, many sources have been tested including microorganisms, wastes, agricultural and forestry residues, energy crops and even used frying oils (UFO) or animal fats. However, among the wastes, economic-crops and algae, it is still not clear who would prevail. Economic-crops are able to be planted in marginal lands but the planting of dedicated energy crops often leads to the carbon stock change, known as land use change (LUC) (Laborde, 2011), which sheds doubt on the predicted positive GHG balance (positive CES). In contrast, microalgae are able to be cultivated in ponds or photobioreator (PBR) located on mudflats or deserts with near zero carbon stock. Meanwhile, other advanced biofuels such as hydrotreated vegetable oil (HVO) (Arvidsson et al., 2011), FT-diesel distillate (green diesel), FT-jet-fuel distillate (green jet fuel) are also candidates produced by Fischer-Tropsch synthesis and hydro treatment, which may be more compatible with existing fuel infrastructures or offer other technical benefits and be prepared with wider feedstock's.

In Francesco's (Cherubini and Stromman, 2011) comprehensive literature review on biofuel development, it was found that there are now an increasing number of papers dealing with lignocellulosic biomass, sugarcane, or palm oil in developing countries in South-Eastern Asia. By contrast, few studies are currently available on the promising feedstock of *Jatropha curcas Linnaeus (JCL). JCL* is a shrub and toxic tree with a smooth gray bark and an average height of 4 m (up to 6 m), belonging to the family Euphorbiaceae. This native species to Central America was introduced to the Cape Verde islands by Portuguese sailors in the 16th century, then to Guinea Bissau from where it spread across Africa and Asia. Its natural habitat is arid and semi-arid zones but it has also been found in damp tropical regions such as North Vietnam and Thailand. *JCL* starts producing seeds within one year of growth, but the maximum productivity is after 4 or five years (typical *JCL* yields in the first 5 years are 0.5, 1.5, 3.0, 5.0 and 6.0 t/ha). Its average life span is over 20 years (up to 50 years) (Cherubini and Stromman, 2011; Kalam et al., 2012).

In this paper, the *JCL* demonstration cases implemented in Thailand, India and West Africa are reviewed and compared with several laboratory works of microalgae, such as *Chlorella,* to further contrast their life-cycle inventory (LCI) of culturing, extracting, producing and processing. The prospective productivities (annual yield) of both feedstocks were compared and discussed to readdress the exaggerated results often found in the literature.

EXPERIMENTALS

Boundaries, functional units and allocation

LCI analysis involves creating an inventory of flows. Inventory flows include inputs of water, energy, feedstock, fertilizer etc. and outputs of CO_2 emission, biofuel products, land and water. The input of water and fertilizer are converted into power, which can be used in manufacturing and irrigating, whereas the output of land and water has not been considered. To develop the inventory, a flow model of the technical system has been constructed using data from the inputs and outputs, and it has given a clearer picture of the technical system boundaries. LCI results would be very different if different boundaries (1: biomass-system; 2: transport fuel system; 3: well (culturing) to wheel system, or 4: by-product included system) were accepted, as shown in Figure 1.

The data used in LCI must be related to the functional unit (FU) defined in the goal and scope. There are four types of FU identified in the LCI of bioenergy systems to compare: 1. given feedstock; 2. different feedstock; 3. dedicated energy crops; 4. Multiple final products, that is, input, output, agricultural land or year unit. The output unit and energy basis (GJ or MJ) were selected as functional units in this paper. All the outputs of the bioenergy systems expressed through other energy units were converted with the conversion factor (1 kg biodiesel = 37.8 MJ or 1kg fossil diesel = 42.8 MJ) to compare the results published in different literatures. The FU of the power input was also converted with the conversion factor (1kWh=3.6MJ).

Allocation in life cycle assessment (LCA) is carried out to attribute the total environmental impact to the different products of a system. This concept is extremely important for bioenergy systems, which are usually characterized by multiple products (e.g. electricity and heat from CHP application, rape-cake and glycerin from biodiesel production), and has a large influence on the final results (Ndong et al., 2009).

Energy balance and fossil fuel saving

The net energy ratio (NER) of a system is defined as the ratio of the total output energy utilized from produced liquid biofuel and residual biomass (produced energy output) over the input energy required in the "production stage," which includes photobioreactor (PBR) construction and materials, nutrition production, and planting (culturing) operation (primary energy input). NER is also called the energy yield. The net energy balance (NEB) is the difference between the effective energy produced and that required in the "production stage." If the bioenergy system is economically viable, then NER and NEB would be larger than one and zero, respectively (Pandey et al., 2011).

NER = Total energy output / Total energy input
NEB = Total energy output - Total energy input.

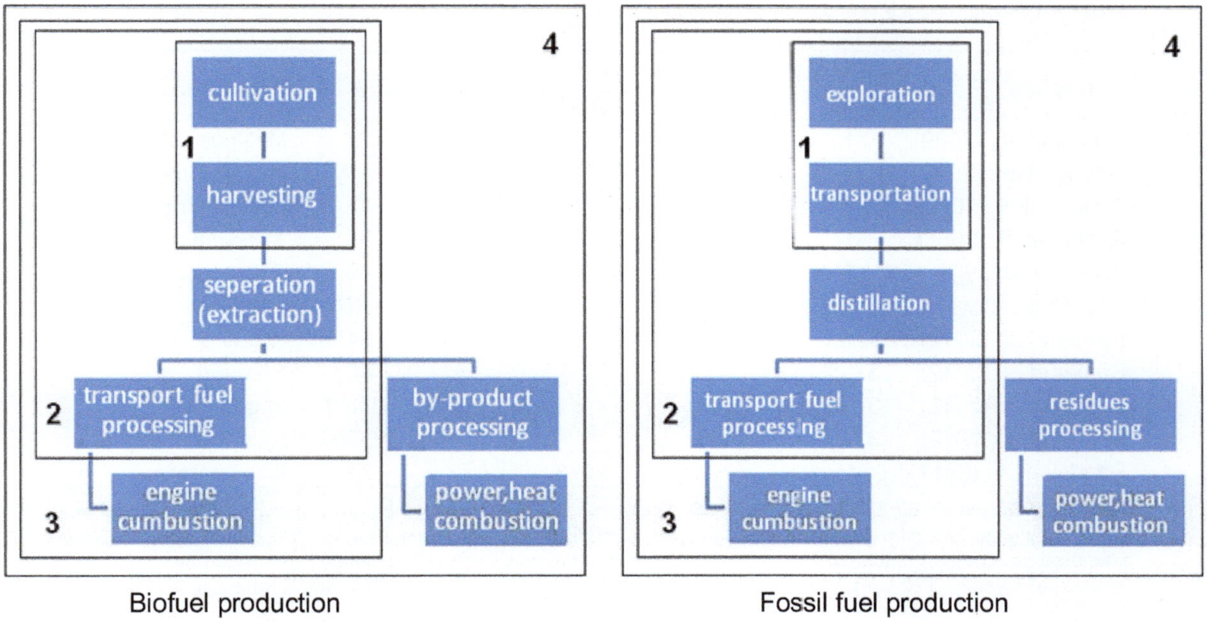

Figure 1. LCI flow model of biofuel and conventional fossil fuel with different boundaries.

Environmental balance and GHG saving

CO_2 was the only green house gas (GHG) considered in this paper and CEB (CO_2 emission of biofuel) in combustion of biofuel was calculated with either kg or MJ as the functional unit.

CEB (CO_2kg/kg) = mass (kg of biofuel combusted) × C content (normalized) × 44/12 = 2.86
CEB (CO_2g/MJ) = 1000 × 2.86 / energy (MJ producing from 1 kg biofuel) = 1000 × 2.86/37.8 = 75.7

Where, 0.78 was used as the carbon content of biodiesel and assuming all of the carbon in biodiesel was converted to CO_2; 44/12 is the ratio of molecular weight of CO_2 and atomic weight of C. CO_2 emission in the "production stage" is calculated as equivalent CO_2 emission from coal-fired electricity generation (0.83 kg CO_2/KWh), which is much greater than from natural gas (0.11 to 0.24 kg CO_2/KWh) but close to that from wood chips (0.82 kg CO_2/kwh) (Kumar et al., 2012). The CO_2 emission from coal-fired electricity in China (~1 kg CO_2/KWh) is even higher due to the use of low-grade coal.

CO_2 emission saving (CES) was used to show the CO_2 emission balance of biofuel and compare the CO_2 emission of fossil fuel used in the production of biofuel.

CES (%) = 100 × [1-(CEF+CEB+CEU)/CEF]

Assuming CO_2 emission in upstream of biofuel production (CEU) to be zero, and CEF was taken as 83.8 g CO_2/MJ fossil fuel and the CEB was -75.7 g CO_2/MJ biofuel, the maximum CES will be up to 90%, assuming GEF, GEB and CEU (CO_2 emission in upsteam of biofuel production) are 83.8, 75.7 and 0 g CO_2/MJ fossil fuel, respectively. In fact, CES is closely dependent on the upstream process and boundary shown in Figure 1, which means it would be much less than 90%. CEU is usually large and even larger than CEB (CES becomes negative) depending on the energy consumed in the upstream process and the electricity source of coal, fuel oil or natural gas-fired power station.

RESULTS AND DISCUSSION

The industrial production of *JCL* is a fairly recent development. Almost 0.9 M ha, 0.765 M ha (0.32 M ha in Senegal) and 0.12 M ha of *JCL* farm have been established to date in Asia, Africa and Latin America, respectively, but it is still far away from the targets of 5 M ha by 2010 and 13 M ha by 2015 (Ndong et al., 2009). Meanwhile, the prices of *JCL* seed have increased from 0.10 $/kg in 2005 to 0.34$/kg in 2011(Kalam et al., 2012).

Fertilizer and watering in JCL plantation

In Thailand, a demonstration of *JCL* plantation was conducted by Kasetsart University, in which annual crop cutting was set within an area of 1 ha (hectare) and a crop density of 2 × 1 m or two trees per m^2 was utilized. Land preparation comprised of plowing, harrowing, and a furrowing process using a tractor with an engine of 75 hp to adjust the soil condition for the new cutting set; ternary (N-P-K: 15-15-15) fertilizer was applied with a rate of 650 kg/ha per year; weedicides and insecticides were also used for the general maintenance of the plantation; the pumping rate was 4.5 m^3/m^2 per year for watering and manual harvesting (Pandey et al., 2011).

In India, the yield of *JCL* increased from 1.5 t/ha (rain-fed) to 5.9 t/ha (irrigated) when double fertilizers and an additional 105 kg/ha diesel were consumed in the irrigated mode as compared with rain-fed mode (Kumar et al., 2012).

In West Africa (Mali and Ivory Coast), *JCL* planting was up to 1500 ha in 2007, and at least 2000 ha more in

Table 1. Energy balance and yield (NER) and CO_2 emission saving (CES).

Reference	Energy output/MJ		Energy input/MJ	NER[1]	CES[2]/%	Plantation
	JME	By-product				
Ndong et al., (2009)	1		0.21	4.7	72	Mali/baseline
Ndong et al., (2009)	1			1.8	11	irrigation/motorized
Achten et al. (2008)	1		0.886	1.1	77	50%Faming E
Achten et al. (2008)	1		0.16	6.3	93	17%Faming E
Yale University	1		0.88	1.1	72	Thailand
Yale University	1	17.88	0.88	21.5	72	Thailand
Kumar et al. (2012)[2]	37.27		21.83	1.7	54	Irrigated
Kumar et al. (2012)[2]	37.27		27.6	1.4	40	Rain-fed
Kumar et al. (2012)[3]	37.27	107.8		1.5-8.6	50-107	Irrigated
Kumar et al. (2012)[3]	37.27	107.8		1.2-7.0	40-93	Rain-fed
Pandey et al. (2011)	1		0.578	1.73	23	Five years[3]

[1]NER was calculated by dividing the biodiesel energy as the sole energy output by the total energy consumed, which is oil yield and allocation mode dependent. [2]Conventional diesel emit 83.8 kg CO_2eq/GJ diesel. [3]NER was estimated as a ratio of biodiesel energy and net allocated process energy to biodiesel. [3]Averaged by the total yield (energy) of 5 years (3.92 t JME, 161.65 GJ) and NEB=161.65-93.51= 68.14 MJ. Faming E (energy).

2008. Two 5-ha experimental fields with plantation densities of 1111 plants ha^{-1} were selected with contrasting soil conditions. Ternary fertilizer only was applied during the first three years: 100 kg/ha in 1st year, 150 kg/ha in 2nd year, 200 kg/ha in 3rd year, whereas both 248 kg/ha ternary fertilizer and 201 kg/ha of ammonium nitrate were applied in the 4th year (Pandey et al., 2011).

Productivities of biomass, JME and by-products

The yield hypotheses had a significant impact on the GHG and energy balances of JME. The weight of each fresh fruit and seed is around 10 to 15 and 2 to 4 g, respectively. The annual yield of *JCL* fresh fruit is about 16 t/ha. The yield of seed is widely spread from 0.1 to 10 t/ha (Basili and Fontini, 2012) and an increase of 1 t/ha on seeds resulted in a 10% reduction in fossil energy use compared to the baseline value of 4 t/ha. An increase of 1 t/ha on seed production from the baseline of 4 t/ha results in a 10% reduction in fossil energy usage. Thus, it appears critical to pursue large-scale field cultivation experiments of *JCL*.

The yield of co-products such as, wood, leaves and seed shell, are 4, 2, and 0.8 t (dry weight), respectively from the process of *JCL* plantation, and press cake (91.5 kg, dry weight) is obtained when 1 FU (1GJ of JME) is produced (Prueksakorn and Gheewala, 2006).

Kumar et al. (2012) reported details of inventory requirements for the farming, oil extraction, biodiesel production and transportation stages for the entire JME production process in India. The oil percentage of *JCL* seed ranges from 21.0 to 48.2% and oil seeds were assumed to be sun-dried. The energy required for harvesting, handling and storing of oil seeds, oil and biodiesel, and the separation of husk from the seeds, were neglected due to the cheap and abundant labor force available.

Energy balance (net energy use)

The selected JME projects of energy balance expressed in NER (energy yield) and environmental balance expressed in biofuel CO_2 emission factor per GJ energy are shown in Table 1.

Ndong et al. (2009) proposed a very detailed LCI analysis on the JEM project in West Africa in which an allocation analysis was also included. The energy yield (the ratio of biodiesel energy output to fossil energy input) is 4.7. In other words, 4.7 MJ of energy is produced from 1 MJ of fossil fuel consumed to produce JME. Biofuel production requires direct (electricity, fuels and natural gas) and indirect (manufacturing of agricultural inputs, methanol etc.) energy consumption. In the allocation of energy expense, cultivation accounted for only 12%, which is even less than the 15% in the transport of seeds, oilcake and unrefined *JCL* oil. Transesteification requires 61% of the energy expense. In an alternative scenario, all co-products and JME are considered as output energy sources. The motorization and irrigation for the first three years were also included in *JCL* production with a high consumption of fuel and irrigation water, which resulted in the energy yield being lowered down to only 1.8 compared to 4.7 in the baseline scenario. Meanwhile, GHG savings compared to fossil diesel become only marginal (11%), only one-sixth of the baseline scenario.

In a Thailand demonstration test (Prueksakorn and Gheewala, 2006), the allocation for energy expenses of transesteification, irrigation and fertilization processes was approximately 40% (0.197 GJ for producing steam),

23% (0.205GJ) and 22% (0.198GJ), respectively. The highest energy consumption was in the process of transesteification (61%) and other energy expenses were from using diesel, electricity and producing fertilizer. Others are from 100 kg of fertilizer with the chemical formula 15-15-15 (energy consumption for transportation of fertilizer is excluded). The consumption of diesel for water pumping is a main contributor to energy expense in the irrigation process. *JCL* plantation requires the process of land preparation one time every five years as the stems are cut every year but a new plantation is made every five years. The residue (by-product) produced energy is 17.883 GJ whereas JME produced energy is only 1 GJ. The highest energy by-product is wood, which produces energy of 10.289 GJ.

Environmental balance (GHG emission)

Achten et al. (2008) deducted GHG emissions from the production phase of JME, and considered the energy content of the co-products in calculating the CES (GHG emissions from JME). The result was ~77% compared to the GHG emission using fossil diesel. Totally, the production and combustion of JME emits 23.5 g CO_2e/MJ JME. The allocation of CO_2 emitted accounted for 52% in cultivation, while the shares of the transesteification and final combustion steps were 17 and 16%, respectively. Large shares of the emissions occurring during the agricultural step are due to fertilizers.

It is assumed that the energy consumed in the production stage is from 100% fossil fuel fired power plants. In India, fossil fuel is 64% of electricity generation, in which 52% of electricity is generated in coal-fired power plants, whereas the other 12% are allocated from natural gas (11%), oil (1%). The rest are from hydro (23%), nuclear (3%) and renewable materials (10%) (Pandey et al., 2011). In China, coal is the major share of fossil fuel in electricity generation (80%), resulting in more CO_2 emitted.

Sunil[8] compared energy and environmental balances of irrigated and rain-fed scenarios. Seed annual yields of irrigated (5.9 t/ha, farming energy: 9333MJ/t JME; farming emission: 680kg CO_2/t JME) and rain-fed (1.5 t/ha, farming energy: 15098 MJ/t JME; farming emission: 1114kg CO_2/t JME) scenarios are closely related.

It was found that the utilization of JME saved more energy and emitted less CO_2 (saving 1.2GJ/ha; emitting 80 kg CO_2/ha per year) than direct use of *JCL* oil (saving 1.0 GJ/ha; 67 kg CO_2/ha per year) based on the comparison of energy and environmental balance in Central India (Center for industrial ecology, Yale University, USA, 2010).

Life cycle costs

It was reported that the total cost of JME without externalities is higher than the current market retail selling price of diesel in Thailand (Sampattagu et al., 2007). The cost allocation is 62.62, 25.27 and 12.12% for agricultural, biodiesel production and environmental processes, respectively. The highest expenditure is the operation costs in agriculture stage such as fertilizers, insecticide, and electricity for water pumping system, especially in the dry season.

Generally, production of JME costs less than palm oil, soybean diesel costs more, and the cost of rapeseed diesel is the highest (Clarens et al., 2010). While as a commercially viable option, microalgae will require further improvements in genetic and metabolic engineering to and obtain promising strains and produce higher yield of oil. Moreover, the development of equipment and methodologies for cost-effective culturing, harvesting and processing are also required as year-round production of biofuels requires constant, reliable feedstock supply (Hoekman, 2009). While microalgae are projected as a future feedstock of biodiesel, production cost is much higher than for these terrestrial crops, in the range of US$2 to $22 per liter (Timilsina and Shrestha, 2011). In addition, the entire microalgae biomass has to be used to produce biofuel otherwise the yield of biofuel would be too low to be accepted (Milledge, 2010).

Situation in China

It's widely hoped that JME production will offer a newer, more sustainable energy source; but that is yet to be proven. In 2007, the global output of biodiesel (FAME) was 8.82 Mt but only 0.1 Mt in China, and there has been no obvious change in suffering from arable land limitation since then (Timilsina and Shrestha, 2011). Therefore, there has been more effort to search energy crops in China than EU, US, and even Asia countries, such as India and Thailand. However, the climate is not suitable to plant *JCL* on most of land in China. Although Southwest China is considered as a prospective area to plant JCL, it has been discovered that both prospective planting area and yield of seed had been overestimated after comprehensive consideration using Agro Ecological Zone method. Based on the remote sensing data on land use, meteorological, soil and land slope, and suitable environment for *JCL* plantation in Southwest China, the potential lands to expand *JCL* areas is only 0.15 M ha by 2008, which is far from the government goal (1.667 M ha by 2020). Although the moderately suitable land was increased to 1.433 M ha after softening the terms, the poor yield of seed on the moderate land would definitely destroy the balance of both energy and environment (NER and CES become minus value) (Wu et al., 2009).

In China, the yield of fresh fruit, dried seed and extracted oil would be less than the countries in the tropic zone, and there is little data from farming sites. The prospective yield of *JCL* seeds and JME are often

Table 2. Energy input (expense) and GHG emission for producing 1 kg algal oil and energy yield (NER).

Reference	Yield[1] (t/ha)	Lipid %(DW)	Output (MJ)	input (MJ)	NER	CES[2] (%)	Cultivation	
							Harvesting	**Algal species**
Kalnes et al. (2012)		25.0	83.6	85.8	0.97	-2	Dewatering	*Nannochloropsis*
Kalnes et al. (2012)		25.0	83.6	59.0	1.42	57	HRJ	*Nannochloropsis*
Batan et al. (2010)	91	42.5	1	0.93	0.93	29	PBR bags	*Nannochloropsis*
Lardon et al. (2009)	62	17.5	103.8	106.4	0.98		Dry	*Chlorella vulgaris*
Lardon et al. (2009)	62	17.5	146.8	41.4	3.55		Wet	*Chlorella vulgaris*
Xu et al. (2011)	128	40.0	37.2	27.5	1.35	41	Tubular	*P. tricornutum*
Vasudevan et al. (2012)	76	25.0	9.2	32	0.3	-232	Dry extract.	Algae
Vasudevan et al. (2012)	76	25.0	9.2	3.7	2.5	37	Wet extract.	Algae

[1]Yield of algal biomass per year, 2 minus means net GHG emission but not fixation.

exaggerated and questionable. For instance, the JME yield (5t/ha) and farming (planting) energy (160MJ/t JME) were accepted in a published LCA research. Nevertheless, the energy of soybean farming is as high 2497 MJ/t FAME as usual and the average yield of *JCL* seed in the tropic zone is only 4t/ha (the JME content is less than 40% of *JCL* seed). A misleading NER of JME (2.004) even higher than soybean (0.981) was proposed based on a unconfirmed *JCL* annual yield (5 t/ha) in Hainan province of China which was greatly overestimated and even over the field obtained in the tropic zone (Ou et al., 2009). The reasonable annual yield of JME is 1 to 2 t/ha based on the demonstration test in Thailand (Prueksakorn and Ghewala, 2006), Malaysia (Kalam et al., 2012) and West Africa (Ndong et al., 2009); where, JME content of 30 to 50% is usually accepted.

Comparison with microalgae

It is known that microalgae has attracted the spotlight around the world during past years and was considered as a very hopeful competitor to replace terrestrial plant as feedstock of next generation biofuels. One reason for the superiority of microalgae is that it is unnecessary to use fresh water and arable land in the culturing. Waste, saline, or brackish water and land resources, such as mudflats or deserts, are all usable for the microalgae culturing so that there is no interference in food production as there was for the first generation biofuels (Clarens et al., 2010). Another benefit of microalgae is due to the high expected yield of biomass, which can be as high as over 100 t/ha and oil content as high as 70% of its dry weight in the form of triglycerides (Chisti, 2007). However, these are only speculations based on the excessively optimistic assumptions or laboratory data using minimalistic culturing volume of several milliliters to liters, which largely deviated from the larger scale culturing results either in ponds or PBR. Although *Chlorella*, *Diatom*, *Scenedesmus*, *Tetraselmis*, *Nannochloropsis* and *Haematococcus pluvialis* have been preferred as hopeful candidates, the real potential of their productivity was not clear until their production was realized at a large scale, resulting to the differences of published biomass and oil (lipid) yield potentials which are as much as 16 times (Quinn et al., 2011).

In this paper, NER was selected to qualify the energy balance and closely related to both the boundaries of LCI and the yield of algal biomass and lipid. Meanwhile CES was selected to qualify the GHG balance. Some of the calculated NER and CES based on published LCA results were summarized in Table 2, and some of them were from laboratory data and extremely exaggerated. The results are dependent on the manner of culturing (pond or PBR), harvesting, biomass yield and lipid content and the boundary. NER for *Nannochloropsis* cultivation process is 4.33 for flat-plate PBR but 7.01 for raceway pond, indicating that both processes were energetically favorable for biomass production (boundary 1) (Jorquera et al., 2010). NER became less than 1 when harvesting stage was included (boundary 2) (except for HRJ or wet harvesting) indicating that dewatering is the most energy consuming process in the upstream (Table 2). The NER is closely dependent on the yield of biomass or lipid and varies as much as six times as reported from different sources (Xu et al., 2011). In order to attain the energy benefit (NER>1), the overestimated LCI data (biomass yield or lipid fraction) were often accepted in the published papers (Chisti, 2007; Yanfen et al., 2012).

If FAME is considered as the only product of algal fuel (boundary 3), it is almost impossible for NER and CES to be >1 and >0, respectively (net energy and CO_2 emission reduction are positive) as large amounts of energy are consumed in the dehydration and extraction processes as shown in Table 2 (Xu et al., 2011).

In fact, current commercial microalgae production is only focused on a few high-value products used mainly for human nutritional supplements, including entire algal biomasses, such as of *Spirulina* (Arthrospira) (3000 t/a) and *Chlorella* (2000 t/a) and extracted products, including β-carotene, astaxanthin and docosahexanoic acid (DHA). The total annual yield of microalgal biomass (dry matter basis) around the world is only about 10 Kt. However, the total revenue of the microalgae-containing products is up to several billion dollars per year, with a typical selling price of $5,000 to $100,000 per dry ton of biomass or extracted products (Spolaore et al., 2006). The biomass of microalgae is also used as live feeds in aquaculture, and in waste-water treatment systems with the lower price in the culturing stage (Pulz and Gross, 2004). Over 90% of the world's commercial microalgae production uses shallow, open, paddle wheel mixed raceway type ponds, in addition to open circular ponds for *Chlorella* production in Japan.

In China, the production of *Spirulina* (*Arthrospira*) and *Chlorella* are sold as nutrients and high nutrition feeds have increased rapidly recently. The annual yields of *Spirulina* (*Arthrospira*) and *Chlorella* have attained to 3000 and 1000 t, respectively. The protein content and lipid content of *Spirulina* sold as a nutrient are as high as 60% and less than 10%, respectively. The *Chlorella,* with lower protein content and higher lipid content than *Spirulina*, was recommended to be used as feedstock of biodiesel, although the price makes it too expensive to be used nowadays. A several-fold increase in algal biomass or lipid production is not feasible by cultivation in either low-nitrogen nutrient or rich-CO_2 environments. When low-N nutrients are used in algal cultivation, lipid content increases but the yield of biomass usually decreases (Illman et al., 2000). The improvement of biomass yield with additional (5 to 14%) CO_2 aeration is only 1.5 times and difficult to double as compared with air aeration without additional CO_2. In the lipid fraction, there are not only triglyceride (TAG) and free fatty acids (FFA) but also sterol, terpene and hydrocarbon, which are unable to be transesterified or esterified to form FAME so that the yield of FAME would be lower than the lipid content of *Chlorella* (Yanfen et al., 2012).

Besides FAME (biodiesel), in order to increase the yield of algal biofuels, other hydrocarbon fuels including methane and FT-fuel have to be considered as the algal biofuels produced from lipid extracted fraction (lipid extracted algae, LEA) which is difficult to be esterified to form FAME (Kalnes et al., 2012). Even now, the algal biofuel has not been produced or demonstrated in China due to the low algal biomass productivity and the complexity of the algal biomass. This is also the reason that the species of microalgae are often not specified in the literature of LCA, easily misleading the reader, as the productivity and lipid composition of various species of microalgae are very different and dependent on the area and time of culturing. Robert summarized the worldwide LCA results of the prior literature and investigated the wide variance in predicted environmental impacts from microalgae cultivation in open-air raceway ponds and deduced a very wide range of CO_2 emission (0.1 to 4.4 g CO_2e/g microalgae) (Handler et al., 2012).

The difficulties of commercialization

Jatropha

The *JCL* has been successfully planted on a large scale, especially in Asia and Africa where planting areas have been near to 1 M ha, respectively. The seed yield ranges 4 to 5 t/ha in above area but is relatively area and climate dependent. The oil content ranges from 20 to 50% and typically 30 to 40%, depending on the culturing area and conditions. The energy and environmental balance of culturing, harvesting and processing show *JCL* to be a preferable sustainable feedstock of biofuel, and the use of JME shows more favorable results in energy saving and CO_2 emission than direct use of *JCL* oil.

In the subtropical zone and even in south of China, the feasibility of *JCL* cultivation must be carefully considered. The yield of *JCL* oil is dependent on the climate and planting area. The published data from tropical zones such as south Asia, south-east Asia and Africa are not suitable for the LCI in China. The real, domestic and large-scale planting data strongly suggests the acceptance as LCI instead of published data from different zones of the world.

On the environmental balance assessment of JME, the CO_2 emission due to LUC was not considered in the paper. Based on the accomplished production scale in South Asia, Southeast Asia and West Africa, Jatropha may be closer to us as the next generation feedstock of biodiesel if there is enough land with proper climate, and we can avoid the CO_2 emission resulting from LUC at the same time.

Microalgae

The cultivation of microalgae even *Chlorella* and *Spirulina* is still in the beginning stages. The biggest farm for culturing *Spirulina* is only 0.1 M ha scale, which is much less than big *JCL* farm up to 1 M ha scale. Therefore, more demonstration farms or projects need to be created to verify the feasibility for microalgae to become the feedstock of biofuel in the future.

The triglyceride and free fatty acid (FFA) contents in the lipid fraction of microalgae are much less than that in the lipid fraction of *JCL*. Therefore, the biodiesel yield cannot be deduced from the lipid content of microalgae directly.

The lipid in microalgae should be referred to as bio-crude other than biodiesel. All of bio-crude or biomass has to be processed to produce the sustainable bio-energy instead of biodiesel (FAME) only so that biohydrogenated diesel (BHD), FT-fuel, bio-gas, bio-power and bio-heat can be produced and used together. The NER and NEB are hardly larger than one and positive, and CES is also rarely positive if the energy prepared from LEA was not accounted for at the present stage.

Conclusion

The biomass yields of both *JCL* and *Chlorella* per hectare have to be increased further by using the high efficiency irrigation systems for *JCL* plantation and well-designed PBR for *Chlorella* cultivation so as to get economic and environmental benefits. Moreover, it was strongly suggested that not only transesterification or esterification but also Fischer-Tropsch process and bio-electricity generation should be considered as routes to produce biofuels. Otherwise, the yield of biofuel would be extremely low using either *JCL* or *Chlorella* as feedstock. Overall, there is a long way to go for either *Jatropha* or *Chlorella* to become real 2nd generation feedstock at the scale corresponding to 1st generation feedstock.

Conflict of Interests

The author(s) have not declared any conflict of interests.

ACKNOWLEDEMENTS

This work was supported by the National Basic Research Program of China through “Fundamental of Chemistry and Engineering for High-Efficiency and Sustainable Oil Refining Technologies” (project number: 2012CB224806).

REFERENCES

Achten WMJ, Verchot L, Franken YJ, Mathijs E, Singh VP, Aerts R (2008). Jatropha bio-diesel production and use. Biomass Bioenergy 32:1063-84.

Arvidsson R, Persson S, Fröling M, Svanström M (2011). Life cycle assessment of hydrotreated vegetable oil from rape, oil palm and Jatropha. J. Clean. Prod. 19:129-37.

Basili M, Fontini F (2012). Biofuel from Jatropha curcas: Environmental sustainability and option value. Ecol. Econ. 78:1-8.

Batan L, Quinn J, Willson I, Bradley T (2010). Net Energy and Greenhouse Gas Emission Evaluation of Biodiesel Derived from Microalgae. Environ. Sci. Technol. 44:7975-80.

Center for Industrial Ecology, Yale University USA (2010).[Internet]. Life-Cycle Assessment of Jatropha Production in Central India. Available at: is4ie.org/Content/Documents/ Document.ashx?DocId=7763

Cherubini F, Stromman AH (2011). Life cycle assessment of bioenergy systems: state of the art and future challenges. Bioresour. Technol. 102:437-51.

Chisti Y (2007). Biodiesel from microalgae. Biotechnol. Adv. 25:294-306.

Clarens AF, Resurreccion EP, White MA, Colosi LM (2010). Environmental Life Cycle Comparison of Algae to Other Bioenergy Feedstock's. Environ. Sci. Technol. 44:1813-19.

Handler RM, Canter CE, Kalnes TN, Lupton FS, Kholiqov O, Shonnard DR (2012). Evaluation of environmental impacts from microalgae cultivation in open-air raceway ponds: Analysis of the prior literature and investigation of wide variance in predicted impacts. Algal Res. 1:83-92.

Hoekman SK (2009). Biofuels in the U.S. - Challenges and Opportunities. Renew. Energy 34:14-22.

Illman AM, Scragg AH, Shales SW (2000). Increase in Chlorella strains caloric values when grown in low nitrogen medium. Enzyme Microb. Technol. 27:631-635.

Jorquera O, Kiperstok A, Sales EA, Embirucu M, Ghirardi ML (2010). Comparative energy life-cycle analyses of microalgal biomass production in open ponds and photobioreactors. Bioresour. Technol. 101: 1406-1413.

Kalam MA, Ahamed JU, Masjuki HH (2012). Land availability of Jatropha production in Malaysia. Renew. Sustain. Energy Rev. 16:3999-4007.

Kalnes T, Lupton S, Handler RM, Shonnard DR, Blowers P, Canter CE (2012). Fossil Fuel Consumption and GHG Emissions for Algal Biofuels. In: 2nd International Conference on Algal Biofuels and Bioproducts. San Diego, California, June 10-14.

Kumar S, Singh J, Nanoti SM, Garg MO (2012). A comprehensive life cycle assessment (LCA) of Jatropha biodiesel production in India. Bioresour. Technol. 110:723-29.

Laborde D (2011). Assessing the Land Use Change Consequences of European Biofuel policies and its uncertainties. Final report No SI2 580403. The International Food Policy Institute (IFPRI).

Lardon L, Helias A, Sialve B, Steyer P, Bernard O (2009). Life-Cycle Assessment of Biodiesel Production from Microalgae. Environ. Sci. Technol. 43:6475-81.

Milledge JJ (2010). The challenge of Algal Fuel: Economic Processing of the Entire Algal Biomass. Mater. Eng. Newsl. 1(6):4-6.

Ndong R, Montrejaud-Vignoles M, Saint Girons O, Gabrielle B, Pirot R, Domergue M (2009). Life cycle assessment of biofuels fromJatropha curcasin West Africa: a field study. GCB Bioenergy 1:197-210.

Ou X, Zhang X, Chang S, Guo Q (2009). Energy consumption and GHG emissions of six biofuel pathways by LCA in (the) People's Republic of China. Appl. Energy 86:S197-208.

Pandey KK, Pragya N, Sahoo PK (2011). Life cycle assessment of small-scale high-input Jatropha biodiesel production in India. Appl. Energy 88:4831-39.

Prueksakorn K, Gheewala SH (2006). Energy and Greenhouse Gas Implications of Biodiesel Production from JCL curcas L. In: The 2nd Joint International Conference on “Sustainable Energy and Environment ”, Bangkok, Thailand, November 21-23. P E-053 (O).

Pulz O, Gross W (2004). Valuable products from biotechnology of microalgae. Appl. Microbiol. Biotechnol. 65:635-648.

Quinn JC, Catton K, Wagner N, Bradley TH (2011). Current Large-Scale US Biofuel Potential from Microalgae Cultivated in Photobioreactors. BioEnergy Res. 5: 49-60.

Sampattagu S, Suttibut C, Yucho S, Kiatsiriroat T (2007). Life Cycle Management of JCL Bio-Diesel Production. In: Thailand. 3rd International Conference on Life Cycle Management, Irchel, Zurich, August 27-29, P326.

Spolaore P, Joannis-Cassan C, Duran E, Isambert A (2006). Commercial applications of microalgae. J. Biosci. Bioeng. 101: 87-96.

Timilsina GR, Shrestha A (2011). How much hope should we have for biofuels? Energy 36:2055-69.

Vasudevan V, Stratton RW, Pearlson MN, Jersey GR, Beyene AG, Weissman JC (2012). Environmental performance of algal biofuel technology options. Environ. Sci. Technol. 46: 2451-59.

Wu W, Huang J, Deng X (2009). Potential land for plantation of Jatropha curcas as feedstocks for biodiesel in China. Sci. China Ser. D Earth Sci. 53: 120-27.

Xu D, Slade R, Bauen A (2011). The Sustainability of Algae Biofuels: LCA, Environmental and Economic Impacts. In: AquaFUELs Project FINAL CONFERENCE. Imperial College, UK, June 30.

Yanfen L, Zehao H, Xiaoqian M (2012). Energy analysis and environmental impacts of microalgal biodiesel in China. Energy Policy 45:142-151.

18

Empirical modeling of solar radiation for selected cities in Nigeria using multivariate regression technique

Adeyemi B. and Ojo O. S.

Department of Physics, P.M.B. 704, Federal University of Technology, Akure, Ondo State, Nigeria.

In this paper, the data for solar radiation, minimum and maximum temperatures, wind speed and evaporation for the years 1970-1995 and for fourteen stations taken over Nigeria were obtained from the archives of the Nigerian Meteorological Agency (NIMET), Oshodi, Lagos. The distributions of solar radiation with each of the four meteorological variables were observed. It was found that solar radiation was adequately monitored by each of the variables indicating that the linear combination of the variables could be used to developed model from where solar radiation can be evaluated. Consequently, a multivariate linear regression model was developed for each of the stations. The statistical indicators such as R^2, MBE, RMSE and MPE were calculated to estimate the efficiency of the developed models. For instance, the values of the R^2, MBE, RMSE and MPE calculated for Sokoto, a Sahelian station are 0.5349, 0.0273, 0.4154 and 0.0811, respectively, which show that the model for Sokoto is significant. Those for other stations have also been calculated with all showing high level of signicance at 0.05 alpha level.

Key words: Solar radiation, meteorological variables, correlation coefficient, multivariate regression, evaporation, significance.

INTRODUCTION

Solar radiation received at the earth's surface is essential for the development and utilization of solar energy devices. It is needed for designing collectors for solar heaters and other photovoltaic equipment that depend on solar energy. Incoming solar radiation has a significant role in hydrological and crop growth modelling. For instance, it is a key input for estimating potential evapotranspiration which play a major role in the design of water supply storage reservoirs and irrigation systems (Sanusi and Abisoye, 2011). The solar radiation reaching the earth's surface depends on the climatic condition of the specific location, and this is essential for accurate prediction and design of a solar energy system (Burari and Sambo, 2001). Technology for measuring solar radiation is costly and has instrumental hazards (Alam et al., 2005); this has led to paucity of it in many stations in the developing countries (including Nigeria). Thus, alternative methods for estimating these data are required (Al-Salihi et al., 2010). One of these methods is the use of empirical models. Accurate empirical modelling depends on the quality and quantity of the measured data used in developing it, and is a good

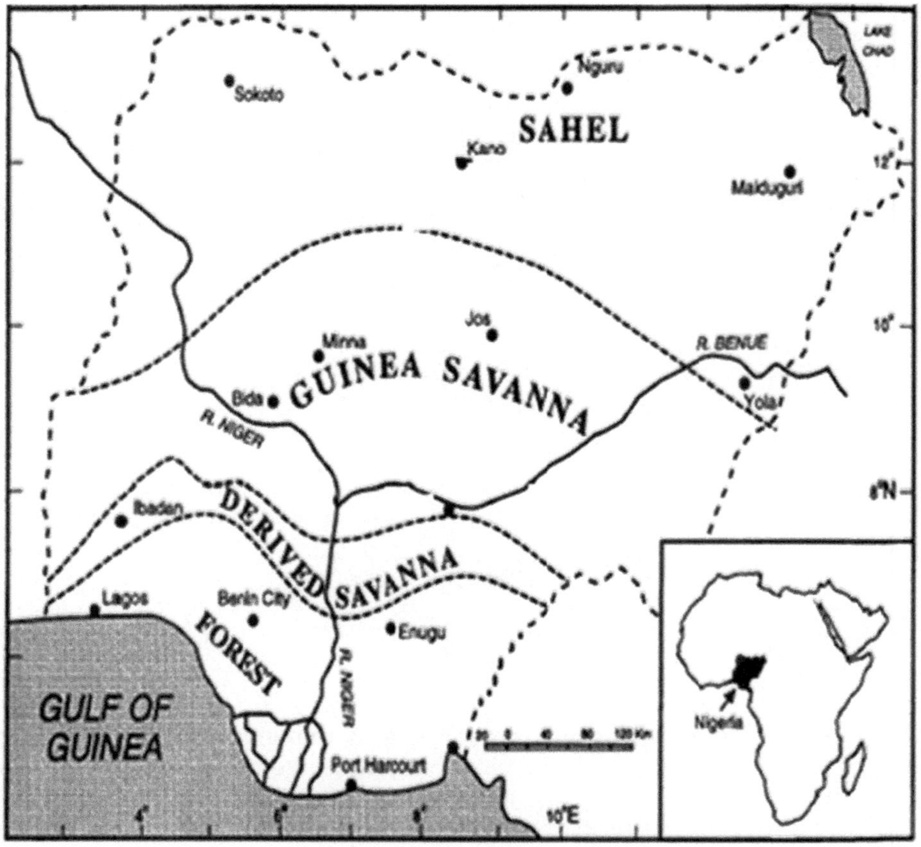

Figure 1. A map of Nigeria showing four climatic regions and the studied stations.

tool for generating solar radiation of any kind (whether global, direct or diffuse) at locations where measured data are not available.

When solar radiation enters the Earth atmosphere, a part of the incident energy is removed through the processes of scattering, absorption and reflection. The scattering of solar radiation is mainly by atmospheric molecules and aerosols while the absorption of solar radiation is by stratospheric ozone, water vapor, oxygen, carbon (IV) oxide, as well as clouds. The reflection of solar radiation on the other hand, is mainly by clouds and this plays an overriding part in reducing the energy density of the solar radiation reaching the surface of the Earth (Exell, 2000).

Several models have been proposed to estimate global solar radiation. Angstrom (1924) was the first scientist known to suggest a simple linear relationship to estimate global solar radiation. Badescu (Badescu, 1999) studied existing relationships between monthly mean clearness index and the number of bright sunshine hours using the data ob- tained from Romania. Trabea and Shaltout (2000) studied the correlation between the measurements of global solar radiation and the meteorological parameters using solar radiation, mean daily maximum temperature, mean daily relative humidity, mean daily sea level pressure, mean daily vapour pressure and hours of bright sunshine data obtained from different parts of Egypt; while Sfetsos and Coonock (2000) used artificial intelligence techniques to forecast hourly global solar radiation. Okogbue and Adedokun (2002) estimated the global solar radiation at Ondo, Nigeria.

The main objective of the present study is to develop predictive models, using multi-variate regression analysis technique to predict solar radiation using four meteorological parameters, that is, minimum and maximum temperature, wind speed and evaporation for the study area and other locations having similar climatic characteristics in Nigeria. This will be useful in the development and the applications of solar energy technology.

METHODOLOGY

The monthly mean solar radiation, minimum and maximum temperatures, wind speed and evaporation data were obtained from the Archives of Nigerian Meteorological Agency, Oshodi, Lagos. The data obtained covered a period of twenty-six years (1970-1995). The stations were grouped according to their respective weather condition as shown in Figure 1. The distributions of the solar radiation with each of the atmospheric parameters were observed. It should be noted here that the solar radiation data were measured using Gun-Bellani distillate. Measurements were taken in millilitres (ml) which were converted to useful format

(MJ/m^2) using a conversion factor proposed by Folayan (Folayan, 1988), which is 1 ml = 1.357 MJ/m^2

Relevant multivariate linear regression analysis theory

Multivariate regression analysis is a technique for modeling and analyzing several variables when the focus is on the relationship between a dependent variable and one or more independent variables. For instance, let z1, z2 zr be a set of r predictors believed to be related to a response variable Y. The multivariate linear regression model for the jth sample unit has the form:

$$Y = \alpha + \beta_1 z_{j1} + \beta_2 z_{j2} + \dots + \beta_r z_{jr} + \varepsilon_j \qquad 1$$

where ε_j is a random error, the β_i where i = 1, 2, r are unknown parameter estimates and α is the intercept. With n independent observations, one model can be written for each sample unit or can be organized into vectors and matrices so that the model now becomes

$$Y = Z\beta + \varepsilon \qquad 2$$

where

$$Y = \begin{pmatrix} y_1 \\ y_2 \\ \cdot \\ \cdot \\ \cdot \\ y_n \end{pmatrix}, \quad Z = \begin{pmatrix} z_{j1} \\ z_{j2} \\ \cdot \\ \cdot \\ \cdot \\ z_{jn} \end{pmatrix} = \begin{pmatrix} z_{11} & \cdot & \cdot & \cdot & z_{j1} \\ z_{21} & \cdot & \cdot & \cdot & z_{j2} \\ \cdot & \cdot & \cdot & \cdot & \cdot \\ \cdot & \cdot & \cdot & \cdot & \cdot \\ \cdot & \cdot & \cdot & \cdot & \cdot \\ z_{n1} & \cdot & \cdot & \cdot & z_{nr} \end{pmatrix}, \quad \beta = \begin{pmatrix} \beta_1 \\ \beta_2 \\ \cdot \\ \cdot \\ \cdot \\ \beta_n \end{pmatrix}, \quad \varepsilon = \begin{pmatrix} \varepsilon_1 \\ \varepsilon_2 \\ \cdot \\ \cdot \\ \cdot \\ \varepsilon_n \end{pmatrix} \qquad 3$$

Least squares estimation

Ordinary least squares (OLS) estimates are commonly used to analyze both experimental and observational data. The OLS method minimizes the sum of squared residual, and leads to a closed-form expression for the estimated value of the unknown parameter β:

$$\hat{\beta} = (Z'Z)^{-1} Z'Y = \left(\frac{1}{n}\sum z_j z'\right)^{-1}\left(\frac{1}{n}\sum z_j y_j\right) \qquad 4$$

where $\hat{\beta}$ denote the least squares estimate of β and Z^0 denotes the transpose of Z.

RESULTS AND DISCUSSION

Distribution of solar radiation with meteorological variables

Figures 2 to 5 show the distribution of solar radiation with meteorological variables, which are, the minimum and maximum temperatures, wind speed and evaporation for four stations, that is one from each of the four climatic regions in Nigeria.

In Sokoto, it was observed that minimum and maximum temperatures and evaporation increased with solar radiation between January and June and also between September and November. A downward trend was discernible between July-September as shown in Figures 2a-b. The increasing trends between January and June may be due to the fact that during these months, dry season condition is predominant in this station. The decreasing trend in July-September may be due to the presence of disturbances like cumulous cloud and cloud cluster which are significant enough to cause variation in weather (Adeyemi, 2004). During this period, all regions in Nigeria will be experiencing intense rainfall because the intertropical discountinuity (ITD) would have reached its maximum northern position. Wind speed on other hand shows an opposite trend to solar radiation during dry season. For instance, between January and March, solar radiation increases as wind speed decreases. However, wind speed and solar radiation show the same downward trend between June and August, the rainy season in the Sahelian zone (Figure 2c). This shows that intense rainfall decreases the intensity of solar radiation and the rate of wind flow in the zone. Meanwhile, there is increasing trends between September and November, another phase of dry season. This may due to the fact that the dry season period in the zone is characterized by mostly turbulent flow of wind. Solar radiation and evaporation show the same increasing and downward trends (Figure 2d). The increasing trends are discernible between January-May and September-November, the dry season period and downward trends between June and August, the rainy season period. This is in agreement with Graham et al. (2004) argument that the physics of evaporation shows that the evaporative demand of the atmosphere is directly dependent on the net radiation which is dependent on solar radiation.

In Yola, there are increasing trends in variations of solar radiation with temperatures, wind speed and evaporation between January-April and September-December, the dry season period (Figures 3a-d). This may due to the topography of the zone that is characterized by short grasses and scattered drought-resistant tree which aids incessant surface heating that caused water vapour to be transported to the higher layers of the atmosphere through buoyancy (Adeyemi et al., 2004; Aro, 1975). Also, solar radiation shows downward trends with the four meteorological variables between April and August, the rainy season period in the zone. This may likened to the presence of the localized convection due to the usual long period of humid condition in the zone. In Ibadan, the dry and the wet season occurrence are greatly influenced by its latitudinal location and solar radiation has considerable seasonal variations. There are downward trends between March and September in the distribution of solar radiation with temperatures and evaporation as shown in Figures 4a, b and d, that is, during the rainy season period. This may due to the fact that these months being the core rainy months, are characterized by incessant cloud formation and thereby depleting the amount of solar radiation reaching the

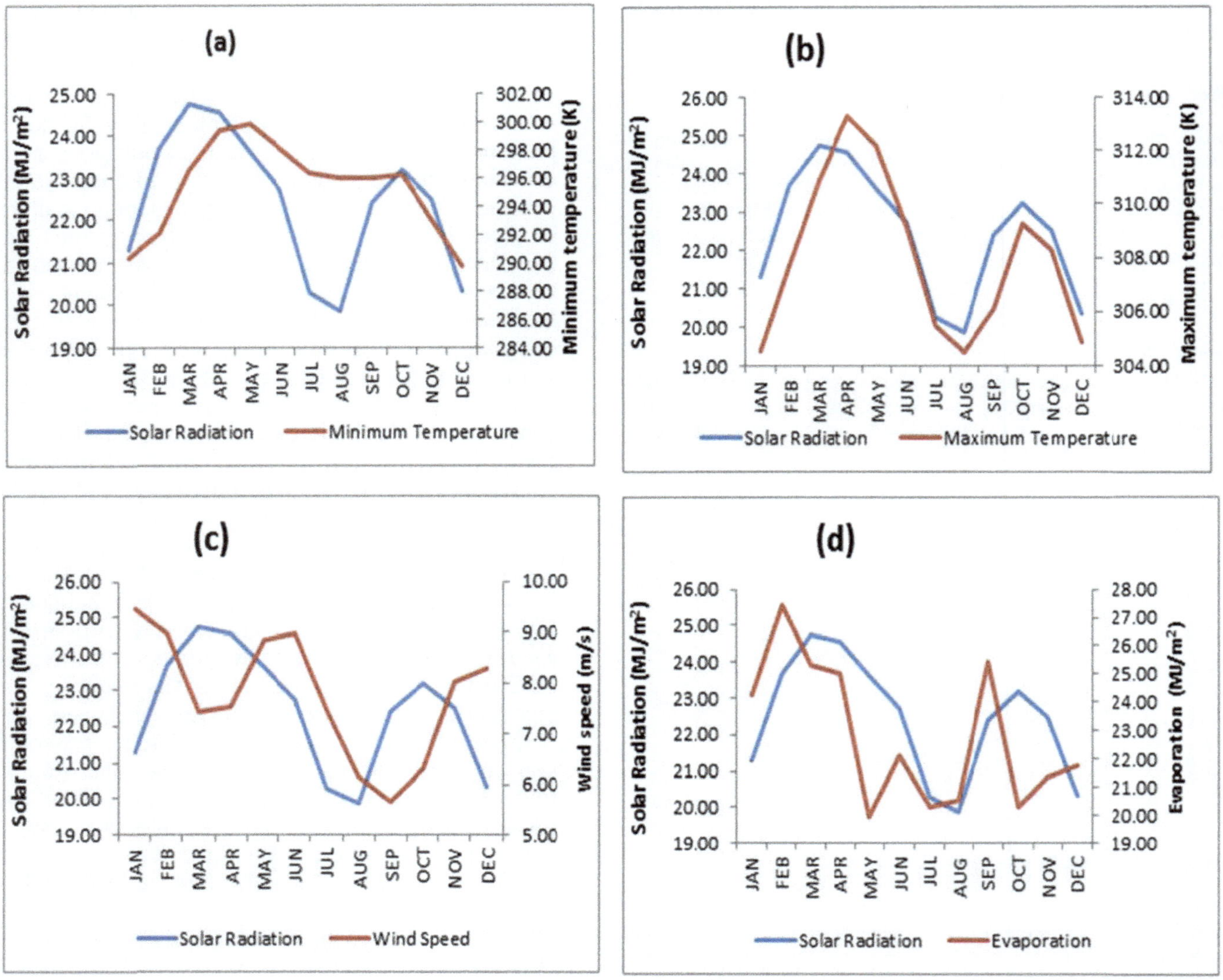

Figure 2. Distribution of (a) solar radiation and minimum temperature (b) solar radiation and maximum temperature (c) solar radiation and wind speed (d) solar radiation and evaporation at Sokoto.

Earth's surface (Ajayi and Adeyemi, 2009). The increasing trends are observed between October and March. This may due to the prevalence of the dry CT air which characterizes the dry season. The period is characterized by low humidity and high rate of evaporation aided by the increase in solar radiation and in air temperature in the zone. On other hand, there is a downward trend in the distribution of solar radiation with wind speed between June and August (Figure 4c). This may be due to the fact that the zone is under the influence of moist maritime south-western monsoon wind which blows inland from the Atlantic Ocean. The dry season occurs from November to February during by the dry dust laden winds blow from the Sahara desert. This period is characterized by low humidity and high rate of evaporation (Ajayi and Adeyemi, 2009). There is a general decreasing trends in solar radiation between November and January in Sokoto, Yola and Ibadan respectively (Figure 2 to 4a-d). These are expected because November and January are harmattan months when aerosol mass loading greatly reduces the intensity of solar radiation (Babatunde and Aro, 2000).

In Port-Harcourt, solar radiation has two peak values in February and November (Figures 5a-d). This is expected because of a very high sunshine hour which is obtainable in these months due to high clearness index (Aro, 1975). Least solar radiation is observed in the months of July and August, respectively. This is expected because heavy rainfall characterizes these months in the station. Therefore, the total solar radiation recorded is quite low because of the wet atmosphere and the presence of heavy clouds. If the weather is cloudy, the solar radiation value would be largely affected.

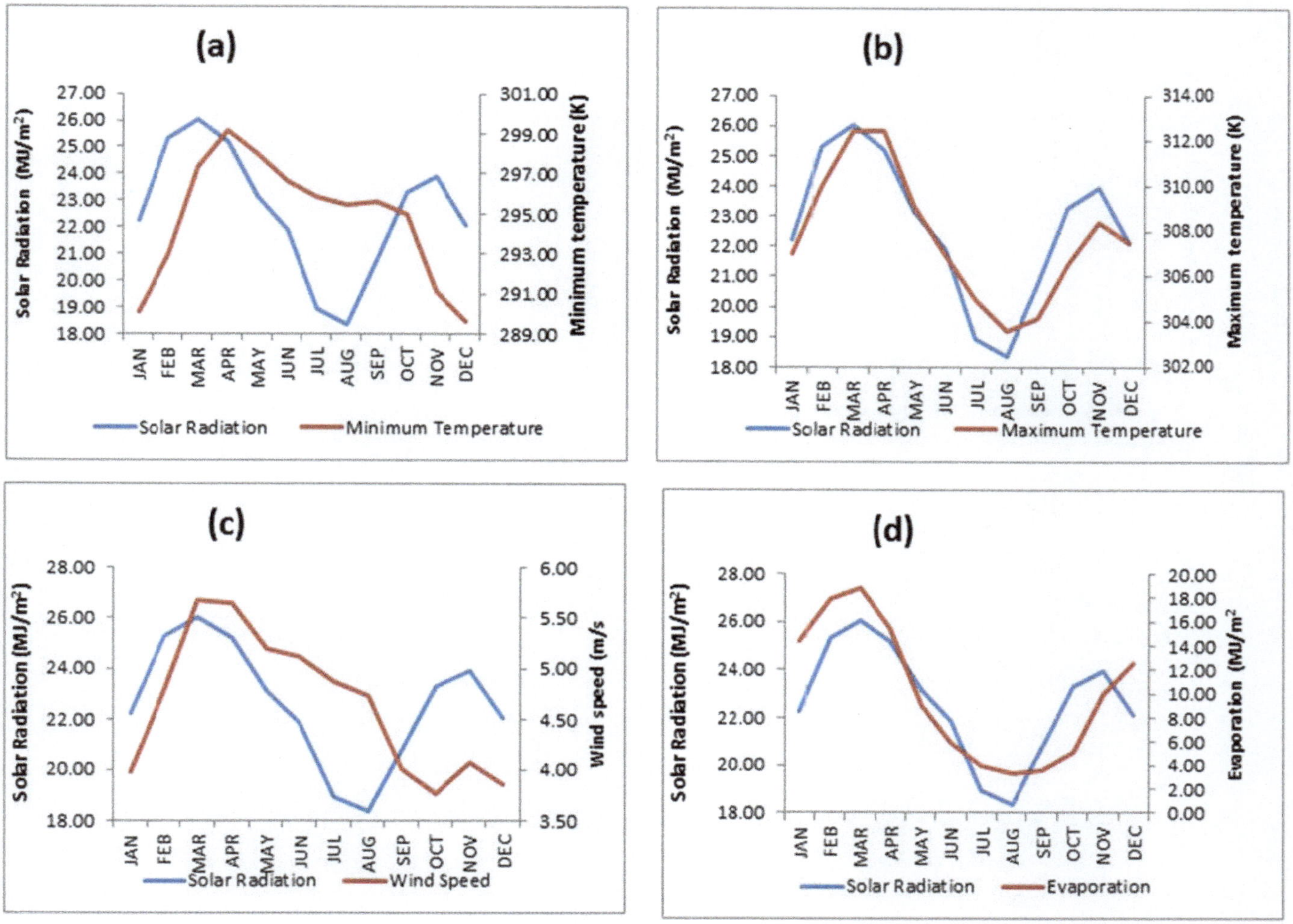

Figure 3. Distribution of (a) solar radiation and minimum temperature (b) solar radiation and maximum temperature (c) solar radiation and wind speed (d) solar radiation and evaporation at Yola.

This is because the solar radiation at normal incidence received at the surface of the earth is subjected to variations due to change in the extraterrestrial radiation and also the two significant phenomena which are the atmospheric scattering by air molecules, water and dust and the atmospheric absorption by O_2, H_2O, CO_2 (Lam et al., 2002). Wind speed increases with solar radiation between November and February while it shows downward trends between March and August in the station as shown in Figure 5c. This may due to excessive rainfall in these months in the zone.

Model development

The multivariate linear regression (MLR) modeling technique was then applied and a model of the form as shown in Equation 6 was developed for all stations using 1970-1990 data.

$$H = \alpha + \beta1Tmin + \beta2Tmax + \beta3Ws + \beta4\,Ep + \varepsilon \qquad 5$$

where α is the regression constant, ε is the error term, β_1, β_2, β_3, and β_4 are parameter estimates of minimum temperature (T_{min}), maximum temperature (T_{max}), evaporation (E_p) and wind speed (W_s) respectively for each of the stations.

The parameter estimates for the empirical model for evaluating solar radiation for each of the fourteen stations are shown in the Table 1. For instance, Sokoto on Lat. 13.5°, Long. 5.25° model is:

$$H = -83.783 - 0.184Tmin + 0.526Tmax - 0.064Ws + 0.079Ep \qquad (6)$$

Model testing and assessment

The models were then validated using surface data of1991-1995 for all the parameters. The performance of the developed models was tested with coefficient of determination (R^2), root mean square error (RMSE)

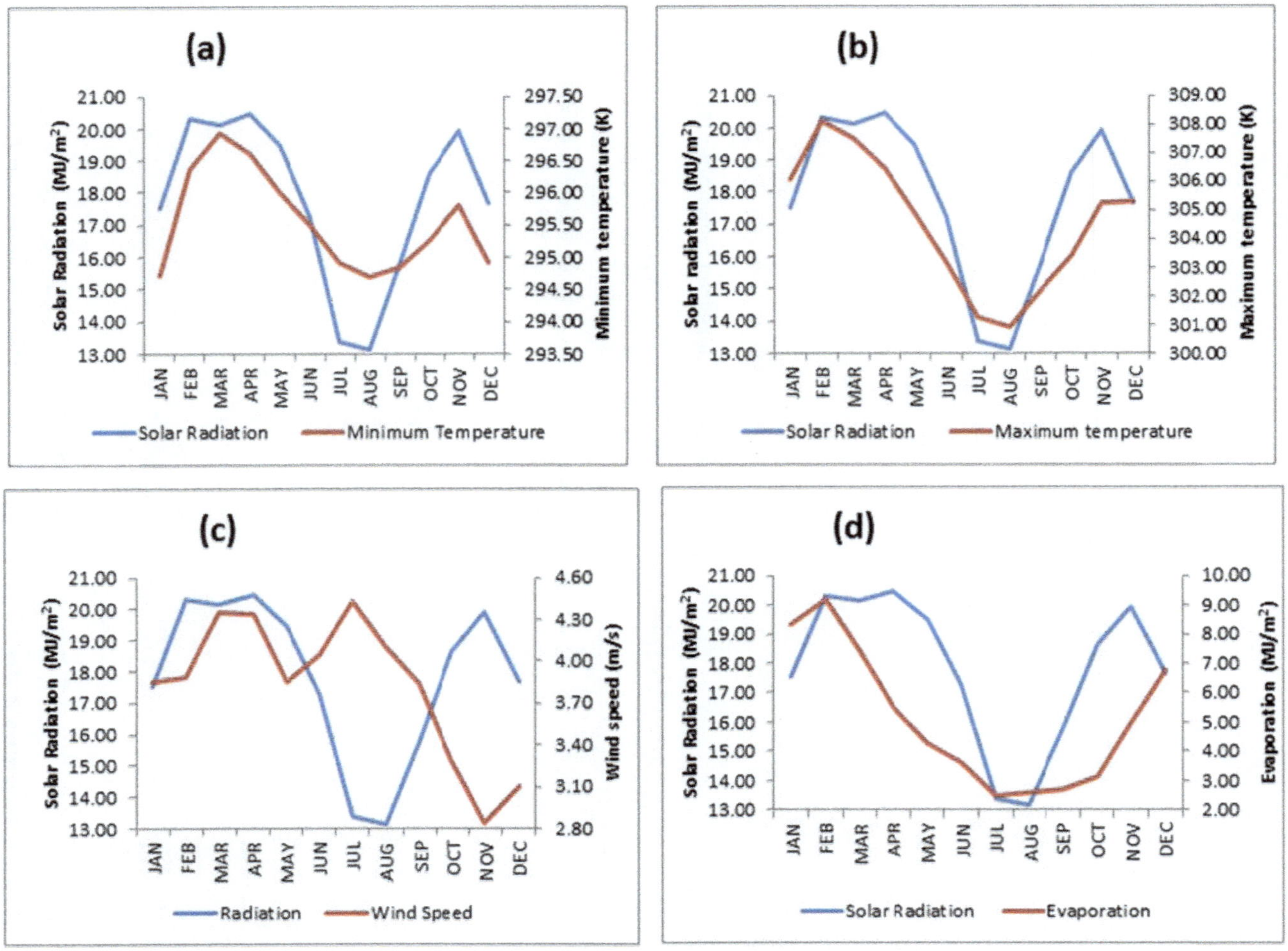

Figure 4. Distribution of (a) solar radiation and minimum temperature (b) solar radiation and maximum temperature (c) solar radiation and wind speed (d) solar radiation and evaporation at Ibadan.

and mean bias error (MBE), mean percentage error (MPE) and their values are shown in Table 2.

The expressions for evaluating the value of MBE, RMSE and MPE as stated by Okundamiya and Nzeako (2010) are shown in Equations (8-10):

$$MBE = \frac{1}{N}\sum_{n=1}^{n}(H_p - H_a) \quad (7)$$

$$RMSE = \sqrt{\frac{1}{N}\left[\sum_{n=1}^{n}(H_p - H_a)^2\right]} \quad (8)$$

$$MPE = \frac{1}{N}\sum_{n=1}^{n}\left[\frac{H_a - H_p}{H_a} \times 100\right] \quad (9)$$

$$t-test = \sqrt{\frac{(P-1)(MBE)^2}{(RMSE)^2 - (MBE)^2}} \quad (10)$$

where H_p and H_m is the predicted and measured values of solar radiation and N is the total number of observations. These indicators are mainly employed for adjustment of solar radiation data (Halouani and Ngguyen,1993; Falayi and Rabiu, 2005; Okogbue and Adedokun, 2002; Al-Salihi et al., 2010). The models developed were applied to evaluate solar radiation at the stations. The result of these was compared with the actual values obtained for all stations which are the average as shown in Table 2. The agreements are remarkable.

From Table 2, the values of the coefficient of determination (R^2) which show the proportion of variance accounted for by the correlation measured values with predicted values and the significance of the developed models to predict the solar radiation vary from 0.3017-0.7655 for the selected stations. The graphical demonstration of the correlation of measured and predicted values are shown in Figures 6 to 8. There are good agreement between the predicted values obtained from the models and the actual measured values of the surface data of solar radiation. The values of the mean bias error (MBE) indicate the average deviation of the predicted values from the actual measured values.

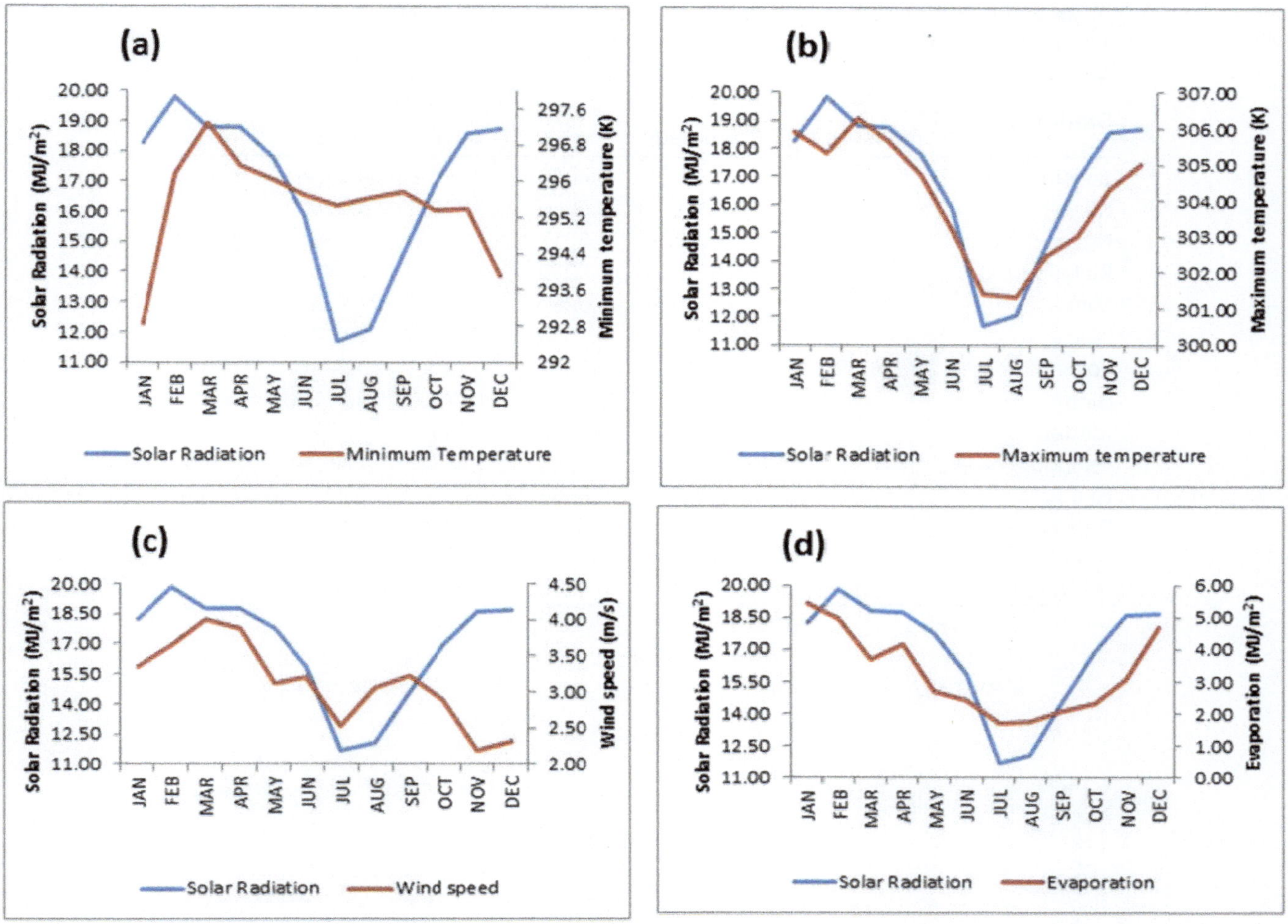

Figure 5. Distribution of (a) solar radiation and minimum temperature (b) solar radiation and maximum temperature (c) solar radiation and wind speed (d) solar radiation and evaporation at Port-Harcourt.

Table 1. Multivariate parameter estimates of the models for the studies stations.

Station	Location		Models' parameters estimates					
	Lat (°N)	Lon (°E)	α	β1	β2	β3	β4	p-value
Sokoto	13	5.25	-83.7827	-0.1837	0.5256	0.0642	-0.0797	0.0000
Kano	12.07	8.43	-120.731	0.0149	0.4593	0.0863	-0.0196	0.0000
Nguru	12.91	5.30	-99.1411	-0.1119	0.5013	0.4478	0.0358	0.0000
Maiduguri	11.91	13.17	-51.7748	0.4025	-0.1448	-0.3399	0.3799	0.0000
Minna	9.62	6.53	-204.529	-0.4197	1.1483	0.2715	-0.21688	0.0000
Jos	9.87	8.52	-192.513	-0.2533	0.9523	0.0370	0.08928	0.0000
Bida	9.08	6.02	-165.524	-0.3573	0.9558	0.3479	-0.06878	0.0000
Yola	9.23	12.47	-148.911	0.0379	0.5248	-0.4690	0.099828	0.0000
Osogbo	7.73	4.48	-225.887	-0.3327	1.1239	0.4135	-0.0383	0.0000
Ibadan	7.43	3.9	-208.052	-0.5119	1.2417	0.1476	-0.0944	0.0000
Ikeja	6.58	3.33	-227.319	-0.4350	1.2284	0.3887	-0.0968	0.0000
PH	4.85	7.02	-212.561	-0.3019	1.0498	0.3169	0.0216	0.0000
Benin	6.32	5.6	-205.085	-0.5191	1.2330	0.4416	-0.0476	0.0000
Enugu	6.37	7.55	-216.977	-0.4366	1.2001	0.0569	-0.0753	0.0000

Table 2. Application of the multivariate proposed model for each of the fourteen stations using 1991-1995 data.

Station	Solar radiation		Performance estimates			
	Measured	Predicted	R^2	MBE	RMSE	MPE
Sokoto	23.20	22.93	0.4086	-0.2717	1.3186	0.8668
Kano	24.74	24.96	0.5600	0.2248	1.2282	-1.1148
Nguru	25.98	24.37	0.3017	-1.2092	2.2361	4.3238
Maiduguri	24.20	25.78	0.5350	1.5808	2.2751	-6.4854
Minna	22.59	23.30	0.6209	0.7120	1.9751	-3.8081
Jos	22.63	21.72	0.7655	-1.9094	2.0708	3.0847
Yola	24.35	23.30	0.6321	-0.4261	1.9346	0.9346
Bida	21.51	22.59	0.6533	1.0763	2.1929	-6.1701
Ibadan	18.12	19.26	0.7132	1.1458	1.9122	-7.2537
Osogbo	17.77	19.15	0.4148	1.3834	2.5899	-9.5505
Enugu	17.03	19.30	0.6604	2.6714	3.1111	-17.150
Benin	15.92	18.45	0.5409	2.6234	3.0702	-17.265
Ikeja	15.19	18.45	0.3225	-2.6345	3.4697	-18.682
Port-Harcourt	16.75	18.98	0.5314	2.2307	3.0028	-15.489

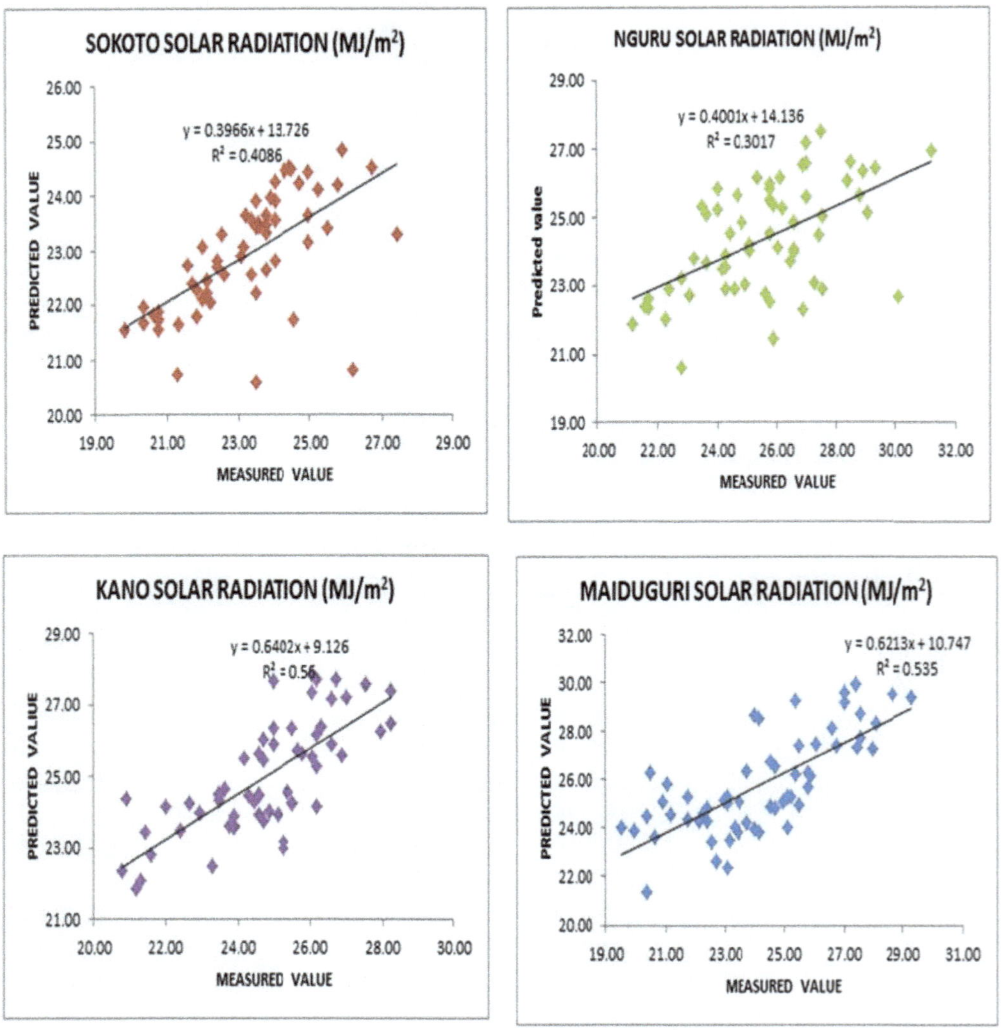

Figure 6. The correlation between the predicted and measured values of the monthly solar radiation Sokoto, Kano, Maiduguri and Nguru Stations in Nigeria.

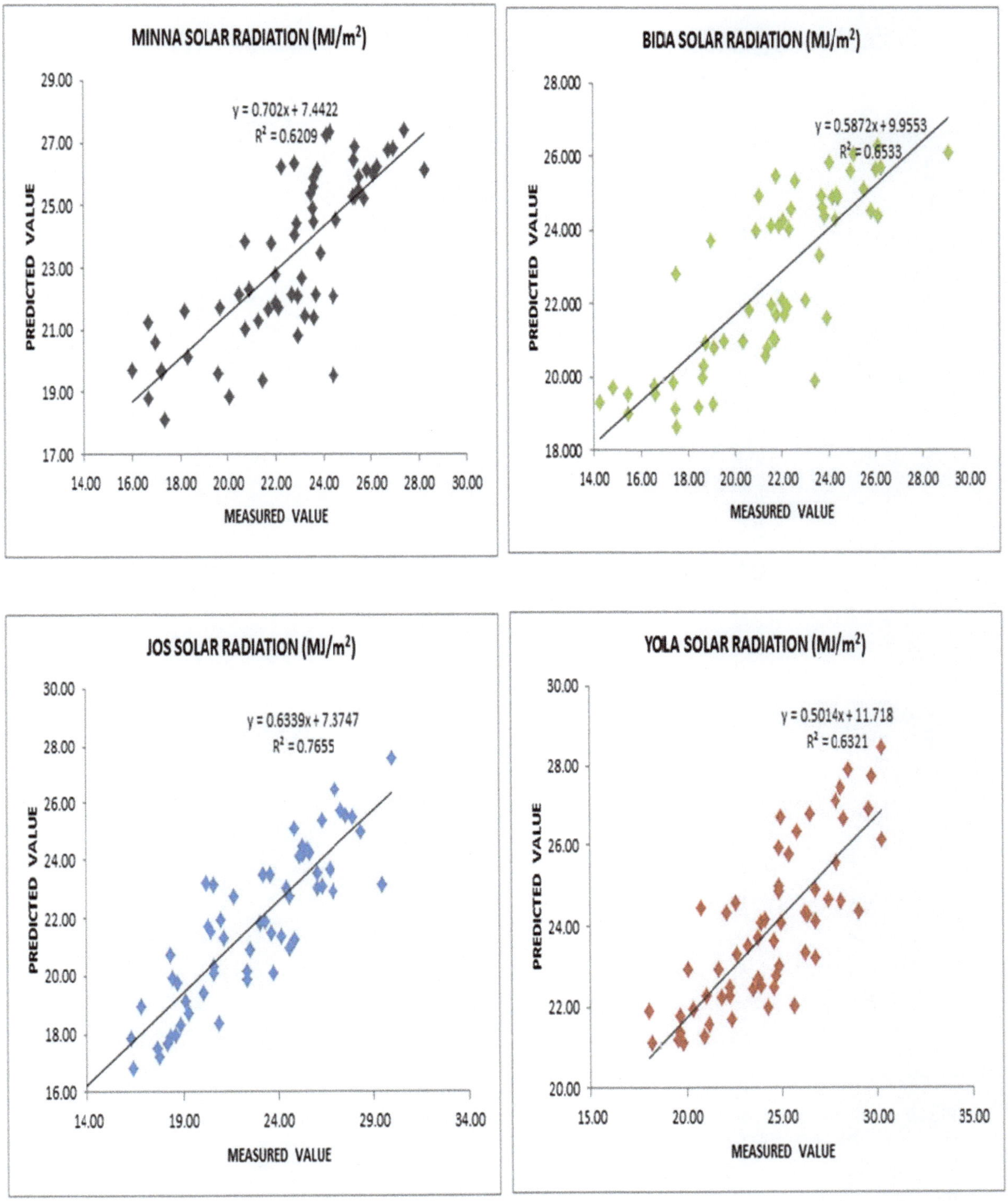

Figure 7. The correlation between the predicted and measured values of the monthly solar radiation for Bida, Minna, Jos and Yola Stations in Nigeria.

The MBE values vary from negative values to the positive values (lowest for Ikeja and highest for Benin). The low values of MBE generally show that the developed models are significant for predicting solar radiations. The MBE values are suitable for the models for each of the fourteen stations (Okundamiya and Nzeako, 2010).

The root mean square error (RMSE) values indicate a measure of the variation of predicted values from the

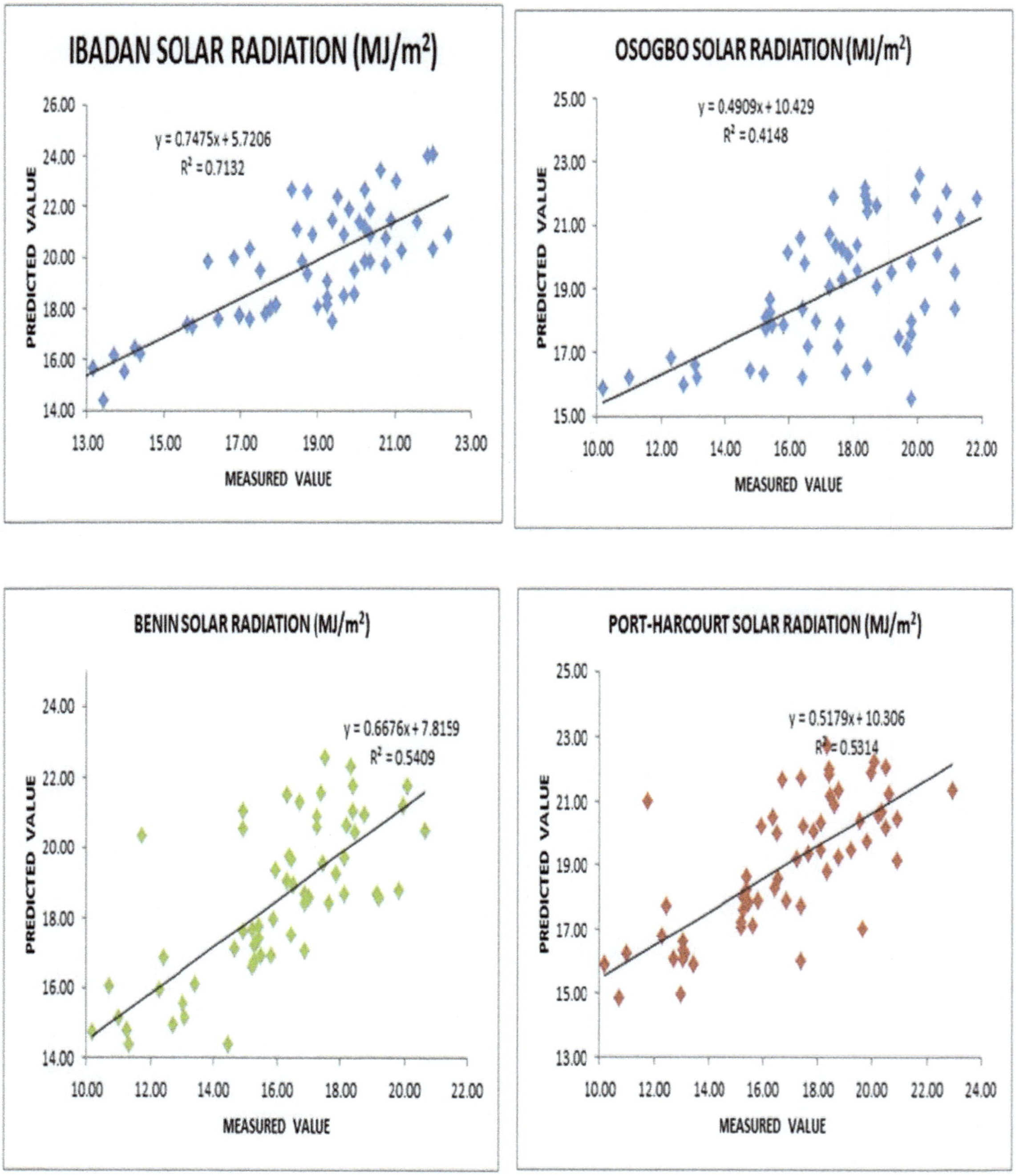

Figure 8. The correlation between the predicted and measured values of the monthly solar radiation for Osogbo, Ibadan, Port-Harcourt and Benin Stations in Nigeria

actual measured values. The RMSE values vary from negative values to the positive values (lowest for Kano and highest for Ikeja). The low values of RMSE show that term by term comparison is allowed for the developed models. The RMSE values are suitable for the models for each of the fourteen stations. Therefore, term by term comparison is allowed for all the developed models (Sanusi and Abisoye, 2011).

The values of mean percentage error MPE give long term performance of the developed models. They also vary from negative values to the positive values (lowest for Ikeja and highest for Nguru). The MPE values are suitable for the models for each of the fourteen stations. Therefore, all the developed models give long term performance (Babatunde and Aro, 2000).

Added to these, the graph of variations plotted for the measured solar radiation for each of the four climatic regions along with their predicted values showed that solar radiation was well monitored at all the regions by the models (Figures 9 to 11).

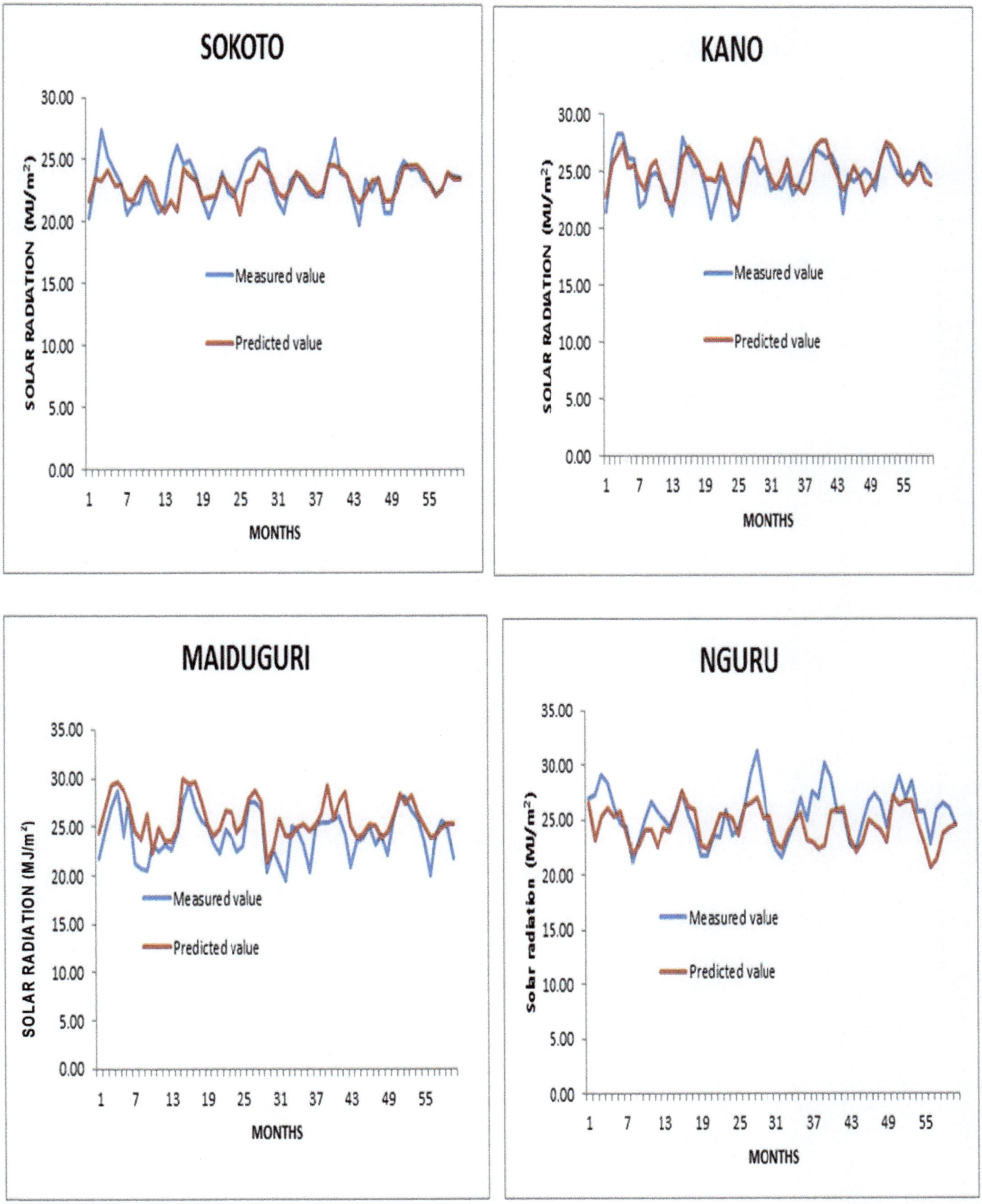

Figure 9. The variations of the predicted and measured values of the monthly solar radiation for Sokoto, Kano, Maiduguri and Nguru stations in Nigeria.

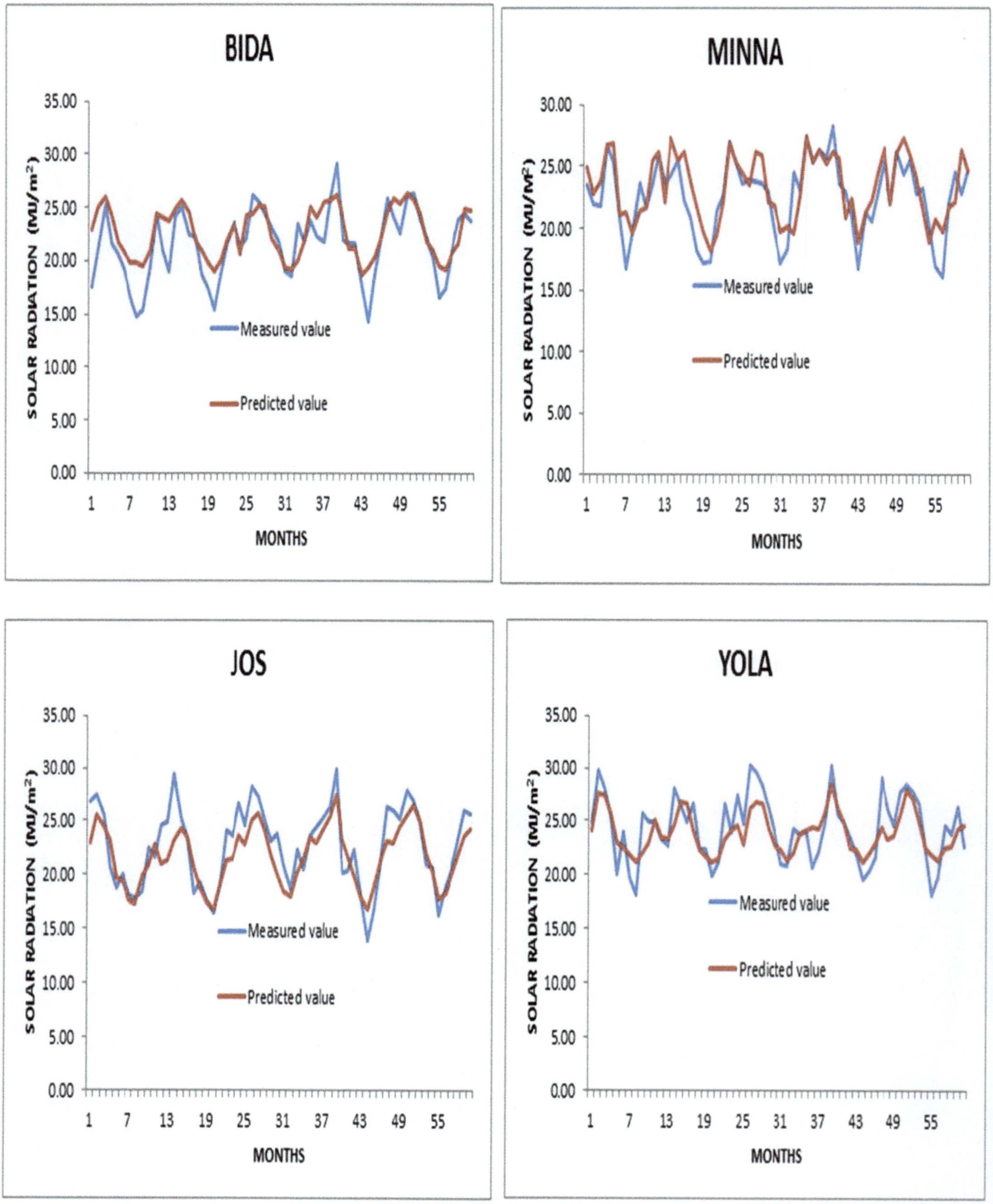

Figure 10. The variations of the predicted and measured values of the monthly solar radiation for Bida, Minna, Jos and Yola stations in Nigeria.

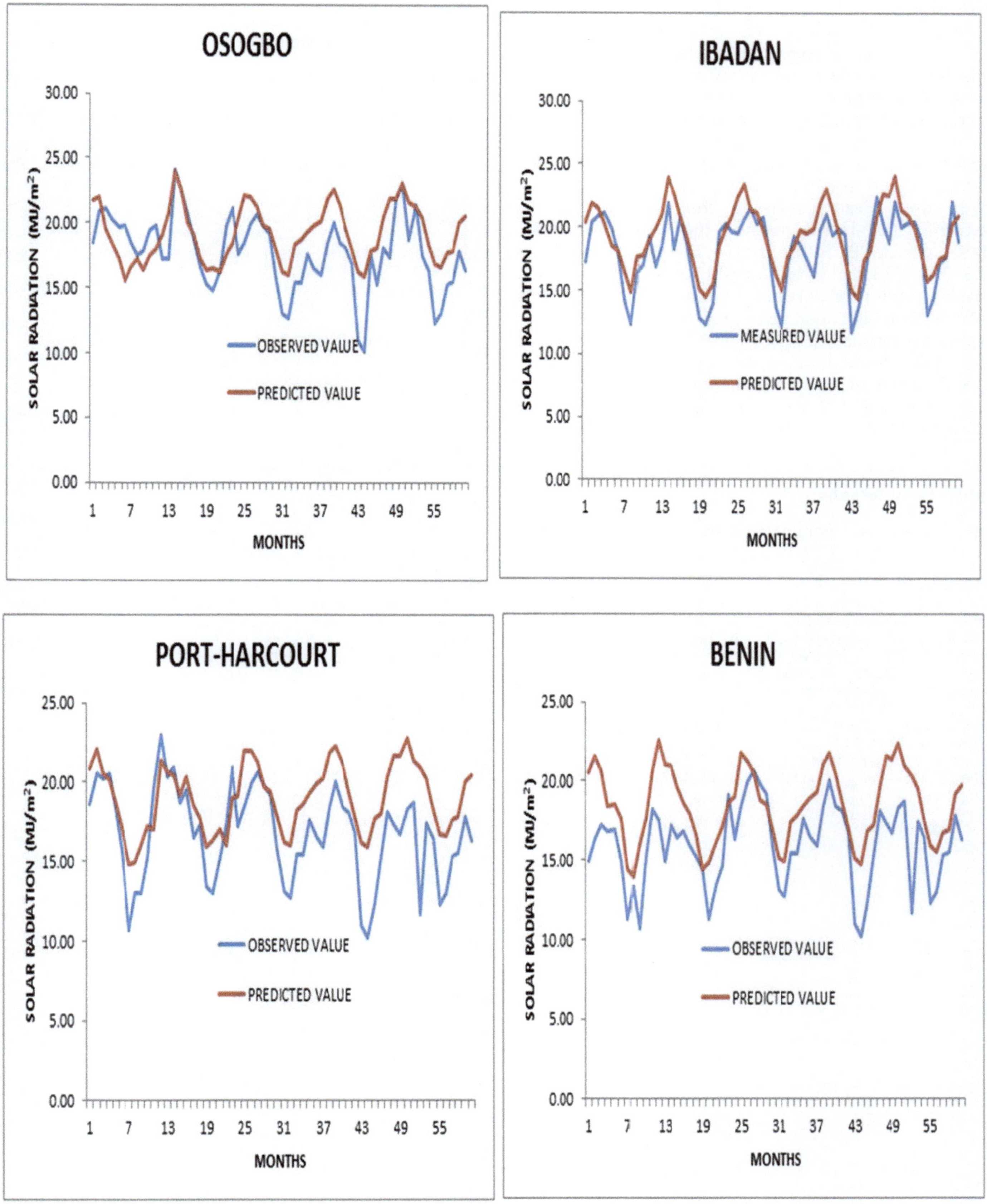

Figure 11. The variations of the predicted and measured values of the monthly solar radiation for Osogbo, Ibadan, Port-Harcourt and Benin Stations in Nigeria.

Conclusion

The distribution of solar radiation with meteorological variables for fourteen stations which represent climatic conditions in Nigeria shows that the linear combination of these variables can be used to predict solar radiation. Analysis have shown that strong correlation exists between solar radiation and all the meteorological parameters at the selected stations. The statistical indicators have also confirmed the efficiency of the developed models. For instance, the values of R^2, MBE, RMSE and MPE for Sokoto, a Sahelian station are 0.5349,-0.0273,0.4154 and 0.0811 respectively, showing that the model for Sokoto is significant. Those for other stations have also been calculated with all showing high level of significance at 0.05 alpha levels. Therefore, the multivariate linear models developed in this research work for each of the fourteen stations is suitable for estimating solar radiation data in the stations and their proximate cities.

Conflict of Interests

The author(s) have not declared any conflict of interests.

REFERENCES

Adeyemi B (2004). Modelling of ITD over Nigeria using Radio meteorological parameters of the troposphere. PhD Thesis, University of Ilorin, Nigeria.

Adeyemi B (2004). Tropospheric radio refractivity over three radiosonde stations in nigeria. Ife J. Sci. 6(2):167-176.

Ajayi OS, Adeyemi B (2009). Empirical modelling of surface refractivity over Akure using ground electrical conductivity. Nig. J. Pure Appl. Phys. 5:24-27.

Alam MS, Saha SK, Chowdhury MAK, Saifuzzaman M, Rahman M (2005). Sim- ulation of solar radiation system. Am. J. Appl. Sci. 2(4):751-758.

Al-Salihi AM, Kadum MM, Mohammed AJ (2010). Estimation of global solar ra- diation on horizontal surface using meteorological measurement for different cities in iraq. Asian J. Sci. Res. 3(4):240-248.

Angstrom A (1924). Solar and terrestrial radiation. Quart. J. Roy. Meteorol. Soc. 50:121-125.

Aro TO (1975). Analysis of data on surface and tropospheric water vapour. J. Atmos. Terr. Phys. 38:565-571.

Babatunde EB, Aro TD (2000). Variation characteristics of diffuse solar radiation at a tropical station, Ilorin, Nigeria. Niger. J. Phys. 12:20-24.

Badescu V (1999). Correlations to estimate monthly mean daily solar global irradiation. Appl. Romania Energy 24:883-893.

Burari FW, Sambo AS (2001). Model for the prediction of global solar radiation for bauchi using meteorological data. Nigeria J. Renew. Energy 91:30-33.

Exell RHD (2000). The Intensity of Solar Radiation. King Mongkuts University of Tech- nology Press, Thonburi, Thailand.

Falayi EO, Rabiu AB (2005). Modelling global solar radiation using sunshine duration data. Niger. J. Phys. 17S:181-186.

Folayan CO (1988). Estimate of global solar radiation bound for some nigerian cities. Nig. J. Solar Energy 7:36-43.

Graham D, Farquhar GD, Michael LR (2004). Pan evaporation: An example of the detection and attribution of trend in climate variables. Proceedings of a workshop held at the shineome, Australian academy of Science, Canberra, pages 20-21, November 2004.

Halouani N, Ngguyen C (1993). of monthly average global solar radiation on horizontal surface using daily hours of bright sunshine. Solar Energy J. 50:247-258.

Lam JC, Li DHW, Lau CCS (2002). A new approach for predicting vertical global solar radiance. Renew. Energy 25(4):591-606.

Okogbue EC, Adedokun JA (2002). On the estimation of solar radiation at ondo, Nigeria. J. Phys. 14:97-99.

Okundamiya MS, Nzeako AW (2010). Empirical model for estimating global solar radi- ation in the six geopolitical zones in Nigeria. J. Appl. Sci. Eng. Technol. 2(8):805-812.

Sanusi YK, Abisoye SG (2011). Estimation of solar radiation at ibadan, nigeria. J. Emerg. Trends Eng. Appl. Sci. 2(4):701-705.

Sfetsos A, Coonock AH (2000). Univariate and multivariate forecasting of hourly solar radiation with artificial intelligence techniques. Solar Energy 68:169-178.

Trabea AA, Shaltout MA (2000). Correlation of global solar-radiation with meteoro- logical parameters over Egypt. Renew. Energy 21:289-291.

Evaluation of groundwater suitability for drinking, domestic and irrigational purposes: A case study of Makurdi metropolis and environs, Benue State, North Central Nigeria

T. I. Akuh, S. A. Alagbe and A. A. Ibrahim

Department of Geology, Ahmadu Bello University, Zaria, Kaduna State, Nigeria.

A study of groundwater samples from Makurdi metropolis and environs was carried out to assess its suitability for drinking, domestic and irrigational purposes. The chemistry of groundwater within the study area with respect to pH, total dissolved solids (TDS), NH_3^-, NO_3^-, Cl^-, Na^+, K^+, Ca^{2+}, Mg^{2+}, HCO_3^-, SO_4^-, Fe^{3+}, Cu^{2+}, Mn^{2+} and Zn^{2+} was used to determine its suitability for drinking. Some parts of the study area have elevated concentration of TDS, magnesium, nitrate, sulphate, copper, manganese and iron and which are considered unsafe for drinking and some domestic uses based on the standard organization of Nigeria (SON, 2007) and the World health organization (WHO, 2011) guideline concentration for drinking water. Sodium adsorption ratio (SAR), percentage sodium (Na%) and bicarbonate hazard have been studied to evaluate the suitability of the groundwater for irrigation purpose. This study show that groundwater within the area is safe for irrigation with respect to sodium adsorption ratio but a good number of the samples had high values of bicarbonate hazard and percentage sodium, and are unsuitable for irrigation.

Key words: Irrigation, groundwater, Makurdi metropolis, suitability.

INTRODUCTION

Groundwater is the largest source of fresh water in the world. It is of major importance to civilization, being the largest reserve of natural drinkable water that can be used by humans. Groundwater occurs in almost all geological formations and the natural filtering effect of these geologic materials often makes groundwater preferable to surface water for many purposes.

The chemical composition of groundwater is related to the soluble products of rock weathering, decomposition and dissolution of organic matter and changes over time (Raghunath, 2007). The chemical quality of groundwater depends on the characteristics of the soil and rocks through which it has percolated and also on its resident time in these rocks. Some constituents of groundwater are also derived from the atmosphere as rain takes some atmospheric gases and ions into solution. Others have resulted from anthropogenic sources such as agriculture, industrial and improper disposal of refuse and sewage which often results in groundwater pollution. Water pollution in Makurdi metropolis and environs results mainly from human activities which results in the deterioration of the physical, chemical and biological properties of water (Akuh, 2014). The population depends mostly on groundwater abstracted through hand-dug wells and boreholes as their source of domestic water supply, farmers in the rural areas also depend on groundwater from wells to irrigate their small farms during the dry season. For effective management of the available water resources, an evaluation of the quantity as well as the quality of water for various purposes has to be carried out. Irtwange and Sha'ato (2009) reported 0.22 mg/l TDS, 60 mg/l Cl^-, 28.81 mg/l SO_4^{2-}, 1.36 mg/l Zn^+ and 200 mg/l Na^{2+} for groundwater from a borehole located at a solid waste disposal site in Makurdi. Idoko (2010) reported iron concentration ranging from 0.03 to 2.38 mg/l for groundwater in Benue State during the rainy season and values ranging from 0.01 to 5.02 mg/l during the dry season. He attributed the elevated iron concentration to geology and confirmed higher concentration in areas with lateritic soil. Edet et al. (2011) reported concentrations of 6136.75 mg/l Na^+, 243.15 mg/l K^+, 390.58 mg/l Ca^{2+}, 70.80 mg/l Mg^{2+}, 10278.75 mg/l Cl^- and 406.05 mg/l HCO_3.

This study was conducted to determine the physicochemical properties of groundwater in Makurdi metropolis and environs and evaluate its suitability for drinking, domestic and irrigational purposes.

The study area

Makurdi is located in Benue State, north central Nigeria. It is bounded by latitudes 8°30'00" to 8°41 '00"N and longitudes 7°40 '00" to 7°50' 00"E (Figure 1). It is located within the middle Benue Trough and covers an area of about 370 km^2. It is accessible by the Makurdi-Lafia road and intra state roads such as the Makurdi-Otukpo road, Makurdi-Gboko road as well as footpaths with many rural road networks linking the rural areas to Makurdi town. The annual rainfall depth ranges from about 1,200 to 1,500 mm with an average depth of about 1350 mm. Temperatures are generally very high during the day, particularly in March and April. Along the river valleys, these high temperatures together with high relative humidity produce debilitating weather conditions. Makurdi records average maximum and minimum daily temperatures of 35 and 21°C during the rainy season and 37 and 16°C during the dry season, respectively (Lower Benue River Basin Development Authority).

Geology and hydrogeology of the study area

Much of Benue State within which Makurdi is located falls within the intra cratonic sedimentary basin known as the Benue Trough which is believed to be structurally developed. The Benue Trough is a linear sedimentary basin which is filled with Cretaceous rocks whose ages range from Middle Albian to Maastrichtian. It is bounded on either side by granite and gneisses which make up the crystalline basement (Cratchley and Jones, 1965). Many others like Najime (2010), Obaje (1994) and Zaborski (1998) described the Benue Trough as a NE-SW trending intracratonic basin ranging between 150 km wide and 800 km long. Lithologies identified in the study area (Makurdi metropolis and environs) include: the Ezeaku shale, the Makurdi Formation, Awe formation, alluvial deposit and pockets of basalt.

The Ezeaku Formation consists of thick flaggy calcareous and non-calcareous shale, sandy or shally limestone and calcareous sandstones. It is reported to be 304.8 m thick in this region and ranges up to 609.6 m towards the south of the state (Offodile, 1989).

The Makurdi formation consists of a thick mass of current bedded coarse grained deposit described as the Makurdi Sandstone (Figure 2) Geological and mineral resources map of Benue State adopted from Nigerian Geological survey agency (NGSA) 2006. Akuh (2014) described this sandstone as well sorted, medium to coarse grained, grain supported and texturally immature subarkose with minor clay rim cement and 3 - 4% porosity. It is highly indurated and cemented mainly by quartz and iron oxide (Adanu, 1981). The Makurdi Sandstone is the most important aquifer in the middle Benue trough. This aquifer is almost impermeable in places; especially the western part of Makurdi metropolis where its compact and indurated nature makes it difficult for the inhabitants to sink hand dug wells into it as the wells are terminated when the sandstone is encountered. Where fractured or less indurated, the formation is usually less compacting, more permeable and has better prospect as an aquifer (Offodile, 2002). Adanu (1981) obtained hydraulic conductivity values of 0.132, 1.73, 3.5 and average porosities of 17.3, 27.8 and 31.3% for various samples of the Makurdi sandstone. These properties influence the rate at which this aquifer is recharged.

The usefulness of the Makurdi Sandstone as a potential groundwater reservoir depends on its secondary permeability derived from weathering and fracturing (Offodile, 2002). It reported yields of 8.2 L/s from a borehole drilled into the Makurdi Sandstone at the Army cantonment, yields between 2.5 and 3.9 L/s were obtained in Aliade and 8.7 L/s from the borehole at the Lower Benue River Basin Estate. Recharge into the aquifer is both direct and indirect. Where the aquifer is uncovered and permeable, the recharge is direct from precipitation, by seepage and along fractures and joints. This is the case in Yaiko and Airport village, indirect recharge to this aquifer is by infiltration through weathered materials overlying the Sandstone aquifer (Adanu, 1981).

The Awe Formation was deposited as passage

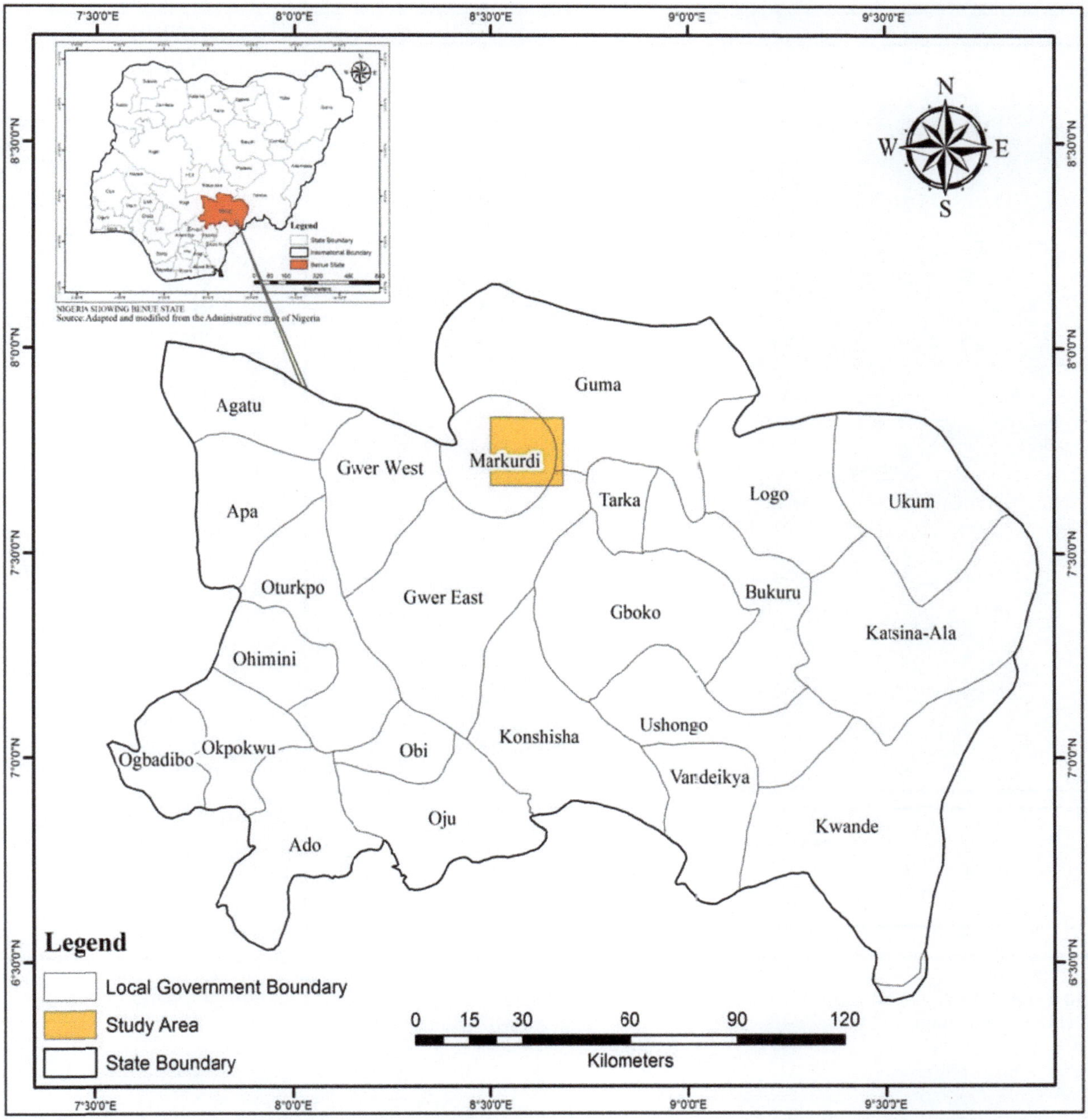

Figure 1. Map of Benue State showing the study area. Source: Adapted and modified from the admistrative map of Benue State State (2012).

(transitional) beds during the Late Albian-Early Cenomanian regression. This formation overlies the Asu River Group. Offodile (1976) estimated the thickness of the formation to be about 1000 m. The formation consists of flaggy, pale-colored medium to fine grained Sandstone with interbedded carbonaceous shale and clay (Offodile, 1976). The Sandstone beds, usually multi-layered appear highly porous and water yielding but the water from these aquiferous beds is contaminated by brines from the interbedded shales. The shale also confines the aquifer and present artesian to sub artesian situations (Offodile, 2002). A few boreholes drilled into this formation have produced good water yields, though often not potable due to the brine contamination.

The major flood plains of Benue State are those of Rivers Benue, Katsina Ala and other smaller rivers. Alluvial deposits, comprising an assortment of clays, sand, gravels and pebbles from these flood plains overlie the metamorphosed sediments and form the superficial geology.

The alluvium aquifer occurs along river and stream channels which are liable to seasonal flooding. It is the least extensive of all the aquifers. The thickness of the

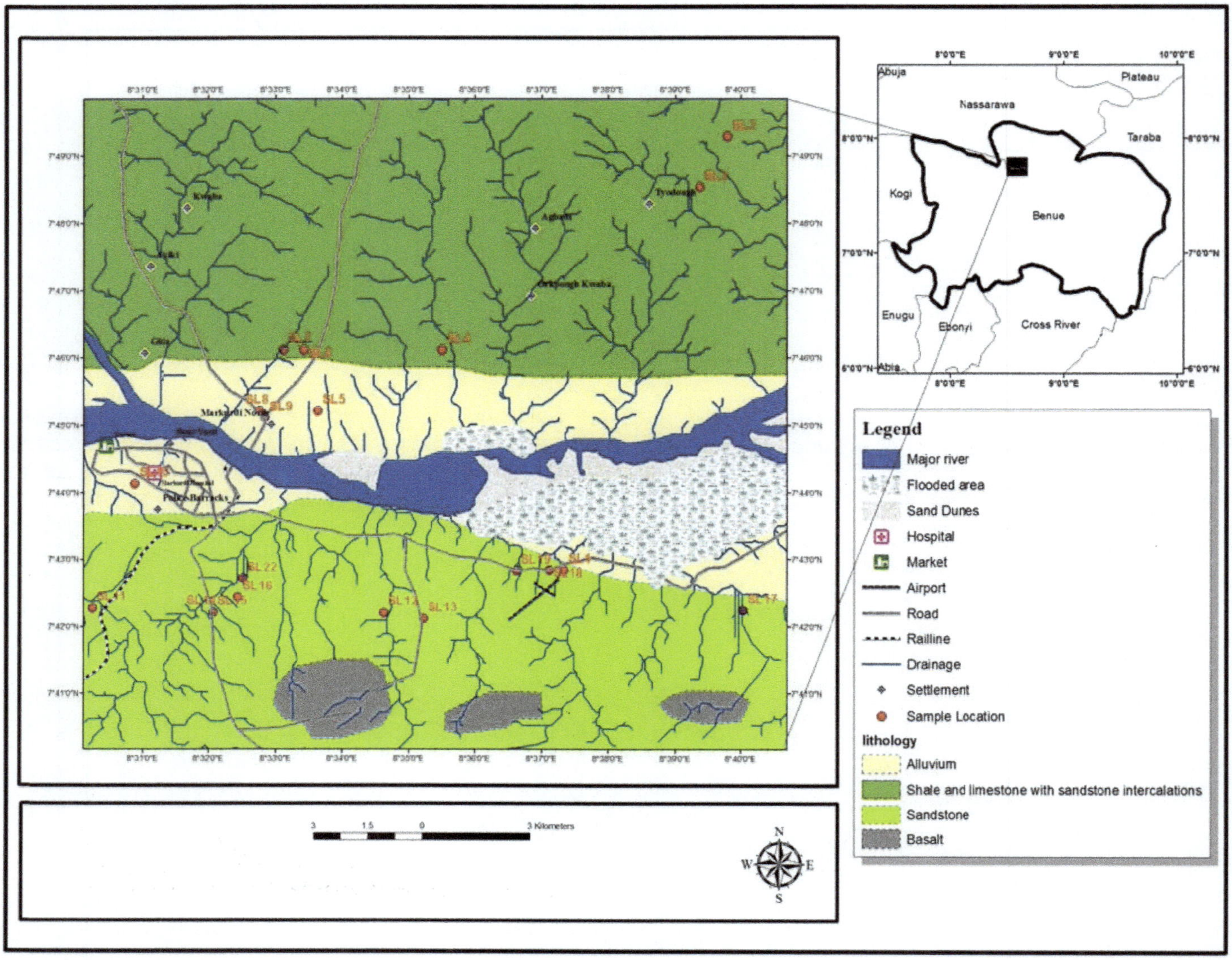

Figure 2. Geological map of Makurdi metropolis and environs adapted from Nigeria Geological Survey Agency (2006) Geological and Mineral Resources map.

aquifer ranges from 2 to 15 m, the thickest occur along the Benue River while the least thickness is along the stream channels (Adanu, 1981). It suggested watertable (unconfined) aquifer condition for the alluvium aquifer. Adanu (1981) also reported porosity and permeability coefficient for this aquifer as 37.5% and about 0.26 m/day respectively. The permeability and porosity value favors effective recharge into the aquifer. Recharge into this aquifer is mainly direct from annual precipitation and indirect recharge occurring when the river character is influent, thus supplying water to the alluvium aquifer (Adanu, 1981).

METHODOLOGY

Water samples were taken from four boreholes and 20 hand-dug wells at the peak of dry season (April, 2012) and stored in plastic sample bottles. The sample points are located as shown in Figure 2. The sample bottles were washed twice with the water to be sampled and filled to overflow, sealed tightly to avoid contamination with atmospheric gases. 500 ml of the sample was acidified to pH of 2 with concentrated nitric acid and another 500 ml non-acidified. A portion of the sampled water was acidified to keep the heavy elements in solution while in storage. *In situ* measurement of some physical parameters such as temperature, pH and total dissolved solids (TDS) was done. Chemical analysis of water samples was carried out at the Center for Energy Research and Training (CERT), Department of Water Resources Engineering Laboratory and Multipurpose Science Research Laboratory all in Ahmadu Bello University Zaria. Three analytical techniques were employed in determining the concentration of the chemical parameters of interest. These includes: photometric method (Zn, Fe, Cu, Mn2+, Cl, NH_3 and NO_3) using a Chemetric V – 2000 model multi – analyte photometer, atomic absorption spectrophotometry (Ca and Mg) titremetric techniques (HCO_3) and flame photometric method (Na and K). Values obtained were compared with maximum permissible limit set by the Standard Organization of Nigeria (SON, 2007) and

World Health Organization (WHO, 2011) to evaluate its suitability for drinking and domestic purpose.

The analytical data obtained were processed with aquachem 4.0 software for geochemical and statistical analysis. This software was used to plot Piper and Wilcox diagrams, deduce possible source of some chemical parameters and obtain correlation coefficient (Table 2) showing the relationship between the measured parameters. Basic statistical parameters such as minimum, maximum, mean, standard deviation and averages were also computed using aquachem 4.0 version software (Table 3).

Sodium adsorption ratio (SAR), percentage sodium (Na%) and bicarbonate hazard were used to evaluate the suitability of water for irrigation purpose. These parameters were determined using Equations 1, 2 and 3 respectively. Values obtained were compared with the classification of groundwater (Table 4).

RESULTS AND DISCUSSION

Physico-chemical parameters of groundwater

A summary of the physic-chemical analysis of the groundwater samples is presented in Table 1. pH values ranges from 2.90 to 7.26 with an average of 4.94 indicating acidic to slightly basic type of water. Only third (12.5%) of the 24 samples have values within the Standard Organization of Nigeria (2007) permissible range of 6.50 to 8.50 indicating that groundwater in most part of the study area is acidic. Total dissolved solids (TDS) ranges from 90 to 520 mg/l in boreholes and from 90 to 1250 mg/l in hand dug wells with an average of 489.55 mg/l. Chloride concentration ranges from 0.00 to 13.11 mg/l with an average of 2.92 mg/l. These values are far below the SON (2007) maximum permissible limit of 250.00 mg/l. No health-based guideline value is proposed for chloride concentration in drinking-water. Nitrate concentration ranges from 0.53 to 196.00 mg/l with an average of 42.61 mg/l. Four (16.7%) of the samples, SL7 and SL8 from North Bank, SL10 and SL11 from Wadata and George's quarters respectively have elevated concentrations of 107, 196, 96 and 117 mg/l, respectively above the maximum permissible limit of 50 mg/l (Table 1). Elevated nitrate concentration in groundwater has resulted from agricultural activities (for example, excess application of nitrogenous fertilizers and manures), from waste water disposal and oxidation of nitrogenous waste products in human and animal excreta. Results show a range of 0.34 to 3.20 mg/l with an average of 1.06 mg/l for ammonia concentration in groundwater within Makurdi metropolis (Table 1).

Sulphate concentration in groundwater within the study area ranges from 5.00 to 500.00 mg/l with an average of 28.25 mg/l. Only two (4.5%) of the samples analyzed had sulphate concentration above SON (2007) maximum permissible limit of 100 mg/l. Sulphate in groundwater within the study area as deduced from the aquachem software is from pyrite oxidation. Other sources of sulphate in groundwater includes; atmospheric precipitation and from the solution of sulphate minerals in sedimentary rocks (Davis and Dewiest, 1966). Most part of the study area (about 95.5%) contains groundwater considered safe with respect to its sodium concentration. One location however, SL 8 at North Bank has elevated sulphate concentration (150 mg/l) and is therefore unsafe for drinking and domestic use. Sodium concentration ranges from 10.70 to 36 mg/l with an average of 17.24 mg/l. All samples analyzed had concentrations below SON (2007) maximum permissible limit of 200 mg/l. The source of sodium in groundwater in the study area has been traced to weathering of albite and ion exchange. Magnesium concentration in groundwater within the studied area ranges from 5.68 to 74.20 mg/l with an average of 23.32 mg/l. All the samples analyzed had concentration above SON (2007) permissible limit of 0.20 mg/l.

Guideline values have not been established for bicarbonate, potassium and calcium. Bicarbonate concentration in groundwater within the study area ranges from 90.90 to 1525.50 mg/l with an average of 555.97 mg/l and is derived from carbondioxide in the atmosphere, carbondioxide in the soil and dissolution of carbonate rocks. Potassium concentration ranges from about 1.80 to 94.50 mg/l with an average of 23.46 mg/l and calcium ranges from 2.78 mg/l to 144.74 mg/l with an average of 41.98 mg/l. Potassium results from weathering of orthoclase and microcline. The abundance of potassium and sodium is about the same but potassium is commonly less than one tenth of sodium in natural water because many potassium minerals have higher resistance to weathering (Davis and Dewiest, 1966). Calcium on the other hand, has resulted from the contact of water with sedimentary rocks of marine origin such as calcite, aragonite, dolomite and gypsum.

Concentration of iron (Fe^{3+}) ranges from 0.00 to 0.47 mg/l in boreholes and 0.00 to 0.86 mg/l in hand dug wells with an average of 0.27 mg/l. 15 (62.5%) of the samples analyzed were within the SON (2007) permissible limit of 0.30 mg/l and nine (37.5%) samples had elevated iron concentration above the permissible limit. At concentrations above 0.30 mg/l, iron stains laundry and plumbing fixtures. There is usually no noticeable taste at iron concentrations below 0.30 mg/l, although turbidity and colour may develop (WHO, 2011). Sources of iron in groundwater could be from atmospheric absorption, leaching of the products of rock weathering especially pyrite, the use of galvanized hand pump (Idoko, 2010). Water from most part of the study area is safe for drinking and other domestic uses. Some parts of the study like Wurukum, Mu village, Kanshio, North bank and Airport however have iron in concentrations that make them unsafe for the uses earlier mentioned.

Manganese concentration ranges from 0.00 to 0.16 mg/l in boreholes and 0.00 to 2.43 mg/l in hand dug wells with an average of 0.48 mg/l. 18 (75%) of the samples show elevated manganese concentration above the SON (2007) maximum permissible limit of 0.20 mg/l while five (25%) of the samples had values below this limit.

Table 1. Results of water analysis from Makurdi metropolis and environs.

S/N	ID	TDS (mg/l)	PH	Temp (°C)	Fe^{3+}	Cu^{2+}	Mn^{2+}	Zn^{2+}	NH_3^-	Cl^-	NO_3^-	Na^+	K^+	Ca^{2+}	Mg^{2+}	HCO_3^-	SO_4^{2-}
1	SL 1	130	5.65	29.5	0.47	0.11	2.43	0.22	0.54	0.97	0.53	11.6	2.5	3.18	22.1	90.9	5
2	SL 2	260	7	31.3	0.1	0.18	0.53	0	1.03	1.37	16	13.3	2.1	31.68	12.53	525.2	15
3	SL 3	400	3.1	30.3	0	0.11	0.39	0.02	1.39	0	15	12.3	3	42.33	36.49	858.5	40
4	SL 4	360	7.26	29.1	0.02	0.15	0.31	0.1	0.76	0.34	0.63	12.8	3	29.58	5.68	401.25	10
5	SL 5	320	4	28.6	0.27	0.09	0.36	0.1	0.82	1.75	32	16.7	18.6	8.16	74.2	232.3	25
6	SL 6	260	4.41	31.4	0.29	0.18	0.19	0.13	0.61	0	32	14	45.4	9.92	22.66	313.1	15
7	SL 7	850	2.24	32.2	0.38	0.01	0.28	0.09	0.86	9.28	107	20.9	51.1	41.5	9.71	700.65	5
8	SL 8	1230	4.34	30.4	0.27	1.93	0.26	0.12	3.2	13.11	150	36	2.99	94.5	7.43	1525.5	105
9	SL 9	90	4	30.5	0.47	0.11	0.53	0.16	0.53	1.39	16	10.5	4.7	18.45	10.99	171.7	15
10	SL 10	1190	5.82	29.4	0.2	0.87	0.49	0.12	2.8	11.29	96	35.5	66.5	49.59	9	550.30	6.5
11	SL 11	1250	6.44	29.9	0.25	0.9	0.54	0.09	0.34	10.76	117	36	64	36.25	15.43	450	50
12	SL 12	240	7.01	28.3	0	0.2	0.93	0.08	0.49	0	17	11.8	17.4	15.27	9.93	255	30
13	SL 13	330	5.8	31.2	0	0.14	1.28	0.06	0.89	0	33	10.7	8.3	4.39	7.09	200	25
14	SL 14	450	7.17	32.3	0.11	0.09	0.5	0.03	0.96	0.3	36	17.8	33.2	2.78	24.27	565.6	40
15	SL 15	170	4.33	32.3	0.35	0.29	0.66	0.11	1.11	0	26	10.4	7.3	32.86	38.59	181.8	15
16	SL 16	810	5.01	32.6	0.18	0.39	0.09	0.13	2.34	5.61	21.3	25.7	53.5	14.12	9.77	757.5	80
17	SL 17	180	4.18	29.6	0.86	0.28	0.24	0.16	0.5	0	38	12.1	1.8	25.19	21.86	131.3	5
18	SL 18	90	3.33	29.8	0.11	0.1	0	0.07	0.96	0	26	10.5	2.2	144.74	24.21	111.1	5
19	SL 19	520	5.91	31.7	0.37	0.27	0	0.11	0.83	0.39	18	20	2.5	119.71	22.98	505	20
20	SL 20	630	2.95	29.1	0.22	0.08	0	0.18	0.79	1.91	49	19.4	4.2	169.74	44.5	505	40
21	SL 21	520	5.88	30.4	0.58	0.35	0.13	0.07	0.61	0.88	16	13	2.3	10.63	39.54	767.6	10
22	SL 22	490	2.9	30.4	0.43	0.23	0	0.15	0.9	4.84	29	19.3	28	110.43	44.08	343.4	15
23	SL 23	130	4.18	28.1	0.13	0.13	0	0.06	0.37	1.09	28	11.7	3				
24	SL 24	520	5.88	30.4	0.66	0.97	1.29	0.14	1.58	0.98	21	13.1	2.8				
	Background	180			0	0.09	0	0.02	0.5	0		10.7	1.8	2.78	21.86	131.3	40
	SON MPL	500	6.5 -8.5		0.3	1	0.2	3		250	50	200			0.2		100

Units: mg/l except for pH and temperature (Temp): SON MPL: Standard organization of Nigeria Maximum permissible limit.

Concentration above 0.1 mg/l causes undesirable taste in beverages, stains sanitary wares and laundry (WHO, 2011) and neurological disorder (SON, 2007). Health-based value for manganese according to the World Health Organization is placed at 0.40 mg/l. Manganese present in groundwater is usually absorbed from the atmos-phere or leached from products of rock weathering.

Copper in water supply usually arises from the corrosive action of water leaching copper from copper pipes in buildings (WHO, 2011). This corrosive action of water is often accelerated by high levels of dissolved oxygen in water. The analysis reveals concentration of 0.09 to 1.93 mg/l with an average of 0.31 mg/l for samples analyzed. At concentration above 5 mg/l copper imparts colour and an undesirable bitter taste to

Table 2. Correlation matrix for measured parameters

	pH	Na	Mg	Ca	Cl	K	SO_4^{2-}	TDS	Temp	Fe^{3+}	Mn	Cu	Zn	NO_3^-	HCO_3^-	NH_3
pH	1	0.05	-0.41	-0.42	-0.26	-0.03	-0.01	0.02	0.3	-0.33	0.27	0.09	-0.16	-0.18	0.1	-0.07
Na		1	-0.22	0.03	0.9	0.87	0.59	0.96	-0.16	-0.24	-0.19	0.78	0.1	0.81	0.78	0.63
Mg			1	0.26	-0.37	-0.32	-0.17	-0.25	-0.18	0.18	-0.1	-0.27	0.02	-0.25	-0.36	-0.25
Ca				1	-0.1	-0.24	-0.18	0.01	-0.13	-0.15	-0.24	-0.16	0.22	-0.07	-0.13	-0.09
Cl					1	0.93	0.51	0.9	0.02	-0.05	-0.17	0.77	-0.07	0.9	0.8	0.63
K						1	0.64	0.83	-0.05	-0.23	-0.26	0.74	-0.12	0.84	0.72	0.65
SO_4^{2-}							1	0.57	0.02	-0.23	-0.21	0.76	-0.01	0.64	0.58	0.65
TDS								1	0.15	-0.21	-0.25	0.74	0.01	0.82	0.88	0.61
Temp									1	-0.08	0.15	-0.06	-0.11	-0.12	-0.15	-0.17
Fe^{3+}										1	0.09	-0.03	0.37	-0.07	-0.3	-0.31
Mn											1	-0.12	0.43	-0.2	-0.36	-0.29
Cu												1	0.12	0.83	0.71	0.71
Zn													1	0.01	-0.18	0.03
NO_3^-														1	0.73	0.56
HCO_3^-															1	0.71

Table 3. Statistical summary of measured parameters.

Parameter	Min	Max	Mean	St. Dev.	Dev. Coef
TDS (mg/l)	90	1250	489.545	362.931	74.136
pH	2.24	7.26	4.942	1.538	31.126
TEMP (°C)	28.6	32.6	30.571	1.194	3.906
Fe^{3+} (mg/l)	0	0.86	0.27	0.213	79.152
Mn^{2+} (mg/l)	0	2.43	0.477	0.549	115.096
Cu^{2+} (mg/l)	0	1.93	0.307	0.433	140.963
Zn^{2+} (mg/l)	0	0.22	0.099	0.057	57.532
Ca^{2+} (mg/l)	2.78	169.74	46.136	48.969	106.139
K^+ (mg/l)	1.8	66.5	19.3	22.48	116.481
Mg^{2+} (mg/l)	5.68	74.2	23.32	16.98	72.814
Na^+ (mg/l)	10.4	36	17.741	8.424	47.483
Cl^- (mg/l)	0	13.11	2.918	4.262	146.068
$HCO3^-$ (mg/l)	90.9	858.5	409.973	232.618	56.74
SO_4^{2-} (mg/l)	5	105	26.205	25.481	97.24
NH_3^- (mg/l)	0.34	3.2	1.057	0.751	71.076
NO_3^- (mg/l)	0.53	150	40.521	39.799	98.218

Min: minimum, Max: maximum, St. Dev.: standard deviation, Dev. Coef: deviation coefficient.

water (WHO, 2011). Maximum permissible limit of copper concentration in water for drinking and domestic uses by SON (2007) is set at 1.00 mg/l; consumption of water with concentration above this value causes gastro-intestinal disorder. All except one (SL 8 at North Bank) of the samples analyzed had concentration below the permissible limit. With an exception of the immediate area around SL 8 at North Bank, groundwater in Makurdi metropolis is considered safe for drinking and domestic uses with respect to its copper concentration. All samples analyzed had zinc concentration below the SON (2007) permissible limit of 3.00 mg/l with concentration ranging from 0.00 to 0.18 mg/l with an average of 0.09 mg/l.

Statistical analysis

The correlat on coefficient reveals some relevant hydrochemical relationships, the content of TDS have high positive correlation with Na^+, Cl^-, K^+ and SO_4^{2-} with correlation coefficients of 0.96, 0.90, 0.62 and 0.56 respectively (Table 2). This positive correlation indicates that the ions are derived from the same source. In a similar way, there exists positive correlation between Cl^- – Na^+ (0.90), K^+ – Na^+ (0.63), K^+ – Cl^- (0.63), NO_3 – NH_3 (0.5), NO_3 – Cl (0.92) and Na – Cl (0.90). The high positive correlation (0.90) between Na^+ and Cl^- may represent influence from salty water.

On the other hand, there are negative relationships between Mg – Na, Mg – K, K – Ca, SO_4 – Mg and HCO_3^- - Mg. These relationships could indicate ion exchange process between the groundwater and soil minerals.

Hydrochemical facies

From the Piper plot (Figure 3), an evaluation of the hydrochemistry of groundwater in Makurdi metropolis was done. The groundwater samples fall in the fields of Mg – Na – HCO_3, Mg-K-HCO_3, K – HCO_3 – SO_4, Mg-Ca-HCO_3, Mg-HCO3 and Ca - Mg - HCO_3 in order of increasing dominance.

Table 4. Calculated values of bicarbonate hazard, percentage sodium and sodium adsorption ratio for groundwater in Makurdi metropolis and environs.

Location description	Sample ID	Bicarbonate hazard (meq/l)	SAR	Na%	EC (µS/cm)
Air Port 1	SL 1	0	0.51	22.45	91
Orowa Village	SL 2	5.99	0.51	17.26	182
Miande Akpu Village	SL 3	8.96	0.33	7.18	280
Antsa Village	SL 4	10.31	0.56	24.89	252
University Of Agriculture Road	SL 5	0	0.40	11.38	224
North Bank 2	SL 6	2.77	0.56	41.69	182
North Bank 3	SL 7	12.52	0.76	40.62	595
North Bank 4	SL 8	24.24	2.54	104.49	861
North Bank 5	SL 9	0.99	0.48	24.92	63
Wadata	SL 10	17.98	1.22	45.18	833
Wadata 2	SL 11	11.82	1.26	46.15	875
Georges Qtrs	SL 12	3.39	0.58	40.31	168
Nyon Village	SL 13	6.65	0.73	63.99	231
Welfare Qtrs	SL 14	7.14	0.75	42.53	315
Kanshio	SL 15	0	0.29	8.77	119
Wurukum	SL 16	10.91	1.29	66.48	567
Mu Village	SL 17	0	0.43	13.00	126
Airport 2	SL 18	0	0.21	2.77	63
Airport 3	SL 19	0.41	0.44	5.71	364
Ikyaan Village	SL 20	0	0.34	3.44	441
Wurukum 2	SL 21	8.79	0.41	10.66	364
Wurukum 3	SL 22	0	0.39	10.37	343

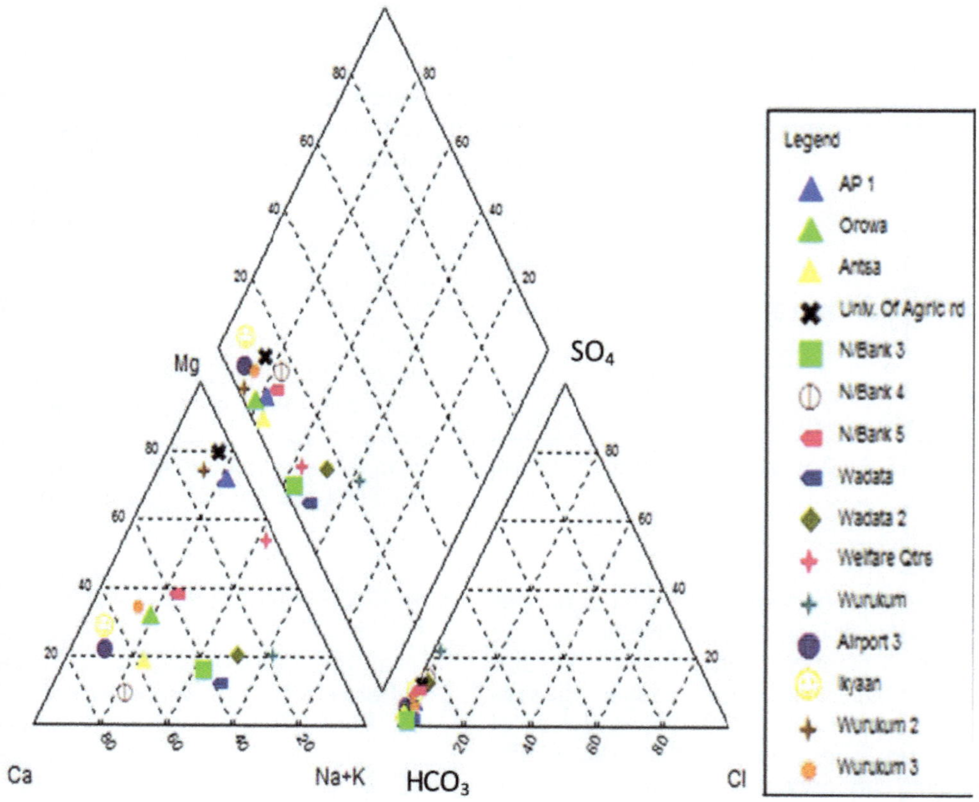

Figure 3. Piper diagram of groundwater in Makurdi metropolis and environs.

Table 5. Classification of groundwater based on Na%, SAR and bicarbonate hazard (source: Tabue et al., 2012).

Parameters	Range (meq/l)	Class
	>200	Above max permissible limit
Na%	20 – 40	Good
	<20	Excellent
	20 - 40	Good
SAR	40 - 60	Permissible
	60 - 80	Doubtful
	>80	Unsuitable
	0.00 – 1.25	Safe
Bicarbonate hazard	1.25 – 2.50	Marginal
	>2.50	Unsuitable

Suitability of groundwater for irrigational purposes

The quality of water used for irrigation purpose is of great importance for crop productivity, soil maintenance and environmental protection (Tabue et al., 2012). The quality of groundwater as earlier explained is influenced by the mineral composition of the aquifers within which water is stored as well as the residence time in these aquifers.

Percent sodium (Na%)

The percentage sodium was calculated using the expression:

$$\text{Na\%} = \left(\frac{Na^+ + K^+}{\mathrm{Ca}^{2+} + \mathrm{Mg}^{2+} + \mathrm{Na}^+ + \mathrm{K}^+}\right) \times 100 \tag{1}$$

Within the study area, Na% varies from 3.44 to 104.50% (Table 4) and are less than the maximum allowable limit of 200% (Table 5). Three (14 %) of the samples fall within the class of 20 to 40 Na% and are considered good for irrigation. Nine samples (41 %) have percentage sodium values above the 20 to 40% class but are still below the maximum permissible limit of 200. Ten (45%) have Na% less than 20% and are considered excellent for irrigation (Table 5).

Higher Na% observed in groundwater from nine locations which constitutes 41% of the study area indicates the dominance of ion exchange and weathering from lithological units of the study area (Tabue et al., 2012). Groundwater with relatively high Na% values above 40 occurs around North bank, Wadata, George's quarters, Nyon, welfare quarters and Wurukum. When the concentration of sodium is high in water, sodium tends to be absorbed by clay particles displacing magnesium and calcium. This exchange of sodium for calcium and magnesium reduces soil permeability and results in soil with poor internal drainage. A larger proportion (59 %) of groundwater within the study area is considered safe for irrigation.

Sodium adsorption ratio (SAR)

This was calculated using the expression (SAR) = ... (2)

SAR for samples within the study area varies from 0.21 to 2.54. The groundwater according to the classification in Table 4 is of excellent quality with respect to sodium adsorption ratio. If water used for irrigation has high Na^+ and low Ca^{2+}, the ion exchange complex may become saturated with Na^+ which destroys the soil structure due to dispersion of clay particles and reduces plant growth (Tabue et al., 2012). High Na^+ and low Ca^{2+} in water used for irrigation results in ion exchange complex which becomes saturated with Na^+ as a result of clay particle dispersion. This causes the destruction of the soil structure and consequently reduces plant growth.

The Wilcox diagram (Figure 4) also shows that all samples analyzed fall within the S1C1, S1C2 and S1C3 regions. This indicates water of good quality for irrigation.

High Na^+ and low Ca^{2+} in water used for irrigation results in ion exchange complex which becomes saturated with Na^+ as a result of clay particle dispersion. This causes the destruction of the soil structure and consequently reduces plant growth.

Bicarbonate hazard

Bicarbonate hazard means excess of bicarbonate over calcium and magnesium, all expressed in milliequivalent per liter. Water with bicarbonate hazard greater than 1.25 meq/l should not be used for irrigation except on very sandy, highly permeable soils or with some soil chemical amendments such as the application of gypsum. Bicarbonate hazard was computed using the expression

$$HCO_3^- - (Ca^{2+} + Mg^{2+}) \tag{3}$$

Only nine (41%) of the samples analyzed (SL 1, SL 5, SL 9, SL 15, SL 18, SL 19 and SL 20) fall within the class of samples considered safe (0.00 to 1.25 meq/l) for irrigation in terms of bicarbonate hazard. Groundwater from other locations should not be used for irrigation purpose except on soils that are sandy, permeable or with some chemical amendments as they are considered unsuitable (Table 4).

Conclusions

Groundwater is of immense importance to both rural and urban dwellers of developing nations. The groundwater of the study area is acidic to slightly basic in nature.

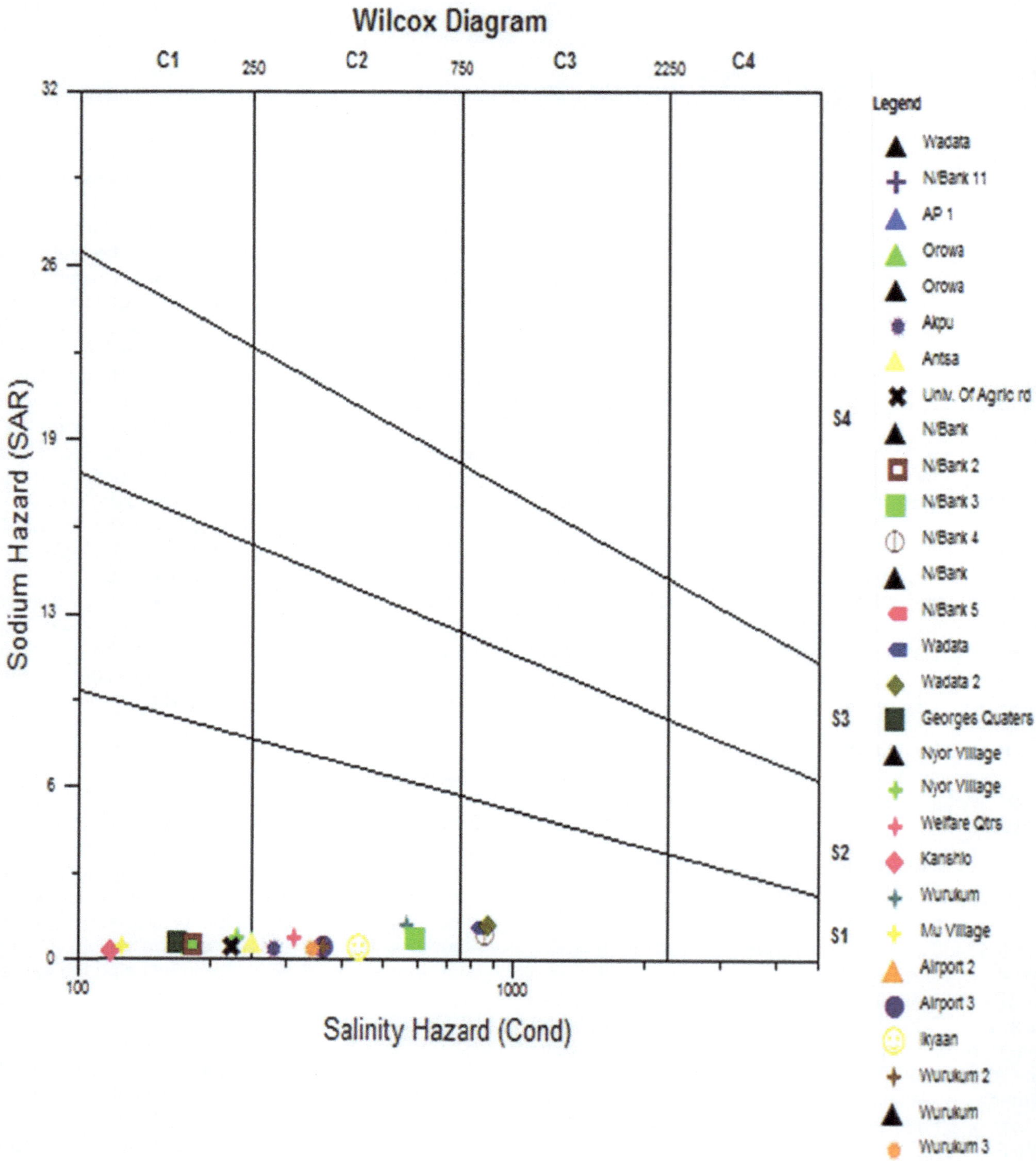

Figure 4. Wilcox diagram of groundwater in Makurdi metropolis and environs.

Groundwater in Makurdi metropolis and environs has been evaluated for its chemical composition and suitability for drinking and irrigation purposes.

The abundance of chemical concentrations in groundwater within the study area is in order of Ca > Mg > Na > K for the cations and $HCO_3^- > SO_4^{2-} > NO_3^- > Cl^-$ for the anions. Groundwater in some parts of the study area have elevated TDS, magnesium, sulphate, nitrate, copper, manganese and iron concentrations above the SON (2007) maximum permissible limit for drinking water and considered unsafe for drinking with respect to these parameters (Figure 5). The acidic nature of groundwater

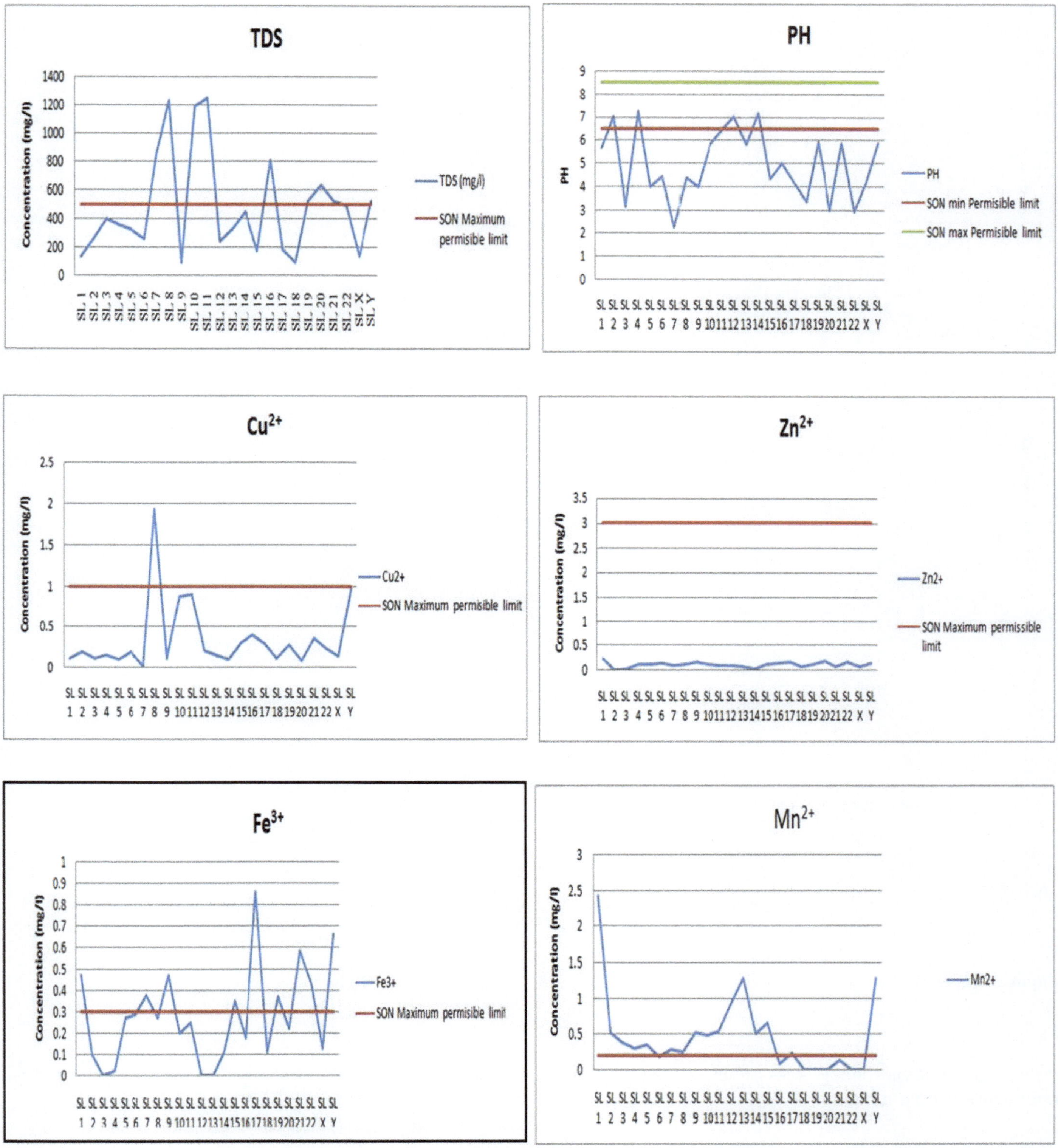

Figure 5. Plots of concentrations of measured parameters and Standard Organization of Nigeria's 2007 permissible limits.

in the study area has also impaired its quality. Northbank and Wadata areas are identified to be areas where groundwater is polluted the most and this is attributed to the poor sanitary conditions in these areas.

Groundwater within the study area is safe for irrigation with respect to its sodium adsorption ratio but a large portion of it has high percentage sodium and bicarbonate hazard and should not be used for irrigation except on

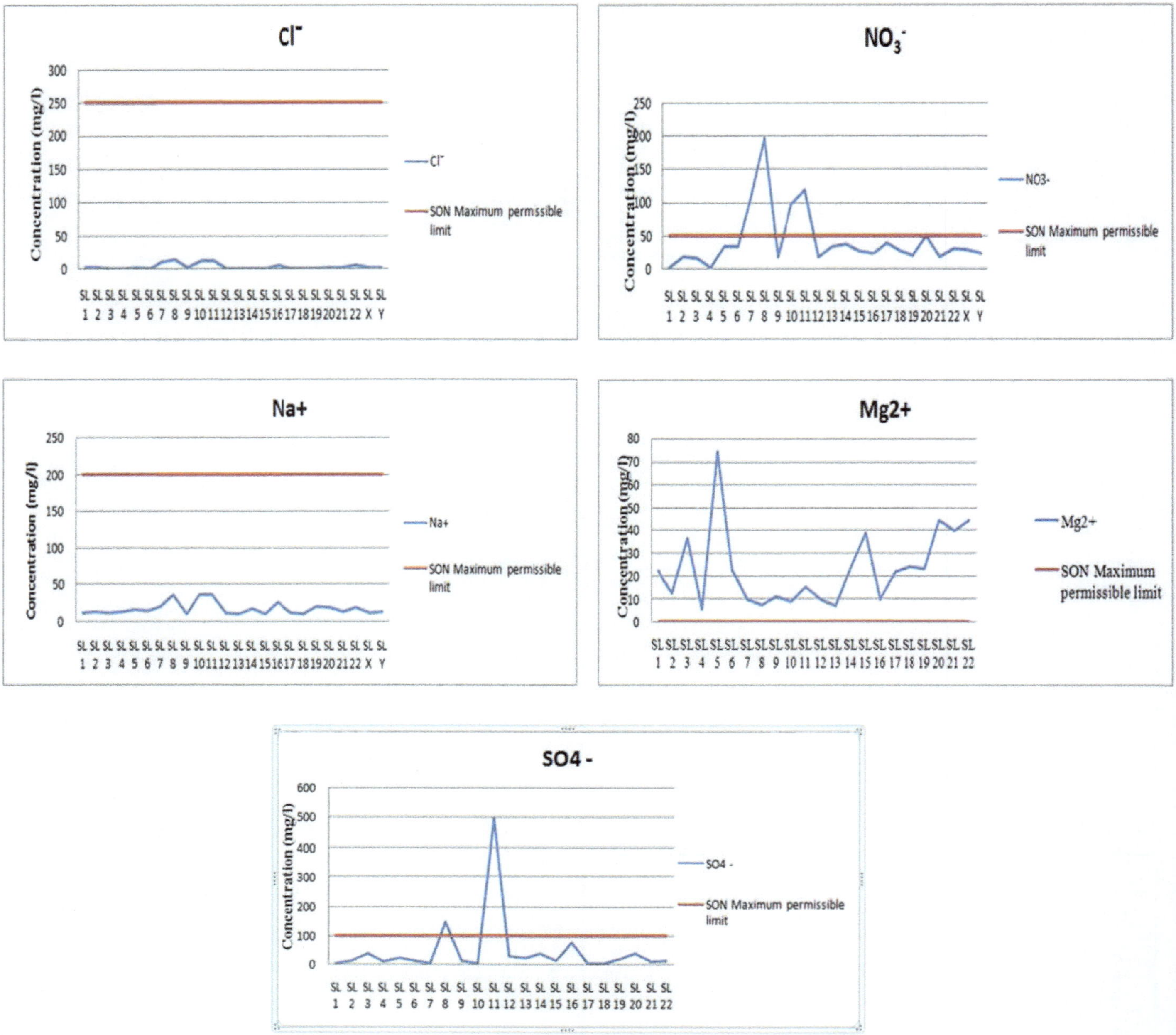

Figure 5. Contd.

very porous and permeable soils or when used with some chemical soil amendments.

REFERENCES

Adanu EA (1981). Hydrogeology of Makurdi and the environ with the General Geology of the Agana sheet (250) and Makurdi sheet (251). Unpublished M.Sc. Thesis, Ahmadu Bello University Zaria.

Akuh TI (2014). Hydrogeology and groundwater quality in Makurdi metropolis and itsenvirons, part of Makurdi (sheet 251), north central Nigeria. Unpublished M.Sc. Thesis, Ahmadu Bello University Zaria, Nigeria.

Davis SN, De Wiest JM (1966). Hydrogeology, John Wiley and Sons, Inc., New York, London, Sydney.

Edet A, Nganje TN, Ukpong AJ, Ekwere AS (2011). Groundwater chemistry and quality of Nigeria: A status review. Afr. J. Environ. Sci. Technol. 5(13):1152-1169.

Idoko OM (2010). Seasonal Variation in Iron in Rural Groundwater of Benue State, Middle Belt, Nigeria. Pak. J. Nutr. 9 (9):892-895.

Irtwange SV, Sha'ato R (2009). Environmental Baseline Characteristics for a Pilot Project Site for Integrated Solid Waste Management in Makurdi, Nigeria. Res. J. Environ. Earth Sci. 1(2):81-98.

Najime T (2010). Cretaceous stratigraphy, sequence stratigraphy and tectono – sedimentary evolution of Gboko area, lowar Benue Trough, Nigeria. Unpublished Ph.D dissertation, Geology department, Ahmadu Bello University Zaria.

Nigeria Geological Survey Agency (NGSA) (2006). Geological and

mineral resources map of Benue State, North central Nigeria.
Obaje NG (1994). Coal petrography, microfossils and paleoenvironment of calcareous coal measures in the middle Benue Trough of Nigeria. Tubinger Mikropalaontolgische Mitteilugen 11, 165 pp.
Offodile ME (1989). A review of the geology of the Cretaceous of the Benue Trough. In: C.A. Kogbe, (ed), Geology of Nigeria (2nd edition). Rock view Nigeria Ltd, Jos. pp. 265-376.
Offodile ME (2002). Groundwater study and development in Nigeria. Mecon Geology and Eng. Services Ltd. Jos, Nigeria.
Offodile ME (1976). The Geology of the Middle Benue Nigeria. Special volume Palaeontological Institution University Uppsala. 4:1-166.
Raghunath HM (2007). Groundwater hydrology, groundwater survey and pumping tests, rural water supply and irrigation systems, third edition.
Standard Organization of Nigeria (SON) (2007). Nigerian standard for drinking water quality. NIS 554:1-30.
Tabue YG, Feumba R, Wethe J, Ekodeck GE, De Marsily G (2012). Evaluation of the groundwater suitability for domestic and irrigational purposes: A case study from Mingoa River Basin, Yaounde, Cameroon. J. Water Resour. Prot. 4:285-293.
World Health organization (WHO), (2011). Guidelines for Drinking-water Quality, fourth ed.
Zaborski PM (1998). A review of the Cretaceous system in Nigeria. Africa Geoscience Review 5:385-483.

Determination of optimum growth conditions and biodiesel production from filamentous algae

Hesham R. Lotfy, Martha N. Amputu and Celestine Raidron

Department of Chemistry and Biochemistry, Faculty of Science, University of Namibia, P. Bag 13301, Windhoek, Namibia.

Petroleum diesel combustion is a major source of greenhouse gas (GHG). It is also a major source of other air contaminants including NOx, SOx, CO and volatile organic compounds. Algae have emerged as one of the most promising sources for biodiesel production. In this study, a higher algae growth rate was observed in the experiments with excess Na_2SiO_4, trace metals, Na_2EDTA and excess vitamin solution, the increase was above 300%. It was also observed that the experiment that was supplied with CO_2 (without simultaneous sunlight exposure) for one hour, for 25 days and the beaker with excess NaH_2PO_4 solution, have shown a slower growth rate than the control. The results of the experiment on the effect of sunlight exposure for certain times daily for 25 days show that the growth rate is directly proportional to increase of sunlight exposure time (during the 90 min). The results of the experiment on the effect of simultaneous exposure to sunlight and CO_2 for certain times daily, for 25 days, show that the growth rate is directly proportional to the increase of sunlight and CO_2 exposure time.

Key words: Algal oil, biodiesel, transesterification, glycerine.

INTRODUCTION

The need of energy is increasing continuously due to the increase in population and industrialization. The continued use of petroleum sourced fuels is now widely recognized as unsustainable because of the depletion of supplies and the contribution of these fuels to the accumulation of carbon dioxide in the environment leading to increase of global warming. The combustion of fossil fuels is responsible for 73% of the CO_2 production (Narendra et al., 2010). With regards to global warming and as dependence on fossil fuels grows, the search for renewable energy sources that reduce CO_2 emissions becomes a matter of widespread attention (Ragauskas et al., 2006; Demirbas and Demirbas, 2007). In recent years, cultivation of microalgae has received renewed attention on account of their utility as a feasible CO_2 sequestration technology (Ono and Cuello, 2006; Hsueh et al., 2007; Jacob-Lopes et al., 2008).

In the last ten years, many studies have been conducted on biofuels for substituting fossil fuels and reduce the greenhouse gas emission (Bastianoni et al., 2008). Algae, especially microalgae, were found to be the only source of renewable biodiesel that is capable of meeting the global demand for transport fuels (Chisti, 2007, 2008).

The idea of using algae as a source of fuel is not new (Chisti, 1980; Chisti 1981; Nagle and Lemke, 1990;

Sawayama et al., 1995), but it is now being taken seriously because of the increasing price of petroleum and more significantly, the emerging concern about global warming that is associated with burning fossil fuels (Gavrilescu and Chisti, 2005). Microalgae can provide several types of renewable biofuels which include, methane, biodiesel (methyl esters) and biohydrogen (Gavrilescu and Chisti, 2005; Kapdan and Kargi, 2006; Spolaore et al., 2006). Oil productivity of many microalgae greatly exceeds the oil productivity of the best producing oil crops (Shay, 1993).

Bioenergy is one of the most important components to mitigate greenhouse gas emissions and substitute of fossil fuels (Goldemberg, 2000). Petroleum diesel combustion is a major source of greenhouse gas (GHG). Apart from these emissions, petroleum diesel is also a major source of other air contaminants including NOx, SOx, CO, particulate matter and volatile organic compounds (Klass, 1998). Biomass is one of the better sources of energy (Kulkarni and Dalai, 2006). Large-scale introduction of biomass energy could contribute to sustainable development on several fronts, environmentally, socially and economically (Turkenburg, 2000; UNDP, 2008). Biodiesel is a nontoxic and biodegradable alternative fuel that is obtained from renewable sources. It is reported that algae were one of the best sources of biodiesel and are the highest yielding feedstock for biodiesel. It can produce up to 250 times the amount of oil per acre from soybeans (Hossain and Salleh, 2008). In fact, producing biodiesel from algae may be the only way to produce enough automotive fuel to replace current gasoline usage. The best algae for biodiesel would be microalgae. Microalgae have much more oil than macroalgae and it is much faster and easier to grow (Shay, 1993).

Algae contain anything between 2 and 40% of lipids/oils by weight (Wagner, 2007). Microalgae have much faster growth-rates than terrestrial crops. The per unit area yield of oil from algae is estimated to be between 18,927 and 75,708 L per acre, per year; this is 7 to 31 times greater than the next best crop, palm oil (around 2,404 L) (Wagner, 2007). In addition, the question “Food or Fuel?” is not raised.

Advantages of microalgal cultivation for biodiesel production over other oleaginous crops are: the former is not only easy to handle and more beneficial from economic point of view but also known for their bioremediation capabilities (Dayananda et al., 2005; Sazdanoff, 2006; Chisti, 2007; Huntley and Redalje, 2007; Li et al., 2008; Schenk et al., 2008; Tan et al., 2009). Some positive points of microalgal cultivation are: rapid growth rates, a high per-acre yield (7 to 31 times greater than the next best crop- palm oil), certain species of algae can be harvested daily, short life cycle (approximately 1 - 10 days), ability to synthesize and accumulate large quantities of lipids per dry weight biomass, algae biofuel contains no sulphur, algae biofuel is non-toxic, algae biofuel is highly bio-degradable, and algae consume carbon dioxide as they grow, so they could be used to capture CO_2 from power stations and other industrial plants that would otherwise go into the atmosphere, potential to grow in saline water and harsh conditions, less fertilizer and nutrient input requirements and it is the most promising non-food source of biodiesel. After extracting oil from microalgae, the remaining biomass portion can also be used as a high protein feed for livestock (Schneider, 2006; Haag, 2007).

The algae used in this study is filamentous algae also known as "pond moss" or "pond scum" and these threadlike algae often occur in huge greenish masses floating upon the waters' surface. They can form dense mats in static water or long, rope-like strands in flowing water. Its filaments consist of series of cells being joined end to end giving a thread-like appearance.

The main aim of the research was to extract oil from a microalgae species and convert it to biodiesel and to determine the optimum growth conditions of the used algae species.

MATERIALS AND METHODS

Stock solutions preparation

The stock solutions were prepared according to the procedures of Guillard and Ryther (1962).

Filamentous algae cultivation

Cultivation of algae in a synthetic medium

The medium was prepared by adding 3.00 ml of $NaNO_3$ solution (150.008 g in 1 L distilled water), 3.00 ml of trace metal solution, 3.00 ml of $NaSiO_3.5H_2O$ solution (10 ml in 1 L distilled water), 3.00 ml of iron citrate solution (9.0008 g $FeCl_3$ and 9.00 g citric acid in 1 L distilled water), 3.00 ml of vitamin solution (Folic acid 0.015 g and 0.0156 g peptone bacteriological in 100 mL distilled water), 1.50 ml of NaH_2PO_4 (11.31 g in 1 L distilled water) and 1.50 ml of $Na_2EDTA.2H_2O$ solution, to a beaker containing 3000 ml of water and 19.62 g of filamentous algae was added.

Optimization of growth condition

The effects of different components concentration of the stock solution on the algae growth (Table 1)

One milliliter of $NaNO_3$ solution, 1.00 ml of trace metal solution, 1.00 ml of $NaSiO_3.5H_2O$ solution, 1.00 ml of iron citrate solution, 1.00 ml of vitamin solution, 0.50 ml of NaH_2PO_4 solution and 0.50 ml of $Na_2EDTA.2H_2O$ solution were added to each of nine beakers filled with 100.00 ml tap water. In the first beaker, 1.00 ml of $NaNO_3$ solution was added in excess, the second beaker, 1.00 ml of trace metal solution was added in excess, the third beaker, 1.00 ml of $NaSiO_3.5H_2O$ solution was added in excess, the fourth beaker 1.00 ml of iron citrate solution was added in excess, to the fifth beaker, 1.00 ml of vitamin solution was added in excess, the sixth beaker, 0.50 ml of NaH_2PO_4 solution was added in excess, the seventh beaker, 0.50 ml of $Na_2EDTA.2H_2O$ solution was added in excess, to

Table 1. The effect of different components of concentration of stock solution on the algae growth.

Stock solution in excess 1.00 ml	Initial mass of algae (g)	Mass of algae after 15 days (g)	Increase in mass (%)
$NaNO_3$	1.00	3.49	249 ↑
Trace metals	1.00	4.47	347 ↑↑
Na_2SiO_4	1.00	4.68	368 ↑↑
Iron citrate	1.00	3.74	274 ↑
Vitamin	1.00	4.04	304 ↑↑
NaH_2PO_4	1.0020	2.75	175 ↓
Na_2EDTA	1.00	4.29	329 ↑↑
CO_2	1.00	2.00	100 ↓
Control	1.00	2.86	186 —

Table 2. Results of the effect of sunlight exposure on growth after 25 days.

Sample	30 min sunlight daily	60 min sunlight daily	90 min sunlight daily
Initial mass (g)	1.00	1.00	1.00
Final mass (g)	2.75	2.87	3.01
Increase in mass (%)	175	187	201

Table 3. The effect of simultaneous CO_2 supplement and sunlight exposure after 25 days.

	30 min		60 min		90 min	
Experimental time	CO_2 and sunlight	Control (sunlight only)	CO_2 and sunlight	Control (sunlight only)	CO_2 and sunlight	Control (sunlight only)
Initial mass (g)	1.00	1.00	1.00	1.00	1.00	1.00
Final mass (g)	4.04	2.75	4.72	2.87	4.82	3.01
Increase in mass %	304	175	372	187	382	201

the eighth beaker, CO_2 was supplied for one hour daily, while the ninth was used as a control, no excess solution was added. One gram of algae was added to each beaker and CO_2 was supplied to each for 10 min.

The effects of sunlight exposure time on the algae growth (Table 2)

Three 250.00 ml beakers were filled with 200.00 ml of water and 1.00 ml of $NaNO_3$ solution, 1.00 ml of trace metal solution, 1.00 ml of $NaSiO_3.5H_2O$ solution, 1.00 ml of iron citrate solution, 1.00 ml ofvitamin solution, 0.50 ml of NaH_2PO_4 and 0.50 ml of $Na_2EDTA.2H_2O$ solutions were added, to each beaker. One gram of algae was added to each beaker and CO_2 was supplied to each for 10 min. The algae were exposed to sunlight at different times, for 30, 60 and 90 min, respectively. The growth rate in each beaker was observed.

The effect of simultaneous CO_2 supplement and sunlight exposure time (Table 3)

Three 250.00 ml beakers were filled with 200.00 ml of water and 1.00 ml of $NaNO_3$ solution, 1.00 ml of trace metal solution, 1.00 ml of $NaSiO_3.5H_2O$ solution, 1.00 ml of iron citrate solution, 1.00 ml of vitamin solutions and 0.50 ml of NaH_2PO_4 and $Na_2EDTA.2H_2O$ solution were added to each beaker.

One gram of algae was added to the beakers. The algae were supplied with CO_2 supplement and exposed to the sunlight simultaneously, for 30, 60 and 90 min, respectively. The growth rate in each beaker was observed.

The effect of available space on the growth rate

Filamentous algae, 34.33 g were collected from a pond and cultivated in two fish tanks. They were filled with 10.00 L of water, 20.00 ml of $NaNO_3$ solution, 20.00 ml of trace metal solution, 20.00 ml of $NaSiO_3.5H_2O$ solution, 20.00 ml of iron citrate solution, 20.00 ml of vitamin solution, 5.00 ml NaH_2PO_4 solution and 10.00 ml of $Na_2EDTA.2H_2O$ solution. To one tank, 22.88 g of algae were added and 11.45 g of algae was added to the second tank. The algae were exposed to sunlight and CO_2 supplement was bubbled into each tank for 90 min daily.

Extraction process

Harvesting

Ninety percent of the algae were collected using a fish net after 10 days of cultivation. The wet algae weighed 56.50 g and they were ground with a pestle in a mortar for 20 min. The ground algae were dried in an oven for 75 min at 80°C to release water. The dried algae weighed 26.85 g.

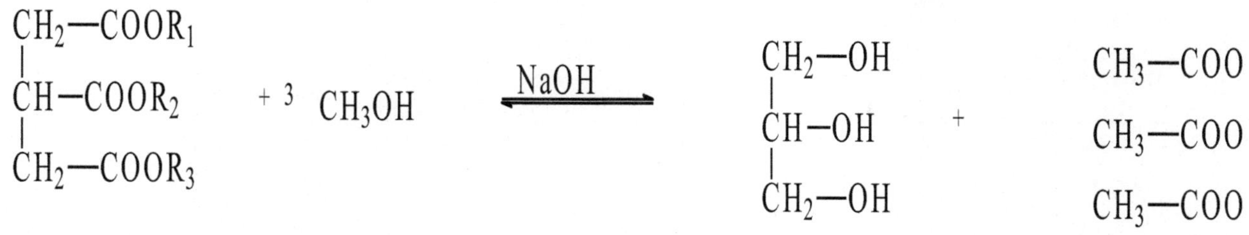

Triglyceride (oil or fat) Methanol Glycerol Methyl Esters

R1–3 are hydrocarbon groups (Biodiesel)

Figure 1. Transesterification reaction to produce biodiesel.

Drying time and temperature

Extraction was done several times on the filamentous algae at different drying times and temperatures in trials to find the best drying time and the optimum temperature. It was found that the best drying time was 75 min at 80°C.

Oil extraction

The two methods used are hexane solvent extraction and soxhlet extraction. The soxhlet extraction did not yield satisfactory results. The hexane solvent extraction used 20.00 ml of hexane and 20.00 ml diethyl ether which were mixed with the dried algae to extract the oil. The mixture was allowed to settle for 24 h. The biomass was collected by gravity filtration and weighed 25.58 g. The extracted oil was collected after filtration. The hexane and the diethyl ether were evaporated using a rotary evaporator.

Transesterification (Figure 1)

Algal oil (0.80 ml) was mixed with sodium methoxide which was prepared by dissolving 0.0122 g of NaOH in 1.20 ml of methanol. The mixture was shaken for one hour, and thereafter transferred to a separatory funnel and allowed to settle for 12 h.

In the separatory funnel, the layers were clearly formed; biodiesel on top and glycerol at the bottom. The bottom layer was drained into a vial and stored. The biodiesel was washed three times with water. The obtained biodiesel was heated at 54°C for 20 min to evaporate all the residual water. Approximately 0.20 ml of biodiesel was obtained and it was then stored in a vial. The extraction and the transesterification processes were repeated several times using the method described by Hossain and Salleh (2008).

RESULTS

Cultivation

The effects of different components of stock solution concentration on the algae growth

A higher growth rate was observed in the beakers with excess Na_2SiO_4, trace metals, Na_2EDTA and excess vitamin solution, the increase was above 300%. It was also observed that the beaker that was supplied with CO_2 (without simultaneous sunlight exposure) for one hour and the beaker with excess NaH_2PO_4 solution showed a slower growth rate than the control.

The effect of the sunlight exposure time, for 25 days, on algae growth rate

The results of the experiment studying the effect of sunlight exposure for certain times daily for 25 days, showed that the growth rate is directly proportional to increase of sunlight exposure time (for 90 min).

The effect of CO_2 supplement and sunlight exposure simultaneously on the growth after 25 days

The results of the experiment on the effect of simultaneous exposure to sunlight and CO_2 for certain times daily, for 25 days, showed that the growth rate is directly proportional to the increase of sunlight and CO_2 exposure time.

The effect of available space on the growth rate

The growth rate in the tank which containing 11.45 g algae was 2.9% after 10 days, while the growth rate in the tank which contained 22.88 g algae was 2.6% after 10 days and under the same experimental conditions. This is in agreement with a different study done by Campbell (2008).

Extraction

Filamentous algae extraction and transesterification results

Extraction was done several times on the filamentous algae at different drying times and temperatures in trials to find the best drying time and the optimum temperature. Although, the dry mass was not constant but based on the ratio between the dry mass and the amount of oil produced it was found that the best drying time, in our experiments, was 75 min at 80°C. Algal oil was obtained as shown in Table 4. With the transesterification process, only some of the algal oil was converted to biodiesel.

Table 4. Extraction and transesterification results of filamentous algae oil.

Dry mass (g)	Biomass (g)	Drying time (min)	Amount of algal oil extracted (ml)	Biodiesel produced (ml)
1.23	0.78	60 min at 80°C	Approximately 0.20	No biodiesel produced
26.85	25.58	85 min at 80°C	0.80	Approximately 0.20 ml of biodiesel produced
13.97	12.70	75 min at 80°C	1.30	Approximately 0.5ml of oil produced
31.01	29.93	180 min at 85°C	1.00	No biodiesel produced
18.83	18.52	170 min at 60°C	0.70	No biodiesel produced
21.42	18.94	30 min at 80°C	0.50	No biodiesel extracted
14.77	13.21	20 min at 80°C	0.70	No biodiesel extracted
25.98	21.35	20 min at 80°C	0.50	0.2 ml biodiesel extracted

Some of the oil solidified during the process. Results in Table 4 shows that the best drying time, in our experiments, was 75 min at 80°C.

DISCUSSION

In the experiment investigating the effect of the sunlight exposure time only and the effect of simultaneous CO_2 supplement and sunlight exposure in 25 days, it was observed that the growth rate is directly proportional to the length of sunlight exposure time and CO_2 supplement simultaneously. This may be attributed to enhanced photosynthesis which in turn increase the growth rate. Space plays a role in the growth of algae, the results of the experiment on the effect of available space indicated that the algae that were sparsely populated in the growth tank had a higher growth rate. In the experiment on the effect of different components' concentrations of media solution on the algae growth, it was found that excess of trace metal and silicate solutions have shown the highest growth rates. This is because they play a structural role in the chloroplast membrane, maintains the green colour and assist in the breakdown of purine, which is in accordance with report of Salisbury and Ross (1992).

Conclusion

The experiment on the effect of different nutrients on the growth of algae showed a higher growth rate in beakers with excess Na_2SiO_4, trace metals, Na_2EDTA and excess vitamin solution, the increase was above 300%. It was also observed that in the experiment which was supplied with CO_2 (without simultaneous sunlight exposure) for one hour and the experiment with excess NaH_2PO_4 solution, there was a slower growth rate than that of the control.

The results of the experiment on the effect of sunlight exposure for certain times daily, for 25 days, showed that the growth rate is directly proportional to the increase in sunlight exposure time (within 90 min).

The results of the experiment on the effect of simultaneous exposure to sunlight and CO_2 for certain times daily, for 25 days, showed that the growth rate is directly proportional to the increase of simultaneous sunlight and CO_2 exposure time (within 90 min).

The results of the experiment on the optimum drying time and temperature, showed that the best drying time was 75 min at 80°C. Algal oil was extracted using hexane and diethyl ether in 1:1 ratio. Through transesterification reaction, some of the algal oil was converted to biodiesel.

Conflict of Interests

The author(s) have not declared any conflict of interests.

REFERENCES

Bastianoni S, Coppola F, Tiezzi E, Colacevich A, Borghini F, Focardi S. 2008. Biofuel potential production from the Orbetello lagoon macroalgae a comparison with Sunflower feedstock. Biomass Bioenergy 10:1-10.

Campbell MN (2008). Biodiesel as a renewable source for liquid fuel. Gueph Eng. J. 1:2-7.

Chisti Y (1980). An unusual hydrocarbon. J. Ramsay Soc. 27-28, 24-6.

Chisti Y (1981). An unusual hydrocarbon. J. Ramsay Soc. pp. 24-26.

Chisti Y (2007). Biodiesel from microalgae. Biotechnol. Adv. 25:294-306.

Chisti Y (2008). Biodiesel from microalgae beats bio-ethanol. Trends Biotechnol. 26:126-31.

Dayananda C, Sarada R, Bhattacharya S, Ravishankar GA (2005). Effect of media and culture conditions on growth and hydrocarbon production by *Botryococcus braunii*. Process Biochem. 40:3125-3131.

Demirbas AH, Demirbas I (2007). Importance of rural bioenergy for developing countries. Energy Convers. Manag. 48: 2386-2398.

Energy Assessment, Preface UNDP (2008). United Nations Development Programme, New York, USA, pp: 219-272. Am. J. Biochem. Biotech. 4 (3):250-254.

Gavrilescu M, Chisti Y (2005). Biotechnology a sustainable alternative for chemical industry. Biotechnol. Adv. 23:471-9.

Goldemberg J (2000). World Energy Assessment, Preface.United Nations Development Programme, New York, NY, USA.

Guillard RRL, Ryther JH (1962). Studies of marine planktonic diatoms. I. *Cyclotella nana* Hustedt and *Detonudla confervacea* Cleve. Canada. J. Microbiol. 8:229-239.

Haag AL (2007). Algae bloom again. Nature 447(31): 520-521.

Hossain ABMS, Salleh A (2008). Biodiesel fuel production from algae as renewable source of energy. Am. J. Biochem. Biotechnol. 4(3):250-254.

Hsueh HT, Chu H, Yu ST (2007). A batch study on the bio-fixation of carbon dioxide in the absorbed solution from a chemical wet scrubber by hot spring and marine algae. Chemosphere 66(5):878-886.

Huntley ME, Redalje DG (2007). CO_2 mitigation and renewable oil from photosynthetic microbes: a new appraisal. Mitigat. Adapt. Strat. Global Change. 12:573-608.

Jacob-Lopes E, Lacerda LMCF, Franco TT (2008). Biomass production and carbon dioxide fixation by *Aphanothece microscopica* Nageli in a bubble column photobioreactor. Biochem. Eng. J. 40: 27-34.

Kapdan IK, Kargi F (2006). Bio-hydrogen production from waste materials. Enzyme Microb. Technol. 38:569-82.

Klass LD (1998). Biomass for Renewable Energy, Fuels and Chemicals, Academic Press, New York. pp. 1-2.

Kulkarni MG, Dalai AK (2006). Waste cooking oil-an economical source for biodiesel: A review. Ind. Eng. Chem. Res. 45: 2901-2913.

Li Y, Horsman M, Wu N, Lan CQ, Dubois-Calero N (2008). Biofuels from microalgae. Biotechnol. Prog. 24:815-820.

Nagle N, Lemke P (1990). Production of methyl ester fuel from microalgae. Appl. Biochem. Biotechnol. 24-5:335-61.

Narendra MV, Shakti M, Amitesh S, Bhartendu NM (2010). Prospective of biodiesel production utilizing microalgae as the cell factories: A comprehensive discussion. Afr. J. Biotechnol. 9(10):1402-1411.

Ono E, Cuello J (2006). Feasibility assessment of microalgal carbon dioxide sequestration technology with photobioreactor and solar collector. Biosyst. Eng. 95(4): 597-606.

Ragauskas AJ, Williams CK, Davison BH, Britovsek G, Cairney J, Eckert CA, Frederick Jr WJ, Hallett JP, Leak DJ, Liotta CL, Mielenz JR, Murphy R, Templer R, Tschaplinski T (2006). The path forward for biofuels and biomaterials. Science 31:484-489.

Salisbury FB, Ross CW (1992). Plant physiology (4th Edition). Wadworth, Inc.

Sawayama S, Incue S, Dote Y, Yokoyama SY (1995). CO_2 fixation and oil production through microalgae. Energy Convers Manag. 36:729-31.

Sazdanoff N (2006). Modeling and simulation of the algae to biodiesel fuel cycle. Report, Department of Mechanical Engineering, The Ohio State University, United States.

Schenk PM, Thomas-Hall SR, Stephens E, Marx UC, Mussgnug JH, Posten C, Kruse O, Hankamer B (2008). Second generation biofuels: high-efficiency microalgae for biodiesel production. Bioenerg. Res. 1:20-43.

Schneider D (2006). Grow your own? Would the wide spread adoption of biomass-derived transportation fuels really help the environment. Am. Sci. 94:408-409.

Shay EG (1993). Diesel fuel from vegetable oils: Status and opportunities. Biomass Bioenergy 4:227-242.

Spolaore P, Joannis-cassan C, Duran E, Isambert A (2006). A commercial application of microalgae. *J. Biosci. Bioeng.* 101, 87-96.

Tan K., Lee K., Mohamed A. and Bhatia S. 2009. Palm Oil: addressing issues and towards sustainable development. Renew. Sustain. Energy Rev. 13(2):420-427.

Turkenburg WC (2000). Renewable energy technologies. In: Goldemberg, J. (Ed). World

Wagner L (2007). Biodiesel from Algae oil. Research report, MORA associates from www.fao.org/uploads.

Local blacksmiths's activity in the west region of Cameroon and their contribution to the development of micro hydroelectric power plants in that region

Kengne Signe Elie B.[1] , O. Hamandjoda[2], Tekounegning[3] and Nganhou. J.[2]

[1]Laboratory of Energy Research, Institute of Geological and Mining Research, BP 4110 Yaoundé, Cameroon.
[2]ENSP University of Yaoundé, I BP 8390 Yaoundé, Cameroon.
[3]Faculty of Agronomy and Agricultures Sciences, University of Dschang, BP 110, Dschang, Cameroon.

As a result of the nature of pollutant and exhaustible of fossils energies, developed countries have made use of renewable energy sources to make effective their energy systems. To stow on that vision in Cameroon, the promotion of micro-hydroelectric powers plants (MHPP) is a priority and any contribution to its ease is beneficial. The objective of this study is to characterize some main forges in the west region of Cameroon and show their technical input in the development of MHPP in that region. The survey was conducted in January, 2014. Socioeconomic data related to blacksmiths and main features were studied by the use of questionnaires, direct interviews, site observations and measurements. Main results show that the first forges was set in the Foto village in 1850 by blacksmiths from Nigeria. The studied forges are of traditional (62.5%) and modern type (37.5%). More than 50 products belonging to the main use sectors are manufactured, adapted and/or repaired. The main problems identified are acquisition of raw materials, the lack of equipment and limited technical knowledge. The blacksmiths has been involved in MHPP facilities since 1997, the installation year of the pilot MHPP with the participation of local craftsmen. Civil engineering structures, hydroelectric unit, transportation and distribution equipment are built by local blacksmiths using as much maximum materials as possible. This is often a source of malfunction. Thus, funding of local craftsman and equipment by the government or other agencies can render the forges useful for sustainable development in the west region of Cameroon.

Key words: Study, blacksmiths, micro hydroelectricity, funding, Cameroon.

INTRODUCTION

All developed countries have made use of renewable energy sources to make effective their energy systems. Energy from micro-hydroelectric powers plants (MHPP) has been an important component of renewable energy for countries which have the potential. For Cameroon to carry out this vision, the promotion of MHPP is a priority and any contributions to its ease are beneficial.

The access to electricity in Cameroon was about 49% in 2012. But this rate is about 23% in rural areas (Lighting Africa, 2012). In many of these areas, there are streams which flow; sufficient development could be interesting alternatives for the production of electricity from MHPP. Several sites favorable for MHPP have been identified, particularly in the Western Region, where without being exhaustive, about 35 sites have already been identified and characterized and fewer than 5 are in operation. The use of MHPP with a power of 153 kW started in 1997 with pilot project funded by World Bank in Bamougoum village. The first MHPP was set up in west region in part by the important contribution of local blacksmiths. Especially for the manufacture and supply of components such as turbines, water wheels, penstocks and many others mechanical accessories. But the activity of the local blacksmith remained much neglected despite its potential to promote development of production activities. Note, however, that so far in many developed countries, craftsmen make a significant contribution to the design, manufacture and repair of equipment and many tools used in rural and in urban areas. In Cameroon, despite their importance, few studies have so far been devoted to a better knowledge of craftsmen to consider improving their technical knowledge and facilities.

The objective of the present study was to characterize some main forges in the west region of Cameroon and show their technical input in the development of production activities generally and particularly MHPP in that region. This is in preparation for the improvement of blacksmiths skill and MHPP development.

The survey took place in the Foto village. It is situated in the west region of Cameroon and geographical coordinates are: 5° 27' 0" North, 10° 4' 0" East. The altitude varies between 1200 (Bamoun plateau) and 2740 m (Bamboutos Mount) with an average of 1500 m. The mountainous relief and the rainfall (1600 to 2000 mm) are suitable to hydroelectricity (Helvetas, 1981). The population is about 1.8 million inhabitants with a yearly growth rate of 2.8%. The average density is 128.5 inhabitants/km^2. The main economic activities are agriculture, livestock breeding, some industrial units of transformation (soap factories and brewery), forges and the commercial exchanges with the other regions.

MATERIALS AND METHODS

Description of MHPP

MHPP is a type of hydroelectric power that typically produces up to 500 kW of electricity using the natural flow of water. MHPP is frequently accomplished with a Pelton wheel for high head, low flow water supply. The installation is often just a small dammed pool, at the top of a waterfall, with several hundred feet of pipe leading to small generator housing (Gagliano et al., 2014; David, 2013). The intake then tunnels water through a pipeline (penstock) to the powerhouse building containing a turbine. The turbine turns a generator, which is then connected to electrical loads; this might be directly connected to the power system of a single building in very small installations, or may be connected to a community distribution system for several homes or buildings (Ramos et al., 2000). Usually, MHPP installations do not have a dam and reservoir. There is a lot of main mechanical component used for building and/or maintenance of the hydropower (turbine, trash rack, boxes, eyebolts, lampshades and screwdrivers).

Power from such a system can be calculated by the equation P=Q*H/k, where Q is the flow rate in gallons per minute, H is the head loss and k is a constant of 5,310 gal*ft./min*kW.

Potentiality for MHPP in Cameroon

In electricity, Cameroon has the second hydroelectric potential in sub-Saharan Africa (19.7 GW fairly technical potential for an energy production of 115 TWh/year). Now, less than 5% are operated. In view to promote the renewable energy in Cameroon, almost 200 sites for MHPP have been identified by ARSEL (the company for regulation of electricity sector), of which 35 were more precisely identified and characterized in the west region (Adam and Associates, 1997; TEKOUNEGNING, 2011).

Methods

This survey was conducted in January 2014. Socioeconomic data related to blacksmiths and main featured were studied by the use of questionnaires, direct interviews and site observations. Workspace dimensions were measured using a tape. During this study, data related to the products were manufactured for main sectors of activities and for development of the MHPP in the region collected. These data concerns history (origin and creation date of forges); socioeconomic features of blacksmiths (sex, age, education level, major activities and forge organization) and forge techniques (equipment, raw materials, manufacturing processes and manufactured products).

RESULTS

Background

Origin

The initial forge (forge mother) was started in Foto village around 1850 by a family of immigrants from West Africa, probably from Nigeria. It appeared that it was first installed in the nearby village: Bafou, in Beng quarter before abandoning this locality to settle in Balefang quarter, in Foto village. Another part of this family left to create another forge in another neighbouring village of Foto, Fongo-Tongo, in Balefang quarter.

Dates of creation and kinship between forges of Foto

The dates of the creation and the kinship between the forges of Foto are indicate in Table 1. It shows that apart from two recent forges, all forges of Foto descended from the same mother, which confirms the strong family tradition of forges. Since 1995, there have been no more new forge in Foto.

Table 1. Dates of creation and kinship between forges of Foto.

Forges	Kinship	Date of creation
I	Mother	1850
II	Sister	1860
III	Sister	1900
IV	Sister	1950
V	Sister	1987
VI	Sister	1989
VII	Independent	1993
VIII	Independent	1993

Table 2. Distribution of number of blacksmiths by forge.

Forges	I	II	III	IV	V	VI	VII	VIII	Total
Number of blacksmiths	45	50	35	26	12	09	08	08	193

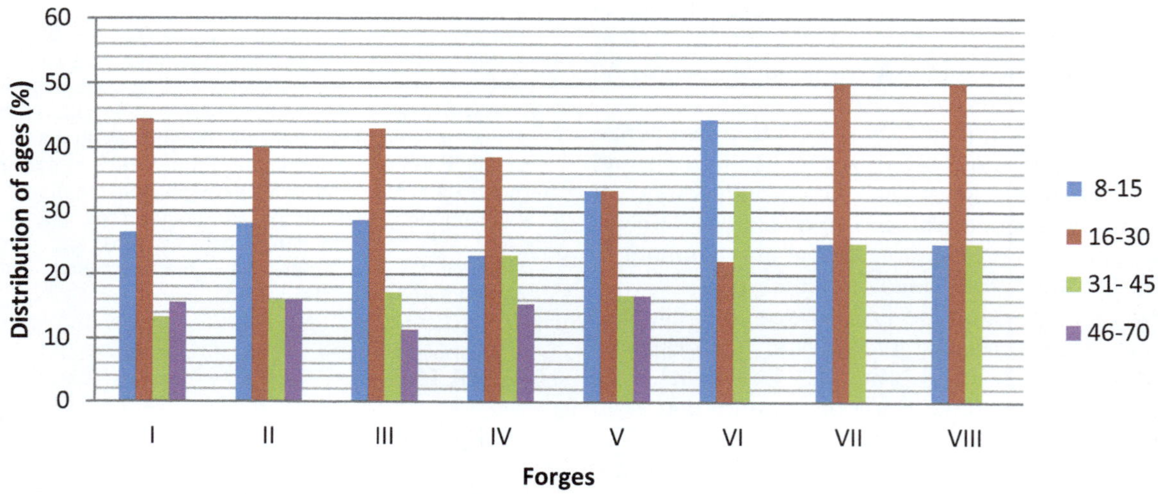

Figure 1. Distribution of blacksmiths by forge and age group.

Socio-economic features of blacksmiths

The socio-economic features of blacksmiths are indicated in Table 2, Figures 1 and 2.

Number of blacksmiths

The number of blacksmiths working per forge varies from 2 to 50 with an average of 24 per forge. About 50% of blacksmiths work temporarily, they are pupil and employees of others sectors. Forges VI, VII and VIII which are more modern possess the smallest number of blacksmiths, and have more efficient equipment. The distribution of number of blacksmiths by forge is summarized in Table 2.

Sex and age

According to the custom, blacksmiths are only men. Age of blacksmiths (Figure 1) varies from 8 to 70 years. These ages are classified in four group: [8-15], [16-30], [31-45] and [46-70]. Nearly 70% of blacksmiths are younger than 30 years since forging requires sustained physical effort. The forges VI, VII and VIII have youth for the sake of profitability. Theses forges have only 13% of blacksmiths but provide more than 95% of the global

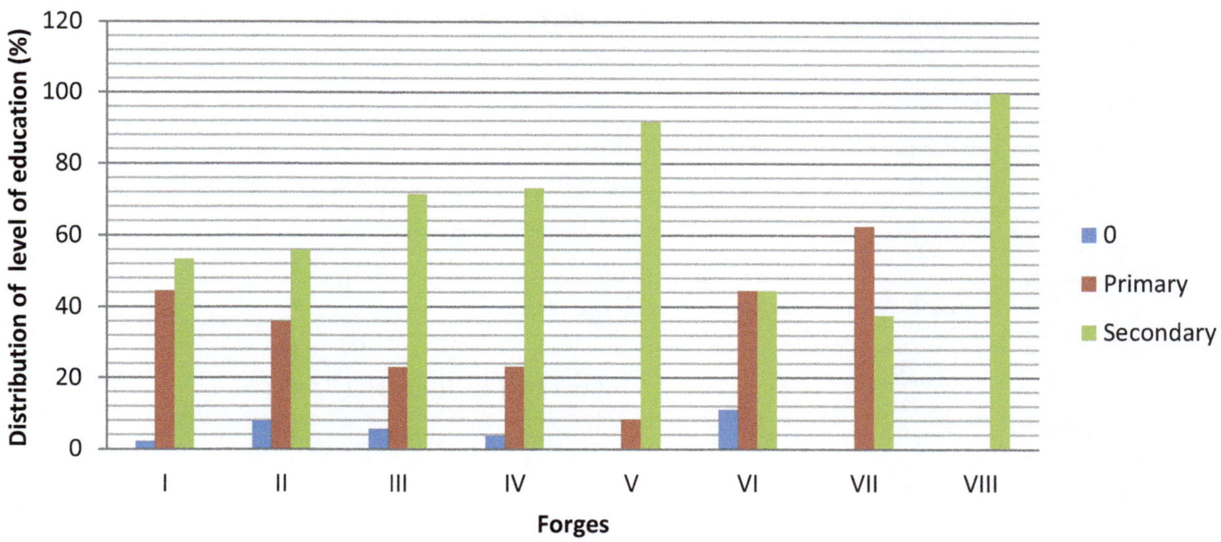

Figure 2. Distribution of blacksmiths by forge and level of education.

production of the forges. This is linked to the dynamism of the young, the productivity of modern equipment used and the organization of work which is the type boss-salaried employees.

Level of education

The distribution of blacksmiths according to the level of education and per forge is summarized in Figure 2. More than 95% of blacksmiths attended school with close to 62% of them having a secondary education. The high proposition of blacksmiths with secondary education is an asset in the context of training, extension and possible technical innovation adoption.

Main activities of blacksmiths

Of the blacksmiths interviewed, 56% them have the forge as the main activity, with agriculture (March-April) the second. The relatively high percentage of temporary blacksmiths is related to the proportion of young students engaged in the forge during school holidays. Other blacksmiths are temporary employees of the liberal sector, public sector or traders.

Techniques of forges

Forge equipment

Repartition of forge according to the type of equipment: Based on UNIDO (United Nations Industrial Organisation) technical standards (Starkey, 1997), one can classify the Foto forges in 2 categories according to the equipment used as shown in Table 3. This table indicates that 62.5% of forges fall in the category of so-called traditional forges. They are less efficient.

Workspaces and blacksmith equipment used: Traditional forges use simple hangars (Table 4) less than 3 m high and opened on all sides. Conversely, the modern forges are housed in real workshop spaces 3 to 4 greater than those traditional forges. The more spaces and comfort found in modern forges arise and are derived from the need to protect a greater number of facilities and also to secure the large volume of storage of raw material and end products. The basic equipment are the same in all forges, although more sophisticated ones are found in modern forges. It is important to note that protective equipment such as safety goggles, gloves and overalls are only in modern forges (CEEMAT, 1974).

Sources of raw materials

Table 5 gives the sources of raw materials for the different categories of forges. Traditional forges get their raw material from two main sources. On the contrary, modern forge materials are from hardware stores.

Manufacturing process

Table 6 summarizes the manufacturing operation. The tasks such as welding and painting are executed only in modern forges. Even if other operations are carried out everywhere, they are faster and more efficient in modern

Table 3. Repartition of forges according to the type of equipment utilized.

Forge categories	Type of equipment utilized	Forges	
		Number	% total
Traditional (I, II, III, IV, V)	Traditional	5	62.5
Modern (VI, VII, VIII)	Traditional and modern	3	37.5
Total		8	100.0

Table 4. Workspaces and blacksmith equipment.

	Forges / *Workspaces*	Area A(m^2)	Number	Total (%)
01	Hangar (traditional forge)	A<20	5	62.5
02	Workshop (modern forges)	A>80	3	37.5

Table 5. Sources of raw materials for the different categories of forges.

Categories of forges	Sources of major raw materials used by forges	
	Recycling materials	**Hardware materials**
Traditional	1 Car wrecks: bodywork, springs, engine part, chassis and tires 2. Saw blades, etc.	Metal sheets, concrete reinforcement steel, angle irons, galvanized pipes, etc.
Modern	0	Metal sheets, concrete reinforcement steel, angle irons, galvanized pipes, welding electrode, etc.

Table 6. Manufacturing operations and equipment used by blacksmiths.

Operation	Equipment or tools used
Heating	Firewood and manual or electric bellows
Cutting	Chisel and hammer, shears; grinder; hack saw
Bending	1. Stone, engine block and other metallic block as anvil 2. Mechanic or hydraulic press and vice
Modelling	Tree-trunk, metallic block, cylindrical iron bar, hammer, press and template
Drilling	Engine block holes; awl; hammer; electric drill; press drill
Riveting	6 or 8 mm iron bar, hammer or press
Welding ^{m}	Arc or welding
Smoothing and sharpening ^{m}	File or grinder
Painting ^{m}	Brushes and paint

m: Operations made only by modern forges.

forges because of their facilities. It is important to note the progressive inter-dependency between the forges; the traditional forges require modern forge services for tasks that their facilities do not permit them to execute (soldering, cutting, etc)

Main manufactured products

Main manufactured products in both types of forges: Table 7 gives the annual production of manufactured products in both types of forges. The hoe is the most manufactured tool. This is because it is highly utilized in all phases of the agricultural calendar and has a low life span. On the other hand, the trident, which is produced less, is manufactured using raw materials of a higher quality and is less frequently utilized. Regarding the MHPP facilities, the World Bank funded the construction of a pilot MHPP in Bamougoum village in 1997 with the participation of artisans (Adam and Associates, 1997). This has created a big impact in this village as it attracts

Table 7. Annual production of main manufactured products and MHPP equipment by categories of forges.

Manufactured tools or equipment	Annual production (number)		
	Traditional forges	Modern forges	Total
Agriculture tools			
Hoe	5 000	150 000	155 000
Jab planter	3 000	8 000	11 000
Axe	700	6 300	7 000
Dabba (tool used to dig)	1 000	4 000	5 000
Trident	160	2 200	2 360
Kitchen knife	4 000	0	4 000
Cooking pot	3 500	0	3 500
MHPP equipment			
Trash rack	0	30	30
Metallic pipe (2 m in length)	0	100	100
Pelton turbine	0	10	10
Cross flow turbine	0	4	4
Water wheel	0	2	2
Boxes	0	300	300
Eyebolts	0	3 000	3 000
Lampshades	0	1 500	1 500
Screwdrivers	0	200	200

development committees of other villages who seek to acquire the same skills in order to replicate that technology in their localities. The MHPP they set up supply electricity to homes and public places. The electricity is also used for other various activities such as food processing. One can expect improvement in the near future with blacksmiths training in the MHPP technology as there is no company in Cameroon making MHPP equipment (Tekounegning et al., 2008). One also notes that blacksmiths dealing with MHPP are from the same modern forge.

Time, period of manufacture and life span of the main manufactured products

The time, period of manufacture and life span of the main forged tools are shown in Table 8. The hoe, the most requested tools was made 30 times faster in the modern forges than in the traditional one. Apart from axes and tridents, the other tools are manufactured according to the agricultural calendar: so hoes are requested during periods of cultivating and weeding; the jab planter during planting, weeding and tuber harvesting period; and the "dabba" (tool to work soil surface) are made during land preparation. The life span of products varies from 1 to 20 years.

Price comparison of forged objects and industrially produced objects

Table 9 lists the forged objects, their sectors of use and sale price as compared to equivalent objects of industrial manufacture. The forges manufacture or repair a very wide range of tools used in various sectors of activities. The main sectors are agriculture and construction. The equivalent products industrially manufactured or imported cost 2 to 10 times higher than those from the forges. This is related to the quality, the functionality and the life of the items. The FCFA was choosen because it is the local currency used (1 Euro = 656 FCFA).

Some important items not manufactured by local industries

Table 10 shows some important items not manufactured by local industries. The forges usually manufacture and repair their equipment. MHPP equipments are not on the market (Adam and Associates, 1997). They are manufactured, on demand, only by one modern forge. Without the contribution of blacksmiths, the development of MHPP in the west region of Cameroon would not have been possible.

Some important photos of local products forged

Photos 1 and 2 show agriculture products, Photos 3 and 4 show turbines (MHPP products), Photos 5, 6 and 7 shows some operation and equipments of forge.

DISCUSSION

With a simple calculation based on data from Tables 7, 9 and 10, in considering the total annual production (Table

Table 8. Time, period of manufacture and life span of the main manufactured products and MHPP equipment.

Manufactured tools or equipment	Fabrication time (minutes)			Life span (years)
	Traditional forges	Modern forges	Period of fabrication	
Agriculture tools				
Hoe	60-120	2-4	Nov-April	1-2
Jab planter	60-180	40-60	Dec-March	1-3
Axe	360-480	40-60	All year	10-20
Dabba (tool used to dig)	120-240	50-60	Dec-Aug	10-20
Trident	480-600	30-40	All year	1-2
Kitchen knife	20-40	-	April-Aug	2-4
Cooking pot	150-200	-	All year	15-35
MHPP equipment				
Trash rack	-	120-240	Dry season	2-4
Metallic pipe (2 m in length)	-	120-240	Dry season	4-6
Pelton turbine	-	720-1020	Dry season	6-8
Cross flow turbine	-	720-960	Dry season	6-8
Water wheel	-	1020-1740	Dry season	4-8
Boxes	-	60-90	All season	4-10
Eyebolts	-	1-2	Dry season	15-30
Lampshades	-	20-40	Dry season	5-8
Screwdrivers	-	35-50	All year	15-25

Table 9. Main forged tools distributed according to the sector of use, their functions and price comparison in the local market.

Sector of use	Name	Functions	Compared unit price (FCFA)		
			Forge (C_1)	Industry(C_2)	$C_2 - C_1$
	Needle	Sew bag after bagging products	60	-*	-
	Crowbar	Unearth stones	5400	30000	+24 600
	Wedge	Split tree trunks	1800	3000	+1 200
	Knives	Work in the kitchen and table	200	1200	+1 000
	Raffia knife	Tap raffia wine	950	-	-
	Trident hoe	Weed, hoe, stir soil surface	2050	3700	+1 750
	Skimmer	Remove food from heated oil	375	600	+225
	pruning	Cut bananas leaves, pick fruit	480	-	-
	Sickle	Cut straw	480	-	-
	pitchfork	Return, load manure	2400	7200	+4 800
	scraper	Grate tuber or fruit	480	4320	+3 840
Agriculture	Axe	Fell, cut, split trees	4800	14500	+9 700
	Hoe	Plough, weed, stumble, dig	1800	3000	+1 200
	Machete	Cut, split	2000	4900	+2 900
	Mould	Give forms to pastry making	960	1920	+960
	shovel	Dig, collect the soil	2300	3700	+1 400
	Pickaxe	Pick, loose stony soil	2400	9600	+7 200
	Planter	Plant seeds, weed, unearth tubers	725	2200	+1 475
	Stove	Fry, warm foods	1450	6000	+4 550
	Rake	Clean, level soil	1000	2200	+1 200
	Flint	Make fire	600	-	-
	Probe	Check the quality of seed bagged	960	-	-
	Sieve	Sift foodstuffs	500	6000	+5 500

Table 9. Contd

	Trident	Break up, dig the soil, weed	4800	18600	+13 800
	Nail puller	Pull mails	2300	3700	+1 400
	Chisel	Cut sheet metal, break the concrete	840	5040	+4 200
	Shears	Cut metals	72000	108000	+36 000
	Art knife	Work bamboos	600	-	-
	Window	Ventilate and secure home	24 600/m^2	-	-
Construction	Scraper	Smooth wood	1200	-	-
	Saw	Saw wood	1800	3600	+1 800
	Hammer	Mail	1700	2200	500
	Hinges	Link post and door wing	1200	1920	+720
	doors	Allow user entrance, secure home	26 400/m^2	-	-
	Awl	Pierce sheet metal	850	1450	+600

*-: not found in local market.

Table 10. Some equipment of animal husbandry, hunting, war, socio-cultural rituals, forges and MHPP and their price on the local market.

Sector of use	Name	Functions	Price FCFA
Breeding, hunting, war	Iron hook	Stabilize attached animal	700
	Bell	Locate the dog during hunt	800
	Lance	Kill the game or the enemy	850
	Trap	Catch the game, mice	2 400
Socio-cultural rituals	Bracelets	Protect against evil spirits	250
	Bells	Rhythm dance	600
	Healer knife	Cure diseases caused by criminals	500
	Razor	Shave hair and beard	700
	Sabre	Perform warrior dances	900
Forge equipment	Anvil	Beat iron	-
	Vice	Tighten iron	-
	Millstone	Grind and refine forged items	-
	Shears	Cut metals	-
	Press	Pierce, give shapes to objects	-
	Bellows	Inflame fire	-
MHPP equipment	Trash rack	Prevent trash getting into conduits	1500 to 6000
	Metallic pipe (2 m in length)	Convey water under the pressure	8000 to 18000
	Pelton turbine	Transform energy of flowing water into mechanical rotary motion transferred to the generator	200000 to 800 000
	Cross flow turbine		140000 to 300 000
	Water wheel		300000 to 1100 000
	Boxes	Secure devices	1500 to 10000
	Eyebolts	Anchor cables on wooden poles	400 to 1000
	Lampshades	Protect outdoor lamps	1000 to 3000
	Screwdrivers	Tie or untie screws	500 to 2000

Photo1

Photo3

Photo5

Photo2

Photo4

Photo6

Photo7

Photos 1 - 7. Some important products of forge.

7) and the unit prize of manufactured main products and MHPP equipment (Tables 9 and 10), an estimation of the turnover of the local workshops studied is about 750 000 000 FCFA (= 1 14 293 Euros). Manufactured items are used both in rural and urban areas. They are cheaper than industrial products but in general, their quality is poorer. The prices of these products correspond to the level of current income and purchasing power of people.

For the same reason, the MHPP can also have a good valorisation in the west region of Cameroon, with the contribution of these local products manufactured by forge.

Conclusion and recommendation

Main results show that the first forges was set in the Foto village towards 1850 by blacksmiths from Nigeria. The studied forges are of traditional (62.5%) and modern type (37.5%). More than 50 products belonging to the main use sectors are manufactured, adapted and/or repaired. Although, there are a few forges, about 193 person work there, and it is appreciable. They produce a wide range of tools for the different sectors of the population activities and at cost compatible with their purchasing power. These products are at least 40% cheaper than products from industries. The value of production is unrelated to investment but is usually high for rural areas.

Actually, the State of Cameroon promotes renewable energies, of which MHPP is an essential component (Law No. 2011/022 on 14th December 2011 governing electricity sector in Cameroon). The contribution of local blacksmiths is very interesting to boost MHPP in Cameroon, with regards to the west region. Local forges have a lot of potential in the production of MHPP components in general and particularly the turbines and water wheels, penstocks and many other mechanical accessories.

More than 95% of blacksmiths attended school with close to 62% of them having a secondary education. The high proposition of blacksmiths with secondary education is an asset in the context of training, extension and possible technical innovation adoption. The areas for improvement should also focus on better organization of forges, equipment and training of blacksmiths. This is left government and non-governmental organizations.

Conflict of interests

The authors did not declare any conflict of interest.

ACKNOWLEDGEMENTS

The authors thank the blacksmiths of Foto, without the collaboration of whom this work would not have been possible.

REFERENCES

Adam H and Associates (1997). Pico-hydro in Cameroon. On behalf of ESMAP, World Bank, Washington DC, USA. 8 South moor Rd, Oxford OX2 6Rd, UK.

CEEMAT (1974). Manual of motorization of tropical cultures. Ministère de la cooperation, Paris France. 1(5281):661 pp.

http://www.inshp.org/Img_Lib/UploadImg/201232712109370.pdf

Helvetas, S. (1981). Manuel technique pour l'Approvisionnement en Eau des zones Rurales MiNAGRi.

Litghting Africa (2012). Report of Cameroon Politic lighting Africa. Yaoundé Cameroon.

Starkey P, Blary D, Le Thiec G (1997). Systèmes d'attelage et matériels à traction animale. Lengericher H Allemagne. 278 pp.

Tekounegning, T. (2011). Modernization of the Production components and transmission Lines of the bapi Village micro Hydroelectric Power Plant. Journal of Applied Science & Engineering Technology, 4.

Tekounegning, O., Hamandjoda, O., Ndzana, B., Lucien, M., & Nganhou, J. (2009). Constraints of micro hydroelectric power plants in the west region of Cameroon. Global Journal of Engineering Research, 8(1 and 2):27-33.

Ramos H, Betâmio de Almeida A (2000). Guidelines for design of small hydropower plants. Western Regional Energy Agency & Network and Department of Economic Development. https://fenix.tecnico.ulisboa.pt/downloadFile/3779576926045/

Gagliano A, Tina G M, Nocera F, Patania F (2014). Technical and Economic Perspective for Repowering of Micro Hydro Power Plants: a Case Study of an Early XX Century Power Plant. Energy Procedia. 62: 512-521.

David Kilama Okot (2013). Review of small hydropower technology. Renewable and Sustainable Energy Reviews. 26 : 515–520.

22

Potential carbon credit and community expectations towards viability of REDD project in Ugalla- Masito ecosystem: A case of Ilagala and Karago villages, KigomaTanzania

Nicholaus Fabian Mwageni, Riziki Silas Shemdoe and Robert Kiunsi

Ardhi University, P.O. Box 35176, Dar esSalaam, Tanzania.

The paper presents results of the potential carbon credit and community expectations towards viability of REDD+ projects in Ugalla- Masito ecosystem using a case of Ilagala and Karago villages whereby REDD+ is being piloted. Various data collection methods were employed and these included focused group discussion, interview, structured questionnaires and document analysis. Results of the study indicate that, there are two suggested payment sharing approaches associated with different preferences at both household and village government levels. These include cash payments to households for compensating the opportunity costs incurred and the second is that funds should be given to village government for community developmental projects. Analysis of the opportunity costs, marginal (incremental) revenues from forest carbon stock as well as the conceptual trend of forest biomass indicates that, there is probability for the project to flop in a very short time. Therefore, in order to ensure the future viability of REDD+ and its associated projects as well as ensuring sustainability of people's livelihoods, any REDD+ associated activities should harmonize community preferences and balance them with project goals by supporting communities in their alternative livelihood activities.

Key words: Carbon stock payments, community preferences and REDD+ project viability.

INTRODUCTION

Carbon exists on the earth's atmosphere primarily as the gas-carbon dioxide (Vashum and Jayakumar, 2012). In the recent past, it has gained a lot of attention as a chief among the greenhouse gases, as it has potential to influence the climate pattern of the world. Why carbon emission drew much attention is because carbon dioxide has long residence time in the atmosphere compared to other greenhouse gases (Brown, 1993).

Forest's ecosystem is one of the most important carbon sinks of the terrestrial ecosystem. It removes the carbon dioxide from the atmosphere during photosynthesis and stores the carbon in the plant tissues, forest litter and soils. This process is more prolific in a relatively new forest where the growth of the trees is still rapid and it decreases as the forest gets old. The carbon stored on the forest trees are mostly referred to as the biomass of the tree or forest, and the carbon concentration of the different parts of a tree is assumed to be 50% (Brown, 1986) or 45% of the dry biomass (Whittaker, 1973). Nevertheless, according to a study conducted by Lugo and Brown (1992), it was revealed that half of the so called matured forests could also sequester carbon in a rate that could be further increased if human pressures like deforestation and forest degradation activities are reduced or removed from the forests.

These activities have tremendous impacts on biodiversity and carbon emissions and hence on the world's climate. IPCC (2007) highlighted that, tropical deforestation and forest degradation accounts for approximately 18-20 % of global greenhouse gas emissions. As a result, current approaches to address climate change include strategies to reduce emissions from deforestation and forest degradation (REDD) through forests conservation practices of which has captured international attention (Stern, 2006; IPCC, 2007). The United Nations Framework Convention for Climate Change (UNFCCC) of 2010 introduced REDD strategy as the financial mechanism compensating countries for the prevention of deforestation and forest degradation that would otherwise occur.

The 2009 Copenhagen accord of the UNFCCC recognized REDD as a valid mitigation strategy and has increased interest in and funding it. Alongside effective greenhouse gas mitigation, the international accord emphasized that REDD may offer other co-benefits like ecosystem restoration, improved land tenure and socioeconomic development due to carbon market that will increase the income of forest communities which has been referred to as REDD+.

Carbon market mechanisms convert emissions reductions from REDD initiatives into carbon credits that industries and countries can use to comply with emissions commitments. The basis for carbon market community-based forestry initiatives, from a business perspective, is a legal contract between the buyer of the carbon credits and the seller (the communities). In many instances intermediaries will pay a key role moving the credits from the level of production (the community forest) to the marketable place (TNRF, 2008). Several payment mechanisms under REDD initiatives do exists. One of them is based on a national level carbon fund that would be the recipient of financial flow for avoided deforestation, carbon credit sales and so on.

In Tanzania, the fund-based approach has been argued to fit with the realities of communal land and forest tenure under village governance and participatory forest management (Burges et al., 2010). The first is an effort-based payment, which rewards communities for improved forest management activities. The second is an output-based payment, which rewards forest managers for empirically verified outputs such as improved forest condition and reduced deforestation (TFWG, 2010). The effort-based approach would reward villages equally even when their ownership of forest resources and utilization of village differs. The output-based approach would benefit more communities who have increased their carbon stocks and are able to demonstrate it by carbon baseline, monitoring and calculation (Bolin et al., 2012).

Kigoma region is one among the areas in Tanzania where REDD project has been piloted. Communities have accepted REDD project implementation with anticipation that the project will be compatible with their local poverty reduction strategies and development goals. In this regard, communities in REDD project areas have been foregoing forestry products despite the fact that their demand is very high compared with what the district or region can supply. These include firewood and charcoal (frequently used for sale or fuel) and timber and building poles (URT, 2009). Thus, there is opportunity cost behind acceptance of REDD project implementation since local communities have been losing income they used to accrue from forestry products. This paper, therefore, confines itself at implication of the potential carbon credit and community expectations towards viability of REDD projects using a case of Ilagala and Karago villages located in Kigoma whereby REDD+ has been piloted. This paper therefore presents the results of the study on the adequacy of potential carbon credit in meeting local priorities and needs as far as the viability of REDD project is concerned using the case of the two above mentioned villages in Kigoma Tanzania.

MATERIALS AND METHODS

Study area

Presence of REDD pilot project and low income communities depending on the forest ecosystem goods and services, were used as the basis for selecting Ilagala ward where Ilagala and Karago villages are situated (Figure 1). The villages are among the seven villages forming Masito-Ugalla Ecosystem where the international Non-Governmental Organization (NGO) (Jane Goodall Institute (JGI)) has been implementing REDD+ pilot project. The villages are located along Lake Tanganyika shoreline within Kigoma rural District. The REDD implementation approach in these villages link with participatory forest management (PFM).

Karago village is bordered to the north by the Ilagala village, to the south by Sunuka village, to the east by Masito-Ugalla forest reserve and southeast by Songambele, and to the west by Lake Tanganyika. It has 8,703 people (about 1600 households) (2012, census) and a land area of 11,218.88 ha has reserved five forests with a size of 5646.49 ha (Figure 2). However, the village has agricultural land of 5137.88 ha and 432 ha for settlement.

On the other hand, the Ilagala village is bordered to the north by Mwakizega village, to the south by Karago and Songambele villages and east by Masito-Ugalla forest reserve, and to the west by Lake Tanganyika. The village has 21, 246 people (about 3500 households) (2012, census) with a land area of 23,840.13 ha and has reserved three forests with a size of 3402.2 ha (Figure 3). However, the village has agricultural land of 14879.91 ha, 4904.85 ha for settlement and a reserved forest for mining activities of 653.17 ha.

Agriculture is the major source of income in Karago and Ilagala villages. If the area under crop cultivation is distributed equally to

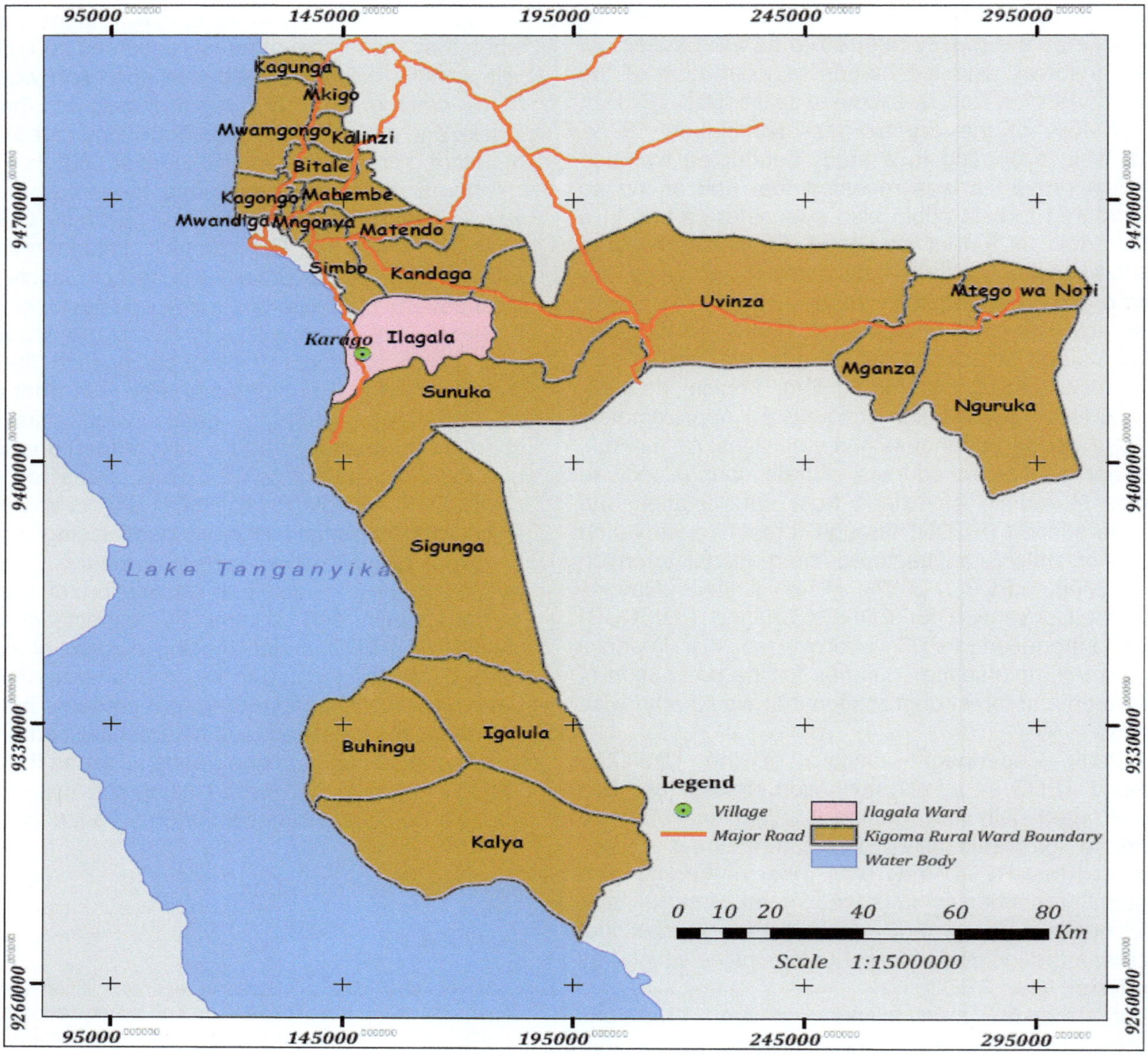

Figure 1. Selected case study villages in Ilagala ward, Kigoma Tanzania.

the total population based on 2012 census every single person in Karago village will be cultivating an area of about 0.6 ha while in Ilagala it will be 0.7 ha.Agricultural production in Karago and Ilagala villages depend mostly on rains for crop growing. Major crops grown include maize, beans, cassava, bananas, groundnuts and oil palm. However, despite of cultivation, there are other socio economic activities like beekeeping, cattle keeping, carpentry and petty business.

Data collection methods and analysis

A combined methodology that involved qualitative and quantitative approaches was adopted. Consequently, multiple methods and techniques for data collection and analysis were used. Data collection methods which were applied included key informants interviews, focus group discussions and in-depth interviews using a standard questionnaire that was structured to obtain the information on potential carbon credit and community expectations towards viability of REDD project. Under the key informants' interviews, different well knowledgeable representatives of the actors at Kigoma rural district level and in the respective villages were interviewed. The interviewee also included the representatives of the NGO which was Jane Goodall Institute (JGI) that is implementing the REDD pilot projects in the respective villages. Other key informants interviewed included ward and village officials, religious leaders and primary school teachers who were regarded as well knowledgeable people at the local level regarding the natural resources issues and again they were useful in advising on the specific community to be interviewed during the in-depth interviews that were carried out in their respective areas. Under the focus group discussions, meetings with village representatives in the selected communities were held to discuss issues related potential carbon credit and community expectations towards viability of REDD project in their respective villages. More information was collected through in-depth interviews using a structured questionnaire which had both open and close ended questions. Under the in-depth interviews, a total of 70 household

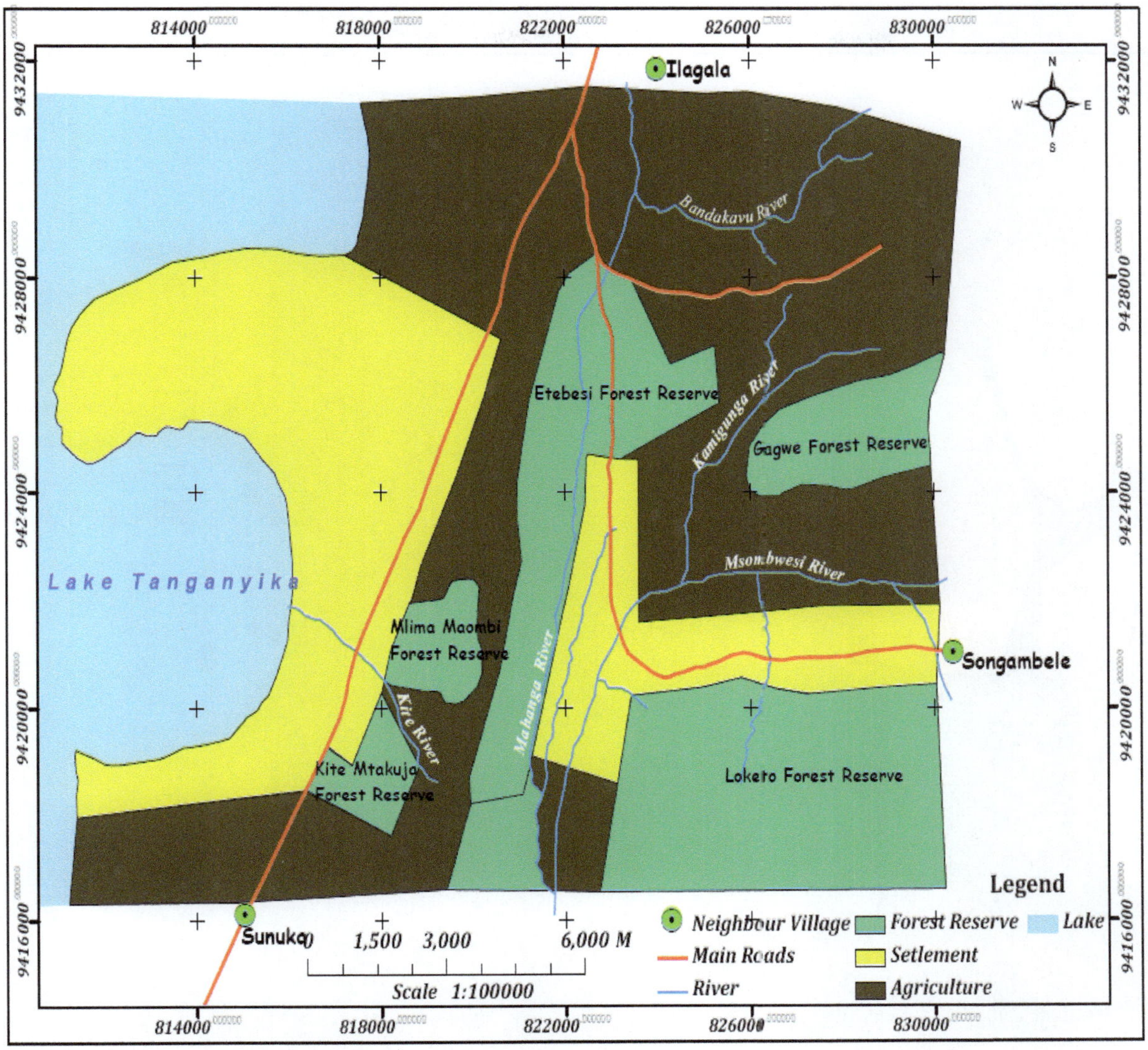

Figure 2. Map showing the land use of Karago Village.

Table 1. Characteristics of the interviewed households.

Village	Households interviewed	Sex		Age (%)			Education level (%)			Residence history (%)	
		%Male	%Female	18-35	36-60	>60	Illiterate	Primary	Secondary	Born in the Village	Born in other village
Karago	30	80	20	30	70	0	3.3	86.7	10	86.7	13.3
Ilagala	40	65	35	25	72.5	2.5	2.5	97.5	0	72.5	27.5

representatives were interviewed. Table 1 gives the characteristics of the interviewed households. However, during the field study, the following criteria were used in selecting respondents especially households: 1) market access – people living near the roads andfar from the roads or people living near or far from the forests; 2) Socio-economic activities for example people who are involved with fishing activities; deforestation rates - areas with high and low deforestation rates.

Most of the qualitative data collected were analyzed using content analysis. Content analysis was carried out for the data that were collected through focus group discussions. Data collected from the questionnaire were analyzed by SPSS whereby descriptive statistics such as mean, standard deviation and percentages were determined.

Opportunity cost estimation for fuel wood accessibility

In order to estimate the opportunity costs for fuel wood accessibility per household in the case study villages, the monetary valuation was done by using the method(s) adopted from Chopra, (1993) as presented in Table 2.

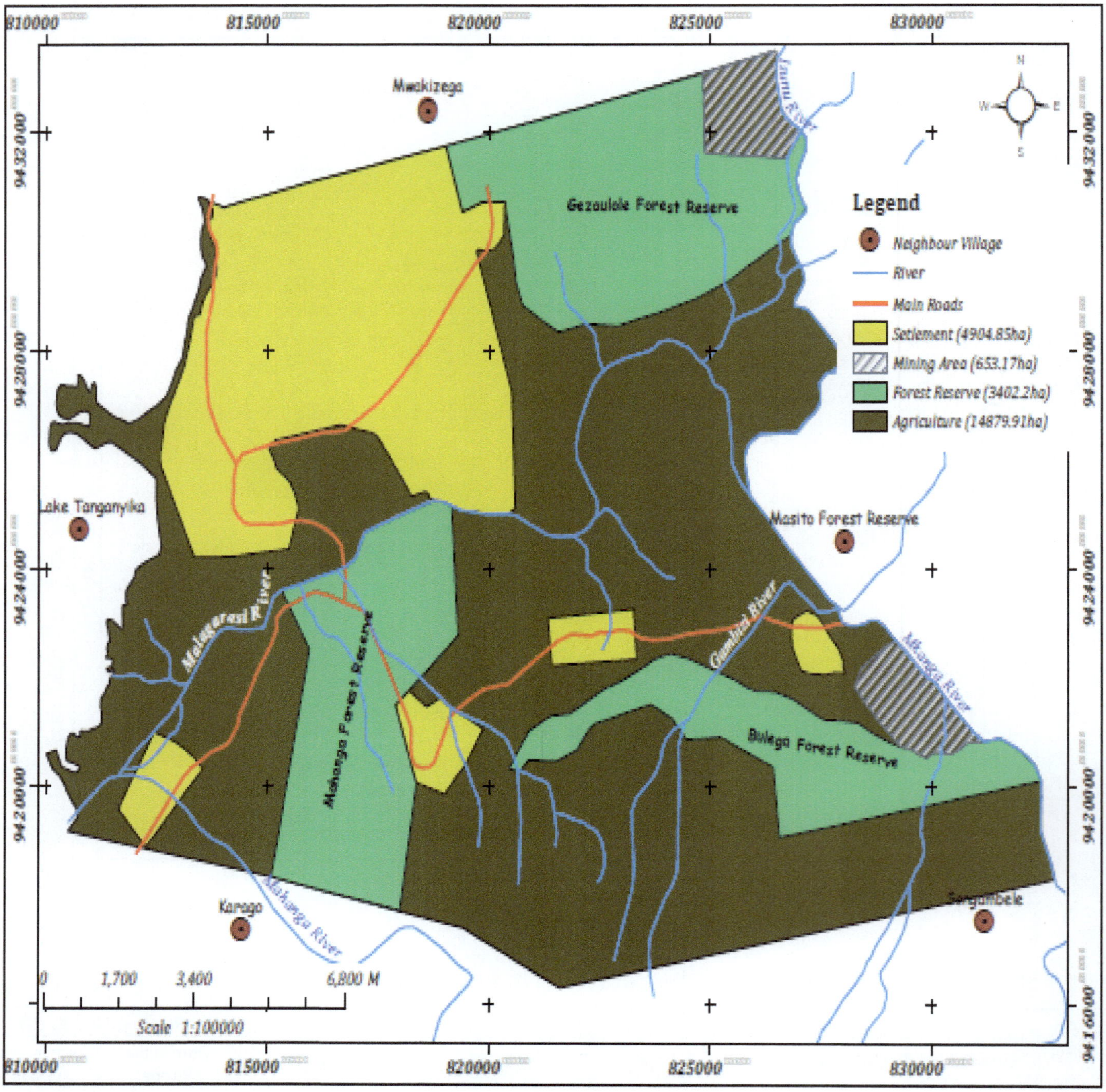

Figure 3. Map showing the land use of Ilagala village.

Table 2. Total Monetary Valuation of foregone fuelwood per household.

Type of Fuelwood	Cost code	Method of Approximation
Firewood	Cost of firewood (PF)	PF= Household consumption per week X UNDP nominal price per m^3 X Number of weeks per year
	Cost of Labor (Time in collection) (CL)	CL=Time taken for firewood collection per day X Firewood Collection Frequency per week X UNDP nominal cost per effective hour X number of weeks per year
Charcoal	Cost of charcoal (MP)	Village current market price

RESULTS AND DISCUSSION

Forests and community livelihoods before and during REDD project

As is the case in many parts of Tanzania, forests have a significant role of securing livelihoods to the surrounding communities. Forest plays three roles to the communities. These include safety netting, supporting consumption, and being an out of poverty pathway (Kajembe et al., 2012). Communities in Karago and Ilagala villages where this study was conducted use surrounding forests for collecting fuel wood, construction materials and non-timber products.

Community representatives interviewed in this study mentioned that during the implementation period of the REDD+ project in the case study area, construction materials and non-timber products have been exploited in sustainable manner but there has been a big concern on fuel wood accessibility. During this implementation period, although forest conservation was highly encouraged to the selected forest reserve, there are no clear identified forests for fuel wood collection and instead, majority (45% -77%) of community members in the case study villages used their agricultural fields for fuel wood collection. This has affected accessibility of fuel wood to residents in terms of space, time and quantity. People who were close to the forest reserves are now obliged to walk a long distance in search of firewood. The average time taken by household for firewood collection per trip was reported to have increased from 5 to 7 h per day and they do that twice a week. Changes in fuel wood accessibility has compelled 43 and 13% of the respondents to buy fuel wood in Ilagala and Karago villages respectively which implied the need for more money to support energy availability for the respective households.

The study has further revealed that, REDD+ project has reduced fuel wood consumption at household level of which is essential for forest ecosystem restoration in the case study villages. The current household consumption of firewood is 1 m^3 per week for Karago and 1.5 m^3per week for Ilagala while for charcoal is 0.75 of a sack (22.5 kg) and 0.7 of a sack (21 kg) per month for Karago and Ilagala villages respectively. The acceptance of REDD+ project has therefore been associated with opportunity costs of which the respondents were willing to take with anticipation that the project will be compatible with their local priorities and needs.

Potential carbon credit and community expectations

Carbon stock estimated in the village forest reserves indicated that Karago and Ilagala villages were required to receive a total $15,625 and$ 11,250 respectively for the first year of the project (phase one). Payments may vary for the next years with respect to amount of the estimated incremental carbon stock of the forest reserves. In these villages, two different payment sharing approaches associated with different preferences at both household and village government level were suggested; these are cash payments to individuals for compensating the incurred opportunity costs which include labour and procurement costs. This was suggested by 20% of the respondents. The second modality was suggested by 80% of respondents. The interviewed community representatives further indicated that funds should be given to village government for development projects and protecting forest reserves (Figure 4).The differences in payment modalities imply that there is need for a harmonized payment mechanism.

Potential carbon credit at household level

Due to spatial and temporal changes of fuel wood accessibility brought by the implementation of the conservation project through REDD+ in the area, some households needed the potential carbon credit for compensating the incurred opportunity cost. People use more time than before the start of the implementation of the REDD+ project. The increased time is used in search of firewood due to the increase in distance. Results revealed that 3% of interviewed households in Karago buy fuel wood while 43% in Ilagala buy fuel wood. In this regard, people may need to be paid $ 1239-$ 1,664 per household per year as compensation for firewood, and $ 40-$ 60 per household per year as compensation for charcoal.

Therefore, in general, the total amount of money to be paid directly to each household in the villages is supposed to be $ 1279-$ 1,724 per year for fuel wood consumption. This amount is far by 92-94 percent compared to the amount that the potential carbon credit which could be divided to households. The amount of money to be given to each household as compensation could not be compared in any case by the forgone opportunity costs.

Potential carbon credit at village government level

In the key informant interviews with the village leaders, it was reported that amount of money to be given to each village government was reported not to be enough to complete even one development project. For instance the amount of money required to complete a water project for the people of more than 1500 households was not less than 200 million (Carlevaro and Gonzalez, 2008). This amount of money is higher by 86 - 88% than the amount of money expected to be given to the households and village government put together. On top of that, the money will be decreasing due to the fact that the incremental carbon stock present in the forests will keep on decreasing too. The achievement of such water

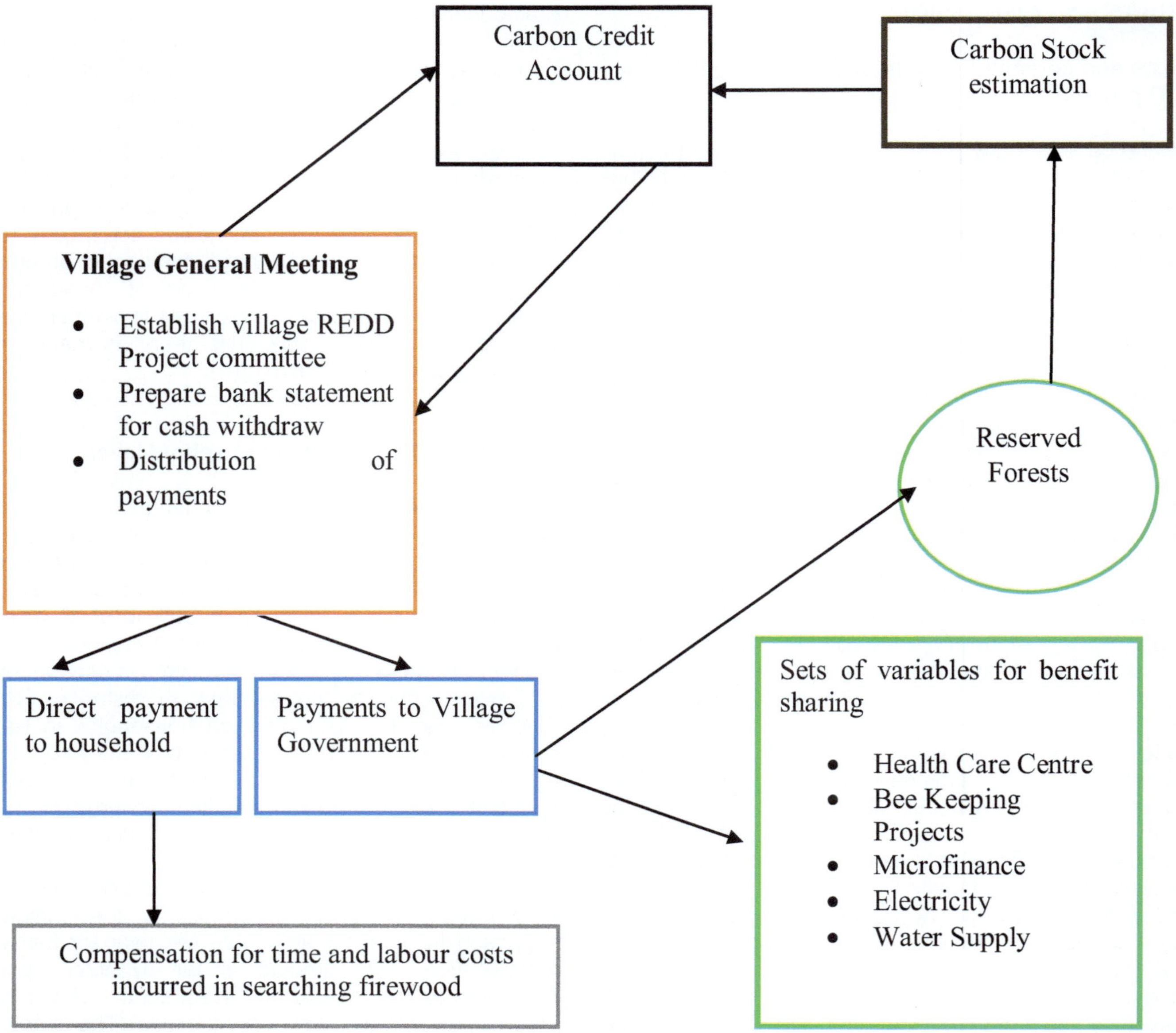

Figure 4. Model for REDD project revenue benefit sharing and payment mechanism for potential carbon credit.

project for example would need an accumulation of money for more than 8 years consecutively under assumption that the amount of money given will not change ($15,625 for Karago and $11,250 for Ilagala) which will not be the case especially when considering the incremental carbon stock.

According to Kabura et al. (2013), the cost of managing a hectare per year is about $6.5. Translating this cost to the case of Karago forest reserve with an area of 5137.88 ha, about $33, 400 per year while for Ilagala forest reserve with an area of 3402.2 ha, about $22, 114 per year is needed for the management cost. These costs are still higher by 45-50% than the financial payments to be received which was $15,625 and $11,250 for Karago and Ilagala respectively.

Trends of opportunity cost, carbon stock and potential carbon credit over time

Analysis of the opportunity cost, marginal (incremental) revenues/potential carbon credit and carbon stock shows that the project will have four major phases (A, B, C and D), each with different behaviour (Figure 5).

In general, there will be a decrease in incremental revenues in each project phase because; the revenues are the function of incremental carbon stock. The incremental carbon stock is also the function of forest age of which increases as the project lifetime increases. Therefore, the project seems to be helpful during the first phase of the project. The probability for the project to flop in near future is high if there will be no means to reduce

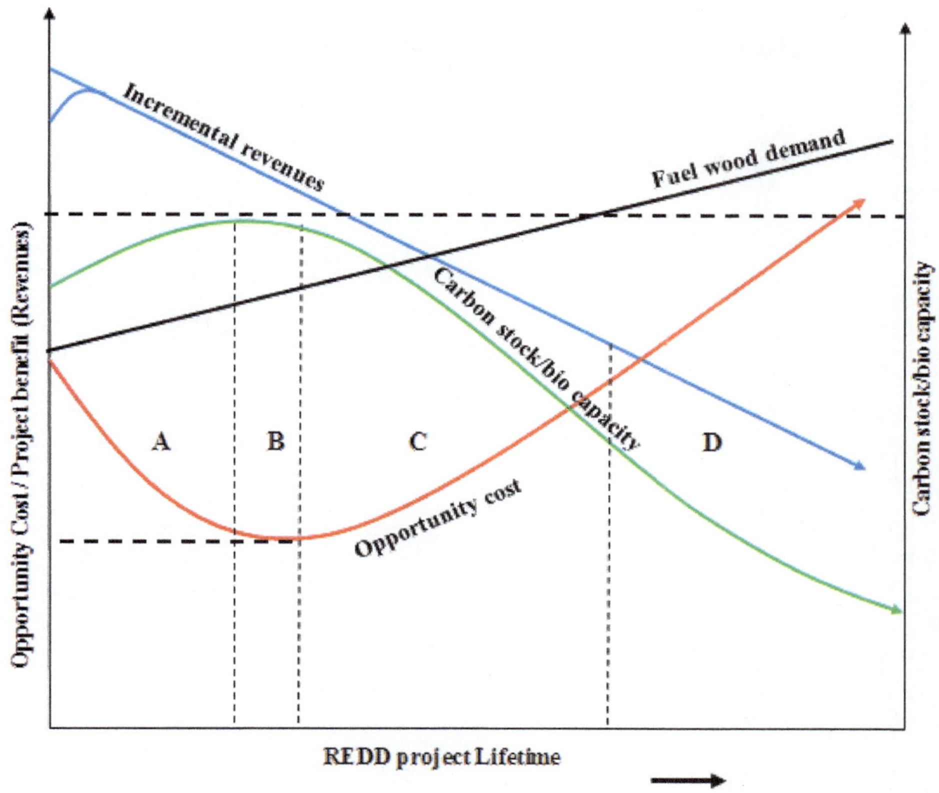

Figure 5. Trends of opportunity cost, carbon stock and potential carbon credit over time.

the opportunity cost and improve forest biomass.

In phase one (A), there will be an increase of carbon credit due to increase of carbon stock in the forest. The bio capacity is slightly higher than the village demand on forest goods. The opportunity cost of forest goods will also have a tendency of decreasing due to the fact that, as the project start operating, there will be additional co-benefits that will lower the opportunity cost. Among the co-benefits from the forest are climate regulation and the increase of food products from the forest such as mushrooms.

Phase two (B) is the positive turning point for the opportunity cost and negative turning point of carbon stock in REDD project forests due to forest age. The opportunity cost will stop decreasing and instead start increasing. The reason is that the forests allocated for fuel wood collection will fail to meet the fuel wood demand and hence raise the labour and procurement costs of fuel wood. It is at this phase when the carbon credit will start decreasing.

Phase three (C) will be associated with high increase of opportunity cost since the revenues from carbon credit will keep on decreasing. On the other hand fuel wood demand will be high if considering no substitute for goods and technologies while bio capacity is low. It is at this time there will be an increase of land degradation and people may start returning to the REDD project forests upon no strict supervision.

Phase four (D) will be associated with unusual increase of opportunity cost which will be far be bigger than the marginal (incremental) revenues from carbon stock increment. This is the time when the ability of forest to sequester carbon will be low due to forest age and hence more less forest revenues to the village. Due to very low bio capacity at constant fuel wood demand, there will be more environmental and social economic problems in the villages. This situation will be worse right from the point (time) when the marginal revenues will have exceeded the opportunity cost. This is the phase when the project will flop.

The role of energy efficient cooking technologies on REDD project viability

The findings from the study indicate that unless energy sources and energy efficient cooking technologies are well addressed, viability of REDD+ initiatives are will still be low. Adoption of high energy efficient cooking technology (HEECT) will reduce pressure in the forests of which is essential for viability of REDD+ project.

According to Tanzania Traditional Energy Development Organization (TaTEDO), the efficient cooking technologies (EECT) include improved charcoal stoves which

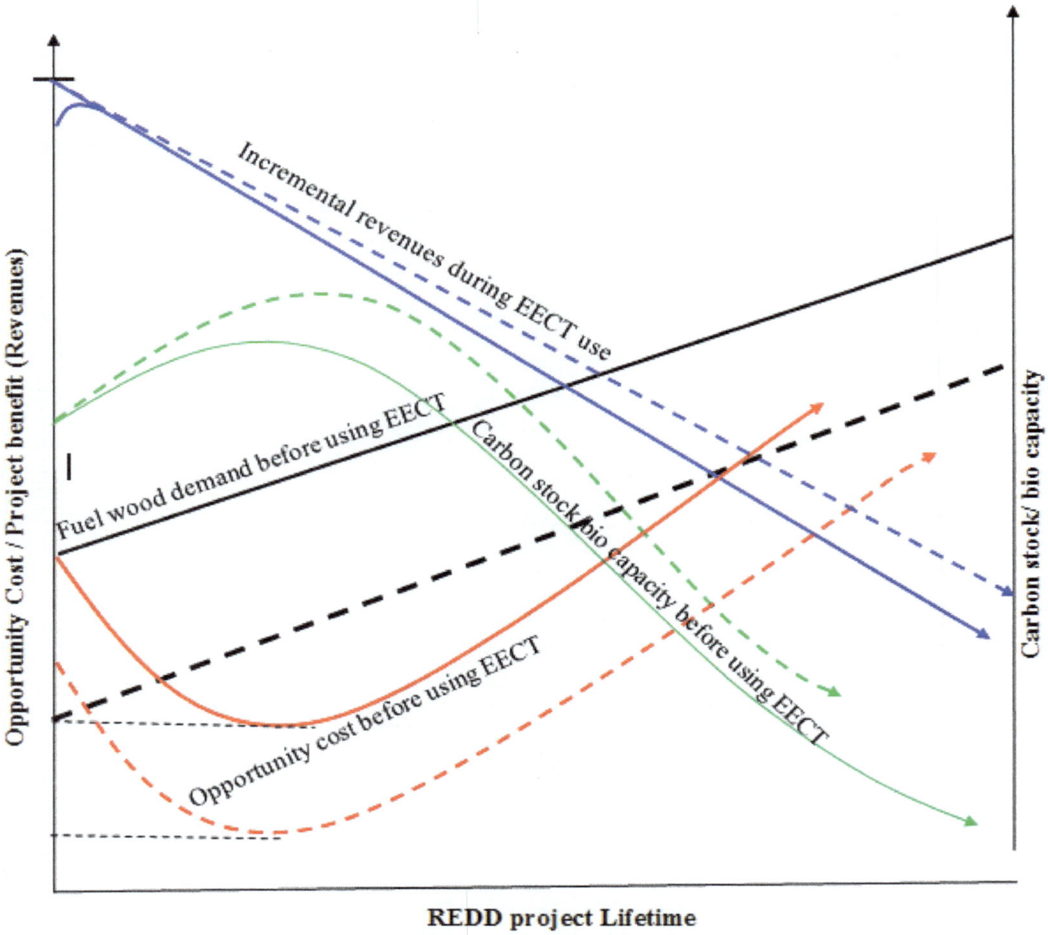

Figure 6. The sketch showing the behaviour of opportunity cost, incremental Carbon stock revenues, fuel wood demand, bio capacity of forest ecosystem before (solid lines)and during EECT application (dashed lines).

have capacity of reducing consumption of charcoal by average of 71% and improved firewood stoves (Improved Okoa Firewood Stove) of which have capacity of reducing consumption by more than 65%. The improved firewood stoves use one or two pieces of firewood. In this regard, by adopting these kinds of technologies, the costs for purchasing firewood will have been reduced significantly for those purchasing firewood as they will consume less fuel wood.

Reduction in fuel wood consumption will in turn reduce further the frequency of fire wood collection. This will help women and children in Karago and Ilagala villages to save around 2,912,000 - 3,328,000 hours per year for collecting firewood with estimated value of $1,456,000 – $1,664,000 per year - at nominal cost of $0.5 per effective hour used for fetching firewood. The time saved can be used for implementing other development activities.

Adoption of EECT will cut down carbon dioxide emissions by 80% since the use of unimproved biomass stove by the 4,000 households in two villages contributes around 1,105,416 tons of carbon dioxide (tCO2) per year at the average of 4tCO2 per cubic meter of firewood but through the use of EECT (improved charcoal technology), the emission could be reduced to 232,137.4 per year.

Moreover, adoption of these technologies may reduce the opportunity costs (labour and procurement costs) for fuel wood accessibility by 67.9 to 70.5%. This is because the opportunity cost for fuel wood accessibility will decrease from $1279-$1724 to $410-$508 per year. The reduction in fuel wood consumption is associated with reduction in frequency of fire wood collection of which has implication on labour costs. Reduction of opportunity cost will increase project viability as there will be a shift of the point at which opportunity cost starts becoming higher than incremental revenues. Also, the decrease in fuel wood demand prompted by the use of EECT will increase project viability in terms of forest biomass as there will be shift of the point at which bio capacity starts becoming lower than the fuel wood demand (Figure 6).The point can be shifted further if communities will be encouraged to have their own woodlots. Thus, application of appropriate measures for opportunity costs reduction and improving forest biomass may make the project more viable in the case villages.

Conclusion and Recommendation

The study reveals that, there is high opportunity cost behind REDD+ project implementation in the case study villages of which 80% of the respondents were willing to take with anticipation that the project will be compatible with their local priorities and needs. The study has further revealed high expectations on the benefits to be gained by the communities in the respective villages. The REDD+ project, local governments and people should therefore, harmonize the community preferences so that project goals become compatible with community priorities and needs. Careful selection of the already proposed alternative livelihood projects like bee keeping project and microfinance must therefore be done in the case villages to ensure sustainability of REDD+ project and people's lives. Notwithstanding the social economic effects, REDD+ project has brought positive changes in the case villages as there is already climate regulation in terms of rainfall, rejuvenation of river water tributaries flowing from the forests and increase of Non wood products in the forest reserves. The study recommends that, in order to maximize the environmental and socio-economic benefits in the case villages, there is a high need for conservation projects such as REDD+ project to cooperate with village government in tree planting campaign in the farm fields and encouraging the use of improved stoves to cut downcosts of fuelwood access and its consumption in the foreseeable future.

Conflict of interest

The authors did not declare any conflict of interest.

ACKNOWLEDGEMENTS

This paper has been prepared as a part of research output of the project on Establishment of Harmonized Modalities and Mechanisms for Community Compensation under Carbon Markets: The Case Study of REDD Pilot Projects in Tanzania funded under the framework of Climate Change Impact Adaptation and Mitigation (CCIAM) programme supported by the Royal Norwegian Government.

REFERENCES

Bolin E, Mustalahti I, Boyd A, Paavola J (2012). Can REDD+ reconcile local prioritiesand needs with global mitigation benefits? Lessons from Angai Forest, Tanzania. Ecol. Soc. 17(1):16.

Brown S (1986). Estimating biomass and biomass change of tropical forests: a primer. Rome: FAO Forestry Paper 134.Food and Agriculture Organization of the United Nations.

Brown S (1993). Tropical forests and the global carbon cycle: The need for sustainable land-use patterns. Agric. Ecosyst. Environ. 46: 31-44. http://dx.doi.org/10.1016/0167-8809(93)90011-D

Burges ND, Bahane B, T Clairs, F Danielsen, S Dalsgaard, M Funder, N Hagelberg, P Harrison, C Haule, K Kabalimu, F Kilahama, E Kilawe, SL Lewis, JC Lovett, G Lyatuu, AR Marshall, C Meschack, L Miles, S AH Milledge, PKT Munishi, E Nashanda, D Shirima, RD Swetnam, S Willcock, A Williams, E Zahabu (2010). Getting ready for REDD+ in Tanzania: a case study of progress and challenges. Oryx 44(3):339-351. http://dx.doi.org/10.1017/S0030605310000554

Carlevaro F, Gonzalez C (2008). Costing Improved Water Supply Systems for Low- income Communities. A Practical Manual. University of Geneva/World Health Organization.

Chomitz KM, P Buys, G De Luca, TS Thomas, S Wertz-Kanounnikoff (2006). At loggerheads? Agricultural expansion, poverty reduction and environment in the tropical forests. World Bank Policy Research, Development Research Group, World Bank, Washington, D.C., USA.

Chopra K (1993). The value of non-timber forest products: an estimation for tropical deciduousForests in india. Economic botany 47(3):251-257 (Institute of Economic Growth, University Enclave, Delhi 110 007, India).

IPCC (Intergovernmental Panel on Climate Change), Aleotti (2007). Assessment of adaptation practices, options, constraints and capacity. In: Contribution of Working Group II to the Fourth Assessment Report of the Intergovernmental Panel on Climate Change. M.L. Pary, O. F. Canziani, J.P. Palutikof, P.J. van der Linden, and C.E. Hanson, eds. pp 976. Cambridge, Cambridge University Press.

Kabura J, Dos SS, Arild V (2013). The cost of managing forest carbon under REDD+initiative : a case of Kolo hills forest, Kondoa, Dodoma, Tanzania.A paper presented during a CCIAM workshop at Palm Tree Village Resort, Bagamoyo January 2013.

Kajembe GC, Silayo DA, Kitula MM, Lyatura.,Mutabuz K.J. and Massawe F. and Vtn A (2012). REDD Realities:Leasons learned from REDD + Pilot projects in Kondoa and Rungwe Districts, Tanzania.

Lugo AE, Brown S (1992). Tropical forests as sinks of atmospheric carbon. For. Ecol. Manag. 54:239-255. http://dx.doi org/10.1016/0378-1127(92)90016-3

Stern N (2006).Stern Review: The economics of climate change. Her Majesty's Treasury, London, UK.

Tanzanian Forestry Working Group (TFWG) (2010). Options for REDD in Tanzania: key design issues for the national REDD strategy. Brief 2. Tanzanian Forestry Working Group,Arusha, Tanzania.

TNRF (2008). Tanzania Natural Resource Forum, Carbon, Communities, and Conservation:Developing a Framework for generating socioeconomic and Environmental Benefits using the voluntary Carbon Market in Tanzania: A Roundable workshop hosted by TNRF 14th March, Arusha.

United Nations Framework Convention on Climate Change (UNFCCC) (2010). Decisions adopted by COP 16 and CMP 6. Cancun Climate Change Conference, UNFCCC, Bonn, Germany.

United Republic of Tanzania (URT) (2009). National Framework for Reduced Emissions from Deforestation and Forest Degradation (REDD). Forestry and Beekeeping Division, Ministry of Natural Resources and Tourism, Dar es Salaam, United Republic of Tanzania.

Vashum KT, Jayakumar S (2012). Methods to Estimate Above-Ground Biomass and CarbonStock in Natural Forests - A Review. J Ecosyst. Ecogr. 2:116.

Whittaker RH, Linkens GE (1973). Carbon in the biota. In Woodwell GM, Pecan EV, Carbon in the biosphere, Proceedings of the 24th Brookhaven Symposium in biology. Upton, New York: United States Atomic Energy Commission 281-302.

Characterization of the solar climate in Malawi using NASA's surface meteorology and solar energy (SSE) model

Senganimalunje, T. C.[1] and Tenthani, C. M.[2]

[1]Malawi Bureau of Standards, Metrology Services Department, Box 946, Blantyre, Malawi.
[2]Physics and Biochemical Sciences Department, Malawi Polytechnic, University of Malawi, Chichiri, Blantyre, Malawi.

This paper presents the characterization of global solar radiation (GSR) for Malawi using NASA's SSE model. The mean monthly daily GSR monthly variation in the three regions of Malawi has been investigated. It has been found that there is a general gradient in GSR in the north-south direction. This gradient correlates well with the variation of the clearness index in the same direction (r = 0.986). It has been observed that the GSR is characteristic of each region. Simple analytic polynomials have been proposed for estimating solar radiation in the traditional Northern, Central and Southern regions of Malawi. There is a strong agreement between the polynomials and the SSE model with R^2 values of 0.988, 0.989 and 0.989 and root mean square errors of 0.061, 0.057 and 0.062 kWh/m^2.day for the regions, respectively.

Key words: Malawi, global solar radiation, solar climate, surface meteorology and solar energy (SSE) model.

INTRODUCTION

Accurate global solar radiation (GSR) data for a particular location is an essential requirement for the design of efficient domestic and industrial solar based systems. Such data is used in project design, cost analysis and in system efficiency evaluation. In spite of the importance of GSR data, direct measurement is challenging due to high cost of maintenance and the need for calibration of the measuring instruments (Mubiru and Banda, 2008) as such models may be found to be handy to come up with GSR values for different locations. Many researchers have used models to estimate GSR. Sanusi and Abisoye (2013) and Oyeleke et al. (2014) used second order quadratic models to estimate the diffuse solar radiation in Lagos and Ogbomoso respectively, and found that they were good enough to be be used for the same. In their study, Safari and Gasore (2011) estimated GSR using Angstrom type polynomials of first and second order. Hassan and Ali (2013) used empirical regression models to predict daily and monthly average GSR for United Arab Emirates based on 9 year measured data. The surface meteorology and solar energy (SSE) section at NASA's Atmospheric Data Center provides GSR data that has been formulated specifically for the needs of photovoltaic and renewable energy system design (NASA, 2008). The

Figure 1. Map of Malawi showing the longitudes used for data extraction.

SSE data used in this paper has a temporal extent of 22 years (July 1983 through June 2005). The validity of NASA's SSE model has been demonstrated elsewhere (Islam, 2009). Thus, the NASA's SSE model eliminates the need for actual GSR measurements for most applications.

Malawi is a tropical country that lies between the latitudes 9° 22' S and 17° 3' S. The country lies entirely in the earth's solar belt, a region of the earth that has an abundance of solar radiation flux (40° N and 40° S) (Islam, 2009). Som (1979) demonstrated the potential for utilization of solar energy in Malawi as early as 1979. Follow up studies by Diabate (2004) categorized Malawi as a solar climate II region without using any local data. This was confirmed by Madlopa (2006) using sunshine data from Chichiri meteorological station in Southern Malawi. Earlier Zingano (1986) found that there is a gradient in the GSR in the north-south direction using sunshine hours data. Tadros (2000) also used a similar approach to estimate GSR for Egypt. However, the accuracy of analytical results from sunshine hours data is inherently limited by the accuracy of the regression coefficients in the Angstrom equation (Amorox, 2004).

The aim of this study was to characterize the mean annual daily GSR flux in Malawi as a function of latitude and to analyse the variation of the mean monthly daily GSR over the year at different latitudes using SSE data. It was envisaged that use of SSE data could provide more accurate results as compared to those obtained using derived parameters such as the clearness index which can also be used to estimate GSR as is shown by Poudyal et al. (2012). The clearness index (which is the ratio of the global solar radiation measured at the surface to the total solar radiation at the top of the atmosphere) is a veritable tool in the characterization of sky conditions (or classification of sky types) over a particular locality (Okogbue et al., 2009). In this study, the cleaness index which was empirically determined by Madhlopa (2006) was compared.

DATA COLLECTION AND ANALYSIS

Malawi is divided into three administrative regions (north, central and south). Most of the land mass in Malawi lies along 33° 30' E longitude in the Northern and Central Regions and along 35° E longitude in the Southern region (Figure 1).

The GSR data extracted from the SSE model was obtained along longitude 33° 30' E for the Northern and Central Regions and 35° E for the Southern Region at latitude intervals of 1° from 9° 30' S through 16° 30' S. This was done by entering the particular geographical coordinates of interest into the model. The data is presented in Table 1 in which the mean monthly daily value of the GSR is given for each month for a particular latitude. The GSR values for each data point in Table 1 are for a 1° × 1° latitude-longitude cell (approximately 100 km by 90 km cells).

The data was then plotted to determine firstly the variation of mean annual daily flux with latitude, secondly the relationship between mean GSR and the clearness index K, and lastly the relation between GSR and altitude for Malawi to ascertain findings by Zingano (1986).

Monthly variation of the mean monthly daily flux at 1° latitude interval along the selected longitudes was analyzed to determine regional variations during the year. Simple analytical polynomials for estimating the GSR have been derived and presented for each of the three regions of Malawi.

RESULTS AND DISCUSSION

The following section presents results of variation of GSR with latitude, monthly variation of GSR and polynomials for estimating monthly GSR in northern, central and southern regions of Malawi in any month.

Global solar radiation and latitude

The insolation on the earth's surface is dependent on the clearness index, K, which is the fraction of the extra-terrestrial radiation transmitted through the earth's atmosphere to the surface. Figure 2 gives a plot of the mean

Table 1. Mean monthly daily global solar radiation (kWh/m^2.day) for Malawi (the data was extracted from NASA's SSE model).

Latitude	Month of the year											
(degrees)	1	2	3	4	5	6	7	8	9	10	11	12
-9.5	4.93	5.11	5.31	5.21	5.33	5.45	5.68	6.23	6.77	6.8	6.01	5.21
-10.5	5.18	5.3	5.57	5.46	5.59	5.54	5.7	6.25	6.81	6.87	6.23	5.52
-11.5	5.22	5.41	5.59	5.52	5.48	5.3	5.43	6.02	6.71	6.87	6.33	5.58
-12.5	5.08	5.33	5.43	5.43	5.32	5.13	5.24	5.87	6.6	6.76	6.24	5.38
-13.5	5.17	5.35	5.53	5.48	5.25	4.96	5.07	5.67	6.48	6.67	6.24	5.38
-14.5	5.18	5.34	5.52	5.33	5.04	4.55	4.65	5.38	6.21	6.51	6.09	5.35
-15.5	5.32	5.46	5.43	5.19	4.84	4.34	4.48	5.26	6.21	6.49	6.12	5.43
-16.5	5.67	5.61	5.59	5.14	4.7	4.2	4.33	5.19	6.12	6.38	6.26	5.71

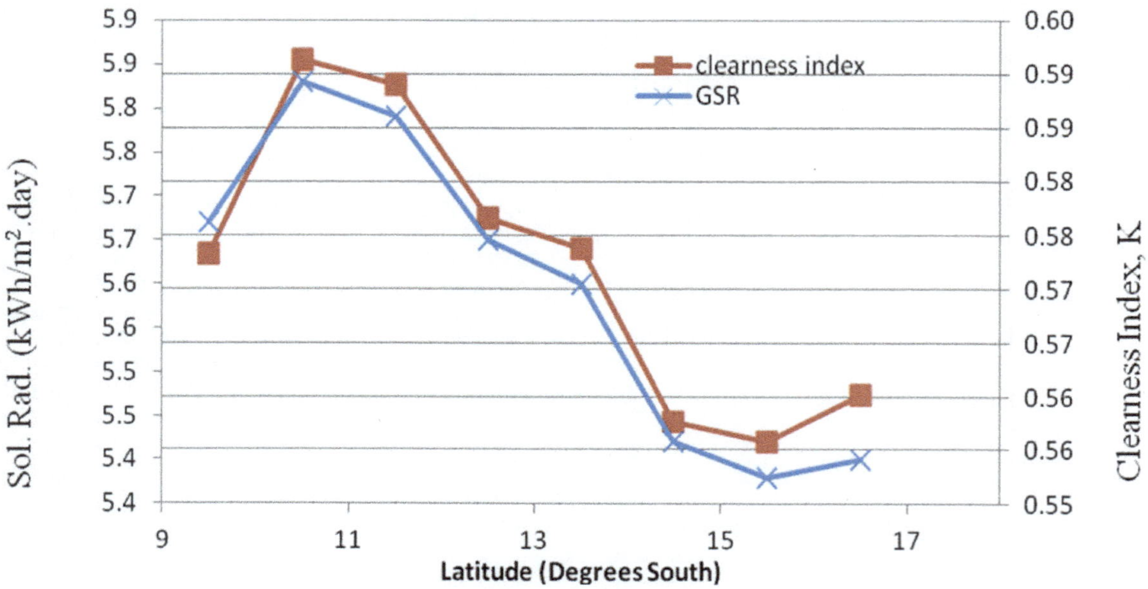

Figure 2. Latitude dependency of the global solar radiation and the clearness index.

annual daily GSR and K against latitude across Malawi in the north-south direction. The values of the clearness index used in the plot were empirically determined by Madhlopa (2006). There is a general gradient in the north-south direction with the maximum value (5.83 kWh/m^2.day) at latitude 10° 30' S and the minimum value (5.38 kWh/m^2.day) at 15° 30 S. The GSR closely matches the clearness index, as expected, since the top-of-atmosphere insolation is slowly varying as a function of latitude with the Pearson product-moment *correlation coefficient r = 0.986.*

Malawi has an altitude gradient in the north-south direction and it might be tempting to attribute the GSR gradient to the altitude. The GSR only drops from the peak by about 8% while the altitude drops by as much as 70% (Figure 3). This indicates that at a national scale, the dependency of GSR on altitude is weak.

Monthly variation of the mean monthly daily GSR

An analysis of the dependence of GSR on the time of the year reveals two general patterns. The first is that there are two peaks in the insolation, a minor one in March a major one in October. The minimum amount of insolation occurs in June. In October, the atmospheric air is dry resulting in minimal attenuation. March is the end of the rainy season in Malawi resulting in reduced cloud cover. The dry air mass and reduced cloud cover account for the observed minor and major peaks, respectively. Malawi being a southern hemisphere country is farthest from the sun in June and has a high air mass and frequent cloudy days resulting in the observed minimum.

The difference between the yearly maximum and minimum of the mean monthly daily insolation values was seen to increase in the southern direction. For example,

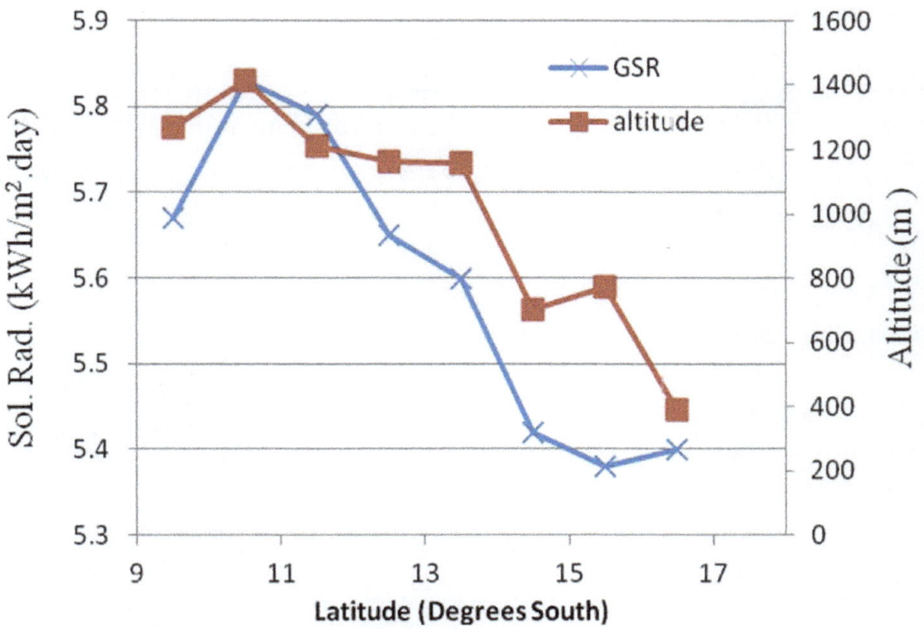

Figure 3. Global solar radiation and altitude.

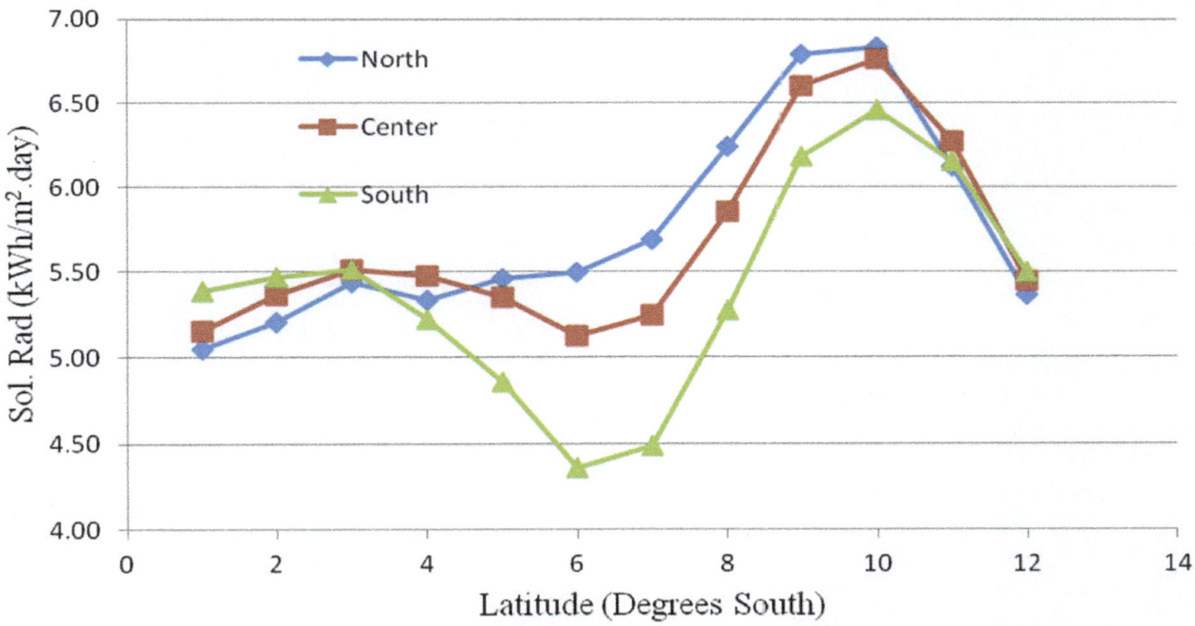

Figure 4. Region averages for mean monthly daily GSR.

the differences between the insolation for the months of October and June for latitudes 10° 30' S, 13° 30' S, and 16° 30 S are 1.33, 1.71 and 2.18 kWh/m^2.day, respectively. This difference gives the variation in the amount of available GSR with the seasons in the Northern, Central and Southern regions. This means there is greater variation in the amount of solar radiation available in the Southern region as compared to the Central and Northern Regions across the months. For some critical system designs applications like solar photovoltaics, this difference may become important and needs to be incorporated early enough at planning stage. Figure 4 gives the regional averages and their fits for the mean monthly daily GSR, highlighting the differences in the maximum and minimum flux during the year in the three regions.

Table 2. Polynomials for estimating GSR in the three regions of Malawi.

Region	Equation – GSR(n)	R^2
North	$6.4115-2.8714n+2.0732n^2-0.6362n^3+0.0939n^4-0.0065n^5+0.0002n^6$	0.9888
Central	$7.0154-3.8856n+2.7698n^2-0.8394n^3+0.1209n^4-0.0081n^5+0.0002n^6$	0.989
South	$7.8074-4.9542n+3.4757n^2-1.0706n^3+0.1559n^4-0.0106n^5+0.0003n^6$	0.989

Polynomials for estimating global solar radiation in Malawi

Data from Table 2 and the plots in Figure 4 were used to generate empirical relationships between the time of the year in months (n) and GSR. The resulting polynomial is of the form

$$GSR(n)=a_o+a_1n+a_2n^2+a_3n^3+a_4n^4+a_5n^5+a_6n^6$$

Where, *n* an integer satisfying $1 \leq n \leq 12$.

The polynomials are given in Table 2. The R^2 value gives a measure of how close the polynomial agrees with SEE data and the agreement is good in all the three cases.

It was noticed that for the months October through April, including all the three polynomials give very close results indicating the insolation is similar throughout the country during this time. That is, it is only in the cold months of May through September that there are important regional variations in the amount of solar flux energy in Malawi.

Conclusion

The SSE model was used to analyze solar insolation in Malawi. There is a north-south gradient in the GSR with an annual mean monthly daily peak of 5.83 kWh/m^2.day at latitude 10° 30' S and an annual mean monthly daily minimum of 5.38 kWh/m^2.day at latitude 15° 30' S. The radiation flux varies greatly over the course of the year with the mean monthly daily maximum of 6.87 kWh/m .day and a mean monthly daily minimum of 4.20 kWh/m^2.day. These spatial and temporal variations may need to be considered in critical applications. The dependence of the GSR data on altitude has also been shown to be weak.

An empirical polynomial model to estimate GSR in the three regions of Malawi has been presented. However, there is a need to assess the accuracy of the SSE model and the derived polynomials using real data to determine their accuracy.

Conflict of interests

The authors did not declare any conflict of interest.

REFERENCES

Amorox J (2004). Global solar radiation estimation using sunshine duration in Spain. Energy Conver. Manage. 45:1529-1535.

Diabate LB (2004). Solar climate in Africa. Renew. Energy 76:733-744.

Hassan ANH, Ali HA (2013) Global and Diffuse Solar Radiation in the United Arab Emirates. Int. J. Environ. Sci. Dev. 4:5.

Islam MK (2009). Measurement of solar energy radiation in Abu Dhabi, UAE. Appl. Energy 86:511-515.

Madhlopa A (2006). Solar radiation climate in Malawi. Solar Energy 80:1055-1057.

Mubiru J, Banda EJKB (2008). Estimation of monthly average daily global solar irradiation using artificial neural networks. Solar Energy 82:181-187.

NASA (2008). Atmospheric Science Data Center, Surface meteorology and solar energy. http://eosweb.larc.nasa.gov/sse/.[accessed 03.09.2011]

Okogbue EC, Adedokun JA, Holmgren B (2009). Hourly and daily clearness index and diffuse fraction at a tropical station, Ile-Ife, Nigeria. Int. J. Climatol. 29:1035-1047.

Oyeleke O, Sanusi YK, Fajinmi GR, Awodugba AO (2014) Estimation of Global Solar Radiation Using a Simple Metreological Data for Ogbomoso Area, Nigeria. Int. J. Bus. Appl. Sci. 1:1.

Poudyal KN, Bhattarai BK, Balkrishna S, Berit K (2012). Estimation of global solar radiation using clearness index and cloud transmittance factor at trans-Himalayan region in Nepal. Energy Power Eng. 4:415-421.

Safari B, Gasore J (2011). Estimation of Global Solar Radiation in Rwanda using Empirical Models. Asian J. Sci. Res. 2(2)

Sanusi YK, Abisoye SG (2013). Estimation of Diffuse Solar Radiation in Lagos, Nigeria, 2nd International Conference on Chemical, Environmental and Biological Sciences (ICCEBS'2013) March 17-18, 2013 Dubai (UAE)

Som A (1979). Solar Utilization Potential in Malawi. Malawi J. Sci. Technol. 3:103-104.

Tadros M (2000). Uses of sunshine hours to estimate the global solar radiation over eight meteorological stations in Egypt. Renew. Energy 21:231-246.

Zingano B (1986). An appraisal of solar water heating in Malawi, M.Sc. Thesis . Zomba: University of Malawi.

24

Assessing soil erosion risk in the Tillabery landscape, Niger

Mansour Mahamane

Faculty of Science, Tubingen University, Tübingen, Germany.

To combat desertification, it is indispensable to understand soil erosion processes in the Sahel zone. The objective of this study was to develop a simple model that uses soil texture data from the field and Digital Elevation Model parameters to predict soil texture. This study also assesses soil erosion research in the Sahel region, illustrated by a case study from Tillabéry landscape (Niger). The most sensitive areas and trends of land degradation processes were identified using the Revised Universal Soil Loss Equation and Unit Stream Power-based Erosion Deposition models. The models used depict a rational evolution of soil loss distribution during the study period from 1973 to 2007. The results show that soil erosion output scenarios predict greater soil erosion in the study area from 2070 onwards. They suggest that human disturbance and topographic factors are the main impact factors in the affected areas.

Key words: Tillabéry landscape (Niger), sheet and rill erosion modelling, data mining.

INTRODUCTION

The Sahel region is considered one of the most vulnerable ecosystems in the world and the majority of the population is highly dependent on soil resources to ensure adequate crop production or to maintain their crop production and livelihoods (Ben Mohamed et al., 2001). Also, it is characterised by erratic precipitation and water shortage due to semi-arid climatic conditions. Speikerman et al. (2015) showed that the degradation of environmental condition was accelerated by droughts and an overall decrease in precipitation in the Sahel.

Soil erosion is one of the principal environmental problems that affects food production in the Sahel (Gandah et al., 2003); it is one of the manifestations of desertification processes. Cobo et al. (2010) explained that soil erosion is one of the most ordinary and extensive forms of soil degradation and is closely related to processes of nutrient reduction and desertification in semi-arid regions. It affects agricultural production, contributes to the contamination and quality of water resources. It further decreases the soil fertility content, the fine grained soil content and the water holding capacity and the depth of the top soils. Michels et al. (1995) reported that erosion processes have a significant impact on land productivity in the West African Sahel, a semi-arid region where both wind and water take action as erosive forces.

Soil texture is influenced by drainage condition, permeability and water holding capacity. It directly affects

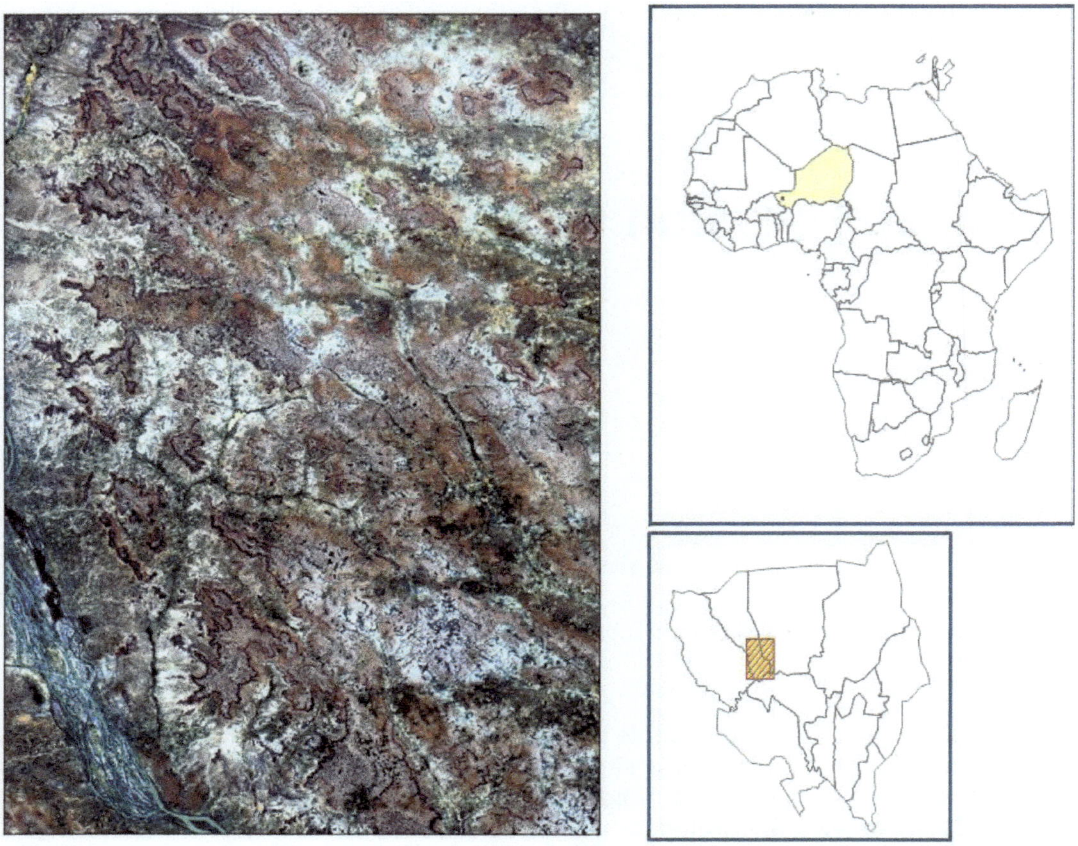

Figure 1. Location of study area in Southwestern Niger and Africa.

the porosity of the soil and its long-term soil fertility. According to Wischmeier and Smith (1978), soil texture determines the soil erodibility and affects the risk of soil erosion. Therefore, the knowledge of soil texture is essential for studying soil erosion. The TreeNet Model (Salford Systems Implementation, cf. Friedman, 1999) was used to estimate the spatial distribution of soil texture. Such techniques have been successfully used by Henderson et al. (2005), Ramadam et al. (2005) and Wilcke et al. (2008). However, the approach employed in this paper is based on a data mining technique developed by Friedman (1999).

To combat desertification, it is indispensable to understand soil erosion processes in the Sahel region. This paper assesses soil erosion research in the Sahel region and used the Tillabery landscape as a case study. In achieving this, it used the RUSLE and USPED models. The RUSLE model was originally applied in the USA, and has proven itself as a valuable model for estimation of soil erosion loss in other regions of the world (Wischmeier and Smith, 1978; Renard et al., 1997; Millward and Mersey, 1999; Angima et al., 2003) This model predict the average annual soil loss from rill and sheet erosion. The USPED is an empirical method which determines the spatial distribution of erosion and deposition (Mitasova et al., 1996) and it is one of the most commonly applied models to estimate soil erosion and deposition (Warren et al., 2005; Lòpez-Vicente and Navas, 2010; Zhang et al., 2011) In this paper, RUSLE and USPED models have shown a realistic evolution of soil loss distribution during all the study period.

MATERIALS AND METHODS

Research area

The study area is located as shown in Figure 1 within the Department of Tillabéry and includes parts of the wide valley of the Niger River. The annual precipitation is between 250 and 400 mm. The soil in this area is very infertile and poses enormous challenges for agricultural production. This area was selected as investigation area due to the fact that desertification is the most serious environmental problem and it is located at the core of the Sahel. To compound it, the depth and width of the Niger River is continuously on a decrease, an issue associated with the moving of sand.

The traditional cereals, millet and sorghum, represent about 85% of the total food crop production requirements of the population in the study area. Livestock equally plays an important role in the economy. Therefore, cattle breeding and crop production are always competing for the fertile pieces of land available. This has

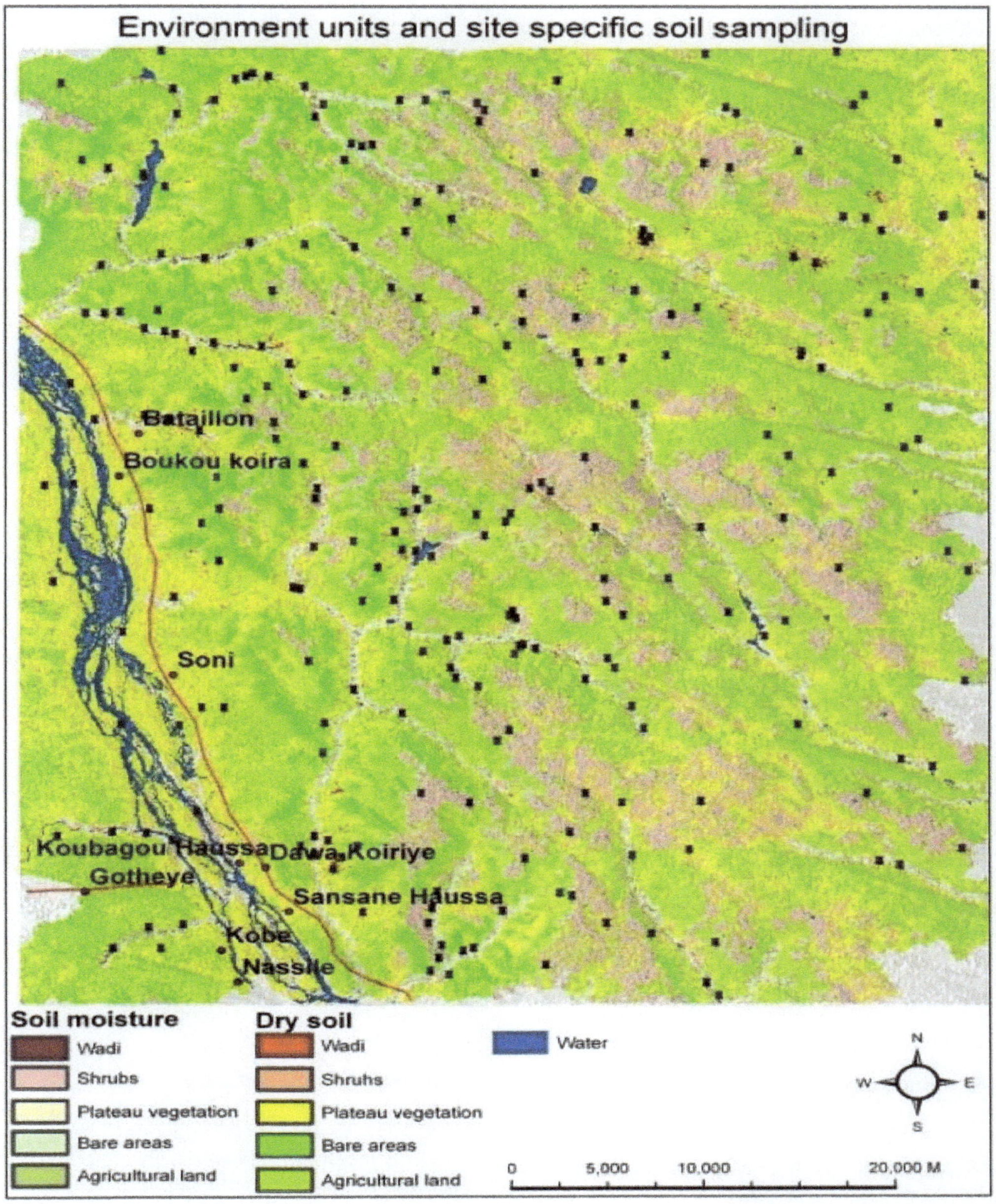

Figure 2. Environment units and site specific soil sampling.

spawned a series of conflicts between the various stakeholders – nomads and sedentary people.

Wind is another climatic factor influencing land degradation. Under the influence of high-speed winds, the process of wind erosion takes place. The combination of wind effect and high soil temperatures seriously affects the establishment of vegetation cover. Wind speeds exceeding 100 km/h have been observed (Sivakumar et al., 1993).

Soil sampling and laboratory methods

To investigate the amount and spatial distribution of the soil structure and soil erosion, a stratified random sample was established based on two-stage sampling plans with topography and land use as the stratifying variables. A topographic wetness index was derived from the Shuttle Radar Topography (SRTM) at a resolution of 90 m with two classes. This index shows areas of potential soil moisture (with value 1) and dry areas (with value 2). The wetness index was combined with land use derived from a Landsat image of 2001. The result displays 12 environmental units within the study area (Figure 2). Two environmental units within the water bodies were not considered because they do not show a classical category of soil attributes. In addition, 25 soil samples were randomly selected within each of the 10 environmental units (250 soil samples). For each point, geographical coordinates were taken using a GPS navigation unit.

The 250 soil samples (soil samples depth 0-30 cm) were collected and tested by feel soil method (Thein, 1979) during rainy season (September, 2012). Four soil samples within the 10 environmental units (40 soil samples) were submitted to laboratory analysis. The following elements were analysed: particle size percentage, USDA classes by pipette method (Gee and Bauder, 1986), organic carbon by combustion method (Read, 1921).

Land forms

A couple of satellite data were used to derive the land form parameters for soil erosion and soil texture regionalisation in the study area. The Shuttle Radar Topography Mission (SRTM) DEM at 90 m spatial resolution was obtained from the University of Maryland (Earth Science Data Interface at Global Land Cover Facility) and an Advanced Spaceborne Thermal Emission and Reflection Radiometer (ASTER) DEM at 30 m spatial resolution was obtained from the National Aeronautic and Space Administration (NASA).

Spatial prediction of soil texture using TreeNet Model

A statistical model was adapted to predict the spatial distribution of soil texture from terrain parameters. For this task, the field survey x- and y-coordinate data together with the z-field representing texture information from 250 observation points were used. SAGA GIS (System for Automated Geoscientific Analyses, Conrad, 2007) was used to overlay terrain attributes derived from digital elevation models. Spatial prediction of soil texture landscape was developed using the TreeNet model (Salford machine implementation, cf. Friedman, 1999).

The dependent variable is categorical (soil texture) and the independent are terrain attributes. A predictive map of soil texture was obtained under SAGA GIS through the application of a spatial interpolation method (inverse squared distance). Inverse squared distance (IDW) was selected based on the error distribution histograms showing that the IDW interpolation method has a superior performance than Kriging and Splines methods. Additionally, from the visualisation point of view, IDW presented the best result.

Soil erodibility (K) and cover management factor (C) calculation

In this study, K was expressed by two empirical equations. A calculating relation was proposed by Renard et al. (1997) in which they applied the percentage of the respective soil texture classes and geometric mean diameter to arrive at the K (without considering organic matter data). This is expressed in this equation.

$$K = 0.0034 + 0.0405 * \exp\left[-0.5\left(\frac{\log Dg + 1.659}{0.7101}\right)^2\right]$$

Where: K: is the soil erodibility (t ha h / ha MJ mm); Dg: is the geometric mean weight diameter of the primary soil particles (mm) expressed by:

$$Dg = \exp\left(0.01 \sum_{i=1}^{n} f_i \ln m_i\right)$$

Where: f_i is the primary particle size fraction in percent, n is the number of size classes in which the distribution curve has been divided and m_i is the arithmetic mean of the particle size limits of that size. The results exposed that the erodibility ranges from 0.05 to 0.40 t ha h/ha MJ mm.

Torri et al. (2002) used a different approach by calculating the percentage of silt plus very fine sand, percentage of sand, percentage of organic matter and soil structure to arrive at the K (with considering organic matter data).

$$K = 0.0293 * (0.65 - Dg + 0.24Dg^2) * \exp\left\{-0.0021 * \left(\frac{OM}{C}\right) - 0.00037 * \left[\left(\frac{OM}{C}\right)^2\right] - 4.02C + 1.72C^2\right\}$$

Where: K is the erodibility factor (Mg ha h/MJ ha mm), Dg is the geometric mean weight diameter of the primary soil particles (mm), OM is organic matter content (%) and C is soil clay content (fraction). After assigning the appropriate equation, the TreeNet was applied. The results show that the equation of the k with organic matter has lowest prediction. This implied the Receiver Operating Characteristic (ROC) integral was less than 0.70 for all the classes due to low amount of soil samples (40 samples). Ließ et al. (2011) explained that "poor model performance is most probably caused by the small size of the dataset". In addition, model performance might be related to the size of the study area (Ziadat, 2005). Finally, both k factors were computed using Co-kriging model with wetness index as independent variable to produce a Co-kriged map of K factors results. Wetness index is selected as independent variable because it is the first important variable for both k factors.

Land use/land cover from Landsat images of 1973, 1989, 2001 and 2007 were converted to C. The agricultural land is assigned to a C of 0.4 as reported by Morgan (1995). The C varies between 0 (for example river), indicating that no erosion occurs, to 1, expressing the maximum of erosion (for example bare areas) and takes under consideration both cover and management variables. For shrub areas, a C of 0.01 was used as reported by Roose and De Noni (2004). For Plateau vegetation, a C of 0.1 was selected, which is also proposed by Roose and De Noni (2004).

Processing for soil erosion

The RUSLE model is supported by five components: rainfall erosivity factor (R), soil erodibility factor (K), topographic factor (LS), cover management factor (C), and support practice factor (P). The average annual soil loss (A, t/ha year) is computed by multiplying these factors.

The spatial patterns of soil erosion and deposition were simulated with the USPED model (Mitasova et al., 1996) to demonstrate the magnitude and spatial variability of soil loss and accumulation and to identify the most severely affected areas in the Tillabéry landscape using land use for the four years 1973, 1989, 2001 and 2007. To determine C factors, the precipitation data, from 1973 to 2007 were taken into consideration to calculate R factor values. K factor without considering organic matter is the same as used in the RUSLE model. However, the topographic component was computed by combining the Profile and the Tangential curvature, Upslope contributing area, Aspect and Slope. In the model output, erosion and deposition areas are shown by negative and positive signs, respectively. The maps obtained using USPED model were divided in the following classes as suggested by De Rosa (2005):

-5 to -0.1 (t/ha year): erosion; -0.1 to 0.1 (t/ha year): stability; 0.1 to 5 (t/ha year): deposition or sedimentation.

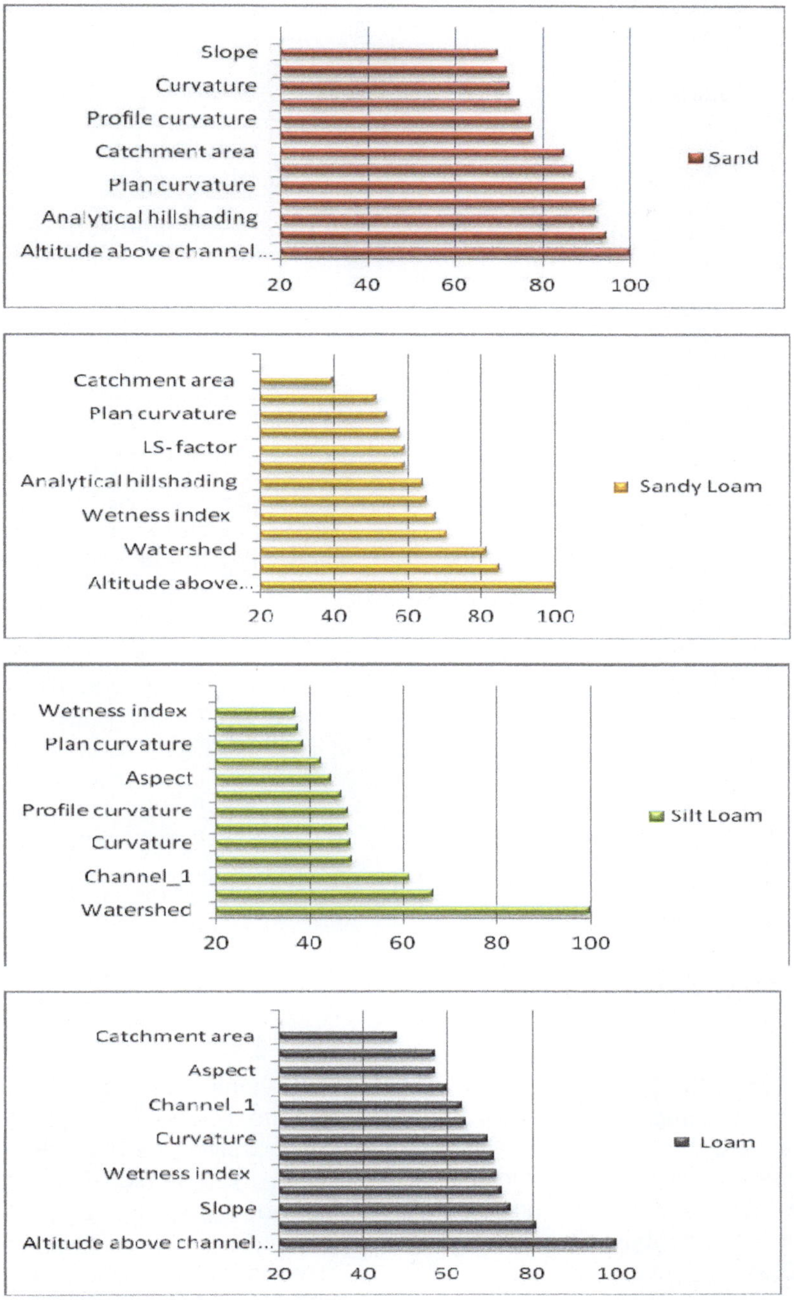

Figure 3. Ranking of the importance variable related to soil texture content regionalization.

RESULTS AND DISCUSSION

Soil texture

Influence of independent parameters on soil texture processes (rankings of important variables)

The quantification of the relationships between topographical elements and soil texture was done in this part. Using TreeNet Model technique, an estimate of important topographical parameters in relation to different soil structure elements is presented in Figure 3. The results show that there is a relatively strong relation between soil structure distribution and topographical parameters.

The global picture is the following: Altitude (100%) has the highest value for soil structure models. Watershed (88%), channel network (76%), profile curvature (75%), plan curvature (74 %), wetness index (72%), analytical

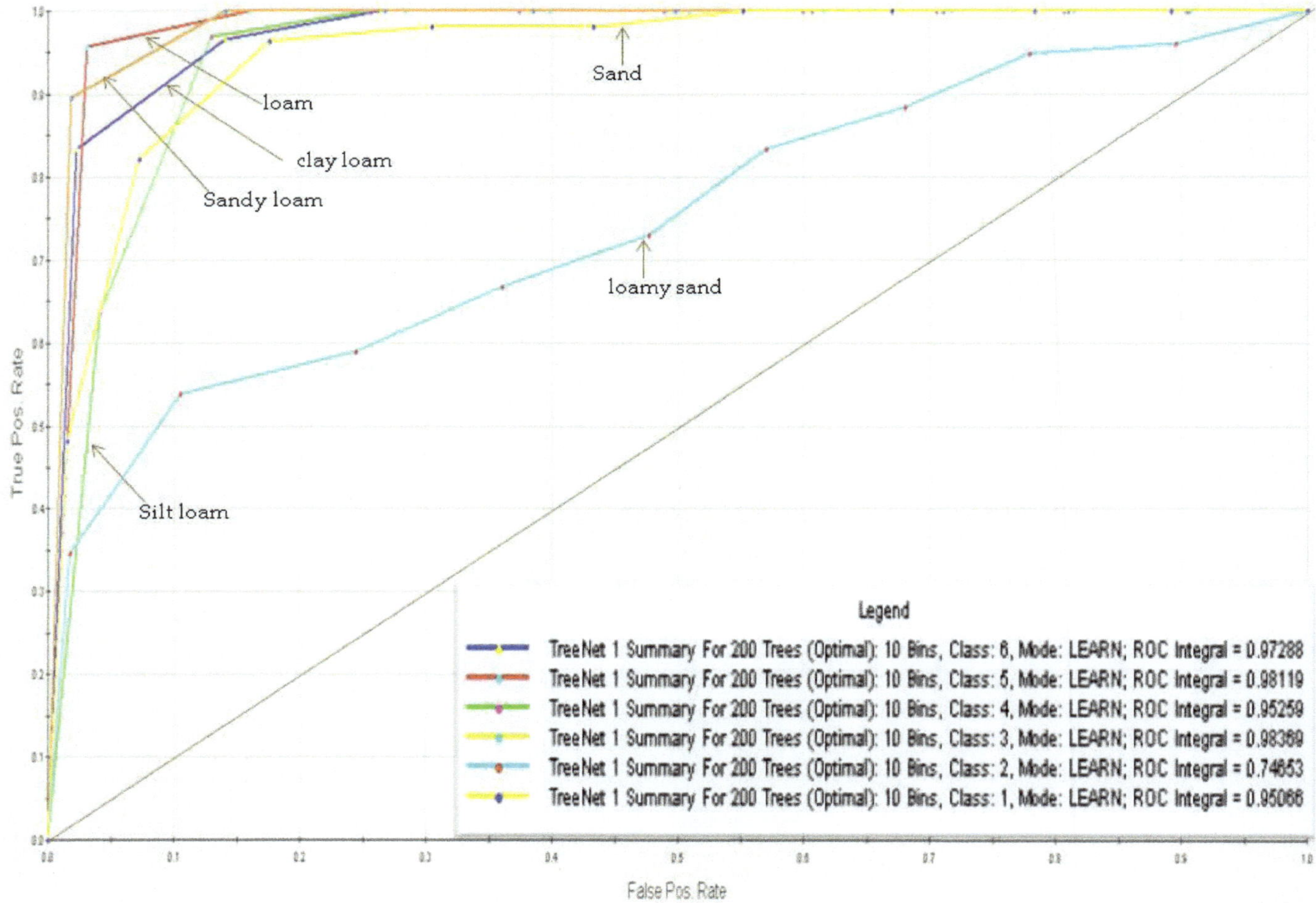

Figure 4. The Receiver Operating Characteristic (ROC) curve for each soil class.

hillshading (72%), curvature (70%), aspect (66%), and LS-factor (6 %) are apparently important as well. Other variables achieved score values of 6 to 56% and their influence cannot be neglected. Sand content displayed the highest score with altitude (100%) and channel network (94%), whereas loam content had the highest score with altitude (100%) and plan curvature (83%). Wilcke et al. (2008) argued that there is a strong dependence of soil texture on altitude, articulated by a good positive correlation between altitude and sand content but a negative correlation concerning clay content. Altitude above channel network was the most important predictor for sandy loam, loam and clay loam in the study area due to erosion and deposition processes.

Model performance evaluation and cross validation

Receiver operating characteristic (ROC) curve was used for characterizing the model performance. The value of ROC is between 0 and 1. In a worse case scenario, the ROC value cannot be less than 0.5 and the best could be close to 1 (Lasko et al., 2005). For the ROC analysis, the true positive rate is plotted on the Y axis and the false positive rate is plotted on the X axis. Figure 4 illustrates the error tradeoffs available with a given model and describes the predictive behavior of the six classifiers, independent of class distributions. The most reliable soil texture predictor was loam and sand (0.981 and 0.983, respectively).

The prediction errors for the spatial texture of the topsoil in the study area were quantified by a ten-cross validation method (Table 1). The model had an accuracy classification rate of 72.80% by ten-fold cross validation. The result from cross validation is unsuitable because there were not enough data sets but with large data sets, the model validation becomes particularly impressive.

Soil texture map

The scoring process of the entire data set of Tillabéry

Table 1. Prediction success from ten-fold cross validation (total 250 samples).

Class	Samples	Percent correct
Sand	56	77.84
Loamy sand	78	39.74
Sandy loam	29	89.66
Silt loam	33	87.88
Loam	24	95.83
Clay loam	30	90.00
Average		80.88
Overall % correct		72.80

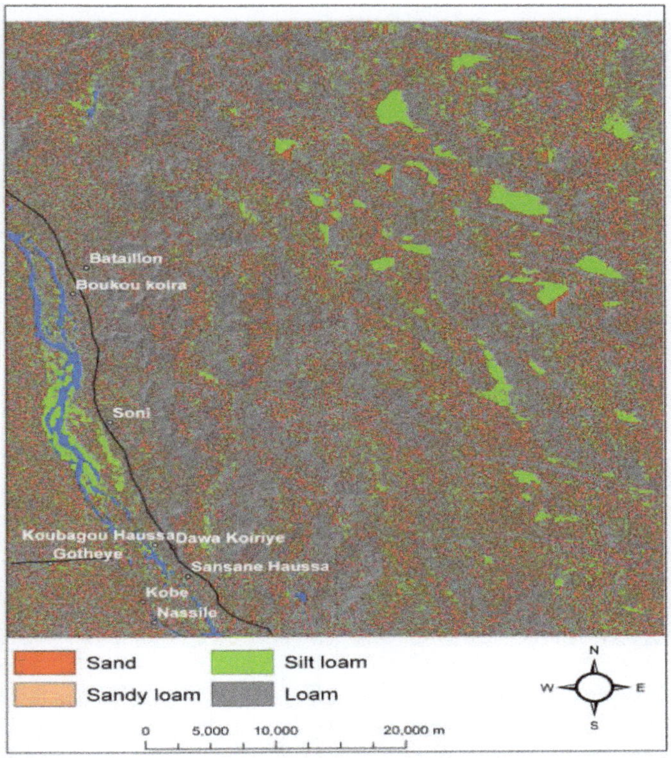

Figure 5. Predicted soil texture for the topsoil layer (0-30 cm).

landscape allowed constructing a map showing the spatially distributed soil texture potential for each class. Soil texture plays a very important role not only for soil fertility, but also for soil stability, water retention capacity and soil biodiversity. Figure 5 shows how the soil texture is spatially distributed in the study area. Soil texture estimation provides a better view of erosion and land degradation processes. The map reflects the high importance of topographic parameters for soil texture distribution. Also the model detected for the following

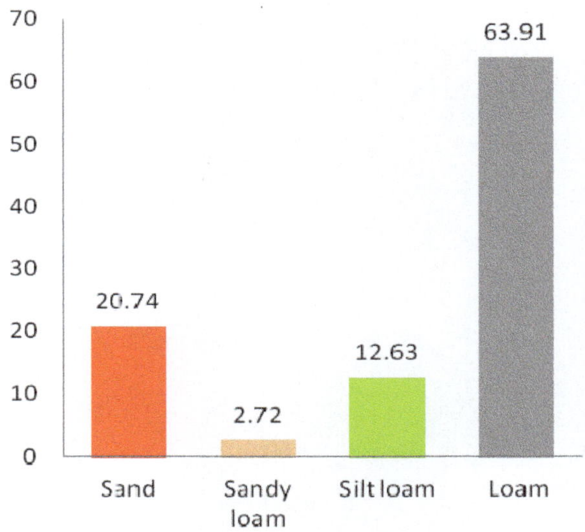

Figure 6. The percentage of soil texture predicted.

terrain attributes a very high prediction potential for soil texture: Altitude above channel network, watershed, channel network, profile curvature, convergence index, plan curvature, wetness index, analytical hillshading, curvature, slope, aspect and LS-factor.

Therefore, the result suggests that topographic elements were important in determining the spatial distribution of soil texture. This can be explained by the fact that all terrain attributes considered in this investigation influenced soil texture due to erosion, transport and deposition processes. This study has demonstrated that loam (63.91%) was the dominant soil texture, suggesting that the study area tends to have high soil loss (Figure 6). The map shows that silt loam was found close to the Niger River and seasonal water bodies.

This result shows that the spatial distribution of soil texture in the study area depends on the wind regime. Sandy soil distribution is indicated by the activity of aeolian processes and act with dry wind direction (October - April) that move northeast and during the rainy season wind direction (May - July) that move westwards through the Sahel in the same time. In addition, the upper part of the study area is mostly dominated by loamy soils under the influence of aeolian processes. In general, the study showed that the spatial distributions of soil texture are influenced by the topographic elements and aeolian activities in the Sahel region.

Spatial distribution of soil loss using RUSLE

The spatial distribution maps of sheet and rill erosion in Tillabéry landscape from 1973 to 2007 are shown in

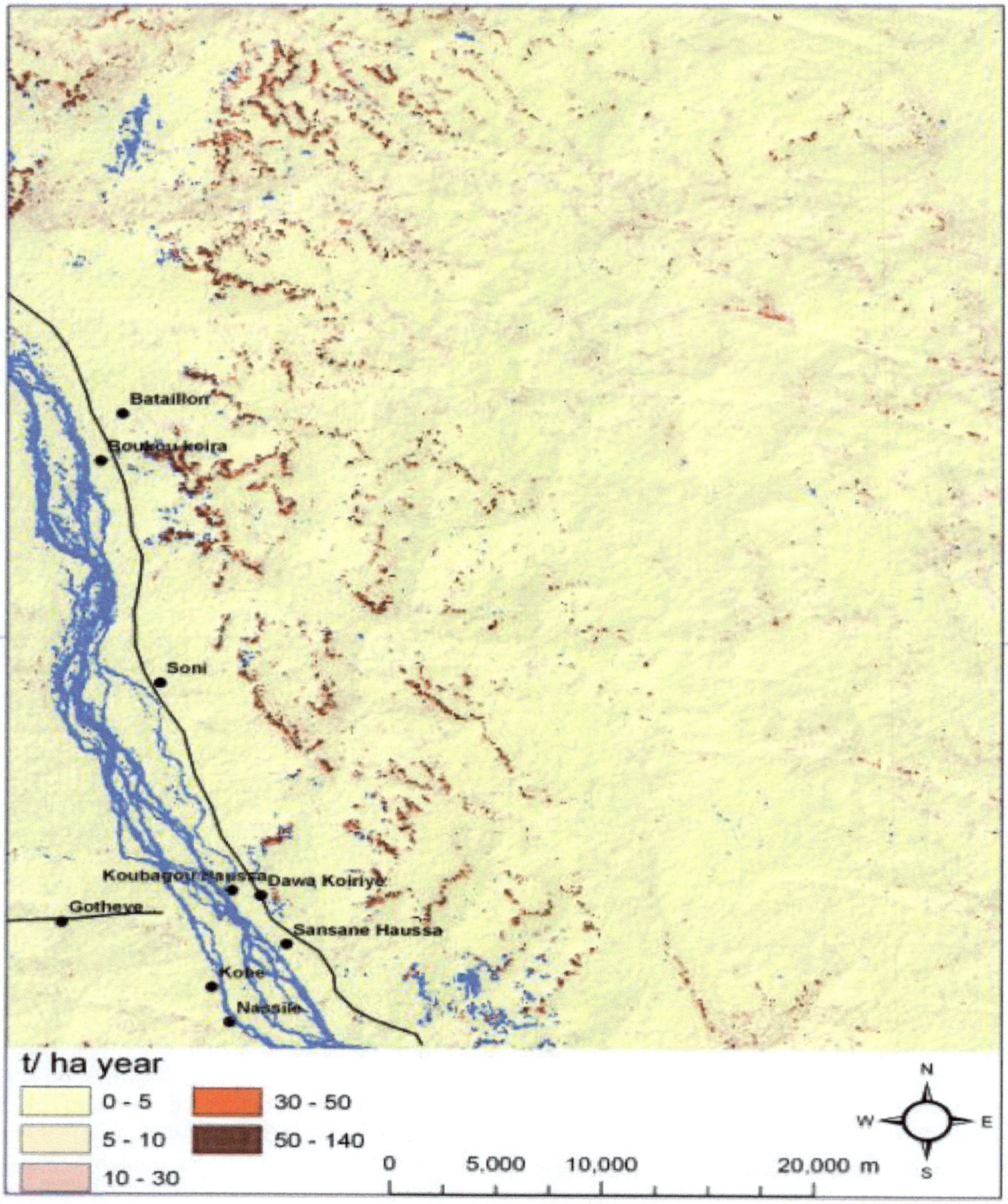

Figure 7. Spatial distribution of soil loss based on K factor without organic matter in 1973.

Figures 7, 8, 9 and 10 based on K factor without organic matter and soil loss based on K factor with organic matter. The results are displayed in a 90 x 90 meter grid cells.

The intensity of erosion is classified in different classifications. Wischmeier and Smith (1978) reported that the maximum tolerable soil loss (12 t/ha year) is described as "the maximum level of soil erosion that will permit a level of crop productivity to be sustained economically and indefinitely". In this study, the K factor, LS factor and P factor values were constant, and R and C factor values were changed according to scenario settings. The results of this analysis demonstrated that rill and sheet erosion during the study periods increased

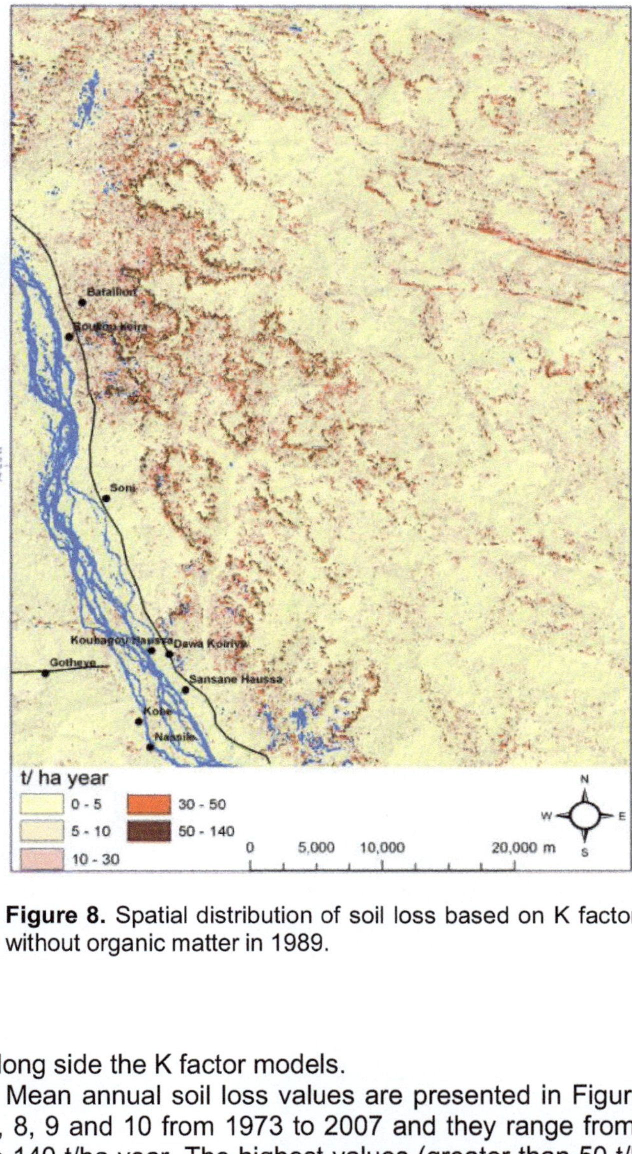

Figure 8. Spatial distribution of soil loss based on K factor without organic matter in 1989.

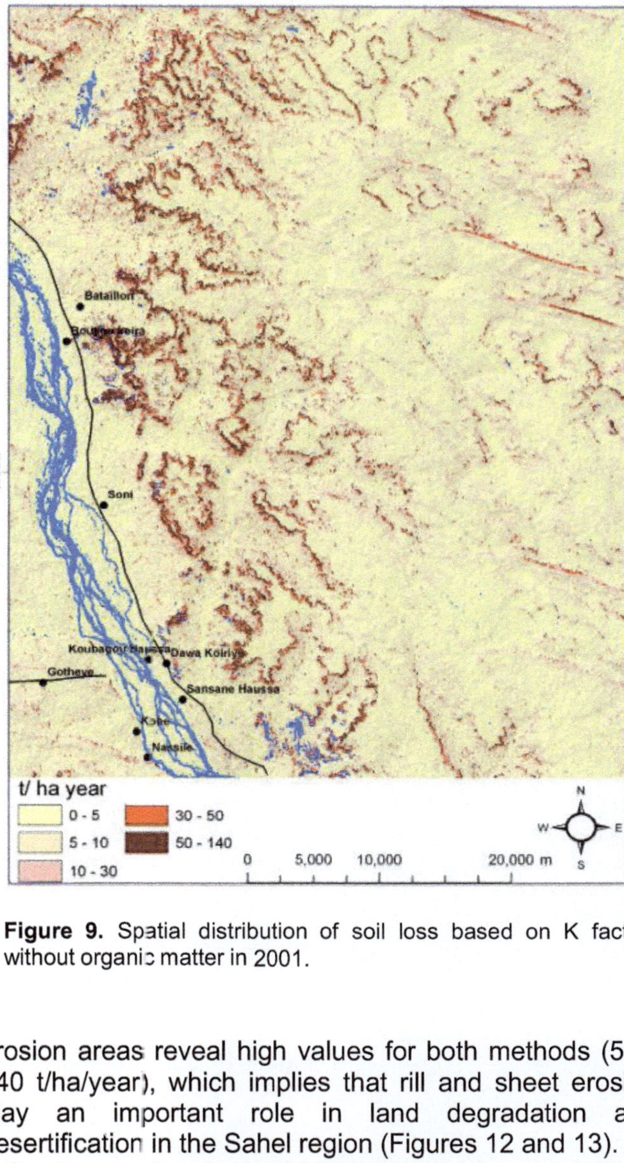

Figure 9. Spatial distribution of soil loss based on K factor without organic matter in 2001.

along side the K factor models.

Mean annual soil loss values are presented in Figures 7, 8, 9 and 10 from 1973 to 2007 and they range from 0 to 140 t/ha year. The highest values (greater than 50 t/ha year) of soil loss were recorded in an infinitesimal amount in the centre of the study area in 1973 and increased with time in the study area for both models. It should be noted that the zone of intensive rill and sheet erosion is situated in the study area with loamy texture soils, bare soil areas and with a slope greater than 7% (Figure 11). Analysis of the RUSLE model series (1973, 1989, 2001 and 2007) show that rill and sheet erosion increased moderately (30 - 50 t/ha/year) and this increase was closely related to an increase in bare areas in the study area.

These results clearly show the existence of interactions between the slope, land cover and soil texture. Amounts of soil loss are higher with the first model because of high values in K factor. Both models indicate low erosion risk close to the river and southwestern part of the study area, but show high erosion risk in the uplands. By comparison of both models, it can be noted that the rill and sheet erosion areas reveal high values for both methods (50 - 140 t/ha/year), which implies that rill and sheet erosion play an important role in land degradation and desertification in the Sahel region (Figures 12 and 13).

On the one hand, sheet erosion is more concentrated in the north east of the study area, dominated by agricultural lands and grasslands. This can be explained by the fact that in the Sahel region, sheet erosion is associated with animal and agricultural production activities which accentuates compaction of the soil and the destruction of vegetation cover. On the another hand, the spatial distribution of rill erosion is more concentrated around the centre and close to major water bodies This is due to the fact that rill erosion in the Sahel region is connected with areas with high altitude (centre of the study area) and with silt loam soil type (close to major water bodies).

Comparison of soil erosion scenarios under rainfall options between two K factors

The rainfall erosivity for the year 2070 in Tillabéry area

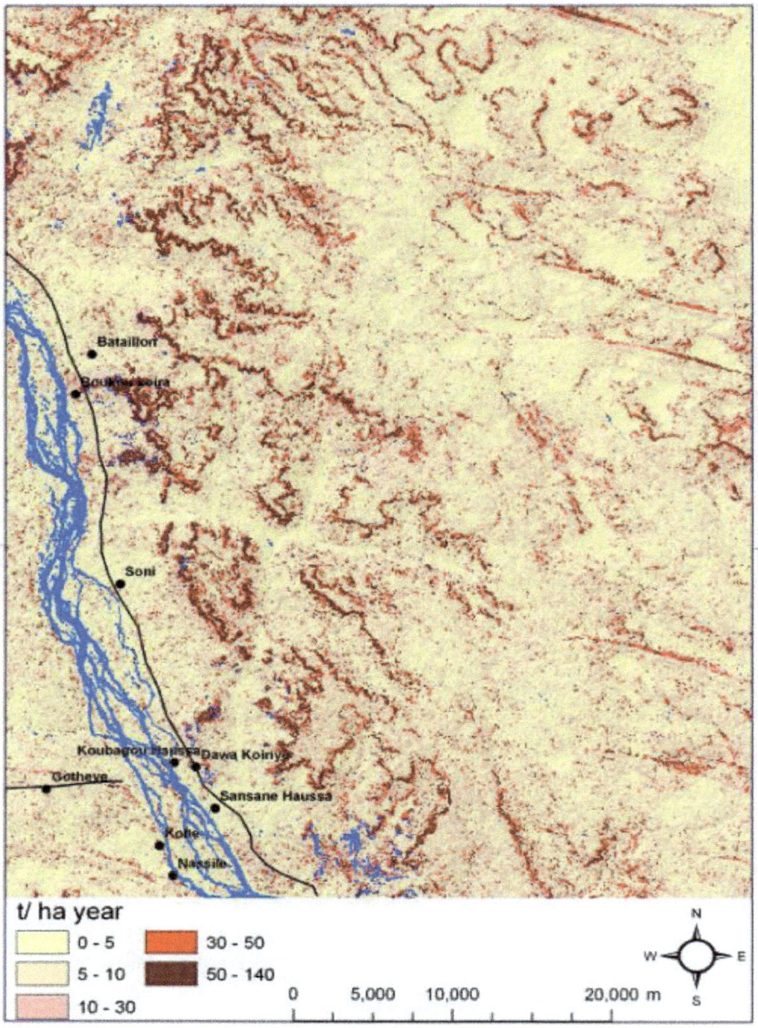

Figure 10. Spatial distribution of soil loss based on K factor without organic matter in 2007.

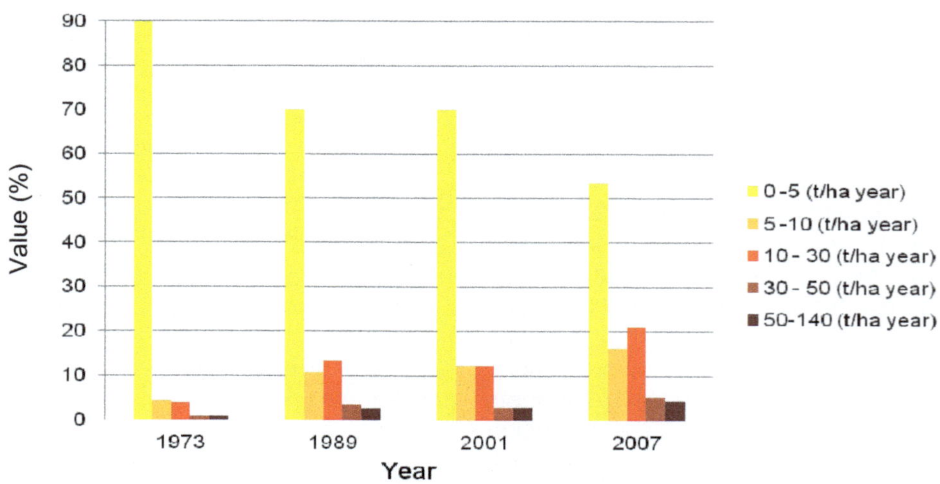

Figure 11. Comparison of soil loss (based on K factor without organic matter) rate (t/ha year) per % of surface affected by rill and sheet erosion in the study area.

Figure 12. Sheet erosion shown in the north east cf the study area dominated by grasslands.

Figure13. Rill erosion shown in the centre of the study area.

using future precipitation predicted from Had CM3 mode under A2 emission scenario (Alcamo et al., 2003; IPCC, 2000) was used. It gives useful information on the erosion risk in a changed climate condition. Figures 14 and 15 illustrate the expected future trend of soil erosion for the Tillabéry landscape up to 2070.

Figures 14 and 15 showed the spatial development of soil erosion up to the year 2070 under two scenarios (Tillabéry I and II). Tillabéry I is a simulated change in soil erosion risk for the scenarios based on rainfall erosivity till the year 2070 in the Tillabéry area using future precipitation predicted from Had CM3 model under A2 emission scenario based on K factor with organic matter. Tillabéry II depicts a simulated change in soil erosion risk for the same area based on rainfall erosivity until 2070 in the Tillabéry area using future precipitation predicted from Had CM3 model under A2 emission scenario (Alcamo et al., 2003; IPCC, 2000) based on K factor without organic matter.

In both scenarios (Tillabéry I and II), an increase in soil erosion is observed. In Tillabéry I, no risk of erosion is in the flat zones of the Niger River Basin. In Tillabéry II, the higher erosion risk (>50 t/ha/year) is visible. As such, these changes affect the rest of the classes. The spatial

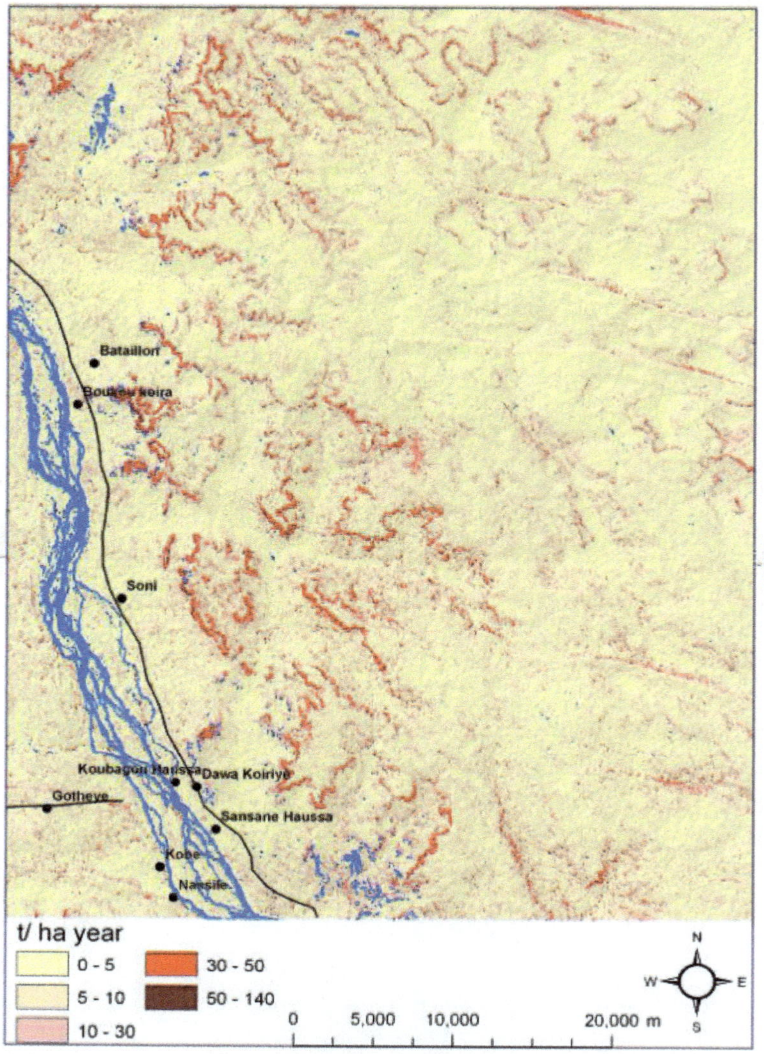

Figure 14. Simulated changes in soil erosion risk scenarios in 2070 (Tillabéry I) with organic matter.

distributions of soil erosion scenarios between the two scenarios show a close relationship in general (Figures 14 and 15). The areas with a high erosion value with Tillabéry I may likely have a high erosion value with Tillabéry II. The spatial distribution of areas with high erosion values with Tillabéry II show a concentration in the central parts of the Tillabéry landscape, whereas the spatial distribution of areas prone to erosion with Tillabéry I are concentrated in the north-eastern part of the study area, precisely close to the Niger River.

Spatial patterns of soil erosion and soil deposition simulated by USPED Model

The results show that high soil erosion was observed and predicted on convex landscape positions and soil deposition was observed and predicted on concave landscape position and confirm the results obtained from the RUSLE analysis (Figure 16). During the four different years (1973, 1989, 2001 and 2007), simulated soil erosion/deposition maps show a low amplitude of simulated values. Indeed, the soil loss values range between 0.1 and 5 t/ha year and deposition values show similar amplitudes in the study area. These values can be elucidated by the specific terrain parameter of the Tillabéry landscape, where areas that have a low slope gradient, produce low soil loss rates and low accumulation.

The soil erosion/deposition maps for 1973 show that most of the study area was very stable. Soil loss was very minimal (0 - 0.1 t/ha/year). Deposition rate was also

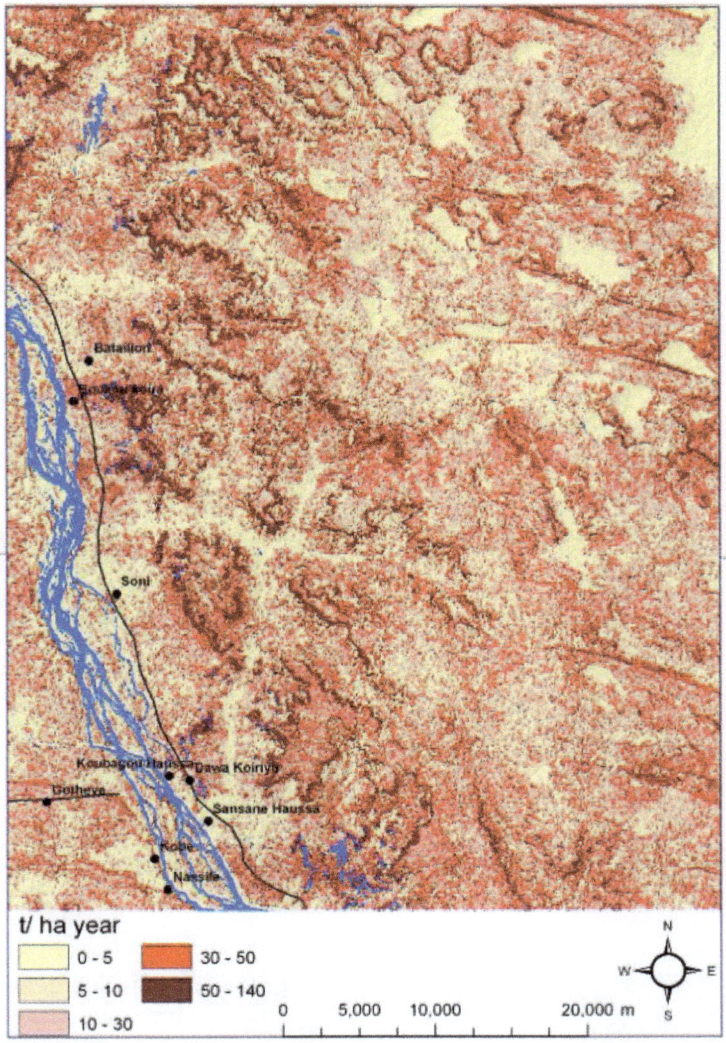

Figure 15. Simulated changes in soil erosion risk scenarios in 2070 (Tillabéry II) without organic matter.

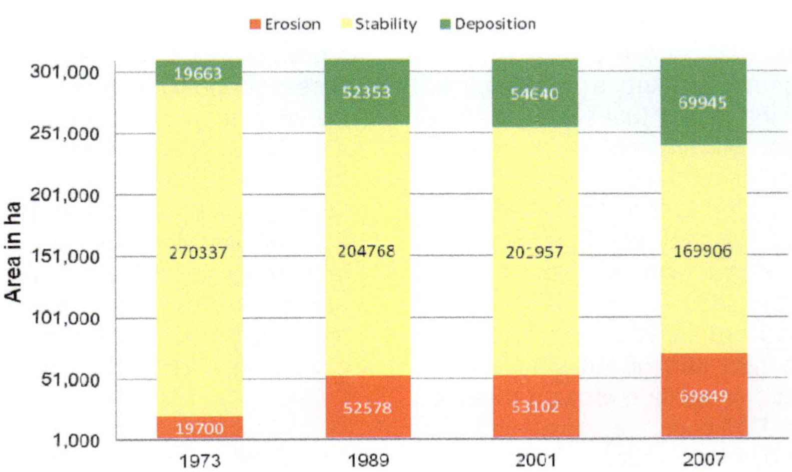

Figure 16. Soil erosion and deposition simulated with the USPED during the study period.

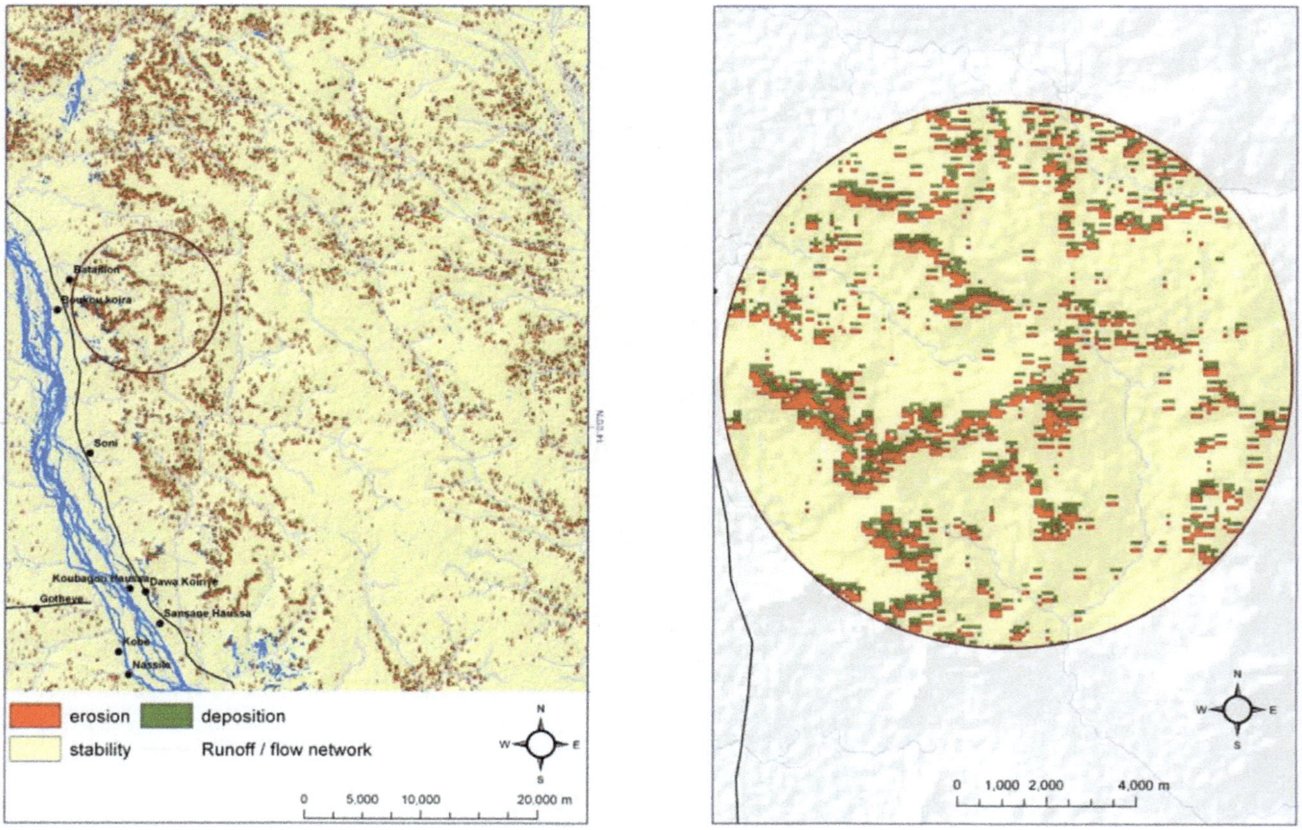

Figure 17. Spatial distribution of simulated soil erosion and soil deposition by USPED model for the study area in 1973.

very low (0 – 0.1 t/ha/year) (Figure 18). Erosion had a significantly lower spatial coverage in 1973. The increase in the erosion endangered zone was caused by a reduction of shrub and grassland, which accentuated desertification and affected landscape stability. This means that rill and sheet erosion were amongst the major contributory factors to land degradation, decreasing soil productivity and an eventual deterioration in the soil and water quality in the Tillabéry landscape and calls for soil conservation and management methods to be applied. The USPED model highlights areas that were potentially affected by water erosion. So erosion areas could be well-known in a short period of time using this model.

The deposition zone is limited to the Niger River prone to accumulation and close to the runoff network as a result of the change in slope gradients. Figures 17 and 18 show a clear correlation between erosion affected zone and bare areas with slope gradients greater that 10%. There are areas in the center part of the study area in 1989, 2001 and 2007 which show high erosion rates, but in 1973 low erosion rates. This could be explained by the fact that during this period, these areas were generally more densely covered by shrubs. Compared with 1973, the stable areas decreased by 100,431 ha in 2007 (Figure 18). An increase in sediments was observed during all the study period due to climatic changes, development of agricultural lands and bare areas. These conversions have intensified the issue of desertification in the study area.

Decreasing soil erosion in the study area is a more cost efficient land management strategy than to deal with removing sediments from the Niger River, which is extremely costly and difficult. Investigating soil erosion/deposition in the Tillabéry landscape can actually be seen as a step to gain information about the spatial distribution of soil erosion and sediment deposition within the area. In addition, the figures could be applied to identify appropriate areas for implementing land conservation assessment. Also, the results show that slope and land cover are the most important factors influencing soil erosion in the study area, which point out at the same time that sound land management practices have a controlling effect on soil loss.

Conclusion and future research needs

The relationship between a set of topographic attributes,

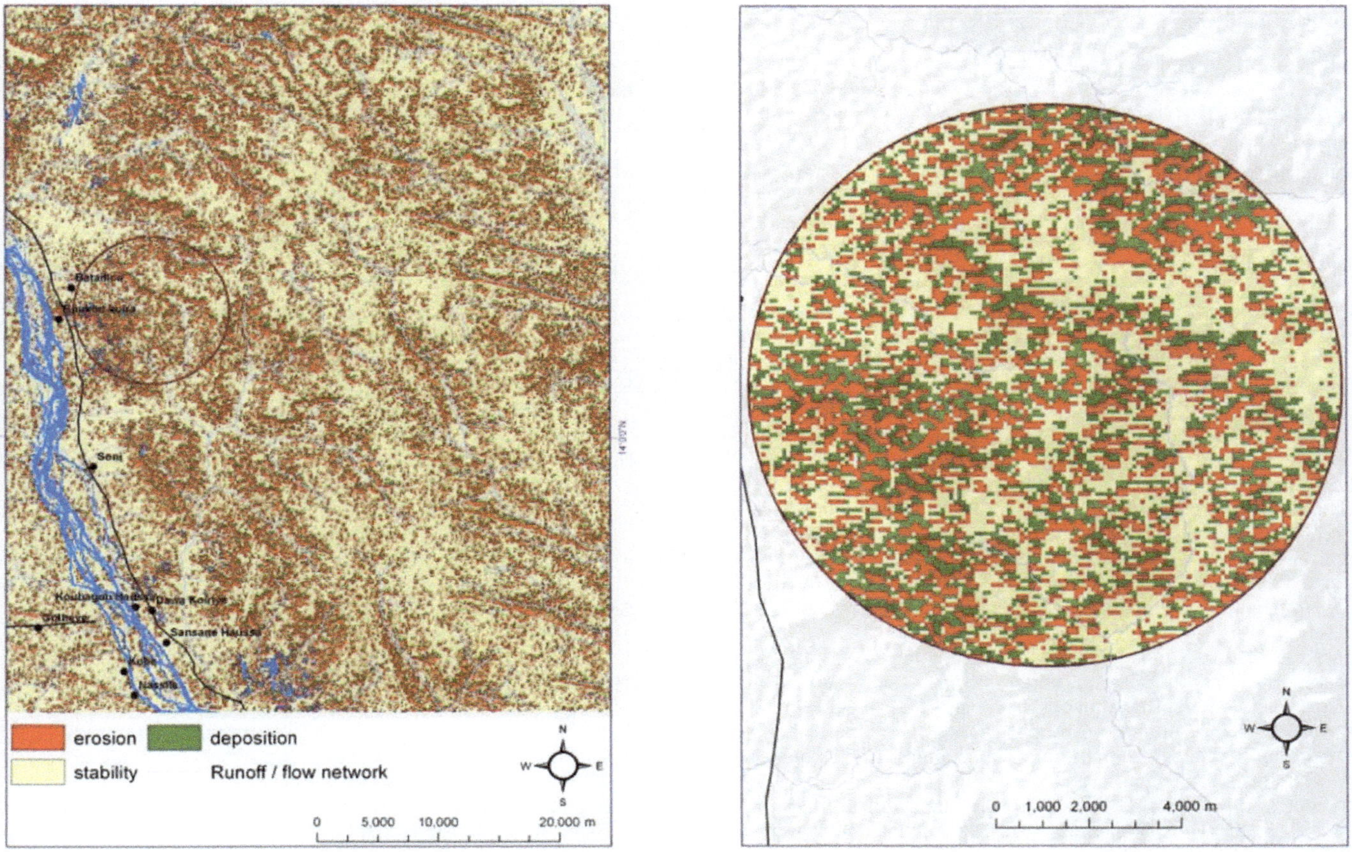

Figure 18. Spatial distribution of simulated soil erosion and soil deposition by USPED model for the study area in 2007.

soil texture and different K factors for soil erosion assessment were investigated by using the TreeNet model in order to improve our understanding regarding soil degradation in the Sahel region and to build up a soil model for the study area. A set of topographic attributes (independent variables) were investigated, which allowed for a better understanding of the major driving factors that influence the spatial distribution of soil texture and erosion processes in the Sahel region. The results show that the altitude above the channel network and the wetness index are the most important factors that play a dominant role in the spatial distribution of soil texture because of their direct impact on runoff and transportation processes.

Soil texture output provided excellent predictive performance, confirming that the Tree Net model was an effective tool. But, the output of K factors provided a bad predictive performance (test data set) due to limited amount of soil sampling (low sample size). Therefore Co-Kriging was used in estimating the K factors in the case of limited available data.

RUSLE and USPED models have shown a realistic evolution of soil loss distribution during all the study period and confirm the need to continue the investigation of soil erosion processes by using data mining tools in the Sahel region. Puerta et al. (2008) reported that soil loss by soil erosion is often noted to play a major role in the desertification process. Both soil erosion scenarios (based on future precipitation predicted from Had CM3 mode under A2 emission scenario) output show greater soil erosion in 2070 in the study area. The scenario based on K factor without considering the organic matter proved to have the highest erodibility.

The present project confirms the importance of organic matter in influencing soil erosion prediction and also as a useful instrument in planning soil degradation control practices. Therefore, reducing soil erosion is more prudent than to deal with removing sediment and sandy soils from the Niger River, which is costly and difficult in the Sahel region.

Erosion models are prone to further research. Further studies must be focused on trying to improve the soil erosion processes by using data mining tools in the Sahel region. However, results indicate that the greater number of sampling sites and using higher quality DEMs derived from LIDAR/RADAR and TanDEM-X are needed in order to improve the result of the soil model.

Conflict of interests

The author did not declare any conflict of interests.

REFERENCES

Alcamo J, Märker M, Flörke M, Vassolo S (2003). Water and climate: A global perspective. Kassel world water series 6, Center for environmental systems research, University of Kassel, Kurt-Wotters-Strasse 3, 34109 Kassel, Germany.

Angima SD, Stott DE, O'Neil MK, Ong CK, Weesies GA (2003). Soil erosion prediction using RUSLE for central Kenyan highland conditions, Agriculture, Ecosystems and Environment 97:295-308.

Ben Mohamed A, Van Duivenbooden N, Abdoussallam S, (2001). Impact of climate change on agricultural production in the Sahel- Part 1. Methodological approach and case study for millet in Niger.

Cobo JD, Dercon G, Cadisch G (2010). Nutrient balances in African land use systems across different spatial scales: A review of approaches, challenges and progress. Agr. Ecosyst. Environ. 136:1-15.

Conrad O (2007). SAGA – Entwurf, Funktionsumfang und Anwendung eines Systems für automatisierte geowissenschaftliche Analysen. Electronic Doktorat Dissertation, University of Göttingen.

De Rosa P (2005). Analisi e Confronti di modelli di erosion del suolo e trasporto di sediment tramite l'uso di sistemi G.I.S. Degree Thesis, Universitá degli studi di Perugia, Italia.

Friedman JH (1999). Greedy function approximation: a gradient boosting machine. Stanford: Statistics Dept., Stanford University, USA.

Friedman JH (1999). Stochastic gradient boosting. Stanford: statistics Dept., Stanford University, USA.

Gee GW, Bauder JW (1986). Particle-size analyisis. In R. Burt(ed), Soil survey field and laboratory methods manual, U.S Depart. of agriculture, Lincoln, Nebraska. Report no 51 pp. 53-54.

Henderson BL, Bui EN, Moran CJ, Simon DAP (2005). Australia-wide predictions of soil properties using decision trees.124, Geoderma pp. 383-398.

Lasko TA, Bhagwat JG, Zou KH, Ohno-Machado L (2005). The use of receiver operating characteristic curves in biomedical informatics. J. Biomed. Inf. 38:404-415.

IPCC (Intergovernmental Panel on Climate Change) (2000). Special report on emission scenarios. Intergovernmental Panel on Climate Change. Cambridge University Press. Cambridge. p.599.

Ließ M, Glaser B, Huwe B (2011). Uncertainty in the spatial prediction of soil texture comparison of regression tree and random forest models. Geodenna 170:70-79.

Michels K, Sivakumar MVK, Allison BE (1995). Wind erosion control using crop residue. I. Effects on soil flux and soil properties. Field Crops Res. 40:101-110.

Millward AA, Mersey JE (1999). Adapting the RUSLE to model soil erosion potential in a mountainous tropical watershed Catena. 32(2):109-129.

Mitasova H, Hofierka J, Zlocha M, Iverson LR (1996). Modeling topographic potential for erosion and deposition using GIS. Int. J. GIS. 10:629-642.

Morgan RPC (1995): Soil Erosion and Conservation. Addision-Westey. Edinburgh, UK. p.198.

Gandah M, Brouwer J, Hiernaux P, Van Duivenbooden N (2003). Fertility management and landscape position: farmers use of nutrient source in western Niger and possible improvements Nutrient Cycling in Agroecosystems 67:55-66.

Ramadam Z, Hopke PK, Johnson MJ, Scow KM (2005). Application of PLS and back-propagation neural networks for the estimation of soil properties. Chemon. Intell. Lab. Syst.75:23-30.

Read JW (1921). Rapid dry combustion method for the simultaneous determination of soil organic matter and organic carbon, J. Ind. Eng. Chem. pp. 304-308.

Renard KG, Foster GR, Weesies GA, McCool DK, Yoder DC (1997). Prediction soil erosion by water: a guide to conservation planning with the Revised Universal Soil Loss Equation. Agricultural Handbook, vol. 703. U.S. Department of Agriculture. p. 404.

Roose E, De Noni V (2004). Recherché sur l érosion hydrique en Afrique-revue et perspectives. Sécheresse 15:121-129.

Sivakumar MV, Maidoukia A, Stern RD (1993). Agroclimatologie de l Afrique de l Ouest Niger, Bulletin d'informations No 5, ICRISAT, Patancheru. India. pp.108.

Thein SJ (1979). A flow diagram for teaching texture by feel analysis. J. Agron. Edu. 8:53-55.

Torri D, Poesen J, Borselli L (2002). Corrigendum to predictability and uncertainty of the soil erodibility factor using global dataset, Catena 46:309-310.

Warren SD, Mitasova H, Hohmann MG, Landsberger S, Iskander FY, Ruzycki TS, Senseman GM (2005). Validation of a 3-D enhancement of the Universal Soil Loss Equation for prediction of soil erosion and sediment deposition, Catena 64:281-296.

Wilcke W, Yasin S, Schmitt A, Valarezo C, Zech W (2008). Soils along the altitudinal transect and in catchments. In: Beck, E., J. Bendix, I. Kottke, F. Makeschin, and R. Mosandl (Eds). Gradients in a Tropical Mountain Ecosystem of Ecuador-Ecological Studies, vol.198. Springer, Berlin, pp.75-87.

Wischmeier WH, Smith DD (1978). Predicting rainfall erosion loss. USDA. Agricultural research Service Handbook 537.

Zhang, SL, Zhang XY, Huffman T, Liu XB, Yang JY (2011). Influence of topography and land management on soil nutrients variability in Northeast China. Nutr. Cycling Agroecosyst. 89:427-438.

Ziadat FM (2005). Analyzing digital terrain attributes to predict soil attributes for a relatively large area. Soil Sci. Soc. Am. J. 69:1590-1599.

Permissions

All chapters in this book were first published in AJEST, by Academic Journals; hereby published with permission under the Creative Commons Attribution License or equivalent. Every chapter published in this book has been scrutinized by our experts. Their significance has been extensively debated. The topics covered herein carry significant findings which will fuel the growth of the discipline. They may even be implemented as practical applications or may be referred to as a beginning point for another development.

The contributors of this book come from diverse backgrounds, making this book a truly international effort. This book will bring forth new frontiers with its revolutionizing research information and detailed analysis of the nascent developments around the world.

We would like to thank all the contributing authors for lending their expertise to make the book truly unique. They have played a crucial role in the development of this book. Without their invaluable contributions this book wouldn't have been possible. They have made vital efforts to compile up to date information on the varied aspects of this subject to make this book a valuable addition to the collection of many professionals and students.

This book was conceptualized with the vision of imparting up-to-date information and advanced data in this field. To ensure the same, a matchless editorial board was set up. Every individual on the board went through rigorous rounds of assessment to prove their worth. After which they invested a large part of their time researching and compiling the most relevant data for our readers.

The editorial board has been involved in producing this book since its inception. They have spent rigorous hours researching and exploring the diverse topics which have resulted in the successful publishing of this book. They have passed on their knowledge of decades through this book. To expedite this challenging task, the publisher supported the team at every step. A small team of assistant editors was also appointed to further simplify the editing procedure and attain best results for the readers.

Apart from the editorial board, the designing team has also invested a significant amount of their time in understanding the subject and creating the most relevant covers. They scrutinized every image to scout for the most suitable representation of the subject and create an appropriate cover for the book.

The publishing team has been an ardent support to the editorial, designing and production team. Their endless efforts to recruit the best for this project, has resulted in the accomplishment of this book. They are a veteran in the field of academics and their pool of knowledge is as vast as their experience in printing. Their expertise and guidance has proved useful at every step. Their uncompromising quality standards have made this book an exceptional effort. Their encouragement from time to time has been an inspiration for everyone.

The publisher and the editorial board hope that this book will prove to be a valuable piece of knowledge for researchers, students, practitioners and scholars across the globe.

List of Contributors

Kabunga NACHIYUNDE
Graduate School for International Development and Cooperation, Hiroshima University

Hideo IKEDA
Graduate School for International Development and Cooperation, Hiroshima University

Kazuhiko TANAKA
Graduate School for International Development and Cooperation, Hiroshima University

Daisuke KOZAKI
Graduate School for International Development and Cooperation, Hiroshima University

Farooq Ahmad
GC University, Sustainable Development Study Centre, Lahore, Pakistan

Amin U. Khanand
GC University, Sustainable Development Study Centre, Lahore, Pakistan

Abdullah Yasar
GC University, Sustainable Development Study Centre, Lahore, Pakistan

Asrat Gebremariam Woldesenbet
School of Chemical and Bio-Engineering, Addis Ababa Institute of Technology, Addis Ababa University, P.O. Box 385, Addis Ababa, Ethiopia

Gizachew Shiferaw
School of Chemical and Bio-Engineering, Addis Ababa Institute of Technology, Addis Ababa University, P.O. Box 385, Addis Ababa, Ethiopia

Bhagwan Singh Chandravanshi
Department of Chemistry, Faculty of Sciences, Addis Ababa University, P.O.Box 1176, Addis Ababa, Ethiopia

EO Akindele
Department of Biological Sciences, Bowen University, Iwo, Nigeria

IF Adeniyi
Department of Zoology, Obafemi Awolowo University, Ile-Ife, Nigeria

Getachew Dagnew Gebreeyessus
Department of Environmental Health Sciences, Haramaya University, P.O. Box 235, Harar, Ethiopia

Berhanu Assefa Demissie
School of Chemical and Bio-Engineering, Addis Ababa Institute of Technology, Addis Ababa University, P.O. Box 385, Addis Ababa, Ethiopia

A. M. Stanley
Department of Building, Ahmadu Bello University, Zaria, Kaduna State, Nigeria

D. M. Stanley
Department of Crop Protection, University of Maiduguri, Borno State, Nigeria

D. W. Dadu
Department of Building, Ahmadu Bello University, Zaria, Kaduna State, Nigeria

A. M. Abah
Department of Building, Ahmadu Bello University, Zaria, Kaduna State, Nigeria

A. O. Ameh
Department of Chemical Engineering, Ahmadu Bello University, Zaria 810261, Kaduna State, Nigeria

I. A. Mohammed-Dabo
Department of Chemical Engineering, Ahmadu Bello University, Zaria 810261, Kaduna State, Nigeria

S. Ibrahim
Department of Biochemistry, Ahmadu Bello University, Zaria, Kaduna State, Nigeria

J. B. Ameh
Department of Microbiology, Ahmadu Bello University, Zaria, Kaduna State, Nigeria

Mengnjo J. Wirmvem
Department of Chemistry, School of Science, Tokai University, Hiratsuka, 259-1211, Japan

Takeshi Ohba
Department of Chemistry, School of Science, Tokai University, Hiratsuka, 259-1211, Japan

Wilson Y. Fantong
Institute of Mining and Geological Research, P.O. Box 4110, Yaounde, Cameroon

Samuel N. Ayonghe
Department of Environmental Science, University of Buea, Box 63, Buea, Cameroon

Justice Y. Suila
G.B.H.S. Atiela, Nkwen, P.O. Box 5147, Bamenda, Cameroon

Asobo Nkengmatia E. Asaah
Department of Earth and Planetary Sciences, Tokyo Institute of Technology, Tokyo, 152-8551, Japan

Gregory Tanyileke
Institute of Mining and Geological Research, P.O. Box 4110, Yaounde, Cameroon

Joseph V. Hell
Institute of Mining and Geological Research, P.O. Box 4110, Yaounde, Cameroon

Engome R. Wotany
Department of Geology, Faculty of Science, University of Buea, P.O. Box 63, Buea, Cameroon

Samuel N. Ayonghe
Department of Environmental Science, Faculty of Science, University of Buea, P.O. Box 63, Buea, Cameroon

Wilson Y. Fantong
Institute of Mining and Geological Research, P.O. Box 4110, Yaounde, Cameroon

Mengnjo J. Wirmvem
Department of Chemistry, School of Science, Tokai University, Hiratsuka, 259-1211, Japan

Takeshi Ohba
Department of Chemistry, School of Science, Tokai University, Hiratsuka, 259-1211, Japan

Ndong Nlo Joseph
Laboratoire Pluridisciplinaire des Sciences de l'Ecole Normale Supérieure de Libreville BP: 17009 Libreville, Gabon

Eba Francois
Laboratoire Pluridisciplinaire des Sciences de l'Ecole Normale Supérieure de Libreville BP: 17009 Libreville, Gabon

Njopwouo Daniel
Laboratoire de Physico-chimie des Matériaux Minéraux, Département de Chimie Inorganique Université de Yaoundé I BP: 812 Yaoundé, Cameroun

Beh Sima Didier
Laboratoire Pluridisciplinaire des Sciences de l'Ecole Normale Supérieure de Libreville BP: 17009 Libreville, Gabon

Ondo Jean Aubin
Laboratoire Pluridisciplinaire des Sciences de l'Ecole Normale Supérieure de Libreville BP: 17009 Libreville, Gabon

P. S. Khisa
Water Resources Management Authority, Lake Victoria South Catchment Area, P.O Box 666-40100 Kisumu, Kenya
UNESCO-IHE Institute for Water Education, P.O. Box 3015, Delft, The Netherlands
Delft University of Technology, Water Resources Section, P.O. Box 5048, Delft, The Netherlands

S. Uhlenbrook
UNESCO-IHE Institute for Water Education, P.O. Box 3015, Delft, The Netherlands
Delft University of Technology, Water Resources Section, P.O. Box 5048, Delft, The Netherlands

A. A. van Dam
UNESCO-IHE Institute for Water Education, P.O. Box 3015, Delft, The Netherlands

J. Wenninger
UNESCO-IHE Institute for Water Education, P.O. Box 3015, Delft, The Netherlands
Delft University of Technology, Water Resources Section, P.O. Box 5048, Delft, The Netherlands

A. van Griensven
UNESCO-IHE Institute for Water Education, P.O. Box 3015, Delft, The Netherlands

M. Abira
Water Resources Management Authority, Lake Victoria South Catchment Area, P.O Box 666-40100 Kisumu, Kenya

James D. Proctor
Environmental Studies Program-MSC 62, Lewis & Clark College, 0615 SW Palatine Hill Road, Portland, OR 97219 USA

Ben Rathbun
Environmental Studies Program-MSC 62, Lewis & Clark College, 0615 SW Palatine Hill Road, Portland, OR 97219 USA

Erin Scheibe
Environmental Studies Program-MSC 62, Lewis & Clark College, 0615 SW Palatine Hill Road, Portland, OR 97219 USA

Samantha Shafer
Environmental Studies Program-MSC 62, Lewis & Clark College, 0615 SW Palatine Hill Road, Portland, OR 97219 USA

D. R. Awoniran
Department of Remote Sensing and Geoscience Information System, Federal University of Technology, Akure, Ondo State, SW, Nigeria

M. B. Adewole
Institute of Ecology and Environmental Studies, Obafemi Awolowo University, Ile-Ife, Osun state, SW, Nigeria

S. A. Adegboyega
Department of Remote Sensing and Geoscience Information System, Federal University of Technology, Akure, Ondo State, SW, Nigeria

A. Y. B. Anifowose
Department of Remote Sensing and Geoscience Information System, Federal University of Technology, Akure, Ondo State, SW, Nigeria

OLADAPO, Michael Ilesanmi
Department of Applied Geophysics, Federal University of Technology, Akure, Ondo State, Nigeria

ILORI, Olugbenga Bashiru
Lagos State Water Corporation, Lagos, Lagos State, Nigeria

ADEOYE-OLADAPO, Oluwakemi Olanike
Department of Physics, Adeyemi College of Education, Ondo, Ondo State, Nigeria

Robert Kirchner
African Development Bank Group, AfDB Temporary Relocation Agency (Tunis),15 Avenue du Ghana, P.O.Box 323-1002, Tunis Belvedère, Tunisia

Adeleke Salami
African Development Bank Group, AfDB Temporary Relocation Agency (Tunis),15 Avenue du Ghana, P.O.Box 323-1002, Tunis Belvedère, Tunisia.

K. R. Sanjay
Department of Biotechnology, Sri Jayachamarajendra College of Engineering, Mysore- 570006, India

M. N. Nagendra Prasad
Department of Biotechnology, Sri Jayachamarajendra College of Engineering, Mysore- 570006, India

S. Anupama
Department of Biotechnology, Sri Jayachamarajendra College of Engineering, Mysore- 570006, India

B. R Yashaswi
Department of Biotechnology, Sri Jayachamarajendra College of Engineering, Mysore- 570006, India

B. Deepak
Department of Biotechnology, Sri Jayachamarajendra College of Engineering, Mysore- 570006, India

Pu Peng
State Key Laboratory of Catalytic Material and Reaction Engineering, Research Institute of Petroleum Processing, SINOPEC, 18 Xue Yuan Road, Beijing 100083, China

Wenguang Zhou
Center for Biorefining, Bioproducts and Biosystems Engineering Department, University of Minnesota, 1390 Eckles Ave., Saint Paul, MN 55108, USA

B. Adeyemi
Department of Physics, P.M.B. 704, Federal University of technology, Akuro, Ondo State, Nigeria

O.S. Ojo
Department of Physics, P.M.B. 704, Federal University of technology, Akuro, Ondo State, Nigeria

T. I. Akuh
Department of Geology, Ahmadu Bello University, Zariya khaduna State, Nigeria

S.A. Algabe
Department of Geology, Ahmadu Bello University, Zariya khaduna State, Nigeria

A.A Ibrahim
Department of Geology, Ahmadu Bello University, Zariya khaduna State, Nigeria

Hesham R. Lotfy
Department of Chemistry and Biochemistry, Faculty of Science, University of Namibia, P. Bag 13301, Windhoek, Namibia

Martha N. Amputu
Department of Chemistry and Biochemistry, Faculty of Science, University of Namibia, P. Bag 13301, Windhoek, Namibia

Celestine Raidron
Department of Chemistry and Biochemistry, Faculty of Science, University of Namibia, P. Bag 13301, Windhoek, Namibia

B. Kengne Signe Elie
Laboratory of Energy Research, Institute of Geological and Mining Research, BP 4110 Yaoundé, Cameroon

O. Hamandjoda
ENSP University of Yaoundé, I BP 8390 Yaoundé, Cameroon

Tekounegning
Faculty of Agronomy and Agricultures Sciences, University of Dschang, BP 110, Dschang, Cameroon

J. Nganhou
ENSP University of Yaoundé, I BP 8390 Yaoundé, Cameroon

Nicholaus Fabian Mwageni
Ardhi University, P.O. Box 35176, Dar esSalaam, Tanzania

Riziki Silas Shemdoe
Ardhi University, P.O. Box 35176, Dar esSalaam, Tanzania

Robert Kiunsi
Ardhi University, P.O. Box 35176, Dar esSalaam, Tanzania

T. C. Senganimalunje
Malawi Bureau of Standards, Metrology Services Department, Box 946, Blantyre, Malawi

C. M. Tenthani
Physics and Biochemical Sciences Department, Malawi Polytechnic, University of Malawi, Chichiri, Blantyre, Malawi

Mansour Mahamane
Faculty of Science, Tubingen University, Tübingen, Germany